Optical Diagnostics for Flow Processes

Optical Diagnostics for Flow Processes

Edited by

Lars Lading

Risø National Laboratory
Roskilde, Denmark

Graham Wigley

Flow Measurement Consulting Service
Kumberg, Austria

and

Preben Buchhave

Technical University of Denmark
Lyngby, Denmark

Springer Science+Business Media, LLC

Library of Congress CIP information goes here.

Optical diagnostics for flow processes / edited by Lars Lading, Graham Wigley, and Preben Buchhave.
p. cm.
"Proceedings of a Summer School on Optical Diagnostics for Flow Processes, held September 26-October 2, 1993, in Roskilde, Denmark"--T.p. verso.
Includes bibliographical references and index.

1. Fluid dynamic measurements--Congresses. 2. Laser Doppler velocimeter--Congresses. 3. Flow visualization--Congresses. I. Lading, Lars. II. Wigley, Graham. III. Buchhave, Preben. IV. Summer School on Optical Diagnostics for Flow Processes (1993 : Roskilde, Denmark)
TA357.5.M43067 1994
681'.2--dc20 94-42621
CIP

Proceedings of a Summer School on Optical Diagnostics for Flow Processes, held September 26–October 2, 1993, in Roskilde, Denmark

ISBN 978-1-4899-1273-2 ISBN 978-1-4899-1271-8 (eBook)
DOI 10.1007/978-1-4899-1271-8

PREFACE

The origin of optical methods for fluid flow investigations appears to be nontraceable. This is no matter for surprise. After all *seeing* provides the most direct and common way for humans to learn about their environment. But at the same time some of the most sophisticated methods for doing measurements in fluids are also based on light and often laser light. A very large amount of material has been published in this area over the last two decades. Why then another publication? Well, the field is still in a state of rapid development. It is characterised by the use of results and methods developed within very different areas like optical physics, spectroscopy, communication systems, electronics and computer science, mechanical engineering, chemical engineering and, of course, fluid dynamics. We are not aware of a book containing both introductory and more advanced material that covers the same material as presented here.

The book is the result of a compilation and expansion of material presented at a summer school on *Optical Diagnostics for Flow Processes*, held at Risø National Laboratory and the Technical University of Denmark in September 1993.

The aim of the course was to provide a solid background for understanding, evaluating, and using modern optical diagnostic methods, addressing Ph.D. students and researchers active in areas of fluid flow research. The disciplines represented by the participants ranged from atmospheric fluid dynamics to biomedicine.

The school was initiated by the European Association for Laser Anemometry, and made possible by a joint grant from the European Commission's Human Capital and Mobility Programme and the Danish Research Academy. Support was also given by Risø National Laboratory, the Technical University of Denmark, Dantec Measurement Technology and AVL List GmbH. In addition we would also like to thank the contributors for both the lecturing and contributions to the book. Ms Bitten Skaarup has been invaluable in the organisation of the course and the preparation of this book. Without her assistance neither the course nor this book would have been possible.

Lars Lading
Preben Buchhave
Graham Wigley

CONTENTS

Whole Field Measurements

Inelastic Methods

INTRODUCTION

Optical methods are increasingly applied to measure fluid phenomena. Use of these techniques implies negligible perturbation of the flow and attainment of high spatial and temporal resolutions. These are the primary reason for the success of optical methods. However, the optical methods are quite different from classical sensor methods like hot wire anemometry or pitot tubes. This applies both to the physics of the interaction with the fluid and to the type of measurement data that is obtained. Optical diagnostics allow localised single-point measurements with laser anemometers and whole field measurements with particle image velocimetry. Laser anemometry is a well established tool and is increasingly being applied to investigate difficult industrially relevant flows. Particle image velocimetry is more in a developmental stage and is attractive for flow processes where it is essential to perform simultaneous measurements of the whole field. In addition to these velocity schemes, diagnostics have been developed for measuring other quantities such as particle size and concentration, temperature, density, and concentration of flow species.

The processing of information can be classified into two areas. In the first type of processing dedicated electronic equipment is used for extracting information, e.g., velocity at a given point in space and time. On the basis of sets of such data statistics characterising spatial and/or temporal dynamics of the fluid process may be inferred. The second type of processing is often performed on a general purpose computer.

The subjects covered in this book incorporate a number of disciplines: fluid mechanics, optics, spectroscopy, and data and signal processing. The book is organised in such a way that it starts by identifying some examples of experiments and the need for measurements (*Flow System Interface*). If models of fluid phenomena were adequate, measurements would be superfluous. However, due to the tremendous problems, both in establishing good models and in solving the equations of the model experiments are essential. The chapter on *Computational Fluid Dynamics Interface* discusses the current state-of-the-art and relates it to measurement needs.

Optics is a classical discipline. After having been almost forgotten - or considered an old-fashioned discipline - optics has attracted a tremendous interest over the last two decades. However, advanced optics is still not a part of the normal engineering or physics curriculum. Thus a chapter on *Optics Toolkit* is included. The physical basis for light is presented with definitions of essential terms. The principles of optical measuring systems are described. Light scattering from small particles is discussed. This is an extremely important area in the context of the present book. A set of mathematical and computational tools for the design and preparation of optical instruments is given.

Data processing here is the manipulation of data from an optical instrument in order to extract information about the flow. In the chapter on *Processing of Random Data* the

Optical Diagnostics for Flow Processes
Edited by L. Lading *et al.*, Plenum Press, New York, 1994

emphasis is on practical methods for estimating mean values, variances, spectra and correlation functions. Methods for evaluating errors and uncertainties are presented.

Laser anemometry is the first of the actual measuring systems to be treated. The basic concepts of laser anemometers are presented. Different types of anemometers are identified as well as the various operational modes. Their spatial structure is analysed, the essential signal statistics are presented, and a number of signal processing schemes are discussed.

Laser anemometry is based on several novel technologies and may be even more so in the future. Fibres have been introduced in order to get a compact measuring head and to make it possible to mount the laser away from the experimental environment. Modulators can be an integral part of a fibre optical system. Diffractive (holographic) elements have a potential for reducing size and enhancing robustness. The chapter on *New Optoelectronic Technologies for Laser Anemometers* presents the state-of-the-art in these areas.

Solid-state devices and in particular semiconductor lasers and detectors are potentially attractive for laser anemometry but so far have not found widespread applications. The chapter on *New Technologies for Laser Anemometers* describes systems using such components. Several novel concepts for simultaneous measuring of different velocity components as well as resolving of the sign of the velocity are shown.

Measuring the size of particles suspended in a fluid is a classical problem. The phase Doppler method has come into widespread use over the last decade for the simultaneous measurement of the velocity and size of spherical particles. Combustion systems are extremely difficult to model; thus extensive measurements are needed. Unfortunately, it is also difficult to perform measurements in such harsh environments. But it can be done. The chapter on *Phase Doppler Anemometry and Its Application to Liquid Fuel Spray Combustion* presents the basic principles of phase Doppler anemometry and expands on the theory of light scattering. Measurements performed in order to investigate reciprocating internal combustion engines are presented.

Many industrially relevant flows involve several phases. Measurements in multiphase flows are particularly difficult. The chapter on *Two Phase Flow Measurements* presents methods for performing measurements in energy conversion systems based on phase transitions.

Particles suspended in the fluid are generally considered to be essential for laser anemometry. However, this is not necessarily true. The chapter on *Collective Light Scattering* presents the theory for coherent detection of collective light scattering from a large number of scatterers - e.g. molecules. The collective motion of the particles induced by turbulence may cause scattering of light from which turbulent transport can be determined. Although this type of laser anemometry is based on the Doppler effect and interference in the same way as the more common type of single-particle Doppler anemometers, the actual layout and implementation are quite different: a CO_2 laser is used in a reference beam mode of operation.

Laser anemometers can perform measurements with high spatial and temporal resolution, but this can only be done simultaneously for a few points in space. In order to obtain whole field measurements particle image velocimetry may be applied. This is a technique based on sequential images of a field of particles. From measurements of their displacements and knowledge of the time between the images the velocity field can be inferred. A pulsed laser and a good imaging system are needed. The computational effort required in order to produce a large number of velocity vectors is quite substantial. Both optical, electronic, and hybrid schemes have been applied. The chapter on *Particle Image Velocimetry* is about the optical configuration and the recording of images. Methods for inferring the velocity pattern by correlation methods are treated in the chapter on *Correlation Methods of PIV Analysis*.

Particle image velocimetry can be substituted by a relatively simple particle tracing scheme based on particle tracking using a video camera. This is described in the chapter on *Visualisation of Coherent Structures*. Novel methods for computer visualisation are also described.

Particle imaging and tracking will only give a two-dimensional flow pattern. However, the schemes may be modified so that three-dimensional structures can be identified. The chapter on *Three-Dimensional Particle Velocimetry* presents several methods that allow for 3D measurements including holographic schemes.

Not all flow patterns are equally likely - and some are physically impossible. The application of a priori knowledge and training can be utilised for the development of sophisticated methods for inferring flow patterns based on, e.g., particle image velocimetry measurements. Such methods are described in the chapter on *Flow Pattern Identification and Neural Network Processing.*

Tomography is a mathematical reconstruction technique used to obtain 3D structures from path averaged measurements. The chapter describes the mathematical background and gives examples of reconstruction.

Laser anemometry and particle image velocimetry are based on quasi-elastic light scattering, i.e. the only frequency shift encountered is caused by the macroscopic motion of the scatterers. Inelastic methods may be applied in order to probe the composition of the fluid. Inelastic methods may also be applied to acquire information about temperature and even velocity. Inelastic methods applied to the measurement of fluid dynamics are in general very expensive. The basic principles and novel schemes are presented in the chapter on *Inelastic Methods.*

The Editors

Fluid Systems and Analysis

THE FLOW SYSTEM INTERFACE

Poul S. Larsen

Department of Fluid Mechanics
Technical University of Denmark
DK-2800 Lyngby, Denmark

ABSTRACT

In every field involving flow processes there is a rising demand for accurate experimental data for temporal and spatial fields of velocity and other physical variables.

The reasons for such demands are many: Advances in computational fluid dynamics, particularly in turbulence modelling, require more complex variables to be resolved on finer scales and with greater accuracy to ensure model validation. Increasing industrial competitiveness in process equipment, power plants, aeronautics, ship building, etc., demands optimal design which can be achieved through reliable methods of prediction that have been verified by model experiments or full-scale experiments. Tighter regulation on particle emission and other forms of pollution demands reliable predictions of design performance and dependable monitoring in operation.

A number of basic and applied flow problems are considered to indicate the kind of deliberations going into the design of an experiment involving the application of optical methods and to illustrate the kind of experimental data that are needed.

INTRODUCTION

Optical methods in flow diagnostics have undergone a remarkable development during the last few decades and are now established as indispensable tools in both research and in routine measurements in industry. From relatively simple, but qualitatively important, visualization techniques employing tufts and streak photography have evolved a number of quantitative techniques, such as

- Laser Doppler Anemometry (LDA)
- Particle Tracking Velocimetry (PTV)
- Particle Image Velocimetry (PIV)
- Particle sizing
- Tomography by light scattering or light absorption
- Flourescence techniques for temperature or concentration.

Optical Diagnostics for Flow Processes
Edited by L. Lading *et al.*, Plenum Press, New York, 1994

In the above list of optical techniques, LDA is by far the most widely used method to get spatial velocity fields and turbulence data. Two- and three-component systems with coincidence filter to assure simultaneous measurements of two or three components of velocity may provide probability density distributions of a single or several components, higher order moments, double and triple velocity correlations, as well as time series for the calculation of autocorrelation or cross-correlation functions in the time domain and the corresponding energy spectra.

Among the contemporary problems of turbulence modelling can be mentioned the shortcomings of the ϵ-equation and the near-wall behavior of both two-equation models of first-order closure, such as the k,ϵ- and k,τ-models, and in differential and algebraic Reynolds stress models (DRS- and ARS-models) of second-order closure. Also, the prediction of swirling flows leaves much to be desired. These models, that are widely used today in solving applied problems, require only one-point statistics. However, two-point statistics will most certainly enter the next generation of turbulence models. LDA-techniqes can, in principle, provide all needed velocity correlations, although it would be desirable to have special optical units for the recording of spatial correlations. Also, probes giving velocity-pressure correlations within the flow are still needed.

In each LDA experiment, the sample size and sample frequency should be chosen carefully to ensure the expected accuracy and resolution for given turbulence scales (see, for example, Durst, Melling and Whitelaw[1], Larsen and Buchhave[2,3], and George[4]). Also, bias corrections for finite size of measuring volume may be necessary, say in thin shear layers and in the near-wall region.

It should be realized that it can be quite time consuming to obtain detailed data over sizeable two- or three-dimensional flow regions. For random sampling, using counter processors, for example, covering 1000 locations in space, each of 10,000 data samples per location, recorded at a mean sampling rate of 100 Hz would require about 30 hours, not counting time for traversing the rig. If all data are kept it would require about 1 Gbyte of storage capacity.

Clearly, the ability of a test rig to maintain steady and reproducible global test conditions is a prerequisite to acquiring reliable data. Drift of a wind tunnel fan, slow changes of the temperature, accoustic pulsations, etc. may not warrant seeking high accuracy of individual measurements. As in all experimental work, data gain in value when qualified by careful estimates of their accuracy.

Particle tracking, using high-speed cinematography or video recording, in either the plane of a light sheet or using a stereo setup, is a classical approach that has been revitalized by use of pulse laser techniques, high-resolution film material, video recorders, holography and efficient computer image analyses. Particle tracking is indepensible, e.g. when the dimensions of the flow system is so small that LDA or other methods are excluded, as in some biological flows[5].

A number of other optical methods based on elastic or inelastic scattering or absorption of radiation may provide properties other than velocity, such as size and concentration of particles, temperture and component concentration in gas mixtures or in solutions. Such techniques, as well as others mentioned briefly above, are elaborated in later section of this volume.

Because of the size of the literature and the variety of flow problems it is not possible to present a systematic and comprehensive summary of the subject. Instead, a number of specific flow problems will be discussed as case studies to illustrate the application of some optical methods. These problems are drawn mostly from our own work. They will illustrate some strong sides of optical methods as well as some difficulties. First, however, we turn to some general considerations of the design of experiments.

ANATOMY OF AN EXPERIMENT

One way to examine a given experiment is to consider the following aspects,

- Physics of process
- Mathematical models
- Objectives
- Experimental method
- Scaling of experiment
- Physical design, including optical access
- Experimental design, strategy, expected accuracy
- Results, processing, interpretation
- Difficulties, corrective measures, achieved accuracy.

Insight into the physical process and possible models (correlations, governing equations, rheological model, turbulence model, etc.) guide in fixing the specific objectives of the experiment. These objectives may be extended or relaxed when selecting the experimental method which, along with the physics, determine if and how the actual process may be scaled up or down.

Geometrical, kinematic and dynamic dimensional analyses, including the consideration of spatial and temporal turbulence scales, are essential tools in deciding on the possible scaling. In most cases true scaling is impossible and the least important requirements must be abandonned and their influence on the results must then be estimated and be part of the documentation.

Following the physical design of test rig and instrumentation (hardware) comes the experimental design (brainware), which includes strategies on spatial sample locations, sample rate and total size, as well as estimates of the expected accuracies.

After data acquisition, processing and display (software) comes the interpretation in light of the physics and models. Various difficulties may be anticipated in advance or, less attractively, be realized after completion of the experiment. The well-known practice of exploratory studies, carried all the way to display and interpretation of processed results, including verification of expected accuracy, is therefore a worth while effort.

SOME CASE STUDIES

To illustrate some of the considerations of the foregoing section we now discuss the following case studies:

1. Turbulent stagnation flow heat transfer (benchmark for model verification)
2. Two-dimensional benchmark test (is it two-dimensional?)
3. Electrostatic precipitator (turbulence and secondary flow)
4. Membrane filtration (weak secondary flow, mass transfer)
5. Jet mixing in parallel or cross-flow (concentration distributions).

Case 1 concerns the turbulent jet impinging perpendicular on a flat heated plate, which was one of two test cases at a recent workshop on refined turbulence modelling[6]. Today's turbulence models, when used with efficient CFD-codes, can often predict velocity and turbulence quite well. This is particularly true for free and wall bounded shear flows, partly because model development relies on empirical input from such flows, e.g. the idea of production-dissipation equilibrium. However, this condition is far from true near a reattachment point or near a stagnation point. This fact, and other shortcomings of most models, shows up in the inability of models to predict the normal Reynolds stress along the stagnation line, see Figure 1. Clearly, accurate data are needed to discover shortcomings and to guide ongoing efforts on turbulence modelling. Axisymmetric flows are held to be the most reliable twodimensional flows in regard to mean fields. Still, the case of the stagnating jet on a flat plate requires careful alignment and axisymmetrical surroundings, and data should confirm such conditions.

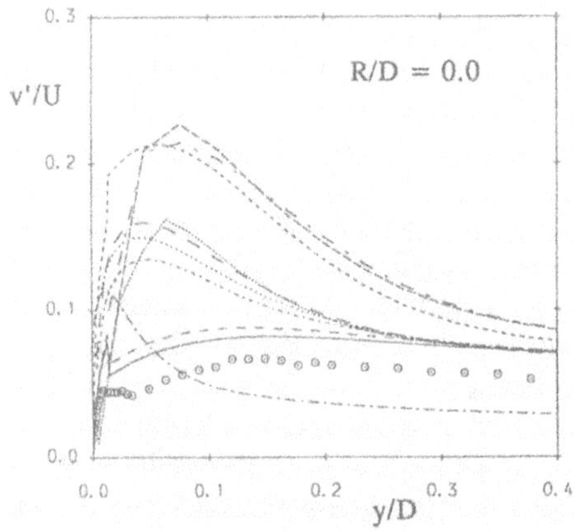

Figure 1. Normal Reynolds stress component (perpendicular to the wall), normalized by mean velocity of jet, v-rms/U, versus distance from wall along stagnation line, y/D. Dashed and solid lines are various Reynolds stress model predictions, and symbols are experimental hot wire data for unheated plate[7]. Jet defined as fully developed turbulent flow (Re = 23000) issuing from pipe of diameter D, at distance y = 2D above plate. Work shop test case[6].

Case 2 concerns efforts to acquire experimental data for twodimensional, low-Reynolds-number, near-wall, turbulent flows, including recirculation regions, also to be used for verification and improvement of turbulence models. Figure 2 shows measurement domain and computational domain of the fence-on-wall test case. The 40 mm high by 10 mm thick fence is mounted on the longer side wall of a 600 x 300 mm wind tunnel section, 760 mm downstream of the contraction and inlet which is fitted with a turbulence generating grid. Using a two-component back-scatter system, LDA-data have been collected over the region shown in Figure 2, in the mid-plane of the wind tunnel. The 2:1 aspect ratio of the wind tunnel and the 600:40 aspect ratio of the fence should ensure two-dimensionality of the near-wall flow in the mid-plane. To test this, and in order to compute the stream function of the mean field, the distribution of divergence of the smoothed LDA-data was computed[8]. The results in Figure 3 show the moderate, but not negligible, departures from two-dimensionality, as well as the resulting streamline plot based on smoothed LDA-data corrected to satisfy zero divergence.

There are several conclusions that can be drawn from this study. Whenever possible, data for mean fields should be checked for consistency with the conservation of mass. It is difficult to establish truly twodimensional test cases for plane mean flows, and the implications of using such data, or smoothed versions hereof, should be studied, particularly in regard to higher order moments. Also, to be useful for comparison to computational results, the experimental data should cover a large domain, both upstream and downstream of the obstacle. Here, it is important to specify as many quantities as possible at the inflow boundary, including the dissipation ϵ which enters the usual k,ϵ-, DRS- and ARS-models. Since ϵ can not be measured directly, indirect methods are

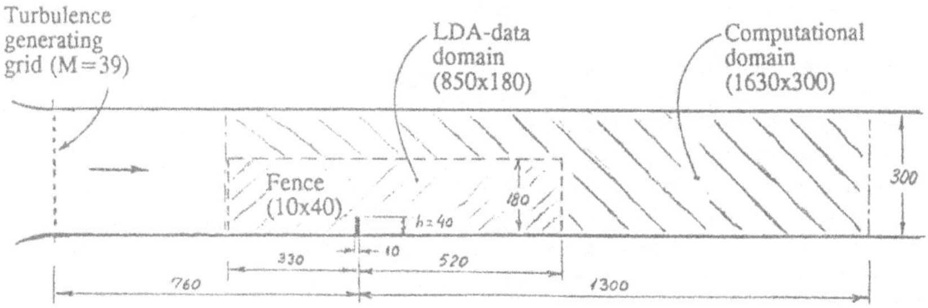

Figure 2. Computational domain and LDA-measurement domain for fence-on-wall test case[8].

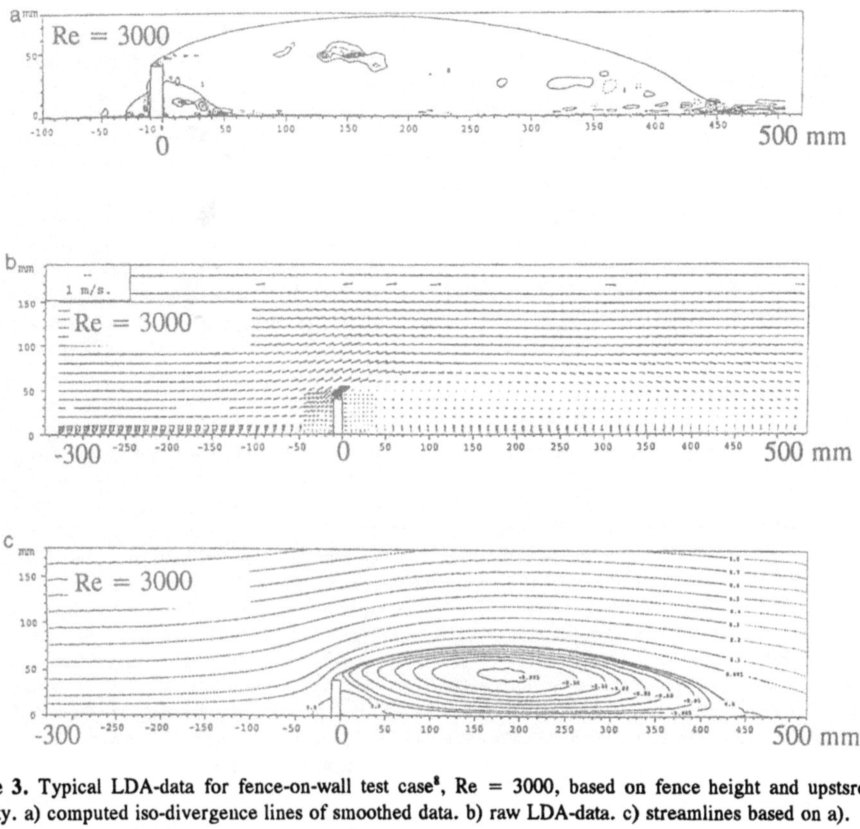

Figure 3. Typical LDA-data for fence-on-wall test case[8], Re = 3000, based on fence height and upstsream velocity. a) computed iso-divergence lines of smoothed data. b) raw LDA-data. c) streamlines based on a).

required. Given the turbulent kinetic energy k from measured normal Reynolds stresses, ϵ can be estimated from k and a length scale L, which can be computed from a measured autocorrelation function. The time-autocorrelation function, using also the Taylor hypothesis, can give an estimate of a spatial length scale. It is still better, although much more difficult and time-consuming, to measure directly spatial correlation functions, say using two LDA-systems or PIV. As mentioned in the introduction, the next generation of turbulence models may be based on two-point statistics. In its simplest form, this may involve use of the Taylor microscale as a variable (Jovanovic, Ye and Durst[9]). Can the spatial distribution of this microscale be measured efficiently for complex flow? - and how?

Case 3 concerns the use of LDA to study velocity and turbulence distributions in a laboratory model of a wire-plate electrostatic precipitator. Here, corona discharge from regularly spaced pins along wires charge the gas and hence the suspended seeding particles which then drift to collector plates under the influence of the electrostatic field (Figure 4a). The nonuniform field also acts on the charged gas and creates an ordered secondary flow in the form of axial rolls superposed the uniform bulk flow. In addition, the flow becomes highly turbulent when the ratio of axial fluid inertia to transversal electrical body force becomes small[10]. Both effects lower the performance of the electrostatic precipitator[11], suggesting that flow management should be an integral part of electrostatic precipitator design.

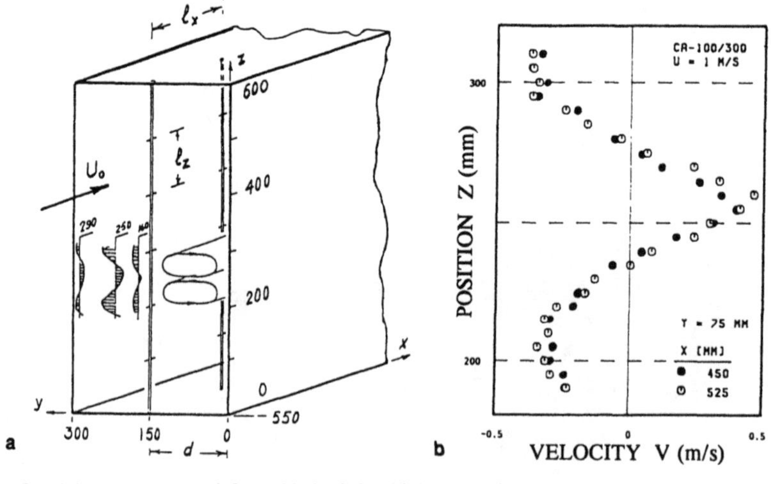

Figure 4. a) Precipitator geometry, 0.3 m wide by 0.6 m high test section with barbed wire electrodes. b) Vertical traverse of transversal velocity v(z), measured by LDA[10] midway between electrode and plate, at two axial locations, indicating axial rolls of secondary flow superposed the axial bulk flow.

Because of the intense electrostatic field, physical probes can not be inserted into the flow and LDA becomes an indispensable instrument. With optical access through the transparent bottom wall of the test facility, a two-component, back-scatter LDA-system has been used to map velocity and turbulence fields, see sample in Figure 4b. Particle sizing and concentration measurements could likewise be employed to verify computational models on precipitator efficiency. Further information on the detailed process of particle precipitation can be obtained by studying the plus-flux and the minus-flux of particles as function of distance from the collector plate. As a simple measure, the probability distributions of transversal velocity component, shown in Figure 5, indicate decreasing rms-values, a nearly constant mean drift velocity that approaches about 0.09 m/s at the plate, and a clear indication that no particles leave the collector plate[12]. The latter is not surprising since the seeding particles were liquid droplets which would adhere to the collector plate once having arrived due to the electrical drift. But the data also suggest that all particles have been charged.

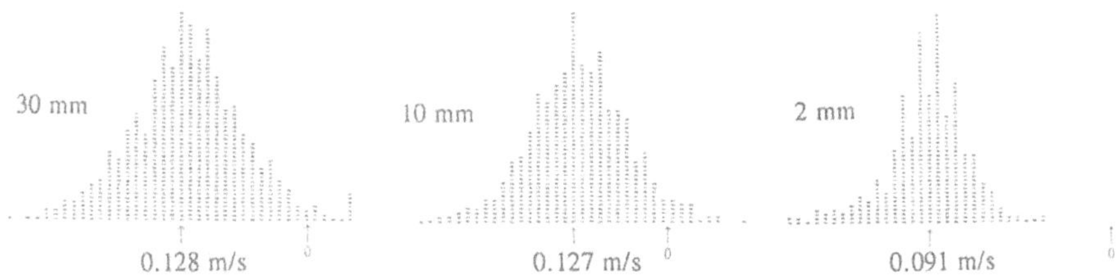

Figure 5. Probability density distributions of transversal velocity component at distances y = 30, 10 and 2 mm from collector plate, showing that no particles move away from the plate[12].

Case 4 concerns the need for velocity distributions of primary and secondary flows of Newtonian and non-Newtonian liquid solutions in narrow (< 1 mm wide) plane and curved ducts of membrane filtration units. Can LDA with frequency shift handle a ratio of 1/200, or less, of secondary to primary flow velocity? The axial mean velocity was about 1 m/s, while the unbalanced normal-stress induced secondary flows attained a maximum of about 0.005 m/s. LDA-results[13] were obtained using a frequency shift of 600-1000 kHz, which should be compared to the velocity Doppler frequency of about 8 kHz. For the highly viscous laminar flow, the LDA-results indicated rms-values of the order of 1-2 times the mean velocity. Nevertheless, acceptable results for mean values of velocity were obtained, see Figure 6. The reason for this is believed to be the everpresent sources of optical and electronic noise, which appeared to be near-Gaussian, centered around the signal frequency, and of a band-width many times the discrete frequency step of digital filtering in the 8-bit counter processor.

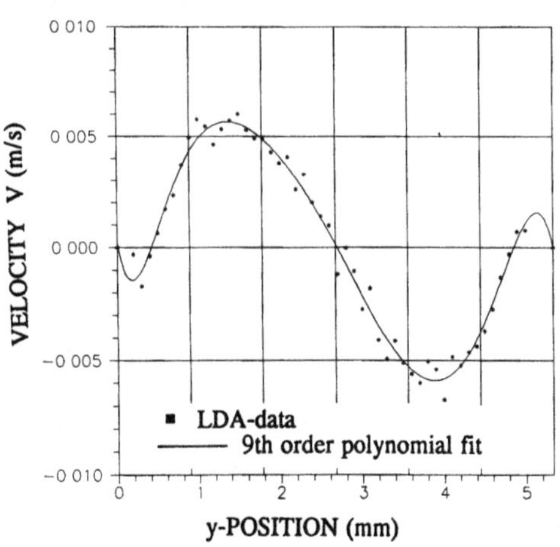

Figure 6. LDA-data of secondary flow profile[13], 1.1 mm from side wall of 3 x 5.33 mm straight duct, for 2% viscarin solution in water at mean velocity of 1 m/s.

Next, the mass transfer through a semi-permeable membrane on the wall causes the development of thin (0.01-0.05 mm thick) concentration boundary layers, having increased viscosity. Local measurements seem to be excluded because of the small dimensions, but integrated concentration distributions through the boundary layer next to a transparent membrane can be readily recorded through simple light absorption measurements[14]. This simple technique requires use of a died solute, such as blue dextran, which absorps light in a narrow band. The calibration is straightforward, using standard cuvettes. When using an observation spot of about 0.01-0.02 mm, the technique also allowed the recording of wavelengths (on the order of 0.1 mm) of the socalled striping phenomenon, appearing as an instability of the process[15], as well as allowing the discovering of some unsteady phenomena.

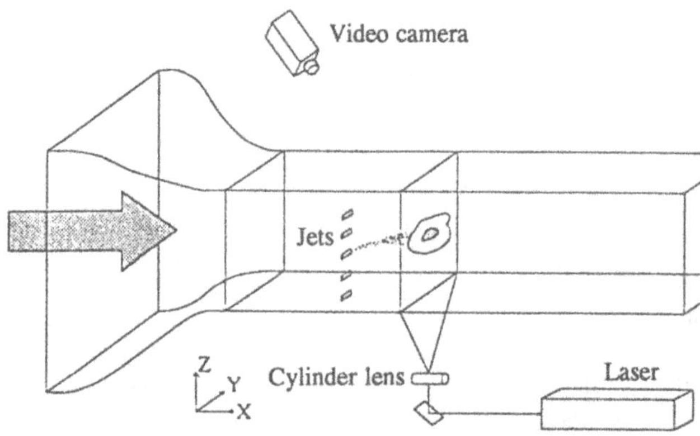

Figure 7. Laser sheet tomography of jet in a cross-flow[16].

Finally, case 5 concerns the mixing of a jet flow with a parallel flow or a cross-flow. Seeding the jet flow with micron-sized particles, quantitative concentration distributions was obtained by laser sheet tomography[16,17], see Figure 7. The necessary calibration may be obtained by making separate recordings of the light sheet in a uniformly seeded channel flow. Video recordings and image analysis, correcting for both the nonuniform intensity obtained in calibration records and for image distortions from geometrical optics, are essential tools in this technique. The exposure time of the recording medium and the flow velocity determine the spatial resolving power in terms of size of coherent turbulent scales that can be registred. To get mean concentration fields, typically the average of 16 single video images, each having an exposure time of about 1/60 s, have been used, Figure 8. However, pulse laser illumination is required to arrest rapid phenomena such as small scale coherent structures of a turbulent flame front[18].

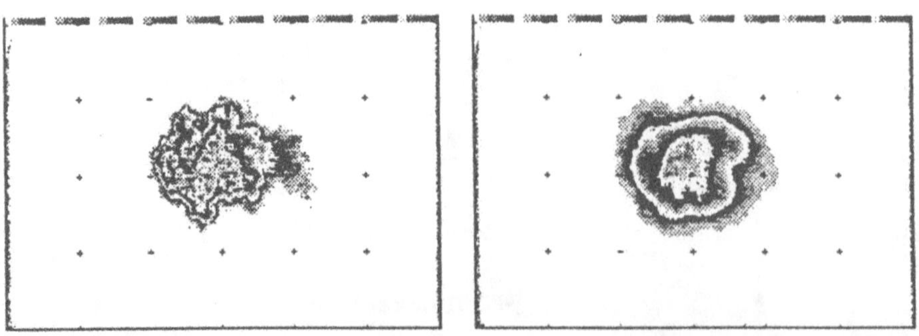

Figure 8. Iso-concentration contours[16] (cyclic scale at top, for relative concentration from 0 to 1.0). Single jet in parallel flow, 37 diameters downstream. Single processed image (left) and average of 16 images (right).

CLOSURE

The case studies discussed above are but a few amongst the many that exist in contemporary research and development. Other cases would bring out some of the same requirements for use of optical diagnostics, as well as other challenges to the experimentalist. Optical techniques are powerful and will continue to play a major role in experimental fluid mechanics.

REFERENCES

1. F.Durst, A.Melling and J.H.Whitelaw, "Principles and practice of laser-Doppler anemometry", 2nd.ed., Academic Press (1981).
2. P.S.Larsen and P.Buchhave, Flow measurements: Why, what and how?, *Dantec Information*, Part 1, no.1:2 (1985).
3. P.S.Larsen and P.Buchhave, Flow measurements: Why, what and how?, *Dantec Information*, Part 2, no.2:2 (1986).
4. W.K.George,W.K., Quantitative measurement with the burst-mode laser Doppler anemometer, *Exp.Thermal and Fluid Sci.* 1:29 (1988).
5. N.F.Nielsen, P.S.Larsen, H.U.Riisgård and C.B.Jørgensen, Particle and fluid motion in the gill system of Mytilus edulis: video recordings and numerical modelling, *Mar.Biol.* 116:61 (1993).
6. M.A.Leschziner and B.E.Launder, eds., 2nd ERCOFTAC-IAHR Workshop on refined flow modelling, 15-16 June 1993, UMIST, University of Manchester Institute of Science and Technology, Manchester, UK (1993).
7. D.Cooper, D.C.Jackson, B.E.Launder and G.X.Liao, Impinging jet studies for turbulence model assessment, Part I: Flow-field experiments, (to appear in *Int.J.Heat Mass Transf.*) (1993).
8. O.A.Jørgensen and U.Marxen,U., Turbulent flow over ribs (in Danish), Report AFM EP 93-06, Master's thesis, Dept.of Fluid Mechanics, Techn.Univ.of Denmark (1993).
9. J.Jovanovic, Q.-Y.Ye and F.Durst, 1992, Refinement of the equation for the determination of turbulent micro-scale, Report LSTM 349/T/92, Univ. of Erlangen-Nürnberg, Germany (1992)
10. J.N.Sørensen, P.S.Larsen and J.Zamany, Experimental and numerical analysis for flow in negative corona precipitator, *in:* Proceedings of 8-th Symposium of Turbulent Shear Flows, München, Sept.11-13, pp.13-3-1 to 13-3-6 (1991).
11. P.S.Larsen and S.K.Sorensen, Effect of secondary flows and turbulence on electrostatic precipitator efficiency, *Atmospheric Environment*, 18:1963 (1984).
12. J.Zamany,J., Modelling of particle transport in commercial electrostatic precipitators, Ph.D. dissertation, EF 316, Dept. of Fluid Mechanics, Techn.Univ.of Denmark, Sept. (1992).
13. B.Gervang and P.S.Larsen, Secondary flows in straight ducts of rectangular cross section, *J.Non-Newtonian Fluid Mech.*, 39:217 (1991).
14. P.S.Larsen, E.M.Christensen and N.F.Nielsen, Optical absorption technique in the study of membrane filtration processes, *in:* Proceedings of the Sixth International Symposium on Applications of Laser Techniques to Fluid Mechanics, July 20-23, 1992, Lisbon, Portugal, pp.31.2.1- 31.2.6 (1992).
15. P.S.Larsen, A simple stability model for the striping phenomenon in ultrafiltration, J.Membrane Sci., 57:43 (1991).
16. E.Andresen, P.S.Larsen and K.Dam-Johansen, Jets in a cross flow: Mixing studies by light sheet visualization, *in*: "Experimental and Numerical Flow Visualization", ed.s Bhalighi, et al ASME 1993, FEC-Vol. 172:293 (1993).
17. L.Böhme, Investigation of turbulent jet with and without chemical reaction (in Danish), Report AFM EP 93-11, Master's thesis, Dept.of Fluid Mechanics, Techn.Univ.of Denmark (1993).
18. Y.Zhang and K.N.C.Bray, Laser tomography of a highly turbulent unconfined premixed flame, *in:* Proceedings of the Sixth International Symposium on Applications of Laser Techniques to Fluid Mechanics, July 20-23, 1992, Lisbon, Portugal, pp. 22.5.1-22.5.5 (1992).

COMPUTATIONAL FLUID DYNAMICS INTERFACE

Simon M. Fraser

Energy Systems Division
University of Strathclyde
Glasgow UK

INTRODUCTION

Computational Fluid Dynamics (CFD) has made tremendous advances during the last twenty years due to the increasing speed and reducing cost of the computer hardware. The software has also been improved substantially with more user-friendly pre- and post-processors and improved solvers. The major difference between solid mechanics and fluid mechanics, using either finite element or finite difference techniques, is that the equations for fluid mechanics are more difficult to represent by the above approximation techniques; in particular when there is no acknowledged equation to represent the fluid flow phenomenon, such as turbulence, the simulations require to be validated before a confidence level can be established. High quality data such as can be obtained with laser anemometry is required in order to validate CFD applications.

There are two main areas which have to be addressed when using validation techniques:-

1) the partial differential equations (PDE's) cannot be solved analytically and either a finite element or a finite difference technique must be used to represent the equations as a set of linearised simultaneous equations which can be solved by various techniques which therefore cannot be an accurate solution to the original PDE's. The way in which the equations are solved may also introduce inaccuracies due to the method by which the schemes allow for the convection of variables from grid point to grid point (false diffusion in finite difference).

2) The main equations for fluid mechanics which can be approximated as above are the equations of Continuity, Momentum and Energy plus any other transported scalar variable such as concentration but there are no equations which adequately represent the effect of turbulence in the flow and all CFD codes model the effect of turbulence in some way. This inevitably leads to errors in the predicted flow pattern.

Optical Diagnostics for Flow Processes
Edited by L. Lading *et al.*, Plenum Press, New York, 1994

EQUATIONS OF MOTION

The essential features of unsteady LAMINAR incompressible isothermal fluid flow can be represented by the following equations in a Cartesian co-ordinate system :-

CONTINUITY

$$\frac{\partial u}{\partial x} + \frac{\partial v}{\partial y} + \frac{\partial w}{\partial z} = 0$$

MOMENTUM

$$\rho\left(\frac{\partial u}{\partial t} + u\frac{\partial u}{\partial x} + v\frac{\partial u}{\partial y} + w\frac{\partial u}{\partial z}\right) = X - \frac{\partial p}{\partial x} + \mu\left(\frac{\partial^2 u}{\partial x^2} + \frac{\partial^2 u}{\partial y^2} + \frac{\partial^2 u}{\partial z^2}\right)$$

$$\rho\left(\frac{\partial v}{\partial t} + u\frac{\partial v}{\partial x} + v\frac{\partial v}{\partial y} + w\frac{\partial v}{\partial z}\right) = Y - \frac{\partial p}{\partial y} + \mu\left(\frac{\partial^2 v}{\partial x^2} + \frac{\partial^2 v}{\partial y^2} + \frac{\partial^2 v}{\partial z^2}\right)$$

$$\rho\left(\frac{\partial w}{\partial t} + u\frac{\partial w}{\partial x} + v\frac{\partial w}{\partial y} + w\frac{\partial w}{\partial z}\right) = Z - \frac{\partial p}{\partial z} + \mu\left(\frac{\partial^2 w}{\partial x^2} + \frac{\partial^2 w}{\partial y^2} + \frac{\partial^2 w}{\partial z^2}\right)$$

Similar equations can be developed for a Polar co-ordinate system.

The above equations are generally referred to as the NAVIER-STOKES equations which were derived on the basis on a linear dependence of the normal and shear stresses on the local rate of deformation eg

$$\tau_{xy} = \mu\left(\frac{\partial v}{\partial x} + \frac{\partial u}{\partial y}\right)$$

This assumption of linearity is not based on physical laws and there might be some doubt as to their validity since there are no general analytic solutions which can be compared with test data; however for the limited range of particular 2-D analytic solutions such as a laminar boundary layer and laminar flow between flat plates etc, the solution given by the Navier-Stokes equations is identical to the experimental data and this gives some confidence in the general validity of the assumptions.

The generally accepted form of these equations for turbulent flow is based on the assumption that the velocity at any particular position in the flow will vary with respect to time but, over a suitably long period of time, the velocity can be characterised in terms of a mean velocity and a fluctuating component which will have a statistically meaningful variance and standard deviation.

Thus $\qquad u = \overline{u} + u'$

where $$\overline{u} = \frac{1}{t}\int_0^t u\,dt$$

and by definition $\overline{u'} = 0$

In order to analyse the statistical turbulence, higher order fluctuating quantities are used which have non-zero mean values eg

Variance $\overline{u'^2}$

Standard Deviation $\sqrt{\overline{u'^2}}$

The fluctuations in the velocity components are due to the passage of eddies through the fluid and the physical effect manifests itself in an APPARENT VISCOSITY increase and this concept of a turbulent apparent viscosity which is considerably larger than the actual viscosity of the fluid is a fundamental feature of turbulence modelling. Applying the concept of a time averaged mean velocity with a superimposed fluctuating component to the momentum equations gives rise to stresses based on the fluctuating components which are additional to the laminar normal and shear stresses indicated above. These additional stresses are referred to as the APPARENT or REYNOLDS stresses and they are composed of normal and shear components in each of the three directions eg in the x-direction there will be one additional normal component and two shear components.

Normal $\sigma'_x = -\rho\,\overline{u'^2}$

Shear $\tau'_{yx} = -\rho\,\overline{u'v'}$; $\tau'_{xz} = -\rho\,\overline{u'w'}$

[**NB** the product terms such as $u'v'$ are negative and hence the shear stresses are positive]

The turbulence statistics of the type indicated above are those which can be of very great importance and are most usefully obtained using laser systems since they, in general, have non-interfering probe volumes and are capable of measuring high quality data.

The nett effect of treating the turbulence as above is that the Navier-Stokes equations for steady flow with no body forces are modified as below :-

$$\rho\left(\overline{u}\frac{\partial\overline{u}}{\partial x} + \overline{v}\frac{\partial\overline{u}}{\partial y} + \overline{w}\frac{\partial\overline{u}}{\partial z}\right) = -\frac{\partial\overline{p}}{\partial x} + \mu\nabla^2\overline{u} - \rho\left(\frac{\partial\overline{u'^2}}{\partial x} + \frac{\partial\overline{u'v'}}{\partial y} + \frac{\partial\overline{u'w'}}{\partial z}\right)$$

$$\rho\left(\overline{u}\frac{\partial\overline{v}}{\partial x} + \overline{v}\frac{\partial\overline{v}}{\partial y} + \overline{w}\frac{\partial\overline{v}}{\partial z}\right) = -\frac{\partial\overline{p}}{\partial y} + \mu\nabla^2\overline{v} - \rho\left(\frac{\partial\overline{u'v'}}{\partial x} + \frac{\partial\overline{v'^2}}{\partial y} + \frac{\partial\overline{v'w'}}{\partial z}\right)$$

$$\rho\left(\overline{u}\frac{\partial\overline{w}}{\partial x} + \overline{v}\frac{\partial\overline{w}}{\partial y} + \overline{w}\frac{\partial\overline{w}}{\partial z}\right) = -\frac{\partial\overline{p}}{\partial z} + \mu\nabla^2\overline{w} - \rho\left(\frac{\partial\overline{u'w'}}{\partial x} + \frac{\partial\overline{v'w'}}{\partial y} + \frac{\partial\overline{w'^2}}{\partial z}\right)$$

When solving problems using this approach there are therefore nine additional terms to be considered for a complete analysis.

As an example of the detailed measurements which experimentalists can provide Figure 1 is based on data obtained by Klebanoff [1]. Even for a relatively straight forward flow such as a turbulent boundary layer it can be seen that the fluctuations are not equal (the turbulence is not isotropic) and very detailed measurements have been made very close to the wall surface (see the inset plot). Note that, as indicated above, the values of u'v' are negative on the plot.

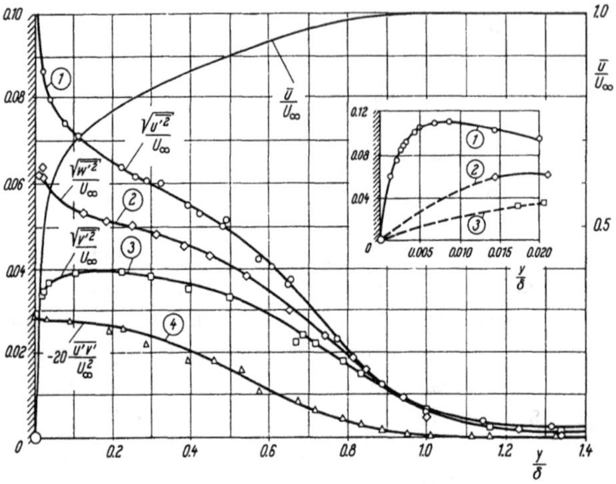

Figure 1 Mean and Turbulent Velocity Data in a Turbulent Boundary Layer
(Courtesy McGraw-Hill - Schlichting Boundary Layer Theory p 534)

NB The above data was obtained using hot-wire anemometry and illustrates the detail that can be obtained with that type of equipment; however the limitation of hot-wire is the relatively low level of turbulence that can be measured accurately and the effect of directional ambiguity on the signal.

TURBULENCE MODELLING

Since the nature of turbulence is complicated many approaches have been made to model turbulence ranging from simple to complete Reynolds stress models. One of the basic problems which has to be faced is that the nature of the turbulent fluctuations is a time dependent phenomenon which cannot be simulated on a normal computer (the latest multi-processor fast computers are now able to simulate simple turbulent structures but at enormous computer cost and it will be many years before they can simulate "real" engineering flows). It is therefore necessary to **MODEL** the effect of turbulence and the simplest of these is the Boussinesq [2] approach which replaces the time-dependent quantities with a relationship based on the local gradient of the MEAN velocity.

$$\textit{Thus} \quad \tau = -\rho\overline{u'v'} = \rho\epsilon\frac{\partial\bar{u}}{\partial y} \quad \textit{where } \epsilon = \textit{eddy kinematic viscosity}$$

and if a value can be found for ϵ then the shear stress can be calculated.

MIXING LENGTH APPROXIMATIONS

This approach has been found to be very useful in boundary layer, jet and wake flows where the eddy viscosity can be related to a typical length scale and the local gradient of the mean velocity; this approach is normally referred to as the Prandtl mixing-length hypothesis.

By making assumptions of this type and dealing with the reduced equations

$$Thus \quad \tau = \rho\epsilon\frac{\partial\bar{u}}{\partial y} = \rho\, l^2\left|\frac{\partial\bar{u}}{\partial y}\right|\frac{\partial\bar{u}}{\partial y}$$

appropriate to boundary layers, jets and wakes it is possible to reduce the equations to Ordinary Differential equations (ODE's) which can then be solved iteratively for known boundary conditions.

eg flow in a plane 2-D shear layer Figure 2 can be reduced to the ODE $F + F''' = 0$ by assuming that the mixing length l is proportional to the local width of the shear layer and by making use of the stream function relationship $\psi = UaxF$ where U = freestream velocity, x= stream direction and a = constant.

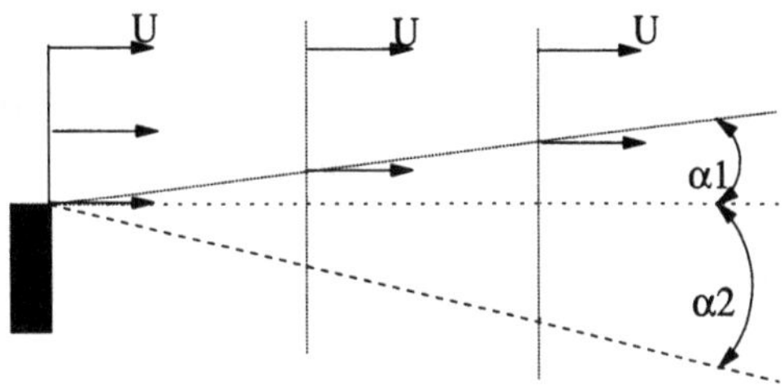

Figure 2 Flow Development in a Mixing Shear layer.

From the solution obtained for the ODE the components of velocity can be calculated as shown in Figure 3.

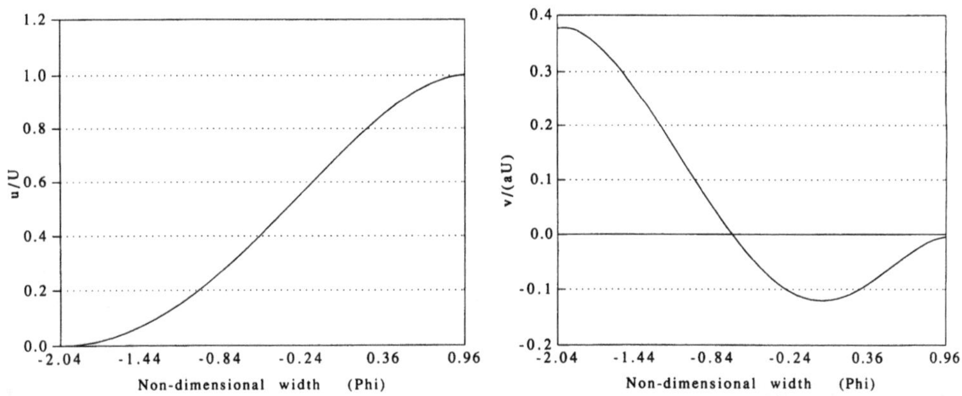

Figure 3 Velocity Profiles Across the Mixing Shear Layer

Equivalent solutions can be found for the types of flow indicated above but when the flow cannot be simplified in this way then more complicated turbulence models must be used.

Most CFD solvers will include an option to use the mixing length hypothesis for solving turbulent flow but the solution must then be restricted to essentially two-dimensional flows in boundary layers, jets and wakes.

TWO EQUATION MODEL $k-\epsilon$

Probably the most widely used turbulence model in current CFD simulations is the k-epsilon model which does not have a very high computer overhead penalty and, in general, can be applied to most fluid flows. k is the turbulence energy due to the fluctuating components :-

$$k = [\overline{u'^2} + \overline{v'^2} + \overline{w'^2}] / 2$$

and epsilon is the rate of dissipation of the turbulence energy to heat.

A Boussinesq relationship is assumed between the mean strain terms and the Reynolds stresses (using the alternating sub-notation for convenience) :-

$$-\rho\overline{u_i'u_j'} = \mu_T \left[\frac{\partial U_i}{\partial x_j} + \frac{\partial U_j}{\partial x_i}\right] - \frac{2}{3}\rho\delta_{ij}k$$

where δ_{ij} is the Kroniker delta (= 0 when i = j and = 1 when i = j).

There is a simple relationship between the apparent turbulent viscosity and the local values of k and epsilon. From dimensional considerations we can show that

$$\mu_T \propto \rho\, V_{scale}\, L_{scale}$$

The typical velocity scale is related to the kinetic energy of turbulence ie V_{scale} is proportional to $k^{1/2}$.

The typical length scale associated with the turbulent eddies can be evaluated from the k and epsilon values

$$L_{scale} \propto \frac{k^{3/2}}{\epsilon}$$

Substituting for these gives $$\mu_T = \rho\, C_\mu \frac{k^2}{\epsilon} \quad \textit{where } C_\mu \textit{ is a constant}$$

The local variation of k and epsilon are modelled in the CFD simulation by the same type of transport equations as the other fluid dynamic equations and as such can only be calculated within known boundary conditions.

REYNOLDS STRESS MODELS (RSM)

Reynolds stress models solve directly for the six additional shear stress components as derived previously and therefore require six additional transport equations to be solved rather than the two required for the k - epsilon model. However the RSM also requires the evaluation of a turbulence length scale and this is usually obtained by solving an epsilon equation as above. Thus the RSM needs to solve seven equations overall compared with the two required for k-epsilon and this is reflected in the additional computer overhead used. It should not be assumed that the RSM is always more accurate than the k - epsilon model and the general rule should be to use the minimum model consistent with the flow analysis.

DATA REQUIREMENTS FROM LDA

CFD solvers can only simulate flows within known boundary conditions; such conditions can be represented by walls, axes of symmetry, inlet and outlet flows etc. The flow domain is sub-divided into a grid of nodes and within each grid volume conditions are assumed to remain constant; one of the problems associated with simulation is called "grid refinement" ie starting with a coarse grid (quick solution) and gradually refining the grid (number of grid nodes in each direction and distribution of the grid) until the simulated values do not change.

Walls

The boundary conditions associated with walls are usually self-evident ie the flow must obey the no-slip condition but the application of this condition will depend on whether the flow is laminar or turbulent.

Axes of Symmetry

An axis of symmetry imposes a simple condition of no flux across the surface and is simple to implement. In all applications the axis of symmetry boundary condition should be used where appropriate to reduce the computational domain required.

Inlet Flow

It is necessary to specify the inlet flow conditions at one or more inlet surfaces; this will normally consist of the velocity profile at inlet and the magnitude of the variables which are transported into the flow domain.

Outlet Flow

It is normal to specify an outlet boundary which is sufficiently far removed from the flow of interest so as not to interfere with the upstream condition.

LAMINAR FLOW

For laminar flow there is a need to provide evaluation data for both velocity vectors and boundary conditions. As stated previously because of the linearisation of the PDE's and the solver scheme used there are inaccuracies in the simulation provided by CFD.

Figure 4 shows the effect of using a "standard" solver compared with a specialist solver to simulate flow over a backward-facing step. In this case the false-diffusion has caused the simulation to under-estimate the length of the re-circulation zone. The specialist solver predicts the correct flow to within 2% as compared with experimental data.

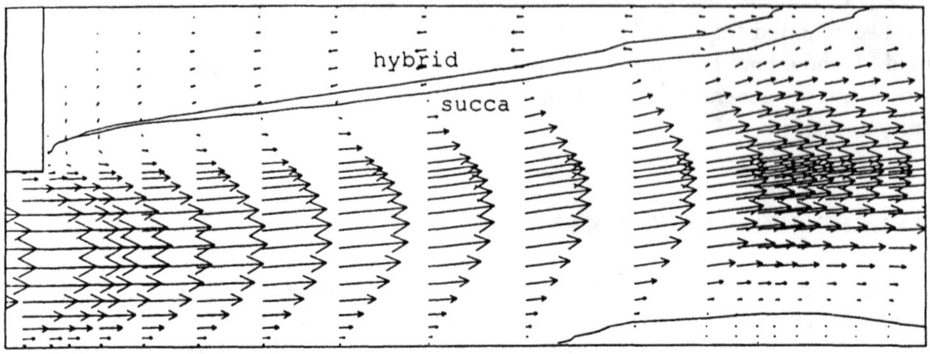

Figure 4 Comparison of Simulated Flow with Standard and Non-standard Solvers

TURBULENT FLOW

The majority of "real" engineering problems involve turbulent flow and it is probably in this area that LDA can help the CFD modellers most. As described above the modelling of turbulence is all that can be done at this point in time and therefore the relevance of the model in predicting flows is crucial to obtaining a good simulation. The same boundary conditions must be addressed as for laminar flow :-

Walls

The wall boundary in turbulent flow will also obey the no-slip condition but because of the more complicated structure of the boundary layer close to the surface (laminar sub-layer, buffer layer etc) it is not easy to represent in the simulation. In this case it is usual to use wall functions such as the log-law of the wall to input the shear stress appropriate to the effect of the wall. Additionally the buffer layer is a region where turbulence generation and dissipation are important and therefore special steps have to be taken to identify these quantities in the near wall grid nodes.

Axes of Symmetry

These are similar to the laminar flow simulation.

Inlet Flow

These are represented as for the laminar flow but in addition the turbulence condition of the flow appearing at the inlet must be specified; in some cases the flow in the main domain is very sensitive to the inlet turbulence quantities.

Outlet Flow

The outlet boundary is specified in the same way as for laminar flow.

Figure 5 shows the comparison between simulated and measured turbulence at various locations in the turbulent flow behind a backward-facing step.

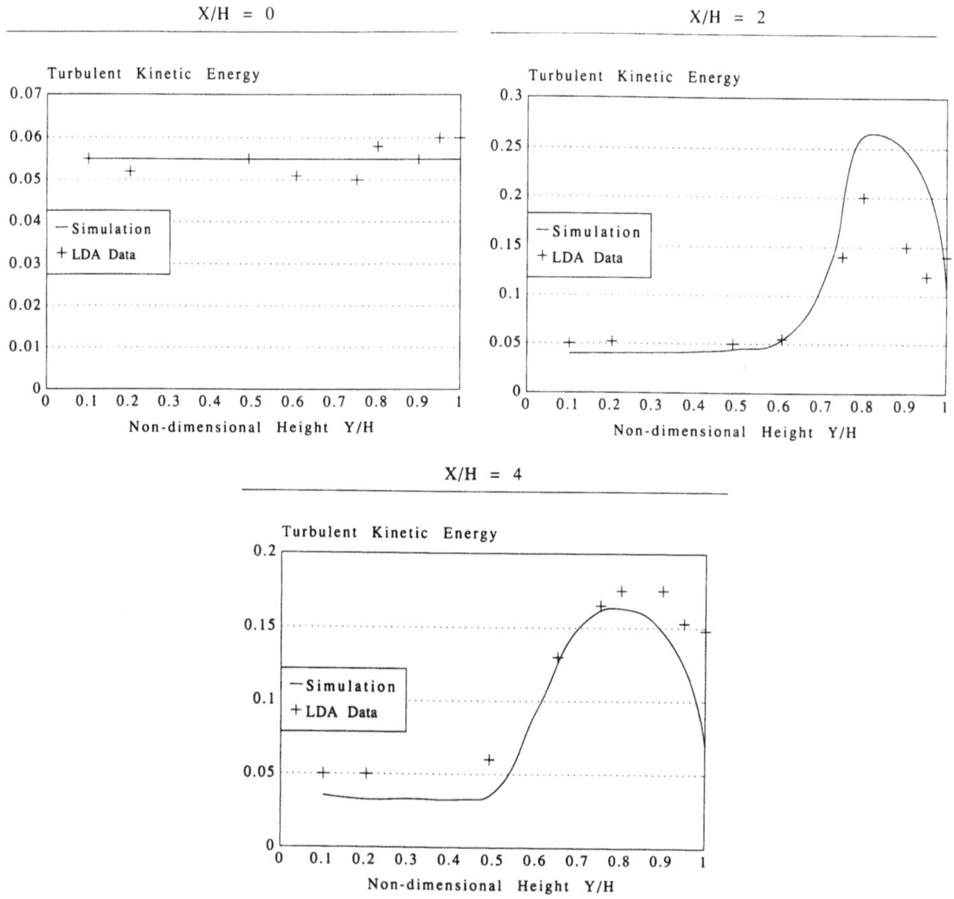

Figure 5 Simulated and Measured Turbulent Kinetic Energy in Flow Over a Backward-facing Step

FUTURE DEVELOPMENT

The future of CFD simulation is to provide better simulations making use of either better models of turbulence or larger multi-processor computers which would not require the modelling of turbulence. In either case there will still be a period during which detailed turbulence measurements of the type provided be LDA will be necessary; these measurements will be required for validation in three-dimensional flows and as such will require to produce point by point data detailing mean velocity vectors, turbulence spectra and indications of turbulent time and length scales.

REFERENCES

1) P.S. Klebanoff Characteristics of turbulence in a boundary layer with zero pressure gradient NACA Report No 1247 (1955).

Background Material

OPTICS TOOLKIT

Preben Buchhave

Physics Institute
Technical University of Denmark
DK-2800 Lyngby, Denmark

FOREWORD

The purpose of this article is to provide the reader with a brief review of optical concepts needed to be able to successfully use optical methods in flow diagnostics. Although one may jump right into optical measurements without more knowledge of optics than provided in basic engineering or natural science courses, optical methods for flow diagnostics are commonly known to be some of the more treacherous and difficult measurements to apply in practice, and often much time is waisted on invalid or erroneous measurements because some common optical principle has been violated or some limits to the applicability of the methods have been exceeded. Worse yet, one may actually in some cases be able to carry out a complete measurement without being aware of some subtle problem or bias, which may infect the data and perhaps cause erroneous conclusions to be drawn from the material.

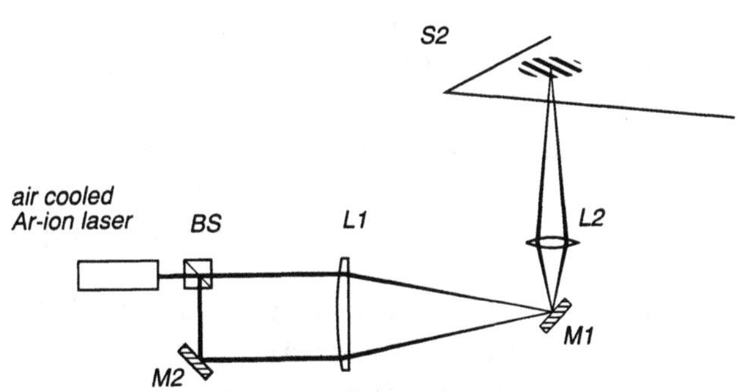

Figure 1. Projection of a laser focal region.

Optical Diagnostics for Flow Processes
Edited by L. Lading *et al.*, Plenum Press, New York, 1994

We cannot in the limited time and space available provide a complete course in optics. Instead we shall attempt to focus on some of the most important areas, which are known to cause troubles and lead the user astray. We shall use a form in which we provide some essential definitions and concepts and in addition some tools such as formulas, tables and other information. Descriptions and derivations will be kept to a minimum, but will will hopefully be sufficient to allow the reader to use the tools in a given situation. To acquire a deeper understanding of the underlying principles, the reader is referred to the up-to-date textbooks and reference material given in the list of references[1-6].

INTRODUCTION

Laser light is still, even for people who work with it on a daily basis, fascinating to observe and in possession of esthetic qualities, which delight and surprise the viewer. As an example look at a laser beam, which is focused to a small spot by a lens, L1 (see Figure 1). A mirror, M1, placed near the focal spot does not change the quality of the beam, but directs it towards the screen, S2, where it can be observed visually or picked up and displayed by a video camera.

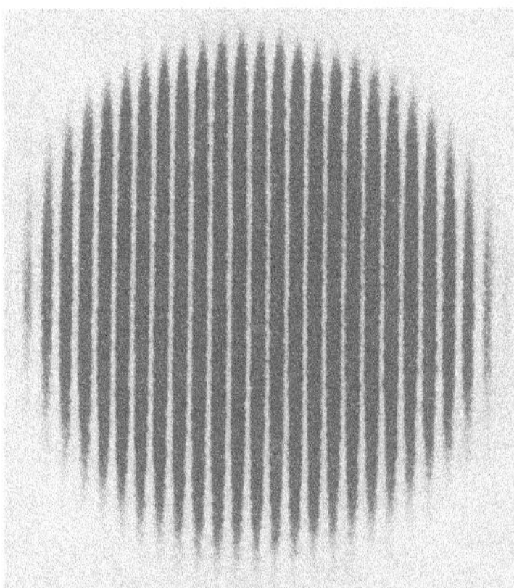

Figure 2a. Interference pattern.

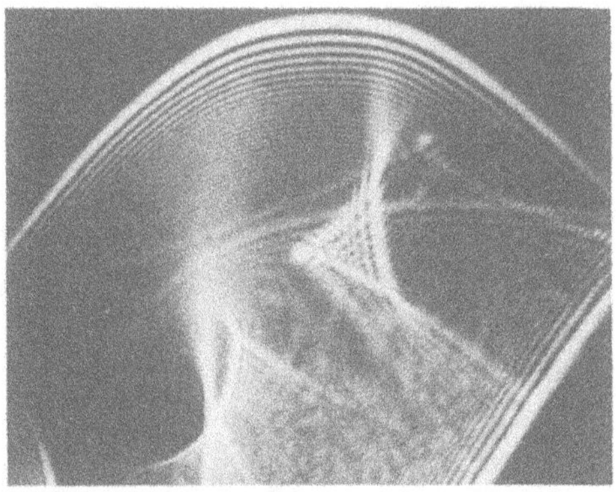

Figure 2b. Distorted interference pattern.

By means of lens L2 we can magnify and project the focus spot onto S2, and we see a nice, circularly symmetric distribution of intensity, which we know to be a Gauss function for a laser in its lowest order mode.

If we split off part of the beam by means of a beamsplitter, BS, and focus this second beam at the same focal spot as the first beam by means of mirror, M2, we see a nice interference pattern forming on the screen, Figure 2a. Clearly, this interference pattern is so regular that it must be useful for something! And indeed, the observation of this pattern or of distortions of this pattern or of light scattered or reflected off this pattern, e.g. by small particles moving through it, forms the basis for many optical measurement principles from interferometry over speckle interferometry to laser Doppler velocimetry and phase-Doppler particle sizing systems.

As a last variation of this experiment let us deform the mirror by applying a thin layer of silicone oil to the surface. The effect is spectacular and beautiful, Figure 2b. The regular fringe pattern is distorted, but it is not totally smeared out. Instead numerous high frequency patterns of different colors are formed. We suddenly see that the blue-green laser light consists of several different laser lines of different colors. Obviously, decoding the information in the new, complex interference pattern in a quantitative way is nearly impossible, but we may still enjoy the qualitative, esthetical impression.

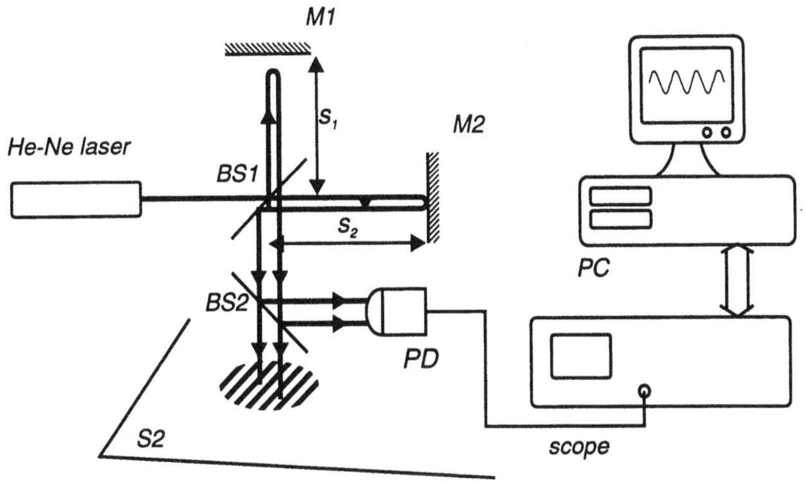

Figure 3. A simple Michelson interferometer.

Consider a second experiment: We now use a small He-Ne laser as a light source in an interferometric setup. This is one of the most widely used optical measurement principles: The laser light beam is divided into two beams by means of a beam splitter, BS1, see Figure 3. Each beam is returned and sent back along the direction from which it came by mirrors M1 and M2 and brought to overlap on the screen S2, where we may view the resulting light intensity. The intensity pattern tells us something about the separate paths of the two beams. We may be able to tell (measure) differences in the optical path lengths due to displacements or vibrations of the mirrors or due to changes in the medium between the beamsplitter and one of the mirrors caused by e. g. changing temperature or composition. If we are clever enough and the setup is stable enough, we may be able to form two identical optical beams or waves on the screen, and we then see the apparently paradoxical result that by adding two strong light beams together we may create total darkness on the screen, a result we may understand as an addition of two

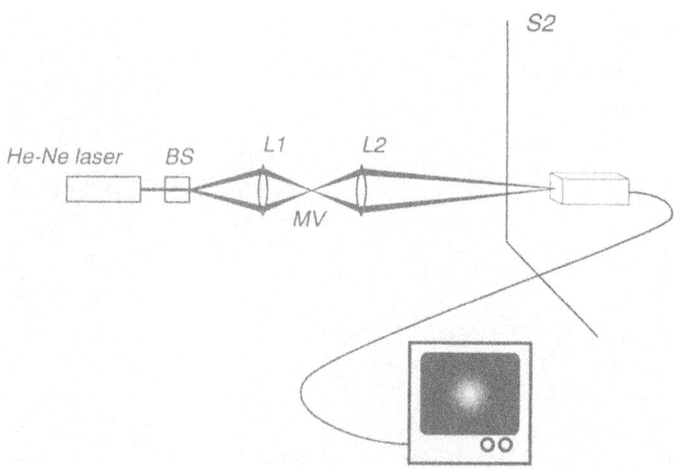

Figure 4. Laser interferometry using a Wollaston beamsplitter.

field strengths of opposite phase, but one which still goes a little against our "common sense". However, the point of this particular experiment this time is not to show the properties of interference, but to show what happens when we adjust the total path length difference between the beamsplitter and the mirrors to some particular value equal to an integer plus one half laser cavity length. What happens then is that the nice interference pattern seems to wash out and a signal picked up by a photodetector will show a very bad signal-to-noise ratio. Had we, unaware of this problem, tried to use the laser in an interference measurement (to measure length or vibration) or in an LDA (laser Doppler anemometer used to measure flow velocity), we would have got exceedingly bad signals, and we might have spent a long time adjusting the beamsplitter angle or the mirror positions without result. Clearly, there are some less obvious properties of gas lasers that we must understand before we can design or use such a system.

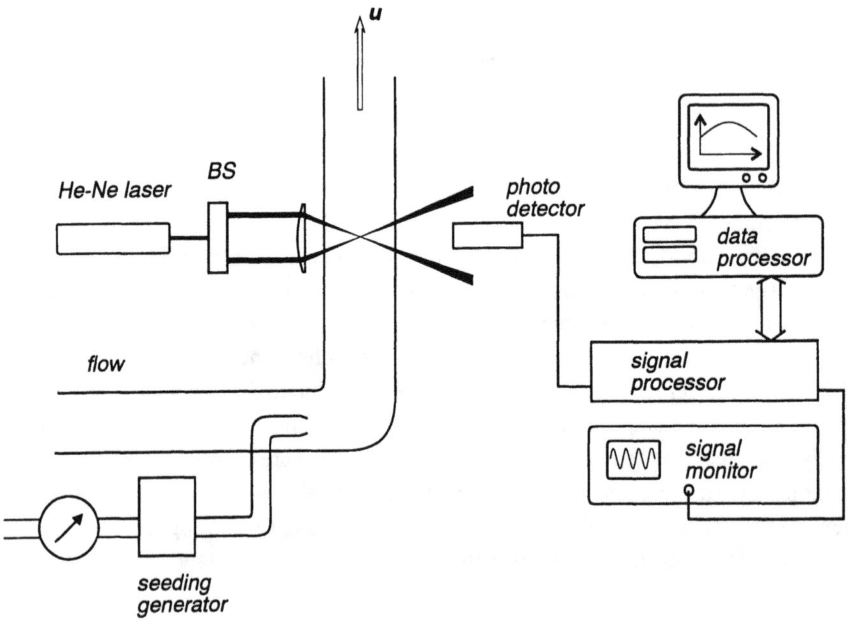

Figure 5. Simple, forward scatter LDA.

As a third example of experimental difficulties consider the simple interference experiment in Figure 4. Here we use a particularly simple beamsplitter called a Wollaston prism, or we could have used a calcite crystal. This type of beamsplitter splits one beam into two beams with slightly different direction of propagation. Again we form a region of overlapping beams by focusing the two beams into a common focal region (which we shall later call a measurement volume). To see what is happening, we project the focal region onto the screen S2 by means of lens L2. We may also use a CCD detector instead of S2 to project the resulting interference pattern to a greater audience on a TV screen. The disappointing result of this experiment is that we seem unable to obtain any interference fringes at all. This setup is not useful for optical measurement purposes, and we have to admit, that there must be some property of laser light, which we did not consider or know about, which prevents us from using this system for optical measurements. In fact, this experiment clearly demonstrates that light must have transverse directional properties (polarization) as we may select one or the other output beam from the beamsplitter by just turning the prism about the laser beam axis. The experiment demonstrates that for two beams to interfere the electric field vectors must oscillate in the same direction - the beams must have the same polarization.

As a fourth and last example in this introduction consider the optical measurement in Figure 5. This is an LDA experiment, commonly used to measure spatially and temporally resolved velocity in a flowing medium. The method relies heavily on the scattering properties of small particles accidentally present in the flow or deliberately added to the flow - so-called seeding particles or tracer particles. Assuming we are measuring in a clear medium such as filtered air or water, we shall find it necessary to add seed particles to the flow upstream in order to get scattered light from the measuring volume, scattered light which in turn can be detected by the photodetector. For the beginner LDA operator one of the more striking things about this measurement is how sensitive the whole measurement is to the type, size and amount (concentration) of the seed particles. Obviously, the first concern is that a single seed particle during its transit through the measuring volume should be big enough to scatter sufficient light into the detector. The received light power must give a detected signal of sufficient strength and signal-to-noise ratio for the signal processor to be able to work with the signal and bring out the information, in this case the Doppler frequency, which will be buried in various forms of noise. However, the particles should still be small enough to follow the flow faithfully. Already here we have an important problem, which is all too often neglected by the user.

Assuming this problem has been carefully dealt with, and the type and size of seed particles appropriate for this particular measurement have been found, we now encounter another important problem: We must have enough seed particles to give a near continuous flow of particles in order to get information about the time fluctuations of the velocity. Let us slowly increase the concentration of particles in the flow and observe the detector signal, which is displayed as the "monitor signal", and the operation of the signal processor, which will be indicated by a display of information such as "data rate" and "pct. validated signals" or even as a "signal-to-noise ratio". We then typically observe the apparently paradoxical course of events that at first (at a very low seed particle concentration) we see very few data per second (low data rate), but high signal quality. As we increase the seed particle concentration, we receive more and more scattered light, and we notice an increased data rate (even for a fixed velocity) and a continuing high signal quality. However, as we increase the concentration of seed particles even further, we suddenly se an abrupt drop in data rate and a deteriorating signal quality, and we can easily reach a point where very few or none at all valid data are detected in spite of the fact that lots of scattered light reaches the detector! Indeed, the signal on the monitor looks large, but the Doppler modulation has nearly vanished. This problem has to do with the concept of coherent and incoherent detection, which is vital for most optical systems using coherent light. Some detection systems use incoherent detection and will be unaffected by this problem. Others use coherent detection and will be strongly affected by a decrease in coherence of the detected light. If many particles are unavoidably present, some optical systems will fare a lot better than others, and this underlines the importance of being able to select the optimum optical configuration for a particular measurement task.

What is Light?

Light is an elastic vibration propagating as waves in a perfect elastic medium, called the ether, which permeates all transparent materials. Surprisingly, this statement was put forth, not as a result of empirical evidence, but as a conclusion deduced from abstract speculation on how nature must work in order to fulfill a principle of divine beauty and perfection. The author was René Descartes in "Principia Philosophiae", Amsterdam 1644. The wave theory was elaborated upon by many "natural philosophers", who tried to explain the various properties of light as they became known during the period from 1600 to the end of the 19th century. The (apparent) rectilinear propagation of light beams and the (apparent) absence of interference phenomena as they were known from e.g. water waves, did not fit the wave picture very well, and this led to an alternative explanation of the nature of light as a stream of small particles. This theory is still associated with the name of Isac Newton, who promoted this explanation in his work on optics. The color dispersion of light beams passing through a prism, which was also described by Newton, was explained by assigning particles of different mass to the different colors. However, also the particle explanation became more and more cumbersome, as more and more properties of light became known. The wave picture finally gained the upper hand after the successful theory of Fresnel, who combined Huyghens' principle of secondary wavelets issuing from a wavefront and the concept of interference put forward by Young. Fresnel's theory very successfully explained observable properties of light propagation, imaging and diffraction, but as to the fundamental nature of light, the situation was basically the same as it was at the time of Descartes: Experimental evidence pointing to the physical nature of light was lacking, and even Fresnel was left with a speculative explanation of elastic waves in a hypothetical medium, the "ether".

Real progress in the explanation of the nature of light had to await the theory of electricity and magnetism presented by James Clark Maxwell in 1867. One of the consequences of Maxwell's theory is the possibility of electromagnetic waves propagating in matter or even in empty space. The speed of propagation of these electromagnetic waves as calculated from Maxwell's theory, exclusively based on the knowledge of material constants, matched the known speed of light, and established for the first time the nature of light waves as electromagnetic waves. All optical phenomena could now be described as solutions to Maxwell's equations, and as it turned out, the theory could even survive the discovery of the theory of relativity.

As we now know, this classical picture had to be modified and expanded as new evidence on the nature of light was discovered. The first problem was the experimentally well established radiation from cavities (blackbody radiation). Only by limiting the possible energy levels of the radiating modes in a blackbody cavity to multiples of a certain minium energy packet, later called a radiation quantum or photon, did Planck succeed in presenting a theory for the blackbody radiation (Planck's law). Later the radiation quantum gained more substance as Einstein associated the exchange of a radiation quantum with the process of absorption and emission of light and in particular with the explanation of the photoelectric effect. Today the wave picture and the particle picture of light coexist as complementary sides of the same phenomenon in quantum optics. However, even today new aspects of the nature of light are being uncovered in the latest experiments on single atom or single molecule lasers and in cavity quantum electrodynamics, and we cannot exclude the possibility that our picture of what light really is will have to be modified or expanded yet again.

However, for the purposes of this course we can safely consider light an electromagnetic wave phenomenon, and the concept of the photon only enters into the picture as an explanation of the noise processes associated with the optical detection process (the semiclassical picture).

A LITTLE E&M

Although we shall not need to solve Maxwell's equations in this course, a little background knowledge of electromagnetism may be useful. The following primarily serves to introduce the basic concepts needed to understand laser optics and the functioning of optical components.

The electromagnetic theory is based on four observable field quantities:

The electric field strength, $\boldsymbol{E}$. This is a vector, which is defined by the force dF exerted on an elementary electric charge, dq, placed in the field: $dF=dq\,\boldsymbol{E}$.

The electric flux density, $\boldsymbol{D}$. This is a vector, which manifests itself by the charge deposited on a small, hypothetical capacitor with area of the plates dA, when the plates are normal to the field: $\boldsymbol{D}=dq/dA\;\boldsymbol{n}$, where $\boldsymbol{n}$ is a unit vector normal to the plate.

The magnetic flux density, $\boldsymbol{B}$. This is a vector, which may be measured by the force dF exerted on a moving charge dq moving with velocity $v=ds/dt$ according to the scheme: $dF=dq\;v\times\boldsymbol{B}=Ids\times\boldsymbol{B}$, where I is the current defined by $I=dq/dt$.

Magnetic field strength, $\boldsymbol{H}$. This quantity results from the presence of currents (moving charges). It may be defined as the magnetic effect of a small loop surrounding an area A carrying the current I: $\boldsymbol{H}=I/A\;\boldsymbol{n}$, the direction of $\boldsymbol{H}$ being along the normal to the plane of the loop, $\boldsymbol{n}$.

In empty space the electric field quantities $\boldsymbol{E}$ and $\boldsymbol{D}$ are proportional and so are the magnetic field quantities $\boldsymbol{B}$ and $\boldsymbol{H}$:

$$\boldsymbol{D}=\varepsilon_0\boldsymbol{E} \tag{1}$$

$$\boldsymbol{B}=\mu_0\boldsymbol{H} \tag{2}$$

In the presence of matter these simple relations may no longer hold, and there may be differences in the directions of the field vectors. In dielectric, nonmagnetic materials we may write:

$$\boldsymbol{D}=\underline{\underline{\varepsilon_0}}\boldsymbol{E} \tag{3}$$

where $\underline{\underline{\varepsilon_0}}$ is a tensor, and the above relation for $\boldsymbol{B}$ still holds.

For light waves in air or transparent liquids we may disregard the currents and the free charges and we may then write the fundamental equations governing the behavior of the field quantities as four coupled first order differential equations (Maxwell's equations).

By elimination these may be written as two uncoupled second order differential equations. Further simplifications may be introduced, if we may assume that the optical fields propagate near to some optical axis, which we shall denote the z-axis. We can then neglect the vectorial coupling of the field equations and look at just one of them, e.g. the equation for $\boldsymbol{E}$:

$$\nabla^2\boldsymbol{E}-\varepsilon\mu\frac{\partial^2\boldsymbol{E}}{\partial^2 t}=0 \tag{4}$$

This is the so-called wave equation. Any function of the general form $f(\boldsymbol{r}\text{-}\boldsymbol{u}t)$, which describes a wave phenomenon linked in space and time by the so-called phase propagation velocity $\boldsymbol{u}$, will be a solution to the wave equation.

The wave equation describes the propagation of an oscillating electromagnetic field. One of the consequences of the wave equations is that the electromagnetic field propagates with a velocity, given by the material constants:

$$c=\frac{1}{\sqrt{\varepsilon\mu}}. \tag{5}$$

In empty space the speed of light is given by:

$$c_0=\frac{1}{\sqrt{\varepsilon_0\mu_0}}=2.998\times10^8\,\mathrm{ms}^{-1}. \tag{6}$$

The wave equation for the electromagnetic field has a multitude of special solutions, depending on the particular boundary conditions. Some of the simplest solutions are the plane wave, the spherical wave and the dipole radiation. We shall here look only at the plane wave, as it displays the essential features of light waves.

We shall assume that the plane wave is a solution to the wave equation. We prove that by inserting the expression for the plane wave into the wave equation and showing that it satisfies that equation. At the same time we find a relation between the parameters describing the plane wave, which must be fulfilled in order for the expression to be a solution.

The electric field of a plane electromagnetic wave polarized in a given direction is written:

$$\boldsymbol{E} = \boldsymbol{E}_0 \cos(\boldsymbol{k} \cdot \boldsymbol{r} - \omega t + \phi_0) \qquad (7)$$

$\boldsymbol{E}_0$ describes the magnitude and the polarization direction of the field strength and ω is the angular frequency of the harmonically oscillating wave. It is related to the ordinary frequency by the expression $\omega=2\pi\nu$. $\boldsymbol{k}$, which is called the wave vector, describes the direction of propagation of the wave by the direction of the unit vector $\boldsymbol{k}/k$, where $k=|\boldsymbol{k}|$, and by the spatial period (wave length) λ through the relation $\lambda=2\pi/k$. ϕ_0 is a quantity, which defines the phase at the origin ($\boldsymbol{r}$=0, t=0).

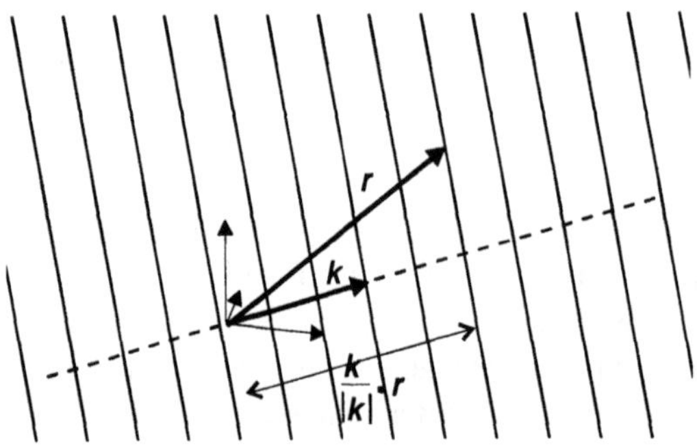

Figure 6. Plane electromagnetic waves.

Equation (7) describes a wave train consisting of an electric field vector, which varies harmonically in time and space. If we consider a fixed point in time, t, then $\boldsymbol{E}$ will have the same phase at all points for which $\boldsymbol{k} \cdot \boldsymbol{r}$ has the same value, i. e. in planes perpendicular to $\boldsymbol{k}$, as shown in Figure 6. Such a plane is called a wave front. We get the same phase for $\boldsymbol{r} \cdot \boldsymbol{k} = p \cdot 2\pi$, where p is an integer, which gives us the relation: $\Delta|\boldsymbol{r}| = \lambda = 2\pi/k$. If $\boldsymbol{r}$ is held constant we see, that $\boldsymbol{E}$ and $\boldsymbol{H}$ will have the same phase for $t = p \cdot 2\pi\omega$ or for $\Delta t = 2\pi/\omega = 1/\nu = T$, where T is the period.

When we insert equation (7) into Maxwell's equations, we find that a condition for solution is that:

$$k^2 = \omega^2 / c^2 \qquad (8)$$

or

$$\lambda \nu = c \cdot \qquad (9)$$

We also find:

$$\boldsymbol{k} \times \boldsymbol{H} = \omega \varepsilon \boldsymbol{E} \qquad (10a)$$

$$\boldsymbol{k} \times \boldsymbol{E} = -\omega \mu \boldsymbol{H} \qquad (10b)$$

$$\boldsymbol{k} \cdot \boldsymbol{E} = 0 \qquad (10c)$$

$$\boldsymbol{k} \cdot \boldsymbol{H} = 0 \qquad (10d)$$

from which we can see that $\boldsymbol{k}$, $\boldsymbol{E}$ and $\boldsymbol{H}$ form a right handed corner, i. e. $\boldsymbol{E}$ and $\boldsymbol{H}$ are always normal to the direction of propagation given by $\boldsymbol{k}$, as long as ε is a constant, not a tensor.

We thus have the following scalar connection between $\boldsymbol{H}$ and $\boldsymbol{E}$:

$$H = \frac{\omega}{k} \varepsilon E = \varepsilon c E. \tag{11}$$

Complex Notation

The complex notation for sinusoidally oscillating, plane electromagnetic waves is more compact and often easier to use in calculations than the real notation. The complex notation is a mathematical tool and is used only temporarily - the real physical solution to a problem is always found by taking the real part of the complex result.

In the complex notation a plane polarized wave looks this way:

$$\boldsymbol{E}^c = \boldsymbol{E}_0 \, e^{i(\boldsymbol{k}\cdot\boldsymbol{r} - \omega t + \phi_0)} \tag{12}$$

where $\boldsymbol{E}_0$ is real.

We arrive at the quantity, which represent the physical field, by taking the real part of the above expression:

$$\boldsymbol{E} = \mathrm{Re}\left[\boldsymbol{E}^c\right] = \mathrm{Re}\left[\boldsymbol{E}_0 \, e^{i(\boldsymbol{k}\cdot\boldsymbol{r} - \omega t + \phi_0)}\right] = \boldsymbol{E}_0 \cos\left(\boldsymbol{k}\cdot\boldsymbol{r} - \omega t + \phi_0\right). \tag{13}$$

Corresponding relations may be used for $\boldsymbol{H}$. Most often we denote the complex field quantities by $\boldsymbol{E}$ and $\boldsymbol{H}$ without the superscript, c.

$\boldsymbol{E}^c$ may be divided into a time independent part and a time dependent part:

$$\boldsymbol{E}^c = \boldsymbol{E}_0 \, e^{i(\boldsymbol{k}\cdot\boldsymbol{r} + \phi_0)} \, e^{-i\omega t} = \boldsymbol{E}_0^c \, e^{-i\omega t}, \tag{14}$$

where $\boldsymbol{E}_0^c = \boldsymbol{E}_0 \, e^{i(\boldsymbol{k}\cdot\boldsymbol{r} + \phi_0)}$. $\boldsymbol{E}_0^c$ is often called a phasor.

We may perform all linear algebraic operations on the complex quantities and then find the value for the physical quantity they represent by taking the real part of the complex result. Such linear operations include addition of optical fields as well as differentiation and integration of field quantities. However, this method is not valid when multiplying or exponentiating fields. In those cases special rules of calculus must be invoked. As an important example we may mention the time average of the product of two optical fields, which we need when computing intensity, correlation or spectrum. We must then use the following special rule:

$$\overline{\mathrm{Re}[A(t)]\mathrm{Re}[B(t)]} = \tfrac{1}{2}\mathrm{Re}\left[AB^*\right], \tag{15}$$

where

$$A(t) = A\, e^{-i\omega t} = |A| e^{ik\cdot r - i\omega t + i\phi_A} \quad \text{and} \quad B(t) = B\, e^{-i\omega t} = |B| e^{ik\cdot r - i\omega t + i\phi_B}. \tag{16}$$

Characteristic Quantities for the Plane Wave

As we have seen, the plane wave may be described by the following variables: The amplitudes $\boldsymbol{E}_0$ and $\boldsymbol{H}_0$, which describe the strength of the electric and magnetic field and their direction of oscillation, the wave vector $\boldsymbol{k}$, which describes the direction of propagation and determines the wavelength of the plane waves, the angular frequency ω, which describes the time variation of the harmonic oscillation, and the phase ϕ, which describes the state of the oscillation in relation to the state at the origin.

The most general expression for the plane wave in complex notation is

$$\boldsymbol{E}^c = \boldsymbol{E}_0 \, e^{i(\boldsymbol{k}\cdot\boldsymbol{r} - \omega t)} \tag{17}$$

where $\boldsymbol{E}_0$ is a complex vector, i. e containing two independent waves:

$$\boldsymbol{E}_0 = \boldsymbol{\varepsilon}_1 E_0 + \boldsymbol{\varepsilon}_2 E_0 \tag{18}$$

If we choose a coordinate system with positive z-axis in the direction of propagation, as shown in Figure 7, we may write for the complex electric field strength (similar considerations apply for the magnetic field strength):

$$\boldsymbol{E}_0 = \left(E_{0,x}, E_{0,y}, 0\right) \tag{19}$$

with

$$E_{0,x} = \left|E_{0,x}\right| e^{i\phi_x} \tag{20}$$

$$E_{0,y} = \left|E_{0,y}\right| e^{i\phi_y} \tag{21}$$

The two components may have a difference in phase:

$$\Delta\phi = \phi_x - \phi_y \tag{22}$$

where ϕ_x and ϕ_y are arbitrary initial phases for the two components along the x- and the y-

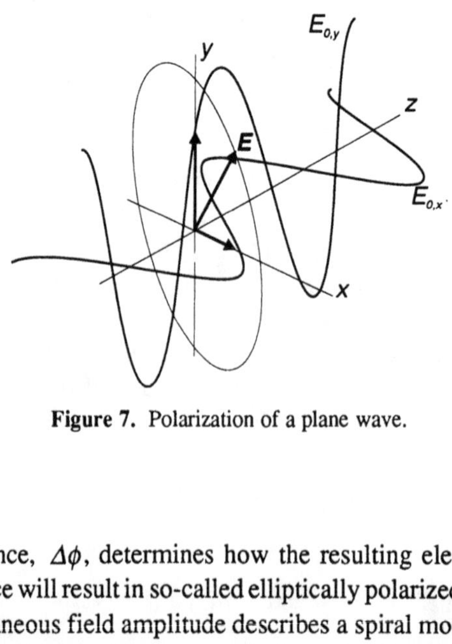

Figure 7. Polarization of a plane wave.

axis. The phase difference, $\Delta\phi$, determines how the resulting electric field oscillates. An arbitrary phase difference will result in so-called elliptically polarized light (see Figure 7). The end point of the instantaneous field amplitude describes a spiral motion in time and space. If the field vector is projected onto a plane perpendicular to the direction of propagation, the end point will be seen to describe an ellipse. If the phase difference is zero, the ellipse will degenerate into a straight line (linearly polarized light). The field vector maintains its direction in space given by the magnitude of the two components. If the phase difference is $\pi/2$ and the two components are equal in size, the field vector will again describe a spiral in space and time, but now the end point projected onto a plane perpendicular to the direction of propagation will describe a circle (circularly polarized light). The state of polarization of the electric and/or magnetic field is thus also a parameter which characterizes the field. Most often linearly polarized light is used in optical measurements because with known polarization direction one has a much better hold on reflectivities and optical path. When polarized light is incident on a plane surface, we often describe the polarization by its component parallel to the surface, which is denoted p-polarization, and its component normal to the surface, which is termed the s-polarization. The general state of polarization may be described by the matrix method or by the so-called Stokes' parameters.

Intensity and Energy

The flux of radiative energy in an electromagnetic wave may be found from Maxwell's equations. One can show that the energy flux into a small volume delimited by the surface A is given by:

$$-\int_A (\boldsymbol{E} \times \boldsymbol{H}) \cdot \boldsymbol{n}\, dA \tag{23}$$

By applying this expression on the plane wave, we get the energy flux per unit area normal to the direction of propagation of the wave, the energy current or radiation, often just called the intensity, denoted by I,

$$I = \overline{|\boldsymbol{E} \times \boldsymbol{H}|}\,, \tag{24}$$

where the overbar indicates time average in a time short in relation to the time scales of interest in the observation (short time average or period average).

By using the relations (11) and (12) for the plane wave, we get:

$$I = \tfrac{1}{2}\sqrt{\frac{\varepsilon}{\mu}}\left(\boldsymbol{E} \cdot \boldsymbol{E}^*\right) = \tfrac{1}{2}\varepsilon c |E_0|^2 . \tag{25}$$

The Quantum Properties of Light

The description of light as electromagnetic waves covers important phenomena such as the propagation of light, scattering of light, diffraction and refraction as well as interference between light beams. However, some physical phenomena concerning light exist, which cannot be described by the wave picture. These phenomena encompass primarily the interaction between light and matter. The classical example is the photoelectric effect, but also other phenomena belong to this group, e.g. the theory of the laser.

It is an experimentally verified fact that the number of photoelectrons, which are emitted from a photosensitive surface, is proportional to the total light energy incident on that surface. It is a reasonable result from a classical point of view, since we assume that the electrons are bound in the material with a certain binding energy, which must be supplied by the incident light. Perhaps one would have expected that the photoelectrons are emitted in an even stream at a rate depending on the intensity of the incident light, assuming that it would take a certain time to absorb the energy needed for the emission of an electron. However, the emission of the photoelectrons happens very unevenly and the fluctuations in the rate of emission of photoelectrons is rather high. Even if the field is very weak, there is a finite probability that a photoelectron will be emitted immediately after the light is turned on. These facts do not fit an explanation, which includes a "saving up" or even transfer of energy from the electromagnetic field to the atom or molecule.

Let us consider a variation of the photoelectric experiment in which we imagine that we can vary the frequency of the incident radiation without changing the amplitude of the light field, which reaches the photosensitive surface (think for example of a tunable laser, where the color of the laser light may be adjusted). How would the number of photoelectrons depend on the frequency?

Under the condition that there are no other competing processes than the photoelectric effect, we would experimentally see that the number of photoelectrons emitted per unit time would increase as the frequency of the incident light decreases, but that the energy of the liberated electron (that is the kinetic energy of the electron as it leaves the surface) would decrease. Thus we get more transitions, but the energy of each transition is reduced. This process may be described by an interaction, where the light in each individual transition delivers an amount of energy (a radiation quantum or a "photon") given by the formula $E = h\nu$, where h is a constant (Planck's constant, $h = 6.63 \cdot 10^{-34}$ Joule) and ν is the frequency of the incident light. The liberation of each photoelectron happens, as mentioned, at an accidental point in time, but the mean emission rate is give by the condition:

$$\nu_p = \frac{\eta \int P dA}{h\nu}, \tag{26}$$

where P is the power of the incident light (the irradiance), η is the quantum efficiency and ν_p is the mean rate of emission. The integral represents the total power on the detector surface, A. Later we shall consider the consequences of this theory for the detection of light in optical measurements and for the noise, which is generated in the detection process.

In conclusion we may state that for our purposes, we may consider light a physical phenomenon, which may be described as a classical electromagnetic wave interacting with matter by the transfer of energy in packets (photons) of a size proportional to the light frequency.

ELEMENTARY LAWS AND CONCEPTS

In this section we shall look at the basic tools for handling the elementary optical concepts described by the classical words reflection, refraction, diffraction, interference and coherence. As stated before, we do not intend to make a textbook presentation, but rather to summarize in a concise form the most useful tools for handling optics problems, which often occur along with work in optical diagnostics of fluid flows. All of these tools may be derived directly from Maxwell's equations, but are stated here without proof.

Reflection and Refraction

Since antiquity the basic facts about reflection have been well known: "Angle of reflection equals angle of incidence". The geometry is illustrated in Figure 8. The plane of incidence is defined by the incident beam and the normal to the surface, $\hat{\boldsymbol{n}}$. The angle of incidence α_i is the

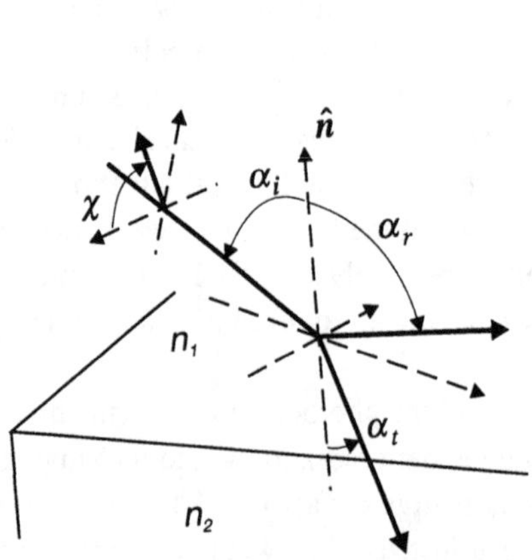

Figure 8. Reflected and refracted rays.

angle between the normal and the incident beam. The reflected beam will appear in the plane of incidence with an angle of reflection α_r between the normal and the reflected ray, which equals the angle of incidence:

$$\alpha_r = \alpha_i . \tag{27}$$

The polarization of the light is described by the angle χ in a plane normal to the beam between a line parallel to the surface and the direction of oscillation of the electric field. The angle $\chi = 0$ corresponds to an electric field oscillating normal to the plane of incidence, often called s-polarization. $\chi = 90°$ corresponds to light oscillating in the plane of incidence, the so-called p-polarization. Finally we should mention that the phase of the reflected light is altered relative to the progression of phase expected from the geometrical optical path length. The actual phase change is a complicated function of material properties of the surface as well as of the polarization and angle of the incident beam.

The refracted ray, see again Figure 8, is also located in the plane of incidence, and the angle of refraction α_t is, regardless of polarization, given by Snell's law:

$$n_1 \sin \alpha_i = n_2 \sin \alpha_t , \tag{28}$$

where n_1 and n_2 are the indices of refraction of the medium of the incident beam and the medium of the refracted beam, respectively.

The coefficients of reflection (fraction of incident intensity, which is reflected) are given by Fresnel's formulas:

$$\left.\begin{aligned} R_s &= \frac{\sin^2(\alpha_i - \alpha_t)}{\sin^2(\alpha_i + \alpha_t)} \\ R_p &= \frac{\tan^2(\alpha_i - \alpha_t)}{\tan^2(\alpha_i + \alpha_t)} \end{aligned}\right\} \quad \begin{aligned} &\text{only for } \alpha_i < \alpha_c \quad \text{when } n_2 < n_1 \\ &\text{If } \alpha_i > \alpha_c \quad R_S = R_P = 1 \end{aligned} \tag{29}$$

where R_p and R_s refer to p-polarized and s-polarized light, respectively.

Two cases are of special interest:

The critical angle

When a ray is incident on a surface from a higher towards a lower index, the angle of refraction is greater than the angle of incidence according to Snell's law, see Fig. 9. Above a certain angle of incidence, α_c, the critical angle, corresponding to an angle of refraction of 90°, all the light is reflected. The critical angle is found from Snell's law:

$$\alpha_c = \sin^{-1}\left(\frac{n_2}{n_1}\right). \tag{30}$$

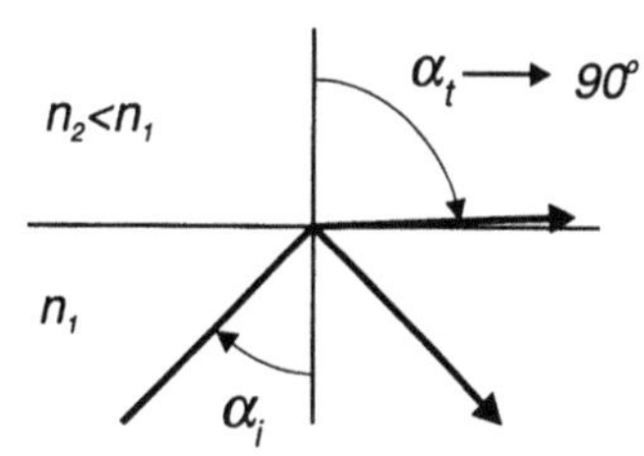

Figure 9. Critical angle.

Brewster's angle

At a certain angle of incidence, α_B, Brewster's angle, all p-polarized light is transmitted, see Fig. 10. This is seen from Fresnell's formula to happen when $\alpha_B + \alpha_t = 90°$ or

$$\tan \alpha_B = \frac{n_2}{n_1} . \tag{31}$$

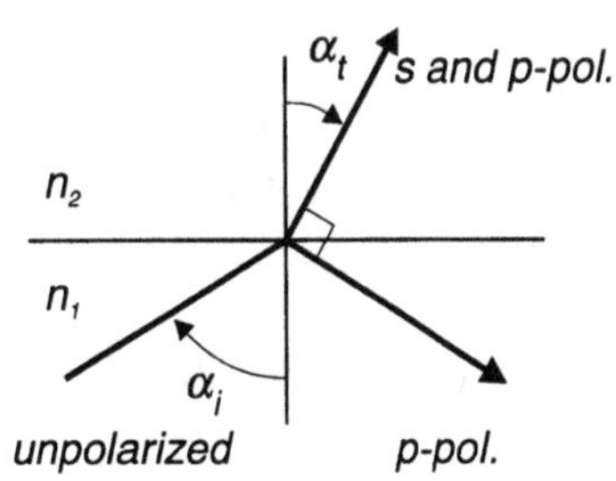

Figure 10. Brewster's angle.

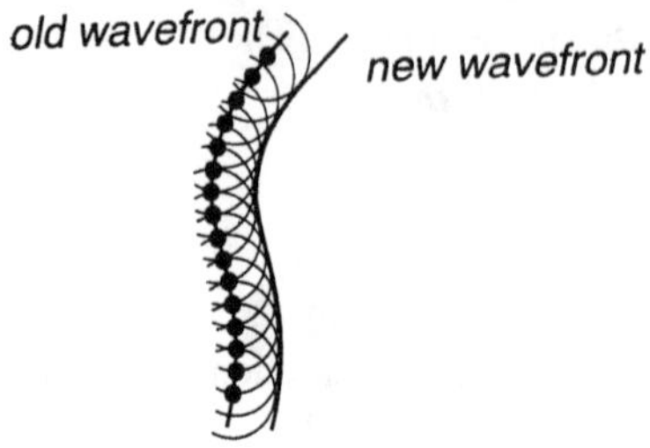

Figure 11. Huyghens' principle.

Brewster's experiment - that no s-polarized light is reflected at Brewster's angle - was instrumental in proving that light is a transversely polarized wave. The fact that total extinction occurs shows that there can be no longitudinal component of oscillation.

Diffraction

The term diffraction describes the fact that a light wave is apparently broken up (diffracted) by objects in its path. The various diffraction theories attempt to construct the diffracted light field by some assumptions about the reemission of light from the incident field - in a way they are all approximations to Maxwell's equations. The earliest diffraction theory is called Huyghens' principle. The propagation of a wave is predicted by assuming that all parts of the incident wave front reemit small wavelets. The wave front at a later time is found as the envelope curve of all the wavelets, see Figure 11.

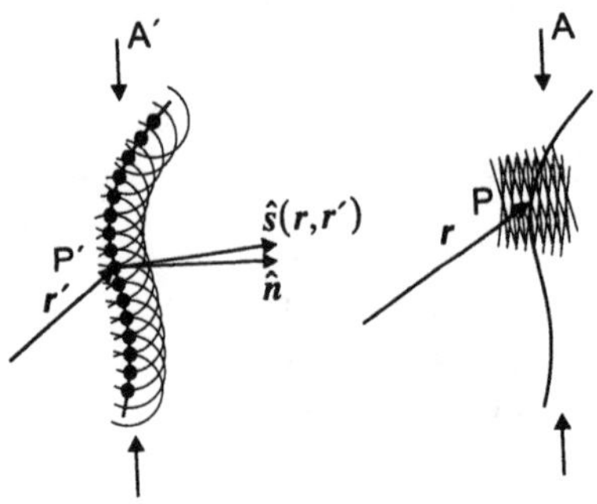

Figure 12. Huyghens-Fresnel principle.

Fresnel's diffraction theory was an extension of Huyghens' principle with the theory of interference. The wavelets from points P′ on the wave front A′ are added at a point P in phase. The strength of the spherical wavelet at P depends on the direction from P′ to P through a direction factor Λ. Today Fresnel's diffraction theory is expressed by the so-called Huyghens-Fresnel diffraction integral, which combined with the Sommerfeld boundary condition, may be written:

$$\Psi(\boldsymbol{r}) = \frac{k}{2\pi i}\int_{A'} \psi'(\boldsymbol{r}')\frac{e^{iks(\boldsymbol{r},\boldsymbol{r}')}}{s(\boldsymbol{r},\boldsymbol{r}')}\cos(\hat{\boldsymbol{n}},\hat{\boldsymbol{s}}(\boldsymbol{r},\boldsymbol{r}'))d^2\boldsymbol{r}, \tag{32}$$

where primed quantities refer to the diffracting aperture, while unprimed quantities refer to the observation plane, see Figure 12. $s(\boldsymbol{r},\boldsymbol{r}')$ is the distance between point $\boldsymbol{r}'$ in the input aperture and point $\boldsymbol{r}$ at the observation plane, while the direction factor turns out to become the cosine of the angle between direction unit vector from $\boldsymbol{r}'$ to $\boldsymbol{r}$, $\hat{\boldsymbol{s}}(\boldsymbol{r},\boldsymbol{r}')$, and the normal to the diffracting aperture, $\hat{\boldsymbol{n}}$.

In the Fraunhoffer approximation, $s(\boldsymbol{r},\boldsymbol{r}') >> k|r'|^2$, the field very far from the diffracting aperture, the far field, has the form of a Fourier transform of the field in the aperture:

$$\Psi(\boldsymbol{r}) = \frac{e^{-ikz}}{i\lambda z} e^{-i\frac{1}{2}k(x^2+y^2)/z} \int_{A'} \psi'(\boldsymbol{r}') e^{-i\frac{k}{z}(x'x+y'y)} d^2\boldsymbol{r}', \tag{33}$$

where we have used $s(\boldsymbol{r},\boldsymbol{r}') \cong z$ and $\cos(\hat{\boldsymbol{n}},\hat{\boldsymbol{s}}(\boldsymbol{r},\boldsymbol{r}')) \cong 1$. Thus, in most cases in optics, where we deal with relatively large F-numbers, we can use this expression to compute the far field diffraction pattern from any given 2-dimensional aperture.

We can get the same simplification of the Huyghens-Fresnel integral as in the Fraunhoffer approximation by multiplying the field in the aperture with a quadratic phase factor, which cancels the quadratic terms resulting from a power series expansion of the optical path length

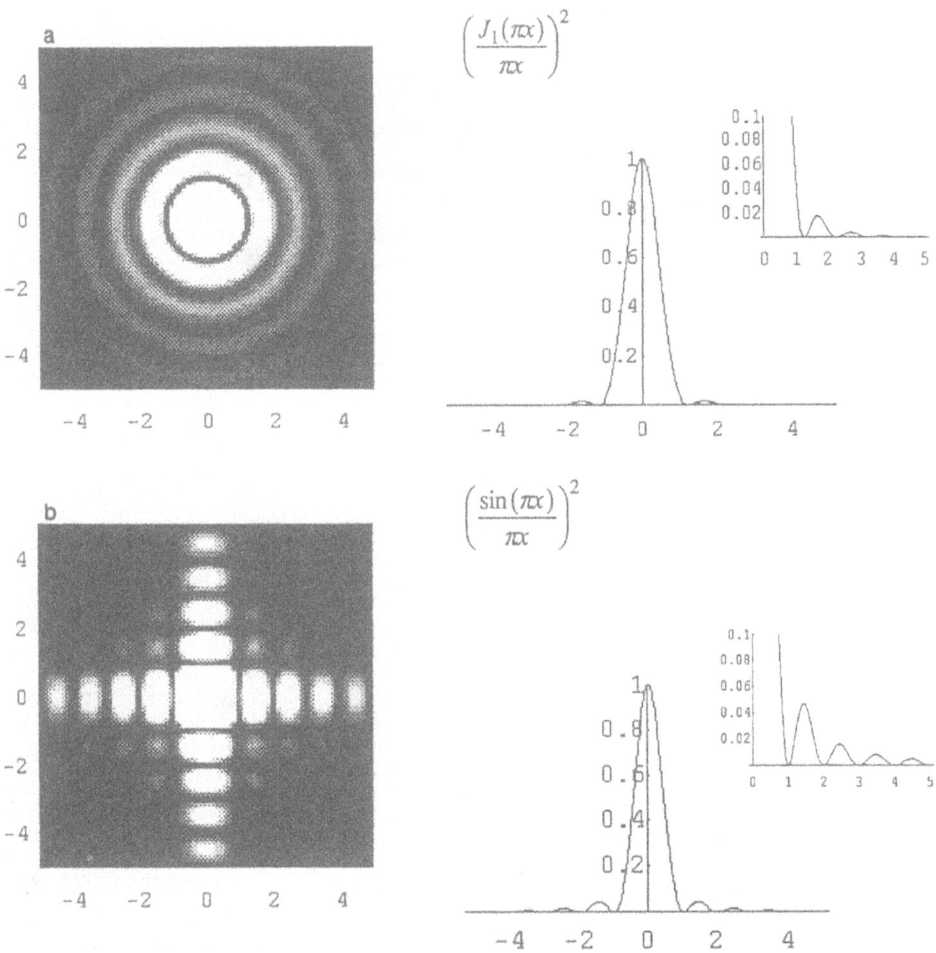

Figure 13. Diffraction patterns. a) circular aperture. b) square aperture.

$s(\boldsymbol{r},\boldsymbol{r}')$. As it turns out, this phase factor is to second order exactly the phase factor, which is introduced by a simple lens inserted into the optical path. A lens placed just behind the diffracting aperture is often called a Fraunhoffer lens. The field in the focal plane of this lens is given by:

$$\Psi(\boldsymbol{r}) = \frac{e^{-ikf}}{i\lambda f} e^{-i\frac{1}{2}k(x^2+y^2)/z} \int_{A'} \psi'(\boldsymbol{r}') e^{-i\frac{k}{f}(x'x+y'y)} d^2\boldsymbol{r}'. \tag{34}$$

Introducing the spatial frequencies in the aperture, $v_x = x'/\lambda f$ and $v_y = y'/\lambda f$, and placing the aperture at a distance $z=f$ from the lens, the 2-d Fourier integral form of the diffraction integral is apparent:

$$\Psi(\boldsymbol{r}) = \frac{1}{i\lambda f} e^{-ikf} \int_{A'} \psi'(\boldsymbol{v}) e^{-i2\pi(v_x x + v_y y)} d^2\boldsymbol{v}. \tag{35}$$

The special position of the lens removes the quadratic phase factor, which is present in Equation (34), but for detection of the intensity in the transform plane, the phase factor is of no significance.

Figure 13 shows the far field diffraction pattern from a circular and a square aperture respectively.

Interference

Optical fields add as vectors. If they are polarized in the same direction, they add algebraically (with the sign) - they interfere. If the two fields are orthogonally polarized, the field strengths cannot cancel each other - they are independent and do not form interference fringes.

The process of observing interference fringes between two or more optical fields, either

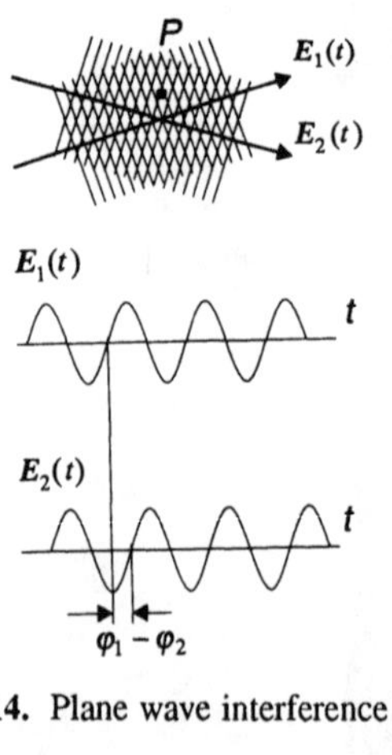

Figure 14. Plane wave interference pattern.

visually or by suitable detectors, is called interferometry. In interferometry we detect the intensity of the total field, since optical detectors including the eye and photographic films respond to optical intensity integrated over the sensitive area of the detector. To make the observation we must observe for a certain minimum of time - we average the visual input to the eye or the signal from the detector over the response time of the detector. The optical fields must stay in the same phase during the observation time in order to obtain a stable interference pattern. We express this by saying that the fields must be temporally coherent over the observation time. We must also demand that the detector is small enough to allow observation of just one interference fringe - otherwise the fringe pattern will appear washed out. We express this by stating that the fields should be spatially coherent over the detector area.

As an example let us take the interference of two intersecting plane waves, Figure 14. The two waves, $\boldsymbol{E}_1$ and $\boldsymbol{E}_2$, are assumed to travel through the point P at $\boldsymbol{r}$, at which a detector will detect the intensity of the total field $\boldsymbol{E}=\boldsymbol{E}_1+\boldsymbol{E}_2$. The intensity is averaged over the detector time constant (long time average as compared to the short time or period time average, which entered into the definition of the intensity). We assume that the detector is small enough to measure the field at the point P. However, we often allow the amplitude and phase to be slowly varying as a function of $\boldsymbol{r}$ and t.

This is what the detector will see:

$$E(\boldsymbol{r},t)=E_1(\boldsymbol{r},t)+E_2(\boldsymbol{r},t)=\left|E_{1,0}(\boldsymbol{r},t)\right|e^{i\varphi_1}e^{-i\omega_1 t}+\left|E_{2,0}(\boldsymbol{r},t)\right|e^{i\varphi_2}e^{-i\omega_2 t}. \tag{36}$$

The intensity at P is

$$\begin{aligned}\overline{I(\boldsymbol{r},t)}&=\tfrac{1}{2}\varepsilon_0 c\overline{\left|E_1(\boldsymbol{r},t)+E_2(\boldsymbol{r},t)\right|^2}\\&=\tfrac{1}{2}\varepsilon_0 c\left\{\overline{\left|E_1(\boldsymbol{r},t)\right|^2}+\overline{\left|E_2(\boldsymbol{r},t)\right|^2}+2\mathrm{Re}\left[\overline{E_1(\boldsymbol{r},t)E(\boldsymbol{r},t)^*}\right]\right\}\\&=I_1+I_2+2\sqrt{I_1 I_2}\cos(\varphi_1-\varphi_2)\end{aligned} \tag{37}$$

where we have defined for short

$$\begin{aligned}I_1&\equiv\overline{I_1(t)}=\tfrac{1}{2}\varepsilon_0 c\overline{\left|E_1(t)\right|^2}\\I_2&\equiv\overline{I_2(t)}=\tfrac{1}{2}\varepsilon_0 c\overline{\left|E_2(t)\right|^2}\end{aligned} \tag{38}$$

and we have used that we know E_1 and E_2 to be real

$$\begin{aligned}\mathrm{Re}\left[E_1E_2^*\right]&=\left|E_1\right|\left|E_2\right|\mathrm{Re}\left[e^{i(\varphi_1-\varphi_2)}\right]\\&=\frac{\sqrt{I_1}}{\sqrt{\frac{1}{2}\varepsilon_0 c}}\frac{\sqrt{I_2}}{\sqrt{\frac{1}{2}\varepsilon_0 c}}\cos(\varphi_1-\varphi_2)\end{aligned} \tag{39}$$

The detector current is

$$i_d=\eta\frac{e}{h\nu}\overline{I(t)}=i_1+i_2+2\sqrt{i_1i_2}\cos(\varphi_1-\varphi_2) \tag{40}$$

where

$$i_1=\eta\frac{e}{h\nu}I_1 \quad\text{and}\quad i_2=\eta\frac{e}{h\nu}I_2. \tag{41}$$

Here η is the quantum efficiency, e is the electron charge, h is Planck's constant and ν is the frequency of the light wave. The overbar indicates long time average.

The case of two stable, plane waves detected at a point is an idealized situation. Real light waves are not stable, but contain fluctuations in amplitude and phase (or frequency), which will influence the contrast of the fringes when we integrate over the detector response time. Likewise, real detectors are of finite size. Again, the contrast of the detected signal may be reduced if the phase of the wave fronts changes across the detector area. The ability to measure interference fringes is described by the term coherence of the field.

Coherence

Definition:

The mutual coherence function: $\Gamma(r_1,t_1;r_2,t_2) \equiv \Gamma_{12}(\tau) = \overline{E(r_1,t_1)E(r_2,t_2)^*}$ (42)

The degree of coherence: $\gamma(r_1,t_1;r_2,t_2) \equiv \gamma_{12}(\tau) = \dfrac{\Gamma}{\sqrt{\overline{|E(r_1,t_1)|^2}}\sqrt{\overline{|E(r_2,t_2)|^2}}}$ (43)

The mutual coherence function measures the covariance of the two fields taken at the space-time points P_1: (r_1,t_1) and P_2: (r_2,t_2). The degree of coherence measures this quantity relative to the product of the rms fields strengths, which would be present at the two points if only one or the other field were present alone.

These functions may be measured if the two fields are somehow sampled at the points P_1 and P_2 and superimposed on a detector. The resulting coherence functions are found as the integrals over the detector averaging time. If the points P_1 and P_2 are part of the same detector aperture, we find the total detector current by summing (integrating) over the total detector area. We thus see that the coherence of an optical field will depend on the detector area and the detector response time. If the measured $\gamma_{12}(\tau)$ is close to one, the fields are said to be fully coherent. If the degree of coherence is close to zero, the fields are incoherent, and if the degree of coherence has a value between zero and one, the fields are said to be partially coherent.

It should be noted that the coherence is always stated as a mutual property of the optical field at two points in space and/or time. If the field is sampled at the same point but at two different times, we speak of temporal coherence, and we may describe the coherence by the time over which the phase of the total field changes at a constant rate (the frequency stays constant). We may also sample the field at the same time, but at two different points in space. The spatial coherence may be described by the area over which the phase of the total optical field is the same, independent of position.

Coherence is essentially the ability of the light to form interference fringes.

Things that may go wrong:

- Light from different parts of the source may change phase randomly within the averaging time.
- Light may change frequency within the averaging time (the frequency fluctuates).
- Light may pass through or be scattered from media, which destroy the coherence by randomizing the phase.

Coherence is not an invariant property:

- Coherence depends on the detector size (and shape) and spectral response
- Coherence depends on the detector averaging time
- Coherence depends on the source spectral width
- Coherence depends on the distance between source and detector

Experiments to Test Coherence

A stable point source would deliver a fully coherent field, but is an idealization. All sources suffer to some extent from some of the imperfections mentioned above. Figure 15 shows three possible ways to measure the coherence of an optical field.

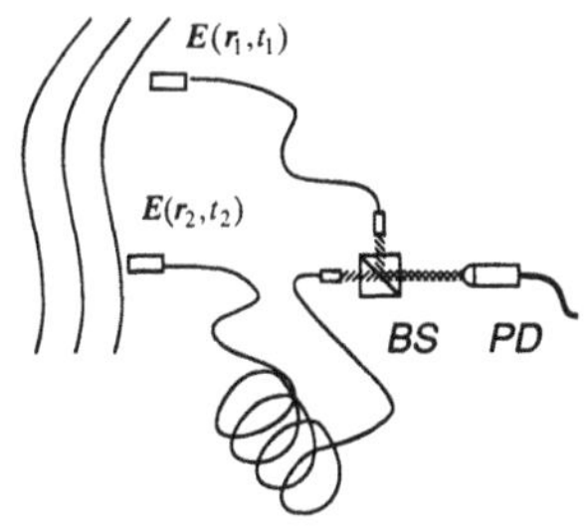

Figure 15a. Coherence experiment with optical fibers.

Fiber optic coherence experiment. A possible way to test coherence could use two optical fibers to sample the field at two different points in space (testing spatial coherence). If one fiber were made longer than the other, the light would suffer a different delay in the two fibers, and the light reaching the detector would have been sampled at two different times (testing temporal coherence).

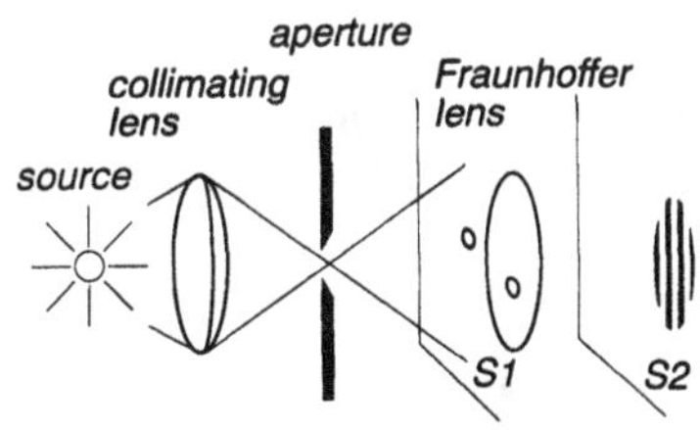

Figure 15b. Coherence experiment with two pinholes.

Young's experiment. Light is sampled by means of two pinholes in a screen, S1. A lens L2 behind the screen brings the far-field pattern into the focal plane of the lens. The light is observed visually or picked up by a scanning detector. Lens L1 and aperture A are used to control the source size. This experiment tests coherence at the same time, but at two different points in space - spatial coherence.

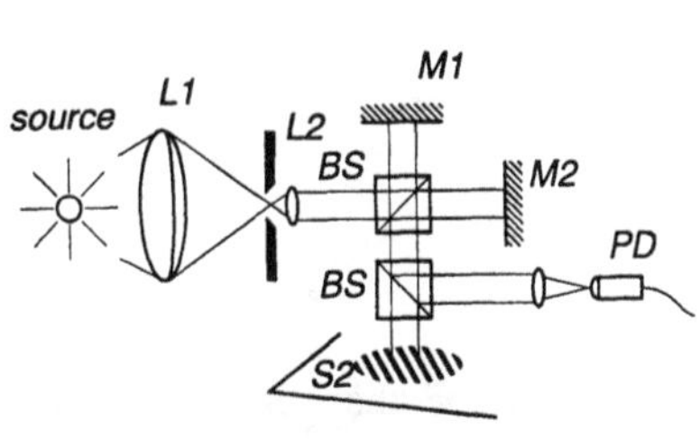

Figure 15c. Coherence experiment with beamsplitter.

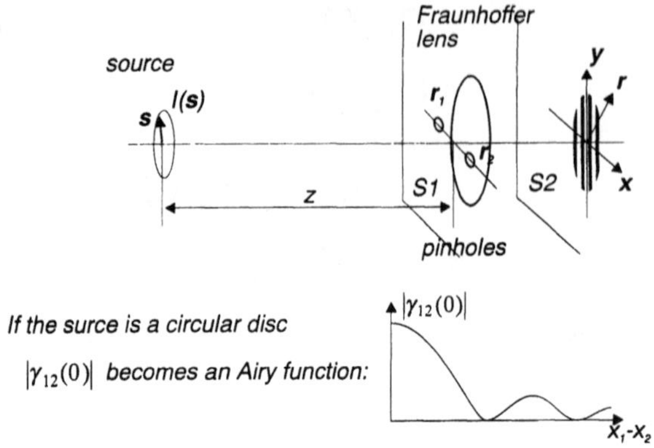

Figure 16. The van Cittert-Zernike theorem.

Michelson interferometer. Light is split in two beams by beamsplitter BS1. The two beams are returned to the beamsplitter by mirrors M1 and M2 and combined. The parallel and overlapping beams are brought to a detector PD or screen S2, where they interfere. The path length difference may be varied to delay one beam relative to the other. This experiment tests coherence of the same field at two different times - temporal coherence.

In general we may test the coherence of the field at two different points in space and time.

Van Cittert - Zernike Theorem

The coherence function behaves like a wave, although it expresses a mutual intensity. However, it can be shown that Γ obeys two coupled wave equations in source coordinates and in detector coordinates. These wave equations can be solved by the Green's function method in the same manner as the diffraction problem is solved for the electric field. The result is the

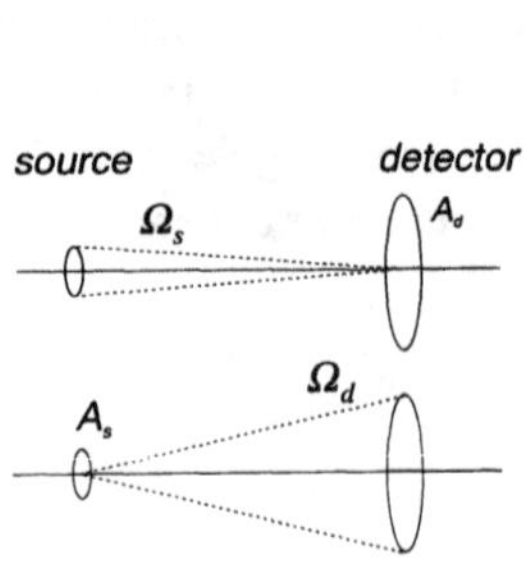

Figure 17. Illustrating detector and receiver area and solid angle.

van Cittert-Zernike theorem, which describes the propagation of the mutual coherence function in space and time. In the Fraunhoffer approximation for an incoherent source the van Cittert-Zernike theorem may be written:

$$\Gamma(r_1, r_2, 0) = \int I(s) e^{i\frac{k}{z} s\cdot(r_1 - r_2)} d^2 s \tag{44}$$

This has the form of a Fourier transform of the source function in the variable pairs s (source coordinate) and (r_1-r_2) (vector from one pinhole to the other in screen S1). This formula is very useful for estimating the coherence properties of an optical field from an incoherent source at a great distance from the source (example: stellar interferometry).

Example: circular incoherent source. The degree of coherence displays an Airy-function dependence on pinhole separation (the visibility curve). The result is analogous to the diffraction pattern from a circular aperture, but there we found the intensity distribution in the far field, see Figure 16.

Coherent Detection

The term coherent detection describes the situation where one interference fringe covers the whole detector area so that no washing out of fringes occurs. In other words we want $|\gamma_{12}(0)| \cong 1$. From van Cittert-Zernike's theorem we see that this in turn requires (see also Figure 17)

$$e^{i\frac{k}{z} s\cdot(r_1 - r_2)} \cong 1 \tag{45}$$

or

$$\frac{k}{z} s\cdot(r_1 - r_2) \cong 0 \quad \Rightarrow \quad \frac{2\pi}{\lambda z} r_s r_d \cong 0 \Rightarrow \frac{r_s r_d}{z} << \frac{\lambda}{2\pi}$$

where r_s and r_d are approximate dimensions of source and detector.

Denoting $\pi r_s^2 \equiv A_s$, $\pi r_d^2 \equiv A_d$, $\frac{A_s}{z^2} \equiv \Omega_s$ and $\frac{A_d}{z^2} \equiv \Omega_d$ we may rearrange the coherence condition:

$$A_s \Omega_d \cong A_d \Omega_s << \frac{\lambda^2}{4}. \tag{46}$$

When this condition is fulfilled, we know that $|\gamma_{12}(0)| \cong 1$, which in turn means that the visibility of the interference fringes is high, bearing in mind that $|\gamma_{12}(0)|$ is a measure of the modulation of the fringes. The optical field is coherent across the detector surface.

Conversely, when we know that $|\gamma_{12}(0)| \cong 1$ and the fields are coherent, we must calculate the intensity according to the formula

$$\overline{I_{tot}} = \tfrac{1}{2}\varepsilon_0 c \overline{\left|\sum_i E_i\right|^2}, \quad \text{(coherent detection)} \tag{47}$$

where we first add the fields together to form the total field and then square and average.

On the other hand, if we know that $|\gamma_{12}(0)| \cong 0$, the field is incoherent and no interference takes place. The intensity of the total field is then found by squaring the separate fields and adding the intensities

$$\overline{I_{tot}} = \tfrac{1}{2}\varepsilon_0 c \sum_i \overline{|E_i|^2}. \quad \text{(incoherent detection)} \tag{48}$$

Example of coherent detection: laser vibrometer: The laser vibrometer is illustrated in Figure 18. A laser beam is expanded and focused onto the target by a transmitter lens with diameter d_t. The scattered light is picked up by the receiver lens with diameter d_r. The scattered light is collimated and mixed with a reference beam. The total field is incident on the

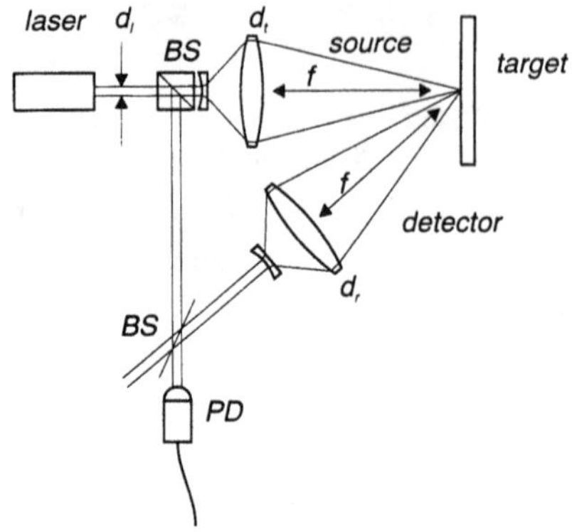

Figure 18. Coherence condition for the laser vibrometer.

photodetector. Basically this instrument is an example of an application of interferometry. The interference between the scattered light and the reference beam is detected as intensity integrated across the total detector surface and averaged by the detector response time. In order for this instrument to function effectively the detection must be coherent, i. e. the total light field scattered by the target and received by the receiver lens should be coherent.

The question is how large we can make the receiver aperture d_r and still get coherence. From the coherence condition we find

$$A_r \Omega_d << \lambda^2 \tag{49}$$

or

$$\frac{\pi}{4} d_r^2 \frac{\frac{\pi}{4} d_f^2}{f^2} \leq \lambda^2 \tag{50}$$

which can be rearranged to express an upper limit for d_r:

$$d_r^2 \leq \frac{\lambda^2 f^2}{\frac{\pi^2}{16} d_f^2} = \frac{\lambda^2 f^2}{\frac{\pi^2}{16}\left(\frac{\pi}{4}\frac{\lambda f}{d_l}\right)^2} = d_l^2 . \tag{51}$$

We thus see that with the same distance between transmitter lens and target as between target and receiver lens, we find that the receiver aperture should be no bigger than the transmitter aperture.

TRADITIONAL OPTICS

In this section we shall give the most elementary definitions and formulas used in traditional geometrical optics and simple diffraction theory. We shall assume that the reader is familiar with standard optical components, which for the use in optical diagnostics for fluid flow would encompass the following list:

Passive Components:

- lenses
- mirrors

- prisms
- coatings
- beamsplitters, polarizing beamsplitters
- gratings
- apertures and pinholes
- polarizers
- polarization dependent components, Glan prism, Glan-Thomson prism, Wollaston prism
- fiber optics

In the following we summarize some of the most useful formulas from geometrical optics and describe two commonly used active components, the Pockels cell and the Bragg cell. Fiber optics is described in the next section.

Tools from Geometrical Optics

- Snell's law: $\alpha_r = \sin^{-1}\left(\frac{n_1}{n_2}\sin\alpha_i\right)$
- critical angle: $\alpha_c = \sin^{-1}\left(\frac{n_2}{n_1}\right)$ $(n_2 < n_1)$
- Brewster's angle: $\alpha_B = \tan^{-1}\left(\frac{n_2}{n_1}\right)$
- first dark ring from circular aperture of diameter d: $d_d = 1.22\frac{f\lambda}{d}$
- first dark ring from rectangular aperture of side d: $d_d = 1.00\frac{f\lambda}{d}$
- first dark ring in interference pattern in Young's experiment, pinhole separation d, distance L: $d_d = \frac{\lambda}{2\sin\frac{\theta}{2}} = \frac{L\lambda}{d}$
- grating formula, grating constant Λ, angle in α_i: $\sin\alpha_n = \sin\alpha_i + N\frac{\lambda}{\Lambda}$
- MTF for perfect lens of diameter d: $MTF \cong 1 - 1.22\frac{\lambda f}{d}\nu$
- image and object position (lens formula): $\frac{1}{l_o} + \frac{1}{l_i} = \frac{1}{f}$
- magnification: $M = \frac{l_i}{l_o}$
- object-to-image distance with principal point distance PP': $OO' = f\left(2 + M + \frac{1}{M}\right) + PP'$

- conjugate distances: $l_o = f\left(1+\frac{1}{M}\right)$ $l_i = f(1+M)$

- chromatic aberration of thin lens: $\delta = \frac{1}{fV}$ where $V = \frac{n_d - 1}{n_f - n_c}$

Active Components

A few active components are significant for the operation of optical diagnostic equipment. We shall briefly describe the Pockels cell and the Bragg cell.

Pockels cell:

The Pockels cell is used for:
- Q-switching pulsed lasers
- reference beam switch in double pulsed holography
- switch in PIV image shifting

The electro-optic crystal acts as a voltage dependent wave plate, Figure 19. At zero applied voltage the crystal is isotropic and no birefringence occurs. The beam passes undeflected through the beamsplitter. When a certain voltage is applied, the crystal acts as a ½-wave plate, i. e. polarization along one crystal axis is delayed 180 degrees with respect to light polarized along the other axis. The result for a beam polarized at 45 degrees to the induced optical axes in the crystal is a 90 degree rotation of the polarization direction. By subsequent passage of the polarizing beamsplitter, the beam is deflected 90 degrees.

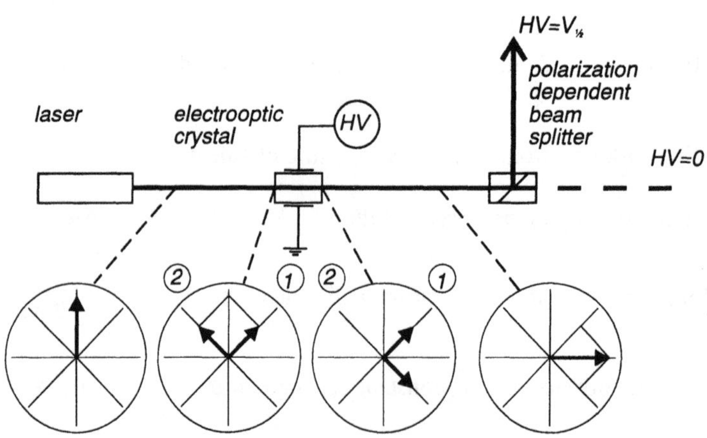

Figure 19. The principle of the Pockels cell.

Bragg cell:

The Bragg cell is used for:
- Q-switching
- cavity dumping
- frequency shift
- deflection

Light is diffracted by a moving ultrasonic grating in an acousto-optic material (flint glass, lead molybdate, quartz, water), see Figure 20.

Thick grating - only one diffracted order, when adjusted for Bragg incidence:

$$\sin\theta_B = \frac{\lambda}{\Lambda}, \tag{52}$$

where λ is the light wavelength and Λ is the wavelength of the ultrasonic wave. The diffracted light is shifted in frequency (Doppler shift):

$$\nu_{-1} = \nu_l - f_s \quad \text{and} \quad \nu_{+1} = \nu_l + f_s. \tag{53}$$

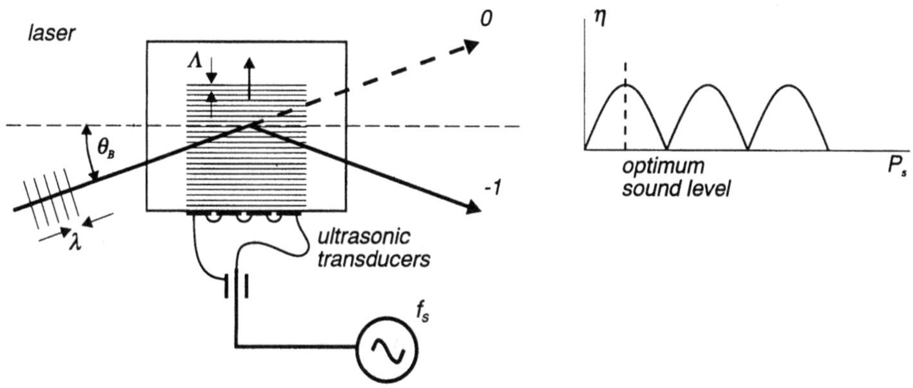

Figure 20. The Bragg cell.

DETECTION OF LIGHT

Basic Facts

A photodetector responds to light intensity integrated over the detector surface, the total incident light power. The photocurrent consists of a train of pulses arriving at random times, each pulse corresponding to a release of one photoelectron (or hole). For each pulse a photon is absorbed from the optical field. However, only some of the incoming photons cause a release of a photocarrier - others are lost due to competing processes. The fraction of photons effective in photocarrier generation is called the quantum efficiency. The quantum efficiency depends on the detector type and the photosensitive material. It is also a function of the wavelength of the incident light.

Types of detectors:

The types of detectors most often used in flow diagnostics are

- vacuum detectors with internal electron cascade gain (photomultiplier tubes)
- semiconductor diodes (Si, Ge) PIN diodes
- semiconductor diodes with internal gain (avalanche detectors)

Detection of Light for Measurement (Signal Carrying Light)

Denoting by P the light power incident on the detector, the mean cathode photocurrent is given by (see e. g. Yariv[7]):

$$\bar{i} = \frac{\eta e}{h\nu}\overline{P(t)} = \frac{\eta \lambda e}{hc}\overline{P(t)} \ , \tag{54}$$

where

η is the quantum efficiency ($\eta \approx 1-20\%$ for vacuum types and 70-90% for Si-based diodes)

$\overline{P(t)}$ is mean incident light power integrated over the detector = $\int_{A_d} I(\mathbf{r}_d,t)dA_d$

h is Planck's constant = 6.63×10^{-34} J

λ is the light wavelength (e.g. 694, 633, 514, 488, 476 nm for ruby, He-Ne and three ar-ion laser lines, respectively)

c is the speed of light = 2.99×10^{8} ms^{-1}

ν is the light frequency $= c/\lambda$

e is the charge of the electron 1.6×10^{-19} C

If the detector has internal gain, G, the cathode current is multiplied by G to give the anode current.

Noise

The following sources of noise are important in optical measurements:

- quantum noise (shot noise)

This is the noise caused by the quantum nature of the optical detection process. The noise is the result of the random superposition of charge pulses, generated by the photodetection process. The spectrum of the fluctuations in the photocurrent depends on the shape of the individual pulse; very short pulses (delta-function) give rise to a constant spectral density (white noise).

The quantum noise may be generated by the signal light itself (noise in signal) or from background light without any signal light reaching the detector (background noise).

- dark current

This is the noise current resulting from random pulses caused by thermal excitation or background radioactivity. The level of dark current and dark current noise is usually provided in the manufacturer's specifications.

- thermal noise

This noise is generated by interaction between the photocurrent carriers and the atomic structure in the electric circuits, especially in the input to the electronic signal processing equipment, the load resistance.

- low frequency noise

This noise is caused by many different processes, e.g. leak currents on surfaces or structural changes in the detector material.

- electron avalanche noise

Some detectors are able to increase the photocurrent before the charge carriers reach the electronic circuits. This may take place as secondary electron emission from special surfaces (dynodes) in a vacuum detector (photomultiplier) or as an avalanche process in a semiconductor detector.

For our applications the dominating noise sources are shot noise and thermal noise.

For detectors with internal gain it is sometimes possible to increase the gain to the point where shot noise dominates over thermal noise. This is possible because the shot noise is generated at the cathode with the signal current, and both may be amplified by the internal gain, while the thermal noise is generated in the electronic circuits at or after the anode. The shot noise dominated detection gives the best S/N.

The noise is measured by its mean square (or rms) fluctuating part: $i_N = \sqrt{\overline{i_N'^2}}$. It may be represented as a current source in an equivalent diagram as shown in Figure 21.

Shot noise and thermal noise are given by the following expressions for the mean square fluctuating noise current:

$$\text{Shot noise:}\quad \overline{i'^{\,2}_{N,S}} = 2e\bar{i}\Delta\nu \tag{55}$$

$$\text{Thermal noise:}\quad \overline{i'^{\,2}_{N,T}} = \frac{4kT}{R_L}\Delta\nu\ , \tag{56}$$

where $\Delta\nu$ is the detector bandwidth
R_L is the load resistor
k is Boltzmann's constant (1.386×10^{-23} JK^{-1})
and T is the absolute temperature.

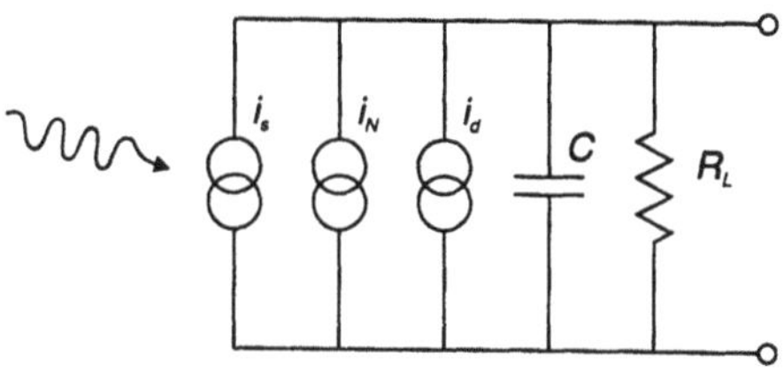

Figure 21. Equivalent circuit for photodetector.

Signal-to-noise Ratio

Formulas for the signal-to-noise ratio are found by dividing the mean square signal current with the mean square noise current. We may distinguish between direct detection, where a single, modulated light beam is incident on a detector, and heterodyne detection, where the signal beam is superimposed with a reference beam. Denoting by $E(t)$ the light field incident on the detector, we may write in the case of direct detection:

$$E(t) = E_0\left(1 + m\cos(\omega_0 t + \varphi_0)\right)e^{i\omega t}, \tag{57}$$

where E_0 is the light amplitude and m is a signal modulation index.

In the case of heterodyne detection of two incident beams of amplitudes $E_{1,0}$ and $E_{2,0}$ and power P_1 and P_2, we may write for the signal:

$$E(t) = E_1(t) + E_2(t) = E_{1,0}e^{i\omega_1 t + i\varphi_1} + E_{2,0}e^{i\omega_2 t + i\varphi_2}, \tag{58}$$

where ω_1 and ω_2 are the frequencies and φ_1 and φ_2 the phases of the two incident beams.

The intensity in the case of direct detection is:

$$I(t) = \tfrac{1}{2}\varepsilon_0 c\,\mathrm{Re}\{EE^*\} \cong \tfrac{1}{2}\varepsilon_0 c E_0^2\left(1 + 2m\cos(\omega_s t + \varphi_s)\right) \tag{59}$$

where the term in m^2 has been neglected, and $\varphi_s = \varphi_0$.

We can then find the cathode photocurrent by integrating over the detector area:

$$i(t) = \frac{\eta e}{h\nu}\int_A \tfrac{1}{2}\varepsilon_0 c E_0^2\left(1 + 2m\cos(\omega_s t + \varphi_s)\right)dA = \frac{\eta e}{h\nu}P_0\left(1 + 2m\cos(\omega_s t + \varphi_s)\right). \tag{60}$$

The mean cathode current is

$$\overline{i_C} = \frac{\eta e}{h\nu} P_0 \tag{61}$$

and the mean square shot noise current at the cathode is

$$\overline{i_{SN,C}'^2} = 2e\bar{i}\Delta\nu = 2e\left(\frac{\eta e}{h\nu}\right)P_0\Delta\nu, \tag{62}$$

while the signal itself is given by

$$i_s(t) = 2m\cos(\omega_s t + \varphi_s)\frac{\eta e}{h\nu} P_0 \tag{63}$$

and thus the mean square signal is

$$\overline{i_s(t)^2} = 2m^2\left(\frac{\eta e}{h\nu}\right)^2 P_0^2. \tag{64}$$

From this we get the power signal-to-noise ratio at the anode in case of internal gain:

$$\left(\frac{S}{N}\right)_P = \frac{G^2\left(\frac{\eta e}{h\nu}\right)^2 P_0^2 2m^2}{2G^2 e\left(\frac{\eta e}{h\nu}P_0 + i_d\right)\Delta\nu + \frac{4kT}{R_L}\Delta\nu} \tag{65}$$

If the internal gain is sufficiently high, and if we may neglect the dark current, we get shot noise limited detection:

$$\left(\frac{S}{N}\right)_P = \frac{\eta P_0 m^2}{h\nu\Delta\nu}. \tag{66}$$

For homodyne detection the intensity becomes

$$\begin{aligned} I(t) &= \tfrac{1}{2}\varepsilon_0 c\,\mathrm{Re}\{EE^*\} \\ &= \tfrac{1}{2}\varepsilon_0 c\,\mathrm{Re}\Big\{E_{1,0}^2 + E_{2,0}^2 + E_{1,0}E_{2,0}e^{i((\omega_1-\omega_2)+(\varphi_1-\varphi_2))} \\ &\qquad + E_{1,0}E_{2,0}e^{-i((\omega_1-\omega_2)+(\varphi_1-\varphi_2))}\Big\} \end{aligned} \tag{67}$$

$$= \tfrac{1}{2}\varepsilon_0 c\left\{E_{1,0}^2 + E_{2,0}^2 + 2E_{1,0}E_{2,0}\cos\left((\omega_1-\omega_2)t + (\varphi_1-\varphi_2)\right)\right\}. \tag{68}$$

The cathode photocurrent is

$$i(t) = \frac{\eta e}{h\nu}\int_A I(t)dA = \frac{\eta e}{h\nu}\left\{P_1 + P_2 + 2|\gamma|^2\sqrt{P_1P_2}\cos\left((\omega_1-\omega_2)t + (\varphi_1-\varphi_2)\right)\right\}, \tag{69}$$

where we have introduced the spatial coherence $|\gamma|^2$ between the beams (possible misalignment).

The mean cathode current is

$$\overline{i(t)} = (P_1 + P_2), \tag{70}$$

the mean square noise current is

$$\overline{i_N'^2} = 2e\left(\frac{\eta e}{h\nu}\right)(P_1 + P_2)\Delta\nu, \tag{71}$$

the signal current is

$$i_s(t)=|\gamma|^2\frac{\eta e}{h\nu}2\sqrt{P_1P_2}\cos((\omega_1-\omega_2)t+(\varphi_1-\varphi_2)), \tag{72}$$

and the mean square signal current is

$$\overline{i_s'^2}=|\gamma|^2\left(\frac{\eta e}{h\nu}\right)^2 2P_1P_2. \tag{73}$$

Thus we find for the signal-to-noise power:

$$\left(\frac{S}{N}\right)_P=\frac{2G^2|\gamma|^2\left(\frac{\eta e}{h\nu}\right)^2 P_1P_2}{2G^2e\left(\left(\frac{\eta e}{h\nu}\right)(P_1+P_2)+i_d\right)\Delta\nu+\frac{4kT}{R_L}\Delta\nu}. \tag{74}$$

Assuming shot noise limited detection, neglecting dark current we find:

$$\left(\frac{S}{N}\right)_P=|\gamma|^2\frac{\eta P_1P_2}{h\nu(P_1+P_2)\Delta\nu}, \tag{75}$$

Finally, if one beam is much stronger than the other ($P_2 \gg P_1$, i. e. a strong reference beam), we find:

$$\left(\frac{S}{N}\right)_P=|\gamma|^2\frac{\eta P_1}{h\nu\Delta\nu}, \tag{76}$$

OPTICS FOR FIBER OPTICS

Types of Optical Fibers for Measurement Purposes

Optical fibers and fiber optic sensors are of increasing importance in optical flow diagnostics. At first fibers were used for carrying light from a detector optics to a remotely located detector. In this type of application the fibers need not preserve the coherence of the received light. However, the need to also transmit coherent light without modifying its coherence properties led to experiments with various types of optical fibers. The so-called SELFOC fiber is a graded index fiber, which essentially acts as a lens. This optical fiber is able to transmit coherently over a distance of 300 - 600 mm. However, the SELFOC fiber as well as other types of graded index fibers are sensitive to mechanical influence and do not preserve the coherence and polarization properties of the light very well. The development of the single mode, polarization preserving fiber for communication purposes provided a component, which is also very useful for measurement purposes. With this type of fiber it is possible to transmit visible laser light over distances of hundreds of meters without loss of coherence or polarization and with only a modest loss of light power.

When discussing optical fibers for measurement purposes, we distinguish between multimode fibers and single mode fibers. The fiber consists of a core, within which the transmitted light is confined, and a cladding, which surrounds the core. The light is confined to the core by an index of refraction difference between the core and the cladding, which causes the light directed away from the core to be deflected back towards the core. To obtain this effect, the index of the core should be higher than that of the cladding, see Figure 22. In practice the fiber may be a so-called step index fiber, in which the index changes abruptly from the value in the cladding to the core value, or a graded index fiber, in which the index changes gradually according to a certain desirable (usually quadratic) function of the radius.

Unless special precautions are taken, a polarized beam travelling through an optical fiber will quickly become depolarized due to stress induced birefringence in the fiber. Certain types of fibers have a built-in high mechanical stress across the core. This stress may have been introduced during the cooling of the fiber in the pulling process by surrounding the core by elements with a different thermal expansion. Referring to the cross sectional pattern of the various elements in the fiber, these fibers are often called bow tie fibers or butterfly fibers. Another term for these fibers is "hi-bi" fiber. These types of fibers are able to preserve the polarization of a laser beam introduced into the fiber with its polarization along one of the optical axes of the fiber over a long distance (several tens of meters). The ability to preserve the polarization is specified by a so-called cross coupling coefficient.

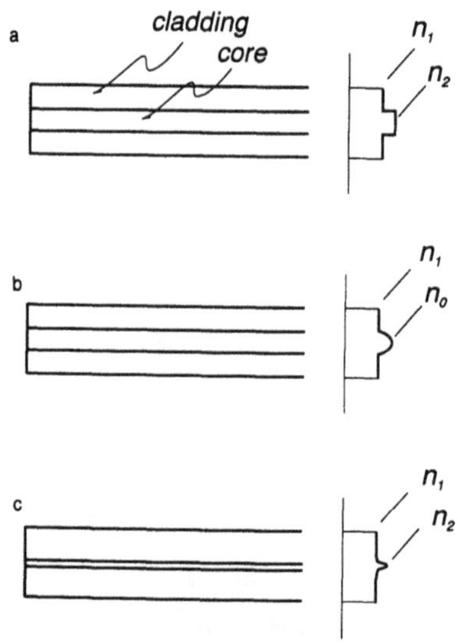

Figure 22. Different types of optical fibers.

Applications

The optical fibers may as mentioned be used as conductors for light. Multimode fibers conduct the light incoherently, i. e. a coherent input will quickly be dissolved into a multitude of modes, and the output of the fiber will appear as a speckle pattern. Single mode, polarization preserving fibers can transmit coherent, polarized light over great distances without significant loss of coherence or depolarization.

However, other applications of optical fibers for measurement purposes are possible. Optical fibers are sensitive to a number of external influences such as pressure, temperature, humidity and many others. The basic mechanism is a change in the fiber index of refraction, which in turn modifies the phase of the light transmitted through the fiber. If the fiber is introduced in one arm of an interferometer (e.g. the Mach-Zhender type), a very sensitive fiber optic sensor results. A sensor may also be based on a change of polarization of the transmitted light. Finally, sensors may be constructed on the basis of reflection from the fiber end, absorption in or just outside the fiber end, fluorescence and other optical effects at the fiber end.

Fiber optic components may also be used for measurement purposes. Such components may encompass fiber optic couplers and beam splitters, fiber optic switches, electro-optic modulators, integrated interferometers and fiber optic phase modulators. A description of these possibilities is outside the scope of this article, and the reader is referred to the literature on the subject of fiber optics and fiber optic sensors[8,9]. In the following we shall concentrate on some useful formulas for the use of optical fibers for transmission of incoherent and coherent light.

Fiber Optic Formulas

Single mode application: coupling a laser to a single mode fiber: The first question to consider is whether the fiber is actually single mode. The important parameter is the so-called mode parameter, V, which controls the number of modes that are able to propagate through the fiber, see Figure 23.

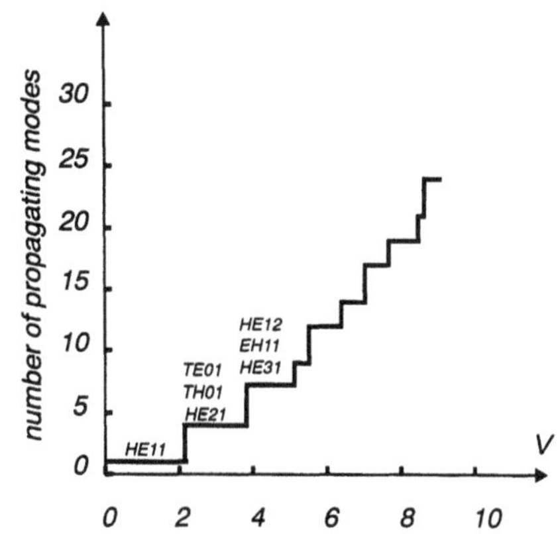

Figure 23. Mode parameter and number of modes for an optical fiber.

$$V = \frac{2\pi a}{\lambda}\sqrt{n_2^2 - n_1^2}\ , \tag{77}$$

where a is the core radius. As it is also indicated in the figure, the condition for single mode operation is that the mode parameter is less than approximately 2.4. Knowing λ, n_1 and n_2 and the condition for V, we may find a from the above formula and thus determine the maximum allowable core diameter for single mode operation, or given the fiber properties we may find the minimum wavelength for single mode operation.

The next step is to determine the optimum beam waist for a focused Gaussian laser beam. A Gaussian field is not a perfect match to the field in the fiber, but we may use the rule of thumb that the e^{-2}-intensity diameter of the focused laser beam should be approximately 1.1 times the core diameter:

$$d_f \cong 2.2\,a \tag{78}$$

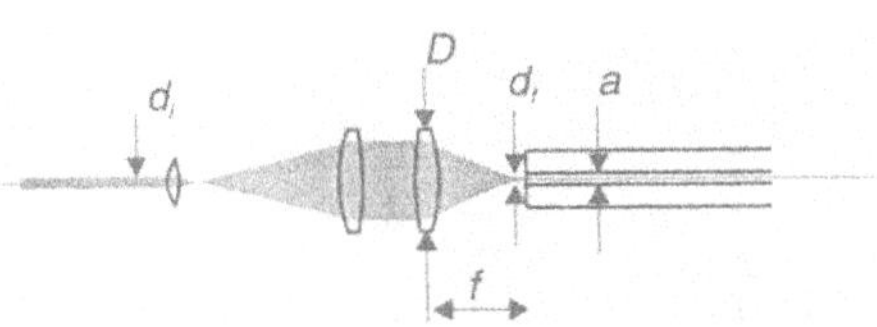

Figure 24. Focusing optics for coupling laser beam to single-mode fiber.

We then have the means for computing the optics needed for coupling a given laser beam into a single mode fiber.

Example, see Figure 24: $\lambda = 530$ nm, $d_l = 0.5$ mm, $n_1 = 1.447$ and $n_2 = 1.450$:

We find:

$$V = 2.4$$

$$a = \frac{\lambda V}{2\pi\sqrt{n_2^2 - n_1^2}} = 2.17\ \mu\text{m}.$$

We may now compute the numerical aperture, F, for the focusing optics:

$$F = \frac{\pi}{4}\frac{d_f}{\lambda} = \frac{\pi}{4}\frac{2.2\,a}{\lambda} = 7.07. \tag{79}$$

Multimode application: Coupling light into a multimode fiber. The problem here is to couple the maximum of light into the fiber. This requires a tight focus with a small F#. However, we also have to consider a possible loss of light from the core to the cladding. To avoid such losses the F# of the incoupling optics must be lower than fiber numerical aperture. We thus have two possibly conflicting requirements: 1) Small F# to image the target inside the fiber core and 2) Large F# to stay within the fiber numerical aperture.

The fiber numerical aperture is given by:

$$NA = 2\sin\theta_{max} \tag{80}$$

with

$$\sin\theta_{max} = \sqrt{n_2^2 - n_1^2} \tag{81}$$

We may again consider an example: $\lambda = 633$ nm, $n_1 = 1.415$, and $n_2 = 1.511$, see Figure 25.

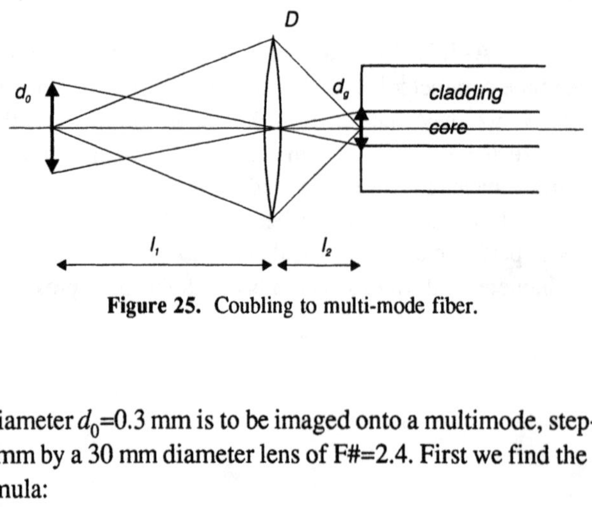

Figure 25. Coubling to multi-mode fiber.

A target of diameter d_0=0.3 mm is to be imaged onto a multimode, step-index fiber from a distance of 600 mm by a 30 mm diameter lens of F#=2.4. First we find the size of the image from the lens formula:

$$d_g = M\,d_0 \tag{82}$$

with

$$f = FD = 2.4 \cdot 30 = 72\text{mm} \tag{83}$$

and

$$M = \frac{f}{l_1 - f} = \frac{72}{600-72} = 0.136 \tag{84}$$

and

$$l_2 = M\, l_1 = 0.136 \cdot 600 = 81.6\text{mm}. \tag{85}$$

This gives

$$d_g = M\, d_0 = 0.136 \cdot 0.3\,\text{mm} = 40.8\,\mu\text{m}.. \tag{86}$$

To be sure we compute the diffraction spot size:

$$d_s = 2.44(1+M)F\lambda = 4.2\,\mu\text{m}. \tag{87}$$

As expected geometrical optics is adequate in this case. As one can see a 50 μm or 100 μm fiber core would be suitable for this application.

The numerical aperture of the light focused onto the fiber is

$$\sin\theta_{lens} = \frac{D}{2l_2} = \frac{30}{2 \cdot 81{,}6} = 0.184 \tag{88}$$

Comparing this to the fiber numerical aperture:

$$\sin\theta_{max} = \frac{NA}{2} = \sqrt{n_2{}^2 - n_1{}^2} = 0.53 > \sin\theta_{lens}\ , \tag{89}$$

we find that the fiber can accept all the light incident on the fiber end.

One final consideration. The fiber should be able to pass the bandwidth of the signal without attenuating some of the frequencies. The attenuation of wide bandwidth signals occurs because of frequency dispersion in the fiber. There are several causes of frequency dispersion in a multimode fiber:

- material dispersion

This is caused by the fact that the index of refraction is a function of the light frequency.

- waveguide dispersion

This dispersion is analogous to the dispersion of microwaves in a wave guide.

- mode dispersion

This dispersion effect may be explained as the different propagation velocities of the different modes.

Only the last one is important for laser light in the visible range.

Example:

We want to pass a signal from an LDA MV to a photodetector through a 100 m long fiber optic cable. Will the dispersion limit the bandwidth of the LDA-signal?

We use the parameters:

$a = 50\mu\text{m}$

$\lambda = 0.5\ \mu\text{m}$

$L = 100\ \text{m}$

$n_1 = 1{,}415$

$n_2 = 1{,}511$

and find:

$$V = \frac{2\pi a}{\lambda}\sqrt{n_2^{\,2} - n_1^{\,2}} = 333$$

$$\Delta\tau_M = 32\,\text{ns}$$

$$\Delta f_{max} \approx \frac{1}{4\Delta\tau_M} \approx 7{,}8\,\text{MHz}$$

We can use a multimode fiber, since the beams are already overlapping in the MV (incoherent detection).

Losses in an Optical Fiber

Losses in optical fibers are normally considered very low, of the order of a few dB per kilometer for near infrared light. However, since our applications are mostly using visible light, losses in fibers for optical measurements cannot be neglected. Figure 26 shows losses in an optical fiber as a function of wavelength. The broken curves are theoretical limits. The ultraviolet absorption edge results from the processes, where photons deliver energy to an electron, which is exited into the conduction band. This energy loss is 8.9 eV for pure quartz corresponding to a wavelength of 0.14 μm. The infrared absorption can take place from 1 μm and up. Impurities such as water can cause absorption in the infrared and near infrared region. Also dopant such as GeO_2 absorb strongly. The curve marked Rayleigh scatter shows light scattering from molecular density fluctuations at the relevant working temperature in the material of the fiber core. This effect is independent of material concentrations and defects and represents a theoretical lower value. Rayleigh scattering goes with λ^{-4} and thus increases strongly towards the blue end of the spectrum. At the Ar-ion wavelength of 488 nm the total absorption is around 33 dB/km, while at the YAG wavelength of 1.06 μm it is around 2 dB/km. The lowest obtainable losses are around 1.1 dB/km.

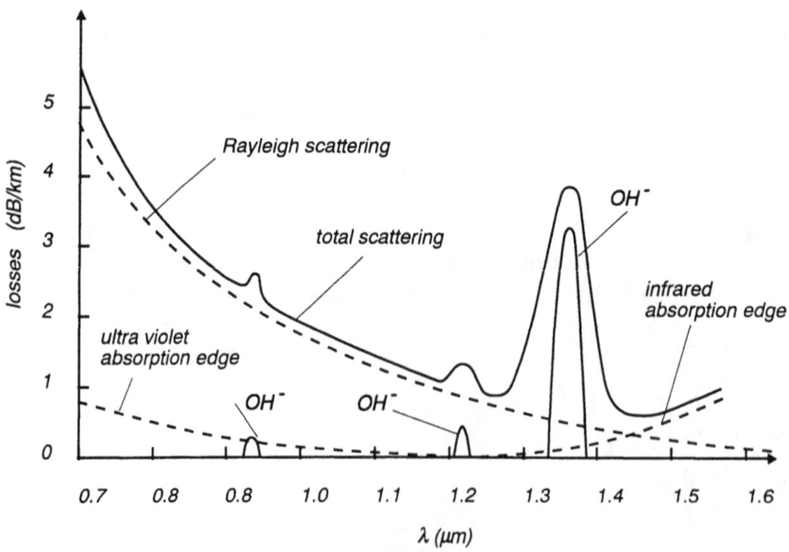

Figure 26. losses in an optical fiber.

SEED PARTICLES

Three requirements are important for seed particles for flow measurements:

- the particles must follow the flow
- the particles should scatter as much light as possible
- the particles should be evenly distributed in the flow

The first condition requires the particles to be as small as possible and to have a density as close as possible to that of the fluid. These conditions are of course most easily fulfilled in liquids, where a good density match may be achieved. As an example, some polystyrene particles have a specific density very close to that of water. Such particles are offered as seed particles by commercial companies specializing in seeding generators and seed particles for fluid flow.

The degree to which the particles actually follow the flow can be estimated from Stokes' law when the acceleration of the fluid is known. The fluid acceleration may result from swirling flow, shock fronts in the flow or turbulence.

Stokes' law is

$$\ddot{u}_p = -18 \frac{\nu}{d_p^2} \frac{u_p - u_f}{(\rho_p / \rho_f)} \tag{90}$$

where

$\ddot{u}_p$ is the particle acceleration
$\nu = \mu / \rho_f$ is the kinematic viscosity
μ is the static viscosity
ρ_f is the fluid density
ρ_p is the particle density
d_p is the particle diameter, and
$u_p - u_f$ is the difference between the particle velocity and the fluid velocity, the velocity slip.

In a practical situation, the fluid and particle properties may be known and the acceleration of the flow can be estimated from fluid mechanics considerations. The formula may then be used to find the maximum particle diameter, which will give a velocity slip (error) less than a given amount.

In Table 1 properties of some often used seed materials are given as well as values for Stokes' parameter and the maximum diameter that will track a harmonic fluid oscillation with a velocity error less than 1 pct.

Table 1. Some commonly used seed particles and their performance in an oscillating fluid.

particle	fluid	density ratio (ρ_p / ρ_f)	viscosity (kg m^{-1} s^{-1}) (μ)	N_s (Stokes' #)	diameter (μm) f=1 kHz	diameter (μm) f=10 kHz
silicone oil	atmospheric air	900	1.8×10^{-5}	19	2.6	0.8
TiO_2	atmospheric air	3.5×10^{3}	1.8×10^{-5}	37	1.3	0.4
MgO	methane/air flame (1800 K)	1.8×10^{4}	5.9×10^{-5}	84	2.6	0.8
TiO_2	oxygen plasma (2800 K)	3.0×10^{4}	1.1×10^{-4}	108	3.2	0.6

a large particle of a refractive index as far as possible from that of the fluid. The scattered light may be computed by Lorenz-Mie computations, which are now available as efficient software packages for PCs and workstations at many universities and commercial companies. Some software packages are able to compute the amount of scattered light power, the visibility and the phase received by a given detector aperture placed in any position relative to the scattering

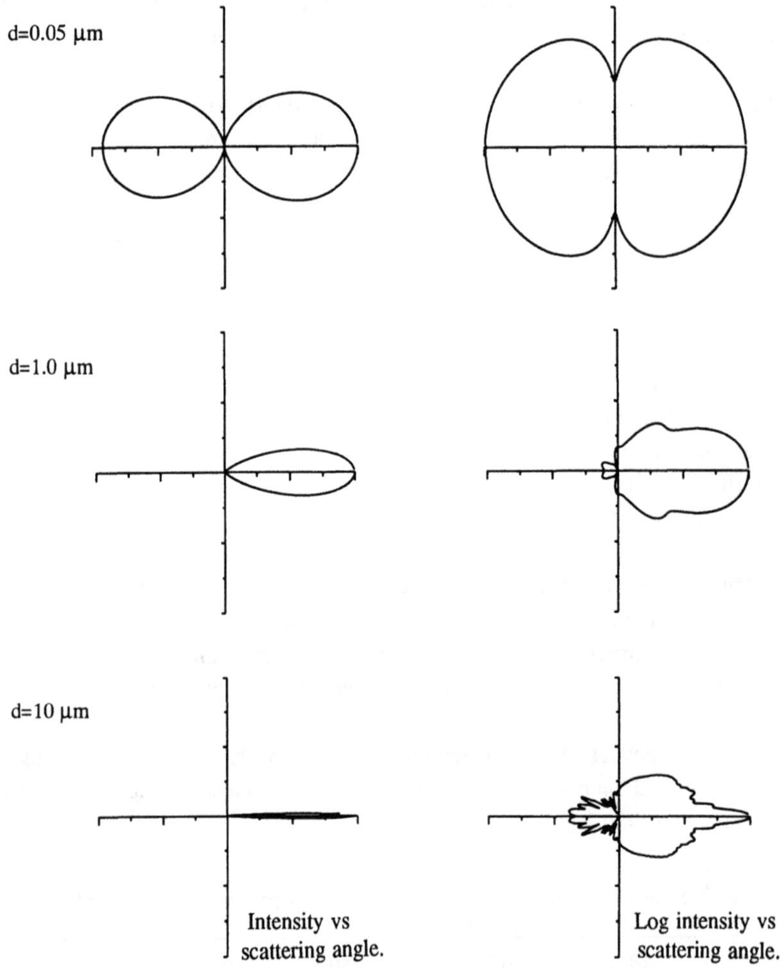

Figure 27. Scattering diagrams for three sizes of particles.

particle and even assuming a realistic laser anemometer configuration with two intersecting beams. Information is also available in the form of Lorenz-Mie scattering tables or curves. In
beams. Information is also available in the form of Lorenz-Mie scattering tables or curves. In
that case it may be necessary to estimate the effect of changes to the system (other wavelength,

Table 2. Properties of aerosol particles.

generation	material	appr. diameter	comments
atomization	water	1 - 2 µm	Evaporation inhibitor added
	DOP	0.35 - 1.2 µm	Commercial atomizer available
	teflon dust	75 % < 2 µm	Atomized from teflon/freon suspension
	silicone oil	1 - 3 µm	Works well
fluidization	titanium dioxide	0.5 - 2 µm	Stable in flames up to 2500 K
	PVC	0.3 - 0.8 µm	Large agglomerates formed; erratic flow
	aluminium, alumin. oxide	up to 8 µm	
chemical reaction	ammonium chloride	1.2 µm	Produced from ammonia and hydrogen chloride gasses (which are toxic and corrosive)
	stannic chloride		Used in flames, but particles consumed
combustion	tobacco smoke	0.1 - 1.0 µm	Dirty unsteady flow
	magnesium oxide		From burning of magnesium powder; used in flames
	smoke bombs	over 3 µm	Corrosive, toxic, uncontrollable production rate
	smoke pellets	0.03 - 1.0 µm	Commercial product, low flow rate
sublimation	ice	0.5 µm	Formed during expansion of humid air through supersonic nozzle

aperture size and position, different medium and particle refractive index etc.) as it is quite impossible to display data for all possible variations of the parameters.

Lorenz-Mie theory is valid for a spherical particle of any diameter in a plane incident wave. However, the computation time increases as the diameter is increased, and beyond a size parameter ($\alpha = 2\pi d / \lambda$) greater than about 10 it is more efficient to use geometrical optics. In geometrical optics we can compute the scattered light based on Fresnel's formulas for reflection

Table 3. Some particles for use in water

material	appr. diameter	comments
polystyrene latex	0.5 - 50 μm	Normal concentration: 1 part in 10^5.
fluorescent polystyrene latex	0.5 - 50 μm	Feasible in close environments.
aluminium powder	< 10 μm	Feasible for systems demanding polarization preservation at scattering.
plastic paint	< 10 μm	Has to be selected. A paint with too small particles cause noise.
bubbles	5 - 500 μm	Can only be used if two-phase flow can be accepted.

and refraction, in which we identify the n'th ray, as the ray, which has had n internal passes through the particle before leaving as a refracted ray. The 0'th ray is the ray initially reflected from the particle. However, it should be noted that not all the scattered light can be accounted for by geometrical optics calculations, even for very large spheres. Diffraction accounts for the light scattered into the very near forward direction, and near the rainbow angles and in the direct backscatter direction geometrical optics is not able to account for all the features of the scattered light.

For $\alpha<1$ we may use Raleigh theory to determine the scattered light. In that case the radiation pattern is the dipole radiation diagram of the molecular dipole radiation induced by the incident field and assumed to radiate in phase due to the small size of the particle.

Recently the Lorenz-Mie theory (LMT) has been extended to spherical particles in a Gaussian incident field, the so-called generalized Lorenz-Mie theory (GLMT)[10]. This theory allows a much more realistic modelling of a phase Doppler system, in which it is possible to take into account the effect of a particle in a narrow, focussed beam and to compute the scattered light from a particle located at different off axis positions in the phase Doppler MV. Using such computations the so-called particle trajectory effect has been investigated, and systems, which reduce or compensate for the beam trajectory error have been proposed.

In any case it is important to take into consideration the very different scattering intensities in the forward, side and backward directions. Figure 27 shows three cases with size parameter much smaller than one (near Rayleigh regime), around one (Lorenz-Mie regime) and much larger than one (geometrical optics regime). As an example we have used polystyrene particles with index of refraction 1.4 in water (index 1.33). These examples show the development of the radiation diagram as the particle diameter is increased beyond the symmetrical Rayleigh regime through the complicated LMT regime and into the geometrical optics regime.

Regarding the third condition, the only general comment is that in order to obtain satisfactory results in an optical flow measurement it is necessary to take into account the requirements to seeding when the experiment is designed. All too often bad results of optical flow measurements are caused by neglecting to take the seeding problem into account in the planning phase.

In Table 2 we have listed some of the relevant properties of aerosol particles and in Table 3 some particles for use in water.

REFERENCES

1. Max Born and Emil Wolf, "Principles of Optics", Pergamon Press (1980).
   Classic reference work on classical optics.
2. Joseph W. Goodman, "Introduction to Fourier Optics", McGraw-Hill (1988).
   Good introduction to Fourier optics, optical filtering and coherent imaging.
3. Eugene Hecht, "Optics 2nd ed.", Addison-Wesley Publishing Co. (1987).
   Good introduction to basic optics.
4. J. R. Meyer-Arendt, "Introduction to Classical and Modern Optics", Prentice-Hall (1989).
   Good introduction to basic optics.
5.. O. S. Heavens and R. W. Dichtburn, "Insight into Optics", John Wiley and Sons (1991).
   Newest textbook on basic optics.
6. G. O. Reynolds, J. B. DeVelis, G. B. Parent, Jr. and B. R. Thompson, "Physical Optics Notebook", SPIE Publications (1989).
   Good book on Fourier Optics and optical imaging, photography and detection.
7. A. Yariv, "Optical Electronics", Holt, Rinehart and Winston (1985).
8. A. H. Cherin, "An Introduction to Optical Fibers", McGraw-Hill (1983).
9. L. B. Jeunhommes, "Single Mode Fiber Optics", Marcel Dekker Inc. (1985).
10. G. Gouesbet, "Generalized Lorenz-Mie Theory: An application to phase-Doppler technique", *in:* Proc.. 5th Int. Symposium on Laser Techniques to Fluid Mechanics, Lisbon (1990).

PROCESSING OF RANDOM DATA

Robert V. Edwards

Chemical Engineering Department
Case Western Reserve University
Cleveland, OH 44106, USA

INTRODUCTION

The processing of random data is always an elaborate method for producing a new random number, or numbers. For instance, estimating a mean of a set of random data means adding up the numbers and then dividing by the number of points used. Every time this process is performed with a different sample from the random data set, a different number is attained. In a similar fashion, estimating the bandwidth of a random signal would vary from estimate to estimate if a new data set is used each time.

The point of this chapter is to examine the *variation* in the estimation of quantities derived from sets of random data. We will examine the variance of the measured mean for correlated and uncorrelated data. In addition, we will look at the variance in the estimate of the autocorrelation and spectrum of random time series. This work will focus on estimates of quantities from the type of data usually encountered in laser anemometry and measurements of turbulence flows.

In every case, the result will resemble the result usually derived for the variance of the mean of a set of random data, e.g.

$$\sigma_m^2 = \sigma^2 / N \ , \tag{1}$$

where σ_m^2 is the variance of the measured mean from the "true" value, σ^2 is the variance of the data set and N is the number of *independent* points used in estimating the mean. The object will be to find out how to estimate the number of independent data points from a time series with correlation.

INDEPENDENT RANDOM VARIABLES

The *expected value* of the function of a random variable can be thought of as the average value of the function taken over infinite time or infinitely many trials. The expected value of a quantity • will be denoted by angle brackets, <•>. The process of computing the expected value is linear in that $<\alpha f(x)+\beta g(x)> \; =\alpha< f(x)>+\beta< g(x)>$, where α and β are any scalars (numbers) and f and g are any functions of the random variable x.

A set of random variables is said to be *independent,* if $<f(x)g(x')> = <f(x)><g(x')>$, when $x\neq x'$, for any functions f and g. Define $\Delta x = x - <x>$. The *expected n^{th} moment* is defined as $<x^n>$. The *expected n^{th} central moment* is defined as $<(x - <x>)^n>$. The *variance,* σ^2, is the second central moment. .

The variance in the estimate of the nth moment, $\bar{x}$, obtained using N independent data points is given by

$$<(\bar{x}^n - <x^n>)^2> \; = \; <\frac{\sum_k x_k^n}{N}\frac{\sum_p x_p^n}{N}> - <x^n>^2$$

$$<(\bar{x}^n - <x^n>)^2> \; = \; \frac{\sum_k\sum_p <x_k^n x_p^n>}{N^2} - <x^n>^2$$

There are $N(N\text{-}1)$ terms where $k\neq p$, and N terms where $k = p$. Using the independence of the variables, we get

$$<(\bar{x}^n - <x^n>)^2> \; = \; \frac{<x^{2n}> - <x^n>^2}{N} \tag{2}$$

Equation (1) is obtained from this equation by setting n =1. A similar expression is obtained for the central moments by simply substituting Δx for x in equation (2). Now, suppose you have a set of 100 random data points from a set you believe to have a Gaussian distribution with zero mean. The third moment is expected to be zero. If you decide to check this by computing the third moment of your data, there will be an expected variance in the measurement of the third moment of $15(\sigma^2)^3/100$, if the distribution was really Gaussian. This result was obtained by using the well-known formula for the even moments of a Gaussian distribution, namely

$$<\Delta x^n> \; = \frac{n!(s^2)^{n/2}}{2^{n/2}(n/2)!}. \tag{3}$$

The odd central moments are zero. A close examination of this expression should reveal the fact that 100 data points is usually a woefully inadequate number to get much information about the higher moments of a distribution.

TIME SERIES

A *time series* is an infinite set of random variables $\{x_n\}$ where the order of the data points is important. You can think of a time series as a regularly sampled random signal. A *realization* of a time series is a finite, contiguous subset of a time series. A *stationary time series* is one where the expected value of any average quantity is the same no matter which realization is used. A typical non-stationary situation would be an attempt to measure the statistics of the fluctuating velocity near the ground during a thunderstorm. The concept of a fluctuation is difficult to define here in that the "mean" velocity will change during the storm, increasing and then decreasing. This is a case where the selection of the averaging time will affect the results. All of the cases examined in this chapter will assume the time series are stationary.

Another type of moment associated with time series is the *autocorrelation function*, $R(m)$, defined as $\langle x_n x_{n+m} \rangle$. The value of m is referred to as the *lag time*. Figure 1 shows a plot of a typical time series. Figure 2 shows the expected autocorrelation function for the type of data in the time series. The first point, $m = 0$, of the autocorrelation function is $\langle x^2 \rangle$.

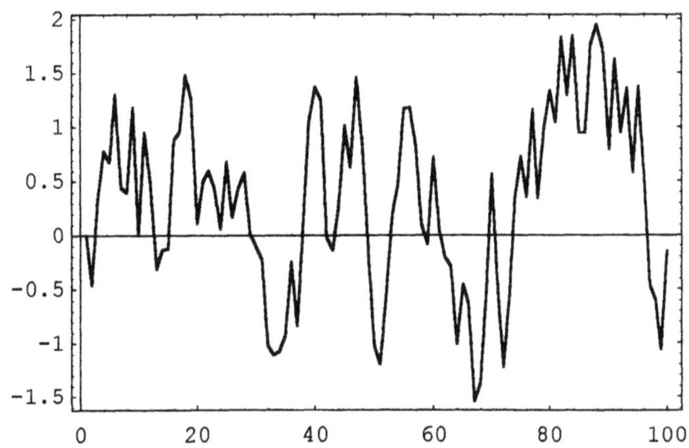

Figure 1. Typical time series.

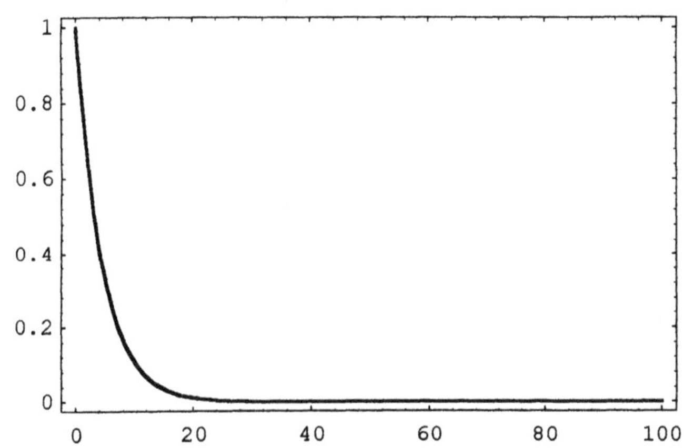

Figure 2. Typical autocorrelation for time series in figure 1.

If $\{x_n\}$ does not change rapidly in the time interval used for sampling, then the values of the autocorrelation for small values of m will be close to the initial point, $m = 0$. In the limit of m going to infinity, the autocorrelation function goes to $\langle x \rangle^2$. The autocorrelation function is also symmetrical, namely, $R(m) = R(-m)$.

The *normalized autocorrelation function*, $RN(m)$, is defined by

$$RN(m) = (R(m) - \langle x \rangle^2) / \sigma^2 \tag{4}$$

VARIANCE OF THE MEAN OF A TIME SERIES

We are now in a position to examine the variance in the measured mean of a time series. Let

$$\overline{x}_N = \frac{1}{N}\sum_{n=1}^{N} x_n, \tag{5}$$

the measured mean. The expected variance of the mean, σ_m^2, is given by

$$\sigma_m^2 = <(\overline{x}_N - <x>)^2> = \frac{1}{N^2}\sum_n \sum_m <x_n x_m> - <x>^2. \tag{6}$$

$$\sigma_m^2 = \frac{1}{N^2}\sum_n \sum_m R(n-m) - <x>^2 = \frac{1}{N}\sum_{q=-N-1}^{q=N-1} \sigma^2 RN(q)$$

Define

$$\Lambda_I = \sum_{q=0}^{\infty} RN(q), \tag{7}$$

the integral time scale. If this were a turbulent flow being measured, this expression would define the Taylor Macro Scale. This scale is a measure of the time the time series data stays correlated with itself. If N is larger than Λ_I, the expression for the variance of the mean becomes,

$$\sigma_m^2 \approx \frac{2\Lambda_I}{N}\sigma^2. \tag{8}$$

The effective number of independent data points is $N/2\Lambda_I$. Please note that although this expression is given in terms of the number of measured points, it really refers to the measurement time. The integral time scale is determined by the process being measured. Thus, equation (8) gives the expected error in the mean in terms of the number of integral time scales measured. If you doubled the number of data points measured in some time interval, you would have to double the number of points measured to get the same accuracy in that time period.

ESTIMATION OF AUTOCORRELATION FUNCTIONS

The autocorrelation function, $R(m)$, is estimated by

$$\overline{R}(m) = \frac{1}{N}\sum_{n=0}^{N-m-1} x_n x_{n+m}. \tag{9}$$

The expected value of this estimate is given by

$$<\overline{R}(m)> \equiv R_N(m) = \frac{1}{N}\sum_{n=0}^{N-m-1} <x_n x_{n+m}> = (1-\frac{|m|}{N})R(m). \tag{10}$$

The expected measured autocorrelation function is the true autocorrelation function multiplied by the factor $1-|m|/N$. If $N >> \Lambda_I$, this results in negligible distortion.

Naturally, an autocorrelation function estimate will differ from the expected autocorrelation function. In the next section, I will examine the *covariance* of the difference between the estimated and the expected autocorrelation at two different lag times, p and q. The covariance of the estimated autocorrelation is given by

$$\varepsilon^2(p,q) = <(\overline{R}(p)-R_N(p))(\overline{R}(q)-R_N(q))>$$

$$\varepsilon^2(p,q) = \frac{1}{N^2}\sum_{n=0}^{N-p-1}\sum_{k=0}^{N-q-1} <x_n x_{n+p} x_k x_{k+q}> - R_N(p)R_N(q). \tag{11}$$

Note that a fourth moment has to be computed to estimate the measurement error for an autocorrelation function. The details of the calculation depend on the statistics of the time series itself. For instance, if the statistics of the series are Gaussian, it can be shown that (11) becomes

$$\varepsilon^2(p,q) = \frac{\sigma^4}{N}\sum_{k=-N}^{N} w(p,N)w(q,N)[RN(k)RN(k+p-q)+RN(k-p)RN(k+q)$$

$$+\frac{<x>^2}{\sigma^2}(RN(k)+RN(k+p-q)+RN(k-p)+RN(k+q)], \quad p>q, \tag{12}$$

where

$$w(p,N) = 1-\frac{|p|}{N}. \tag{13}$$

The terms on the second line arise from the errors in the estimate of the average of the series. It affects the estimated baseline of the autocorrelation function. Note also that the errors in the estimate of the autocorrelation function are expected to correlate with each other. In other words, if the error in the pth point is high, the error in the points around it are also expected to be high. This fact is usually overlooked when the autocorrelation function is used to estimate parameters such as the bandwidth of the time series.

This expression is somewhat hard to visualize, but it can be shown that a reasonable approximation is given by

$$\varepsilon^2(p,q) \approx \frac{2\sigma^4\Lambda_I}{N}[RN(p-q)+RN(p)RN(q)+4\frac{<x>^2}{\sigma^2}]. \tag{14}$$

The approximation above is valid no matter what the detailed statistics of the time series.

DIGITAL ESTIMATES OF THE SPECTRUM OF TIME SERIES

The complex digital spectrum of a time series, $\bar{s}_N(m)$, is defined through the digital Fourier transform as follows:

$$\bar{s}_N(m) = \frac{1}{\sqrt{N}} \sum_{k=0}^{N-1} x_k \exp(-i\frac{2\pi\, km}{N}). \tag{15}$$

The ***average power spectrum*** is defined as

$$\bar{S}_N(m) = |\bar{s}_N(m)|^2 \tag{16}$$

$$\bar{S}_N(m) = \frac{1}{N}\sum_{k=0}^{N-1}\sum_{l=0}^{N-1} x(k)x(k+(l-k))\exp(-i\frac{2\pi\, m(k-l)}{N}) \tag{17}$$

Inverting the order of summation and renaming the summation variable, we get

$$\bar{S}_N(m) = \sum_{p=-N-1}^{N-1} \bar{R}(p)\exp(-i\frac{2\pi\, mp}{N}) \tag{18}$$

The expected spectrum is given by

$$S_N(m) = \sum_{p=-N-1}^{N-1} R(p)\exp(-i\frac{2\pi\, mp}{N}) \tag{19}$$

However, be careful because there are some problems with the DFT. First, the DFT is usually defined as a one-sided sum, i.e. from 0 to N-1. The expressions above are two-sided sums, from -N-1 to N-1. Since the autocorrelation function is, by definition, symmetrical, you can get the equivalent result from a typical one-sided DFT by

$$S_N(m) = 2\mathrm{Real}[\sum_{p=0}^{N-1} \bar{R}(p)\exp(-i\frac{2\pi\, mp}{N})] - \bar{R}(0) \tag{20}$$

Further, be careful of what normalization factor is used by the package. These days, most packages seem to use a factor of $1/\sqrt{N}$. You will have to multiply by the inverse factor to obtain the correct energies in the spectrum computed as shown above, in (20).

The digital power spectrum is the two-sided Digital Fourier Transform of the estimated correlation function. In fluid mechanics circles, I believe this is called Taylor's theorem. Everywhere else, it is known as the Weiner-Khinchin theorem.

Of particular interest to measurements in fluid mechanics is the fact that *the Digital Fourier Transform is not a digitized version of the Fourier Transform.* In fact the two are related by

$$\mathrm{DFT}(m) = \sum_{k=-\infty}^{\infty} \mathrm{FT}(m - kN). \tag{21}$$

This means that the Digital Fourier Transform at any point is the sum of an infinite number of shifted Fourier Transforms. For a spectrum with a long tail such as those typically found in turbulence velocity fluctuation spectra, the consequent distortion of the spectrum can be quite severe. This phenomena is called *aliasing*. The cure for this problem is to filter out *all* of the frequency content of the signal above a frequency of $1/2\Delta t$, where Δt is the sampling rate. This frequency is the well-known *Nyquist frequency*. The digital spectrum is symmetric about $N/2$, so that only half of the spectrum contains useful information. The effective number of frequency points is $N/2$.

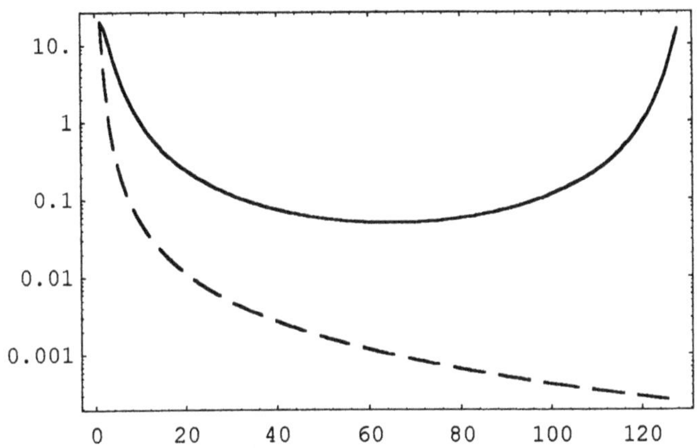

Figure 3. The solid line is the Digital Fourier Transform of a 128 point sample of exp[-.8n] and the dashed line is the Fourier Transform of exp[-.8|n|].

Other interesting properties of the DFT are:

$$\sum_{n=0}^{N-1} f(n)g(n)\exp[-i\frac{2pnm}{N}] = \sum_{k=0}^{N-1} F(k)G(m-k) \quad \textbf{Convolution} \tag{22}$$

$$\frac{1}{N}\sum_{m=0}^{N-1} F(m)\exp[-i\frac{2pnm}{N}] = f(n) \quad \textbf{Inverse} \tag{23}$$

We can now examine the covariance of the spectrum estimates.

$$\varepsilon_S^2(m,n) = <(\bar{S}_N^2(m) - S_N^2(m))(\bar{S}_N^2(n) - S_N^2(n))>$$

$$\varepsilon_S^2(m,n) = \sum_{p=-N-1}^{N-1}\sum_{q=-N-1}^{N-1} e^{-i\frac{2\pi mp}{N}} e^{i\frac{2\pi nq}{N}} (<\bar{R}_N(p)\bar{R}_N(q)> - R_N(p)R_N(q)) \tag{24}$$

Fortunately the term in brackets has already been computed in equations (12) and (14). If (12) is substituted into (24) and the computations carried out, assuming for the sake of simplicity a zero mean signal, one obtains

$$\varepsilon_S^2(m) \simeq S_N^2(m)\delta_{nm}, \tag{25}$$

where δ_{nm} is the Kronecker delta. *The expected variance of the spectrum is the square of the spectrum!* It may not be obvious yet, but the problem is that the Fourier Transform was taken of the entire N data points.

There is another way to compute the spectrum. Take sub records of length of $N_s < N$. Compute the spectrum of this record using (14). Do that for N/N_s times and average the results. It can be shown that the variance of this estimate is

$$e_S^2(m) \simeq \frac{N_S}{N}(\tilde{S}_N^2(m)\delta_{nm}) \tag{26}$$

where $\tilde{S}_N(m)$ is the spectrum obtained from $R_N(m)(1 - \frac{|m|}{N_S})$, i.e. the result is equivalent to computing the entire correlation function and then truncating the correlation function. What has happened here is that the spectrum information has been obtained from a subset of the possible frequencies. This is equivalent to smoothing of the data after computation of the Fourier Transform, e.g.

$$\tilde{S}_N(m) = \sum_{n=-N-1}^{N-1} w(n, N_s)R_N(n)e^{-i\frac{2\pi nm}{N}}. \tag{27}$$

By (22), the Convolution Theorem, this expression can be rewritten

$$\tilde{S}_N(m) = \sum_{k=-N-1}^{N-1} W(m-k, N_s)S_N(k), \tag{28}$$

where $W(\ )$ is the Fourier Transform of $w(\ )$. The function $W(k)$, smoothes the Transform $S(k)$.

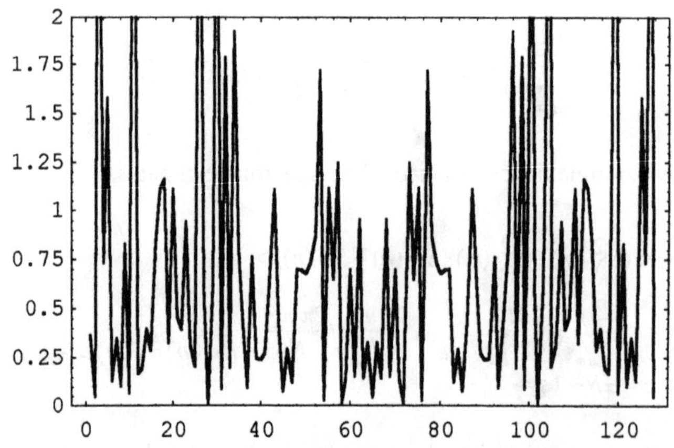

Figure 4. Measured spectrum of 12,800 random data points using equation (17) at the frequencies m/N_S.

A careful reading of the above section should convince you that the smoothing function is necessary in order to get good estimates of the spectrum of random signals. In electrical engineering this smoothing process is called *windowing*. There is an inherent trade-off between smoothness and resolution when measuring the spectrum of a time series. The bandwidth of the smoothing filter is $\approx 1/(\Delta t N_s)$, so the finest feature visible will have to be wider in bandwidth than that. There is no point in having a finer frequency scale than the bandwidth of the smoothing filter, so most users modify the calculation to take place only over the range where $W(k,N_S)$ is not zero and with a frequency step of the bandwidth of the filter, e.g.,

$$\tilde{S}(m) = \sum_{n=0}^{N_S-1} w(n, N_S) R_N(n) \exp[-i\frac{2\pi\, nm}{N_S}] \tag{29}$$

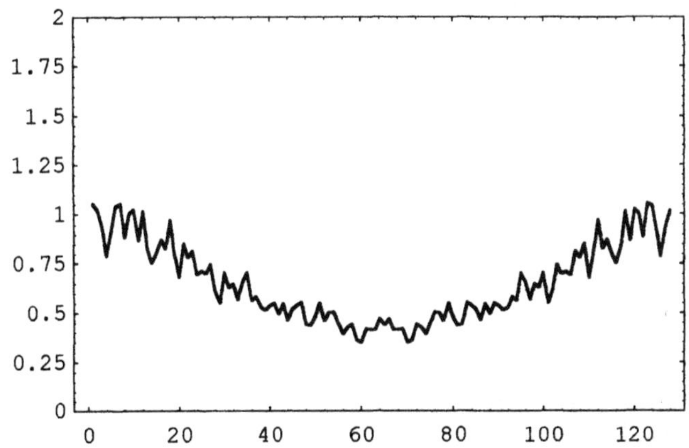

Figure 5. Measured spectrum of same data set as figure 4, using equation (27).

A bonus obtained is that the number of computations is decreased. A FFT routine takes $N\ ln_2 N$ multiplications and additions to compute the DFT. Computation of the correlogram indicated above takes $N_S\, N$ multiplications and additions and then computation of the spectrum takes $N_s\ ln_2\ N_s$ more computations. The latter method usually involves fewer computations than the FFT of the full data set, followed by smoothing.

Many experimenters underestimate the number of points needed to estimate a spectrum to a given degree of accuracy. For instance, suppose you wanted to measure a spectrum with a frequency range of 512 points to an accuracy of 5% at each point. You would need to compute a spectrum of 1024 points to get that range, so the minimum value of N_S would have to be 1024. A 5% accuracy would mean that the relative variance would be 0.0025 requiring that N/Ns be greater than 400. That means that 409,600 data points would be needed.

RANDOM SAMPLING

Laser anemometer typically measure the light scattered from particles randomly placed in the flow of interest. In a turbulent flow, this means that you are randomly sampling a time series. To complicate things more, *the sampling rate for any given velocity depends on that velocity*. Intuitively, you would expect that high speeds would be sampled more often than low speeds because the flow itself convects the particles into the measurement region and high speeds bring more particles per second than do slow speeds. A consequence of this effect is that averages from the randomly sampled flow are *biased* towards the higher speeds. There are many schemes in the literature for correcting this bias. However, most of them depend on uniform particle number density in the flow and on every particle crossing the region being measured. These conditions are often not perfectly met in practice. These correction schemes also depend on ancillary data from the experiment such as the time the particle spends in the measurement region. Again, assumptions are made about the measurement that may not be met in practice.

In this section, I will endeavor to outline the critical concepts needed to deal with random sampling. The essence of the problem is that probability of getting a measurement of a velocity depends on that velocity appearing in the measurement region *and* on a particle actually being measured crossing the region. In general, the probability of a measurement of a velocity v at some time t_i, is given by

$$p_m(v, t_i) = p(v)\,p(t_i | v), \tag{30}$$

where $p(v)$ is the desired probability distribution function for the velocity and $p(t_i|v)$ is the conditional probability of a measurement if the velocity is v. This conditional probability is given by

$$p(t_i | v) = \frac{\rho(v)}{<\rho>}, \tag{31}$$

where $\rho(v)$ is the measurement rate for the velocity v. Note that in all cases the probability of a measurement of v in two different intervals are independent, since there is no assumed correlation in the particle positions.

Independent Sampling

Initially we will examine the situation where the sampling rate does not depend on the velocity measured. Much more is known about this situation than the general one where the measurement rate depends on the measured quantity. The trend and order of magnitude of the independently sampled results should give some insight into the more general situation. Since the probability of a measurement does not depend on the measured quantity, the expected means and the variances in the estimates of the means depend only on the expected number of independent samples obtained. In fact, it can be shown that the expected means for a randomly sampled time series are the same as for a regularly sampled series. However, there is a difference in the expected variance in the measurement. The variance is given by

$$\sigma_m^2 = \frac{\sigma^2}{N}(1 + 2\rho\,\Lambda_I). \tag{32}$$

The term $\rho\Lambda_I$ is the average number of measurements per integral time scale. In the limit of a small measurement rate, the expression becomes that for independent sampling and in the limit of a high measurement rate, it becomes that for continuous sampling.

Autocorrelation functions and spectra can be estimated from randomly sampled data. One method is to create a new regularly sampled series from the data by selecting a time interval Δt and inserting the average of the randomly sampled measurements occurring in the interval $[k\Delta t, (k+1)\Delta t]$ in the corresponding slot in the new series. Usually, zero is inserted in the slot if there are no measurements in that interval. The motivation for making this new series is that estimation of the autocorrelation function depends on averaging the product of the values separated by a fixed time and in a randomly sampled series, the probability of getting two measurements at a given separation in time is negligible.

Again, the expected value of the autocorrelation function calculated from the derived series is the true autocorrelation, provided that each measured set of lag products is divided by the number of actual pairs encountered in the data set, e.g.

$$< \overline{R}(k) > = \frac{1}{N_k}\sum_{i=1}^{N_k} < y_i y_{i+k} > = \frac{1}{N_k}\sum_{i=1}^{N_k} R(k\Delta t) = R(k\Delta t) \tag{33}$$

where N_k is the number of pairs found for the lag time k.

A minor problem is that the autocorrelation derived from the second series is actually a filtered version of the true function because each point of the second series is the average of the data found in the interval Δt. If the interval is picked to be small enough, this distortion is unimportant.

The covariance of the errors in the measured autocorrelation function is the same as before except that N_k is substituted for N. The expected number of such pairs can be estimated from the average measurement rate. If ρ is the average measurement rate, and Δt is smaller than the underlying series' time scales, the probability of finding at least one measurement in any particular slot is given by $1-\exp[-\rho\Delta t]$. Since the probability of a measurement in one interval is independent of that for another, the probability of a pair is the square of the probability for one interval, i.e.,

$$< N_k > \approx N(1 - \exp[-\rho\,\Delta t])^2 \tag{34}$$

It can be seen from this expression that it is important to pick the proper value of Δt since too small a value and there will be few useful pairs for a given lag time, increasing the measurement variance. On the other hand, if it is picked too large the smoothing of the autocorrelation function will be excessive.

The spectrum can be computed from the autocorrelation in the normal fashion. Another popular method for computing the spectrum is as follows:

$$\begin{aligned} \overline{s}_T(m) &= \frac{1}{\sqrt{T}}\sum_k x_k \exp[-\frac{i2\pi t_k m}{T}] \\ \overline{S}_T(m) &= |\overline{s}_T(m)|^2 \end{aligned} \tag{35}$$

where t_k is the time of the kth sample. In this case, the expected measured spectrum is not just the expected spectrum of the underlying time series.

$$\langle S_T(m)\rangle = \rho R(0) + \rho^2 S(m) \tag{36}$$

The $\rho R(0)$ term is added to the spectrum. In the limit of large measurement rates, the expected spectrum goes to that of the underlying series. In the lower limit, it goes to a constant. The expression for the variance is likewise modified from the previous results.

$$\varepsilon_S^2(m) = \frac{1}{BT}\langle S_T(m)\rangle^2, \tag{37}$$

where B is the bandwidth of the smoothing filter.

Non-Independent Sampling

Using Equation 28, it can be shown that the expected value of the arithmetic average of a randomly sampled time series is given by

$$< \bar{x} > = \frac{< \rho \nu >}{< \rho >}. \tag{38}$$

If the measurement rate is assumed to be proportional to the measured quantity ν, then the expected average is

$$< \bar{x} > = \frac{< \nu^2 >}{< \nu >} \tag{39}$$

resulting in a bias of σ^2 / ν above the expected value of ν.

In principle, all bias effects of random sampling can be eliminated if you know the exact form of $\rho(\nu)$. For instance, the unbiased mean of the quantity ν can be computed from

$$< \nu > = \frac{\sum_k \nu_k / \rho(\nu_k)}{\sum_k 1 / \rho(\nu_k)}. \tag{40}$$

Most of the schemes in the literature attempt to correct for errors due to the sampling rate by estimating forms for $\rho(\nu)$, by making some assumptions about the optical setup, assuming uniform particle seeding, and assuming that every particle is measured and stored exactly once. Naturally, they are successful to the extent that the approximations are good. In many cases the approximations are not good because of unforeseen factors like electronics whose efficiency is frequency dependent, data handling systems that are not 100% efficient, faulty methods of rejecting more than one measurement from the same particle, and non-uniform seeding. Other times, the models fail because they are simply incorrect.

Another approach is try to make an averaging scheme that is somewhat insensitive to the exact form of $\rho(\nu)$. One such scheme is the so-called Sample and Hold method. Here, the mean is estimated by

$$\bar{\nu}_{SH} = \frac{1}{T} \sum_k \nu_k (t_{k+1} - t_k) = \frac{1}{T} \sum_k \nu_k \Delta T_k \tag{41}$$

where ΔT_k is the time between the kth and the $k{+}1^{th}$ measurement. The idea is that the average time after a measurement should be roughly inversely proportional to the measurement rate and thus the bias effects should cancel. However, a more detailed look at this method reveals that there are restriction that apply before it is possible that it can eliminate the effect of $\rho(\nu)$.

$$< \nu_{SH} > = \int \nu p(\nu) \rho(\nu) < \Delta T | \nu > d\nu \tag{42}$$

where $<\Delta T|v>$ is the mean time between measurements, if the velocity at the beginning of the interval is v, the conditional inter arrival time. It can be shown that if $<\rho>\Lambda$, the average number of measurements per integral correlation time, is greater than 4, the conditional inter arrival time is approximately inversely proportional to the measurement rate and the two terms cancel out in equation (42), leaving an unbiased measurement. On the other hand, if $<\rho>\Lambda$ is less than 0.1, the inter arrival time is essentially unrelated to the measurement rate and thus the bias remains in the calculation.

Another method is to set the measurement system up to take measurements at regular intervals. This is equivalent to the method discussed above where we made a second regularly spaced series from a randomly sampled one. Typically, only the first data point in a time slot is recorded. If, again, Δt is small compared to the underlying scales, the probability of measuring a value of v is given by

$$p_m(v) \propto p(v)\left(1-\exp[-\rho(v)\Delta t]\right). \tag{43}$$

If $\rho(v)\Delta t << 1$, the probability for a measurement becomes proportional to

$$p_m(v) \propto p(v)\rho(v)\Delta t \tag{44}$$

The probability of a measurement is still conditioned by the value being measured. If on the other hand $\rho(v)\Delta t > 4$, the exponential term is negligible and the probability for a measurement is no longer conditioned by the value being measured. There are other similar schemes that limit the sampling of the random series, but they all have the limitation that they are not effective unless the time interval Δt is small compared to the relevant time scales and $\rho(v)\Delta t$ is large.

BIBLIOGRAPHY

Papoulis, A., 1965, "Probability, Random Variables, and Stochastic Processes," McGraw-Hill Book Company.

Saleh, B., 1968, "Photoelectron Statistics: With Applications to Spectroscopy and Optical Communications," Springer-Verlag, New York.

Bendat, J.S. and Piersol, A.G., 1986, "Random Data: Analysis and Measurement Procedures," John Wiley and Sons, Second Edition,.

Tennekes, H. and Lumley, J.L., 1972, "A First Course in Turbulence," The MIT Press, Cambridge, MA,.

Lumley, J.L., 1970 , "Stochastic Tools in Turbulence," Academic Press, New York.

Gaster and Roberts, 1977. "The spectral analysis of randomly sampled records by a direct transform," *Proc. R. Soc.* London. A, **354**, 27-58.

Laser Anemometry and Particle Sizing

PRINCIPLES OF LASER ANEMOMETRY

Lars Lading

Risø National Laboratory
Optics and Fluid Dynamics Department
DK-4000 Roskilde, Denmark

e-mail: ofd-llad@risrms1.risoe.dk

INTRODUCTION

Laser anemometry is an area characterised by the use of laser light for localised non-perturbing measurements of fluid velocity. The start of what could be called modern laser anemometry goes back to the measurements of Yea and Cummins[1]. They performed a light scattering experiment on small particles suspended in a fluid pipe flow. Light scattered in a given direction would exhibit a well-defined Doppler shift. The shift is generally too small to be detected by direct spectral analysis of the optical signal. However, by coherently mixing the scattered light with a reference beam on a photodetector it was possible to obtain a beat signal oscillating with the frequency of the Doppler shift itself. The light beam illuminating the flow and the reference beam came from the same laser. The reference beam light was frequency shifted by Bragg diffraction in an acousto-optic cell. The principle of frequency shifting in order to overcome the sign ambiguity of the measured velocity did not find widespread use until several years later.

In the late sixties several laboratories pursued the technique. The company *Ratheon* made one of the first systems for high-speed flows applied by *NASA*. The research establishment *Harwell* developed the basis for the first commercial system introduced by *DISA Electronics* (now *Dantec Measurement Technology*). The system was based on the so-called differential Doppler configuration - now often called a *fringe* anemometer - developed independently by several groups in Europe, the United States, and Russia. There are now several manufacturers of sophisticated laser anemometer systems.

The applications of laser anemometry have mostly been in laboratory fluid flows. However, industrial applications in more complicated fluid systems with difficult access have lately attracted a considerable amount of interest. Also in medicine has laser anemometry been applied to investigate both the respiratory system and the arterial flow. At the other extreme in terms of range we find the measurements of atmospheric wind velocity. For these applications systems based on CO_2 lasers in a backscattering reference beam configuration

Optical Diagnostics for Flow Processes
Edited by L. Lading *et al.*, Plenum Press, New York, 1994

have dominated. Measurements on solid surfaces are of interest in connection with the production of paper, metal and plastic sheets and profiles, fabric, etc.

In laser anemometry light scattering is generally from particles. However, in some cases the principles of laser anemometry (which in this context should probably be called velocimetry) are also applied to measure the movements of surface or bulk waves.

The areas of laser anemometry optics, signal processing, data processing, and applications in many very different fields have been investigated by many research groups all over the world and are still subjects of active research and development. Many of the newer developments have appeared at various conferences[2]. The presentation given here is concentrated on work in which the author has been involved. No attempt has been made to give references to all who made important contributions to the development of laser anemometry.

In this chapter we shall outline the principles of the laser anemometer itself. The spatial structure is analysed, the essential statistical properties of the detector signals are given, and the methods of signal processing are investigated.

BASIC PRINCIPLES

The basic principles of a laser anemometer are explained in this section. The essential components of a system are identified and the functions of each of the parts are described. We introduce the concept of the *code* of a laser anemometer.

A Simple Experiment

Let us consider a simple experiment (Fig. 1): a laser beam is weakly focused in a fluid where the velocity is to be measured. A small particle - much smaller than the beam diameter - suspended in the fluid is convected through the beam by the movement of the fluid. The particle scatters light. The light power in a given direction depends on the intensity with which the particle is illuminated. Because of the finite and supposedly well defined beam diameter the intensity exhibits spatial variations. As the particle moves through the beam, the detected light power varies with time. From the temporal variations, e.g. the temporal width, it is possible to deduce the velocity assuming that the parameters characterising the illuminating beam are known.

The scheme outlined in Fig. 1 is simple and does in fact illustrate the basis of laser anemometry but, as shown, it is not advisable for velocity measurements. The width of the beam is often poorly known, small variations of the beam profile are often encountered, and the spatial selectivity is poor. But even if these problems were properly solved, the method would still represent poor utilisation of the available light power (or number of photons). So let us consider other configurations. But before we do that, we shall identify the essential components of a laser anemometer (Fig. 2). The *transmitter* defines the way in which the measuring volume is illuminated. The receiver collects the scattered light from the measuring volume and attenuates light coming from outside the volume by a spatial filtering process. Also in the receiver the collected light power is converted to an electronic signal by the photodetector. The transmitter and receiver jointly define the way in which the velocity information is encoded. In Fig. 1 the encoding is in the form of a pulse of varying width. In a fringe anemometer (often called a laser Doppler anemometer - LDA) it is in the form of a spatial wave packet of varying scale (frequency) (Fig. 2a). In a time-of-flight anemometer (LTA) the encoding is in the form of two spatially separated pulses (Fig. 2b). On the basis of the temporal evolution of the signal the *signal processor* provides estimates of the velocity and, possibly, some auxiliary parameters. We shall name the function that defines the encoding the *code* of the laser anemometer.

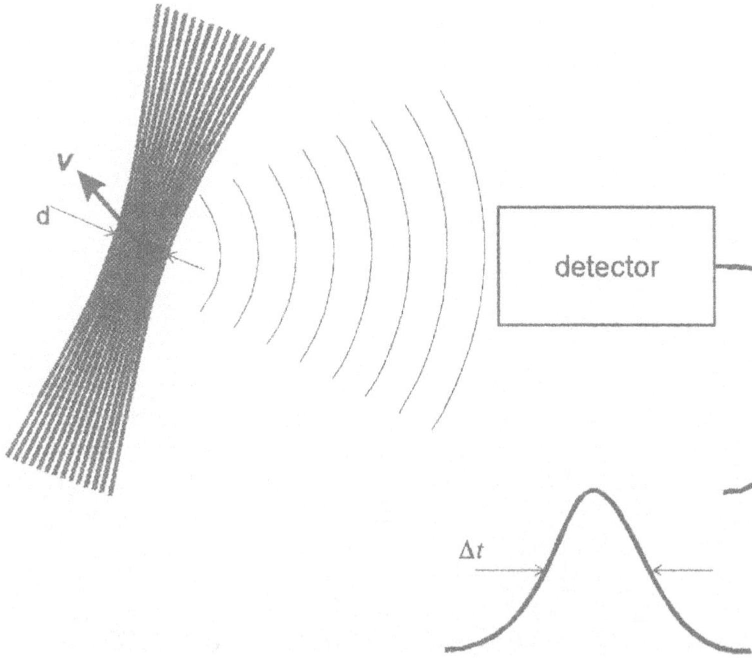

Figure 1. A simple experiment for measuring the velocity of a small particle passing a laser beam. The velocity is given by $d/\Delta t$.

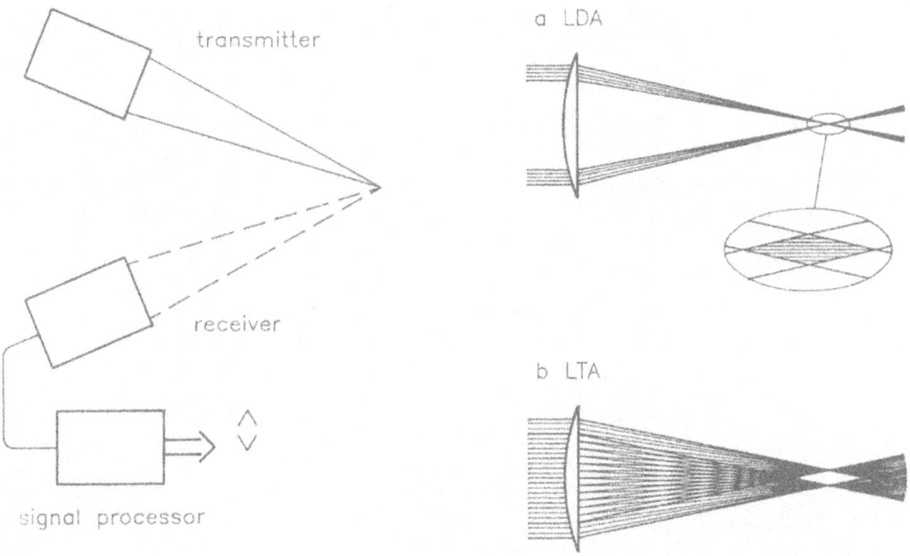

Figure 2. A laser anemometer can be divided into a *transmitter*, a *receiver*, and a *signal processor*. Insert **a** shows the transmitter of an LDA and **b** of an LTA.

It is important to realise that a laser anemometer is based on stochastic processes. The photocurrent (or photon count rate) will even for spatial fixed scatters exhibit temporal fluctuations, the so-called *shot noise* (see the chapter on *Optics*). Electronic thermal noise may also play a role. The scatters - usually small particles suspended in the fluid - are generally randomly distributed in space. As they are convected through the measuring volume, they will provide a random sampling of the fluid velocity. And the velocity itself may be a random process - e.g. turbulence. Therefore, the estimated velocity is in general different from the true velocity. A good laser anemometer is one that provides for estimates that are very close to the true values. Some configurations will certainly do a better job than others. The best configuration depends on the actual conditions - and, of course, on the applied definition of *best*.

FOURIER OPTICAL MODEL

In this section we shall consider the spatial structure of a laser anemometer, i.e. the first two parts shown in Fig. 2. Different *operational modes* are identified and two specific configurations are considered. Fourier optics[3] is used for the analysis because it allows for a relatively simple analysis of the impact of the various components in a system. Rudd[4] was the first to apply Fourier optical methods to the analysis of laser anemometers in what he called a *filter model*. The results given in this section are based on the material published by Lading[5] and the results of Lading, Mann, and Edwards[6].

The spatial intensity distribution in the measuring volume, as seen by the detector when a small particle passes the measuring volume, defines the code of the anemometer. As already mentioned: in a laser Doppler anemometer (LDA) the code is given by a spatial wave packet; in a time-of-flight anemometer (LTA) the code is given by two displaced peaks.

The model is based on the configuration shown in Fig. 3. The transmitter incorporates the light source (a laser is practically always used, but in principle not mandatory), some beam-shaping device, often incorporating a beamsplitter, and a transmitting lens. The spatial pattern of the transmitted light is initially defined in the left focal plane. The receiver incorporates a lens system that generates a spatially filtered image of the measuring volume on the pinhole in front of the detector. The pinhole is matched to, or slightly larger than, the image of the measuring volume. The expected photocurrent is obtained by integrating the intensity over the pinhole area. It is noted that systems where the optical axes of the transmitter and receiver, respectively, are nonparallel can be introduced by mathematically introducing a wedge in the measuring plane. Proper attention will then have to be given to the extra phase shifts as, e.g., utilised for the measurement of particle size by the phase Doppler method (described in the chapter on *Phase Doppler Anemometry*). Performing the analysis for a general system (i.e. without specifying the code or the transmitted field distribution) we find that the photocurrent can be written as a sum of three terms:

$$
\begin{aligned}
i = K\eta^2\Big[& \int p^2 |u_0 *\_h|^2 \mathrm{d}^2 r + \\
& \int p^2 |(u_0 g) *\_h|^2 \mathrm{d}^2 r + \\
& \int p^2 ((u_0 *\_h)(u_0 g) *\_h)^* + c.c.)\mathrm{d}^2 r \Big]; \quad \mathrm{d}^2 r = \mathrm{d}x\mathrm{d}y, \qquad (1)
\end{aligned}
$$

where K is the detector sensitivity (amps/watt), η an optical efficiency factor, and p the pinhole pupil function, which is one where light is transmitted and zero where it is blocked. u_0 is the field in the measuring plane, and $\_h$ is the inverse Fourier transform of the pupil function of the filter H. It is assumed that the directly transmitted beams are unaffected by the scattering (which is of course never exactly true, but a very good approximation in most cases), thus the transmission function of the measuring plane is written as $1 + g$ with $|g|^2 << 1$.

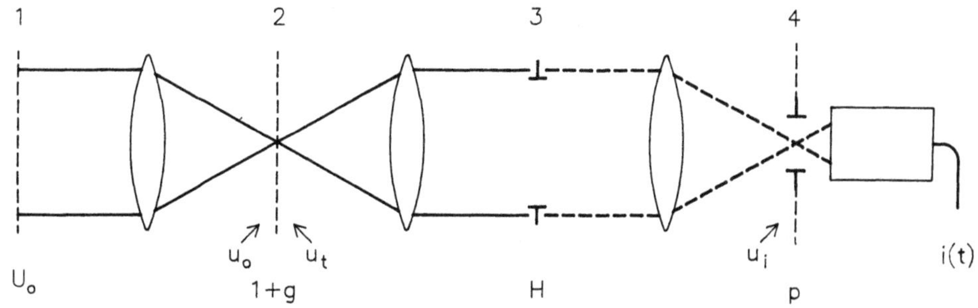

Transmitter — 1: U_0

Measuring plane — 2: $u_0 \propto \Im\{U_0\};\ u_t = u_0(1+g)$

Receiver aperture — 3: $U_t \cdot \_H$

Detector plane — 4: $u_i = u_t * \Im\{\_h\};\ i = K \int_{\substack{\text{det.}\\ \text{area}}} |u_i p|^2 \, dA$

Figure 3. Basic Fourier optical model for the laser anemometer. The spacing between each of the planes 1, 2, 3, and 4 and the nearest lens is assumed to be f, the focal length of the lenses. The optical axis for transmitter and receiver does not necessarily have to be the same. A backscattering configuration is obtained by folding at plane 2 (the measuring plane).

Let us assume that the scattering occurs from randomly distributed particles suspended in the fluid. Each of the particles may be represented by a weighted delta function. This is legitimate if the sizes of the particles are small compared with the fringe spacing (see the chapter on *Phase Doppler Anemometry* for a discussion about scattering from larger particles). The function g is then given by a weighted sum of weighted delta functions, i.e.

$$g = \sum_i m_i \delta(\boldsymbol{r} - \boldsymbol{r}_i), \qquad (2)$$

where $\boldsymbol{r} = (x,y)$ is the coordinate in the measuring plane and m_i is related to the scattering cross-section of the i'th particle.

The different operational modes are defined on the basis of the position and size of the opening in the filter H.

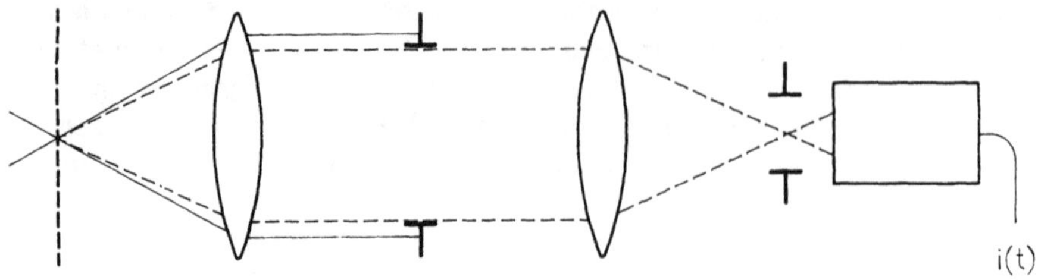

Figure 4. LDV homodyne mode. The directly transmitted beams are blocked by the receiver aperture. If the aperture is much larger than the directly transmitted beams in the plane of the aperture, the system operates with incoherent detection; if the aperture is smaller than the beam diameter, it operates with coherent detection.

The Homodyne Mode

The homodyne mode[7] is characterised by the fact that the directly transmitted beams do not reach the detector. This is illustrated in Fig. 4 for a forward scattering configuration. In this mode the first and second terms of Eq. (1) vanish; only the second term is of importance. Substituting Eq. (2) into Eq. (1) gives that the photocurrent is

$$i = K\eta^2 \int p^2 \left|\sum m_i u_0(\boldsymbol{r}_i) h(\boldsymbol{r}-\boldsymbol{r}_i)\right|^2 \mathrm{d}^2\boldsymbol{r}. \tag{3}$$

Eq. (3) can be written as a sum of squared terms plus a sum of cross-terms. Using the fact that the Fourier transform of $h(\boldsymbol{r})$ defines the receiver aperture we find that

$$\int \left|\_h(\boldsymbol{r})\right|^2 \mathrm{d}^2\boldsymbol{r} = \frac{1}{4\pi^2}\int \left|H(\boldsymbol{k})\right|^2 \mathrm{d}^2\boldsymbol{k} = \frac{A}{(\lambda f)^2}. \tag{4}$$

If $h(\boldsymbol{r})$ is much narrower than the average particle spacing, the cross-terms will vanish and the photocurrent is given by

$$i = K\eta^2 \frac{A}{(\lambda f)^2} \sum_i \left|m_i u_0(\boldsymbol{r}_i)\right|^2. \tag{5}$$

The term $|m_i|^2 / \lambda^2$ can be interpreted as the scattering cross-section of particle i. This mode of operation, characterised by *incoherent detection,* is by far the most common in laser anemometry. The images of two particles separated in the x-y plane by the diameter of the illuminating beam are - potentially - separated.

If $h(\boldsymbol{r})$ is much wider than the width of $u_0(\boldsymbol{r})$, we have *coherent detection.* The photocurrent is then given by

$$i = K\eta^2 \frac{A}{(\lambda f)^2} \left|\sum_i m_i u_0(\boldsymbol{r}_i)\right|^2 \tag{6}$$

We note that in this case the photocurrent consists of a sum of contributions from single particles - as in Eq. (4) - but in addition also a sum due to mixing between light from different particles. The cross-particle terms will of course only appear if there is more than one particle in the measuring volume. The images of two particles separated in the *x-y* plane by the diameter of the illuminating beam are overlapping.

Summarising:

Incoherent detection implies that the receiver diameter is larger than the beam diameter in the measuring plane.

Coherent detection implies that the receiver diameter is smaller than the beam diameter in the receiver plane.

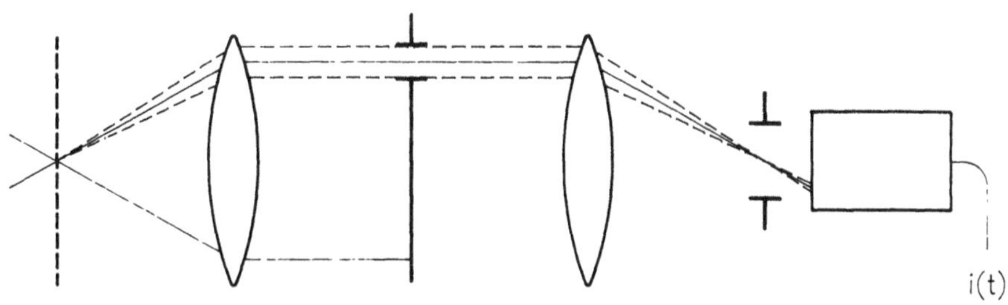

Figure 5. LDA reference beam mode. One directly transmitted beam impinges on the detector.

The Reference Beam Mode

The reference beam mode is characterised by a situation where the dominating dynamic signal is given by heterodyning between scattered light and a reference beam. A possible receiver configuration for this mode of operation is shown in Fig. 5. If the power of the reference beam is much larger than the power of the collected scattered light, then the second term of Eq. (1) vanishes and the dynamic part is given by the last term. This term can be written as follows

$$i = 2K\eta^2 \sum_i m_i |u_0(\boldsymbol{r}_i)|^2 . \tag{7}$$

As in the incoherent detection case no cross-particle terms appear in Eq. (5).

Three-dimensional Effects

The material presented so far assumes that all particles are confined to the measuring plane (Fig. 2). In general, this is obviously not true. A rigorous evaluation of 3D effects is generally rather complicated. This applies both to the spatial filtering effects and to the coherence effects, i.e. from what region are cross-particle beats observed. The most common mode of operation is homodyne with incoherent detection. In that case we can evaluate the length of the measuring volume *seen* by the receiver and find that

$$\Delta z \cong \frac{f}{a} r_p, \tag{8}$$

where a is the radius of the receiver aperture, r_p is the radius of the pinhole, and f is the focal length (see Fig. 2). For the LDA operating in either forward- or backscattering mode this length is much smaller than the intersection length of the two beams. In the derivation of Eq. (8) it is assumed that optical aberration effects cause a smearing of the image of a point source much smaller than the pinhole radius.

Backscattering operation implies that a term $\exp\{jkz_i\}$ must be encountered from each particle. For the homodyne mode with incoherent detection the term has no effect on the signal (the absolute square is 1!). However, mixing with a reference beam this term may give the z-component of the velocity. This essentially corresponds to a *Michelson interferometer-like* configuration, as indicated in Fig. 6 and is the basis for the very long-range laser anemometers used for atmospheric measurements[8].

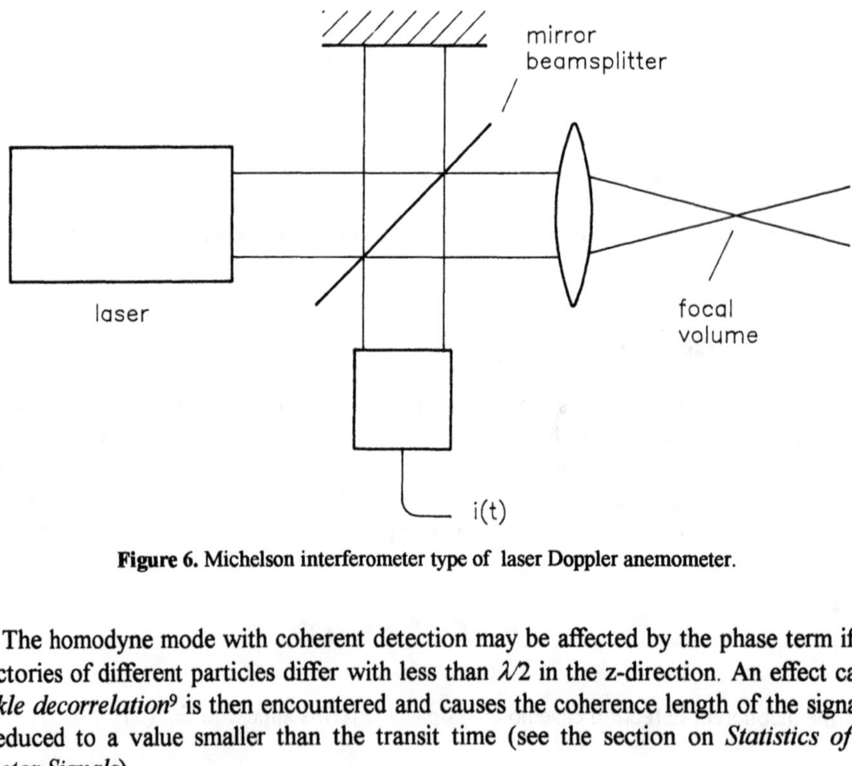

Figure 6. Michelson interferometer type of laser Doppler anemometer.

The homodyne mode with coherent detection may be affected by the phase term if the trajectories of different particles differ with less than $\lambda/2$ in the z-direction. An effect called *speckle decorrelation*[9] is then encountered and causes the coherence length of the signal to be reduced to a value smaller than the transit time (see the section on *Statistics of the Detector Signals*).

The Doppler Configuration

The field in the measuring plane is given by a superposition of two tilted beams, i.e.

$$u_0(x,y) \equiv \frac{E}{\sqrt{\pi} r_0} \exp\left\{-\frac{1}{2}\left(\frac{r}{r_0}\right)^2\right\}\left[\exp\{jk_{x0}\} + \exp\{-jk_{x0}\}\right], \tag{9}$$

where k_{x0} is the spatial frequency corresponding to the angular offset. The transmitted field is the Fourier transform of Eq. (9) and is given by two displaced parallel beams. The intensity distribution that also defines the code is given by

$$I_0(x,y) \equiv \frac{E^2}{\pi r_0^2} \exp\left\{-\left(\frac{r}{r_0}\right)^2\right\}\left[\cos\left\{\frac{k_{x0}}{2}x\right\}+1\right]. \tag{10}$$

Calculating the three-dimensional intensity distribution reveals that the spatial frequency does depend on the z-position as shown in Fig. 7. The relative spatial frequency versus the z-position is given by

$$\frac{k_{x0}(z)}{k_{x0}(0)} = \frac{1}{1+(z/(kr_0)^2)^2}. \tag{11}$$

Incorporating the extension of the measuring volume in the z-direction, it can be shown that the relative error is smaller than the square of $1/(2\pi\times$the number of fringes). The number of fringes is typically in the range of 20 to 100. Thus, the error is in general negligible.

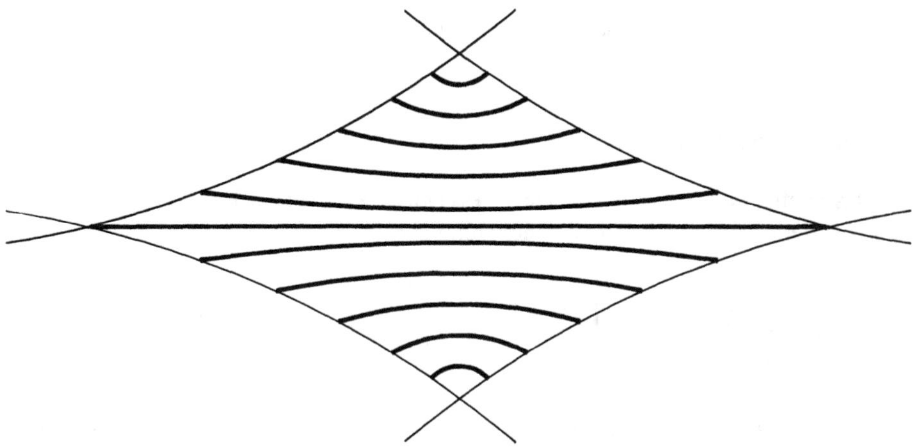

Figure 7. Curvature of the interference fringes of an LDA (exaggerated). The beam waists are at the beam intersection.

A more important error is introduced if the transmitted field has a phase curvature or, equivalently, the beam waists are located in a plane different from the left focal plane of the transmitter lens (see Fig. 8)[10]. We find that

$$\frac{1}{k_{x0}(0)}\frac{\partial k_{x0}(z)}{\partial z}\bigg|_{z=0} = \frac{z_t k}{(z_t/r)^2+(kr_t)^2}\kappa\frac{\alpha}{r_0}. \tag{12}$$

In the same way as for the fringe curvature we find that the error is smaller than $1/(2\pi\times$the number of fringes) and is absent if the beam waists are located in the proper focal planes.

If the beam waists of the two intersecting beams are located in planes at different z-positions, the fringe spacing will not only depend on z but also on x; i.e. a particle passing the measuring volume at a constant velocity will generate a chirped burst - as if it had an acceleration.

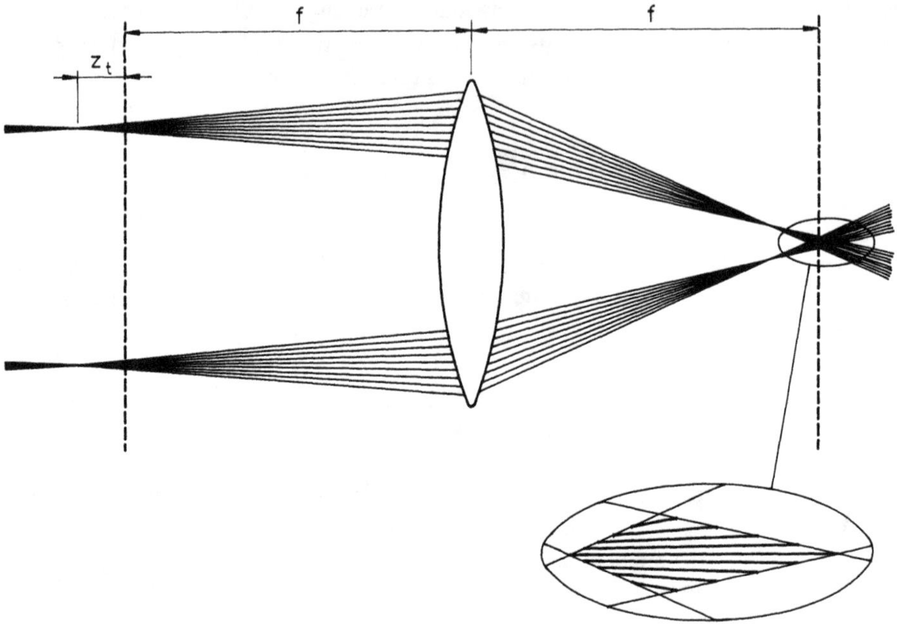

Figure 8. Gradients between the interference planes of an LDA when the beam waists are displaced from the focal plane.

The Time-of-flight Configuration

This configuration can be considered essentially as a spatial Fourier transform of the Doppler system. Thus, the field in the measuring plane is

$$u_0(x,y) \equiv \frac{E}{\sqrt{2\pi r_x r_y}} \exp\left\{-\frac{1}{2}\left(\frac{(x-x_1)^2}{r_x^2}+\frac{y^2}{r_y^2}\right)^2\right\} + \exp\left\{-\frac{1}{2}\left(\frac{(x+x_1)^2}{r_x^2}+\frac{y^2}{r_y^2}\right)^2\right\}. \tag{13}$$

Note that the two spots may have an elliptic shape. In some cases it is an advantage to have $r_y > r_x$ in order to increase the *acceptance angle*.

The transmitted field (the field in the left focal plane of the transmitter lens) consists of two overlapping Gaussian beams with tilted wave fronts. The beam waists of the transmitted beams are supposed to appear in the focal plane. A displacement will not change the beam spacing in the measuring plane or make it z-dependent (as in the Doppler case). But a displacement of the intersection point z_t of the intersection point of the transmitter beam axes will give a spacing

$$2x_1 = 2x_{10}\left(1 + z\frac{z_t}{f^2}\right), \tag{14}$$

i.e. the beams are no longer parallel. Practical systems will often incorporate a lens that images the two spots at the desired measuring distance. In order to make the final beams parallel a displacement of the intersection point along the optical axis is necessary (see Fig. 9).

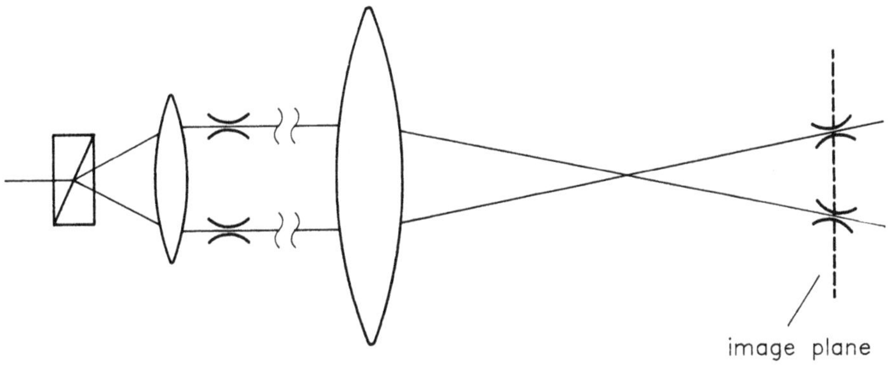

Figure 9. LTA where the two spots are imaged by a relay lens. Note that the beam axes are not parallel.

Comments on the Spatial Configurations

The LDA configuration is by far the most common for general laser anemometry. The LTA is also applied for a number of cases where high intensity of the illuminating beams and a low background are important. However, many other configurations are certainly possible - defined by the spatial code of the anemometer.

STATISTICS OF THE DETECTOR SIGNALS

The statistics of the signal(s) of a laser velocimeter depend on several stochastic processes: the velocity fluctuations are often random; the scatters are supposed to follow the velocity of the supporting medium exactly, but are generally randomly distributed and will thus give a random sampling of the velocity; the number of detected photons within a given epoch $\{t, \delta t\}$ (the time from t to δt) is random, but the expected value is modulated by the movements of the scatters. The way in which the expected photocurrent (or photon count rate) is modulated - the encoding - defines the code of the velocimeter.

Thus we have the following basic stochastic processes in a laser anemometer:

Turbulence → *Particles* → *Photons*

In addition to these processes the refractive index fluctuations along the paths of the propagating light may also play a role. Laser noise can also have an unwanted effect in reference beam systems. The effect is much smaller in homodyne (differential) configurations when the illuminating beams are of equal intensity.

Any measurement will necessarily imply some kind of spatial and temporal averaging. In general, we shall assume that the spatial averaging is done on a scale smaller than the smallest spatial scale of the velocity variations.

The averaging time may be smaller or larger than the characteristic coherence time of the velocity fluctuations. Of course, fluctuations on a time scale smaller than the averaging time cannot be resolved.

Single Bursts

In many applications of laser anemometry the particle concentration is so low that the likelihood of having more than one particle in the measuring volume at any one time is negligible. This essentially implies that most of the time no particles are in the measuring volume. This is generally the case in gas flows. It is assumed that the particle passes the measuring volume at a constant velocity. If the particle is much smaller than any spatial scale of the *code*, then the expected temporal signal is simply a temporal replica of the spatial code. For an LDA it is a wave packet where the frequency is proportional to the velocity in the measuring direction and the width is inversely proportional to the speed of the particle. Such a burst is illustrated in Fig. 10. The expected signal may be written as

$$i(t+t_0) \propto \exp\left\{-(t\omega_s)^2\right\}\left[\cos\{\omega_0 t\}+1\right], \tag{15}$$

where t_0 is the time of occurrence of the burst, ω_s is the bandwidth of the burst (reciprocal transit time), and ω_0 is the Doppler frequency given by

$$\omega_0 = 2\pi f_0 = \frac{1}{2} k_{x0} v_x = 2 v_x k \sin\alpha, \tag{16}$$

where $k = 2\pi/\lambda$ and α is half the angle between the incident beams.

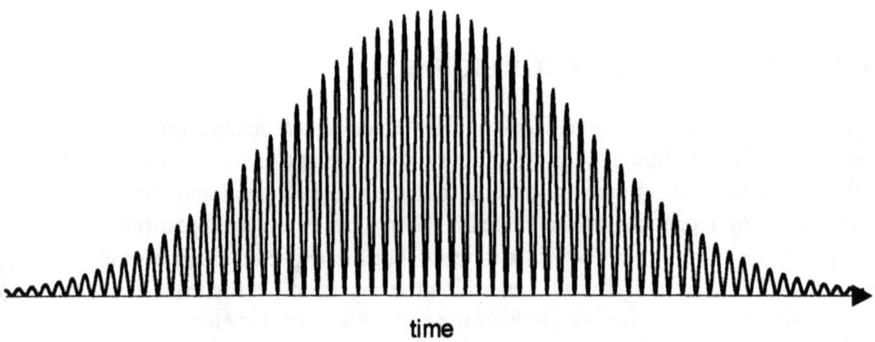

Figure 10. Ideal LDA burst - no noise. The modulation depth is often lower than 1 - as shown here.

In addition to the dynamic signal a background term, *b*, is often present. The term is often caused by light scattering from walls, windows, or other light components that reach the detector. The spatial filtering in the receiver is supposed to minimise the term. The primary difficulty with the background is the shot noise associated with it. In difficult measuring situations it may be much larger than the intrinsic signal noise.

The photodetector converts photons to electrons (electron-whole pairs in *pn*-junctions). Thus, let us briefly consider the photon statistics of the detected photoelectrons. The number of photons emitted from an ideal laser within a given time has a Poisson distribution

$$p(n) = \frac{\langle n \rangle^{n_i}}{n_i!} \exp\{-\langle n_i \rangle\}. \tag{17}$$

The count numbers at different nonoverlapping times are statistically independent. Thus the joint probability distribution for a set of count numbers is simply given by the product of functions of the type specified by Eq. (17). The number of photoelectrons of the photodetector does also have a Poisson distribution, but the expected value of the photoelectron count number is lower than the number of photons impinging on the detector. The ratio between the two expected count numbers defines the quantum efficiency of the detector. The expected value of the photon rate (photons per unit time) impinging on the detector is given by the intensity distribution on the detector integrated over the detector area.

Let us consider the situation where a particle is at a fixed position in the measuring volume. The distribution of the count numbers is given by Eq. (17). The expected count number is proportional to the product of the light intensity at the particle position, the particle scattering cross-section, and the light collection efficiency of the receiver. As the particle moves, it samples different intensities. The expected value of the count number at different times (i.e. particle positions) can be written as

$$\langle n_i \rangle = \langle r(t) \rangle = [as(t_i - t_0, \mathbf{v}) + b]\Delta t, \tag{18}$$

assuming that the counting time is much smaller than any time scale associated with the movement. a is the signal count rate and b is the background. $s(t_i - t_0, \mathbf{v})$ is given by the code of the anemometer - for an LDA by Eq. (15). The noise is independent of the signal if $b >> a$ and is then additive. A burst with additive noise is shown in Fig. 11a. Figure 11b shows a situation where $b = 0$. The noise is then multiplicative - the expected noise power is proportional to the signal amplitude; thus the noise is lowest when the signal amplitude is smallest! Figure 12 shows part of a burst represented by the actual count numbers.

The arrival rate of bursts depends on particle concentration, the velocity, and the cross-section of the measuring volume perpendicular to the direction of the velocity. Thus, although the random distribution of particles gives a random sampling of the velocity, the distribution of the samples depends on the mentioned quantities. This implies that a simple arithmetic mean of the measured velocities will in general be a biased representative of the true mean velocity. For a one-dimensional flow with homogeneous seeding an unbiased estimate of the mean velocity can be obtained by weighting each of the measurements by the measured velocity, i.e.

$$\langle \hat{v} \rangle = \frac{\sum_i \frac{1}{v_i} v_i}{\sum_i v_i}. \tag{19}$$

However, the weighting does not work in the case of a three-dimensional flow or if the seeding is inhomogeneous. Also, a laser anemometer without frequency shift will have an acceptance angle of the velocity direction smaller than +/- π. The velocities of particles moving in directions outside the acceptance angle are not measured. It has been suggested to use a residence time weighting. Each measurement is weighted by the time a particle spends in the measuring volume, i.e.

$$\langle \hat{v} \rangle = \frac{\sum_i \frac{1}{\delta t_i} v_i}{\sum_i \delta_i}. \tag{20}$$

This scheme requires a homogeneous seeding - which is rarely fulfilled - and a reliable measurement of the residence times. For a discussion of the bias problems and various correction schemes see[11].

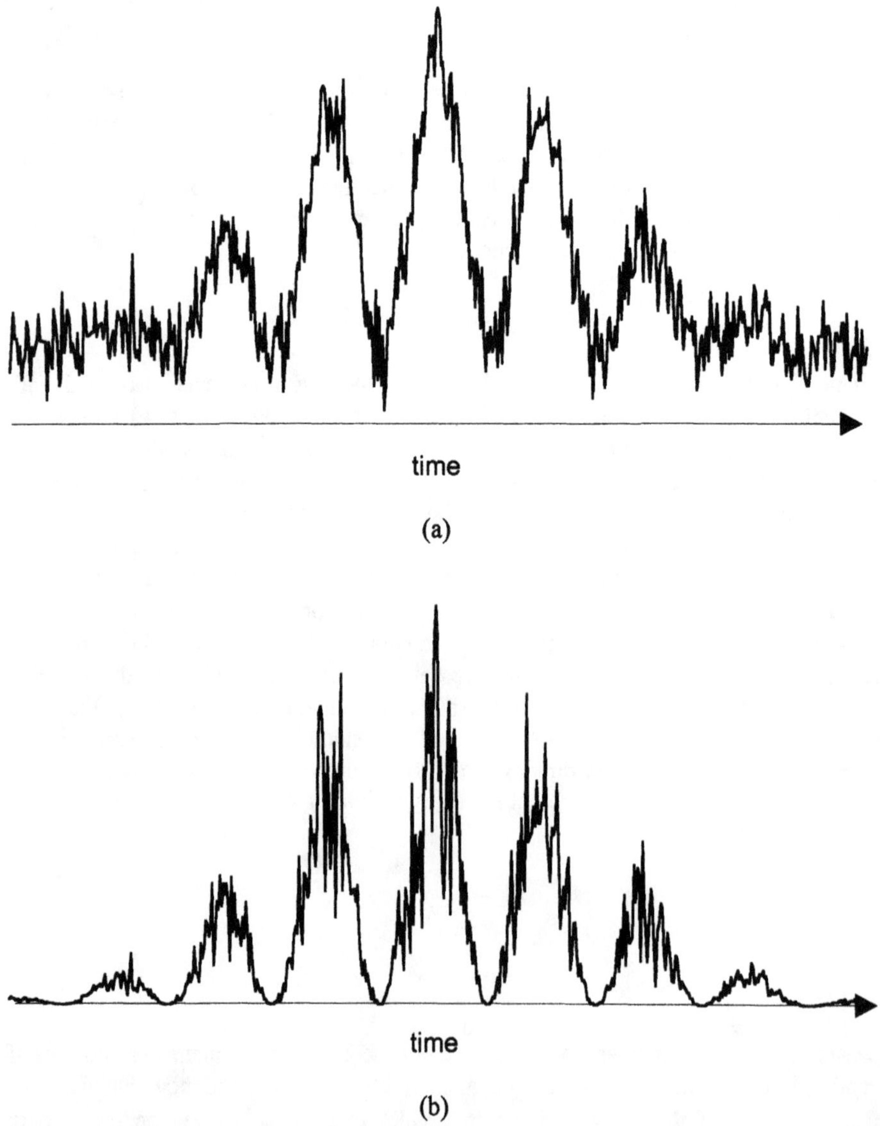

Figure 11. A Doppler burst with a 100 % modulation and with (a) additive noise and (b) multiplicative noise. It is noted that the number of oscillations in a burst is often considerably higher than shown here. The number of oscillations of the signal is generally larger than shown here.

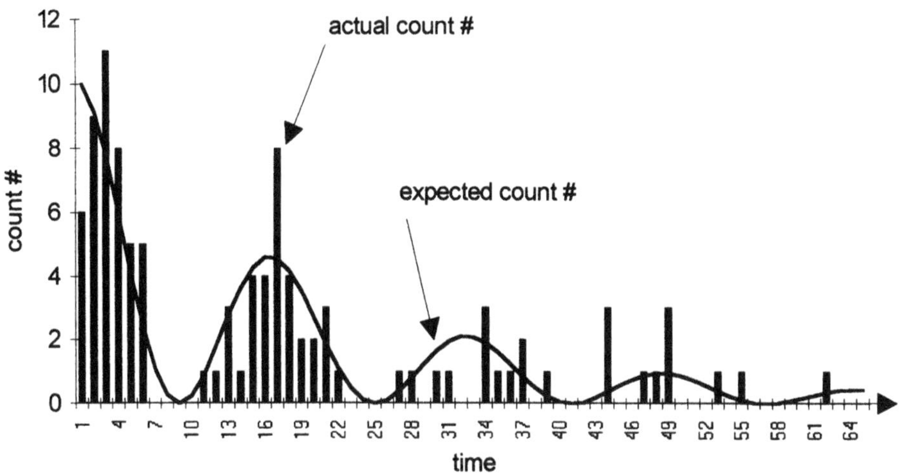

Figure 12. Photon counting: illustrating one half of an LDA burst obtained by photon counting. It is noted that it is possible to recover the Doppler frequency even if the mean number of photons per cycle of the signal is less than 2 (Nyquist sampling only applies strictly for periodic sampling).

Many Particles in the Measuring Volume

The biasing problems could in principle be solved if the velocity samples could be taken with a temporal spacing much smaller than the coherence time of the velocity fluctuations. A replica of the velocity fluctuations can then be produced. However, it is often difficult - if at all possible - to control the particle concentration. If the particle concentration is so high that several particles are in the measuring volume at the same time, then cross-particle interference plays a role. The fact that the particles are randomly distributed in space implies that the phases of the contributions to the signal from the different particles are randomly distributed. Bursts from two particles may be in phase or of opposite phase, which would cancel the dynamic part of the signal. Each time a new set of particles enters the measuring volume the phase of the signal changes. The *instantaneous* frequency may be defined as the derivative of the instantaneous phase and will exhibit fluctuations even in the case of a constant velocity. In the early seventies this fact almost discredited the LDA for turbulence measurements. However, many turbulence measurements have actually been performed with laser anemometers. But it must still be considered a nontrivial task to perform turbulence measurements with a very high spatial and temporal resolution. Let us consider the problem for the LDA[12].

The instantaneous Doppler signal may be written as follows

$$i(t) = A(t)\ \cos\{\omega_0 + \phi(t)\}. \tag{21}$$

With many particles in the measuring volume or a high concentration of surface scatters the statistical properties are somewhat unpleasant: the variance of the derivative of the phase is infinitely large. The distribution for ϕ is rectangular ranging over 2π. $A(t)$ is nonnegative and has a Rayleigh distribution.

Now, any processor will of course give an output with a finite variance. Averaging

$$\dot{\theta} \equiv \partial\{\omega_0 t + \phi(t)\} / \partial t; \; \hat{\omega}_0 \equiv \langle \dot{\theta} \rangle_T \tag{22}$$

over a time larger than zero will give an output of finite variance. If the averaging time T is larger than the reciprocal signal bandwidth, then the variance of the estimated frequency is given by

$$\mathrm{var}\{\hat{\omega}_0\} \cong \omega_s^{\,2} / T. \tag{23}$$

However, it is so that phase and amplitude are correlated. It can be shown that the conditional variance

$$\mathrm{var}\left\{\hat{\omega}_0 \middle| A > A_0\right\} \to 0 \text{ for } A_0 \to \infty \tag{24}$$

as shown in Fig. 13. This implies that if the amplitude is continuously monitored and used for gating the frequency measurements, then an arbitrarily low accuracy can be obtained provided that no photon or thermal noise appears in the signal and that the accepted observation time (which is different from the averaging time) is unlimited. Now, for a specified temporal resolution and a given noise level there exists an optimum threshold for data acceptance. This is illustrated in Fig. 14. The effect of amplitude conditional sampling is that *holes* will be generated in the measured record of the velocity. Raising the amplitude of acceptance implies a transition to the single-burst case.

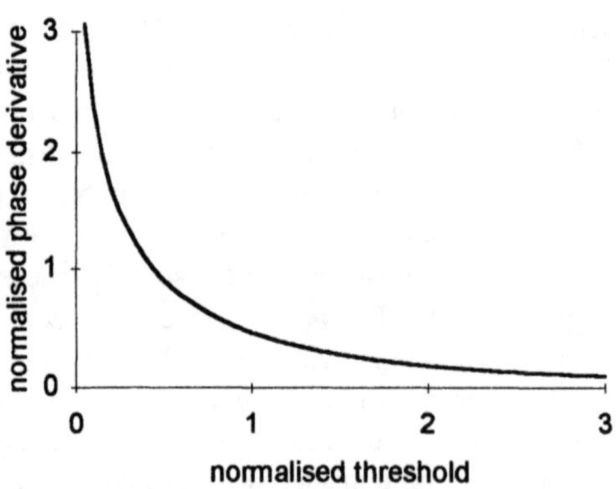

Figure 13. The variance of the normalised phase derivative (instantaneous frequency) versus the normalised threshold for the signal amplitude measurement is conditioned.

The effect of the particle concentration may also be evaluated in terms of a so-called coherence factor defined by

$$\gamma \equiv \frac{\langle v\hat{v}\rangle^2}{\langle v^2\rangle\langle \hat{v}^2\rangle}, \tag{25}$$

where v is the fluctuating part of the true velocity and $\hat{v}$ is the estimated velocity. The coherence factor has been evaluated as a function of the threshold for different temporal scales of the velocity fluctuations (Fig. 14b).

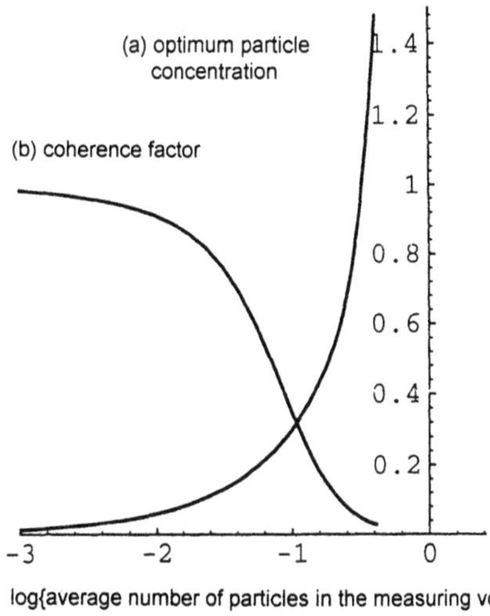

Figure 14. The optimum number of particle concentration in the measuring volume for an LDA (a) versus the logarithm of the ratio of the spatial dimension of the measuring volume to the spatial scale of the velocity fluctuations. Also shown is the corresponding normalised correlation between the measured and the true velocity fluctuations (coherence factor) (b). Instead of controlling the particle concentration an equivalent effect can be obtained by performing amplitude conditioned sampling. The normalised threshold for accepting a measurement depends monotonically on the reciprocal optimum particle concentration. A turbulence intensity of 10 % and 60 fringes are assumed.

Equivalent evaluations of the LTA can be performed. Figure 15 shows the coherence factor versus the particle concentration calculated for different turbulence scales assuming that the spatial resolution in the measuring direction is given. It is worthwhile to notice that there is in general an optimum particle concentration for the LDA. This is not the case for the LTA. Also, for the constraints applied here the LDA provides the best coherence factor for low particle concentrations and large-scale turbulence (large compared with the dimensions of the measuring volume), whereas the LTA is superior for high particle concentrations and low turbulence levels.

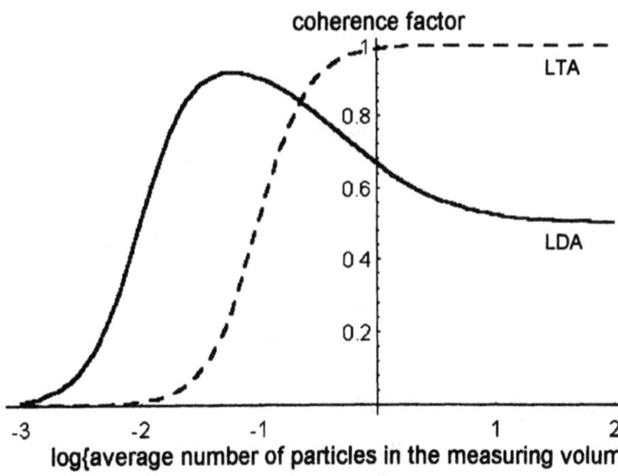

Figure 15. The coherence factor versus the particle concentration (log-lin scale) for an LDA and an LTA, respectively. It is assumed that the number of fringes in the LDA is 10 and that the ratio of the beam spacing to the beam diameter in the LTA is 10. The relative turbulence intensity is 0.1. The ratio of the largest dimension of the measuring volume (length of the measuring volume of the LDA) to the turbulence length scale is 0.1.

Resolving the Sign of the Velocity

The sign of the velocity is not available in a simple laser anemometer. In the experiment shown in Fig. 1 it is not possible to determine which way the particle is passing the laser beam, neither is it possible to do so in the systems shown in Figs. 2a and 2b unless special measures are taken. This can be illustrated by considering the frequency spectrum of the Doppler signal (Fig. 16a). The spectrum is symmetric around the frequency zero. Changing the sign of the frequency does not change the appearance of the spectrum.

For the LDA the introduction of a frequency shift can overcome the sign ambiguity problem. The frequency of one of the transmitted beams is shifted relative to the other beam. This is most commonly done with an acousto-optic Bragg cell (see the chapter on *Optics*). Introducing a fixed shift implies that the phase of the field in one beam changes linearly with respect to the phase of the other beam. In the expression for the intensity (Eq. (10)) a phase term will appear in such a way that the fringe pattern is moving at a constant velocity under a fixed envelope. A particle that is not moving will cause a detected signal oscillating with a frequency given by the velocity of the moving fringe pattern. This frequency happens to be exactly equal to the acoustic drive frequency of the Bragg cell. If the velocity of the particle has the same direction as the direction in which the fringe pattern is moving, the detector frequency is reduced. If the direction of the particle velocity is opposite to the direction of the fringe pattern velocity, the detector frequency is increased. As long as the Doppler frequency is smaller than the Bragg frequency, the sign of the velocity can be unambiguously determined. Frequency shifting corresponds to *single sideband modulation* in communication systems. An ideal frequency shift does not degrade the performance of the system in any way if the detector and signal processor can handle the higher frequency range without degradation. The spectrum of the frequency shifted signal is shown in Fig. 16b. The spectrum is still symmetric, but 0-velocity corresponds to the Bragg frequency. The dynamic range is also increased.

Most Bragg cells used in laser anemometry operate at 40 MHz although both lower and higher frequencies have been applied. A high Bragg frequency gives a small cell. The beam is inevitably deflected in a Bragg cell (< 10 mrad). This necessitates a realignment of the optical system. The efficiency of a Bragg cell is generally not higher than 90 % - generally lower. The Bragg cell may also introduce a small amount of optical aberration and requires careful shielding to avoid electromagnetic interference. In some early laser anemometers two Bragg cells operating at slightly different frequencies were used. One would give a shift up and the other a shift down. In this way a very small optical shift could be obtained making it possible to use the signal processing equipment without modifications. Modern systems use only one Bragg cell in combination with electronic down mixing. It is also possible to use the Bragg cell as a beam splitter. Only the diffracted beam exhibits a frequency shift. The difficulty with this approach is the very small diffraction angle. It is noted that frequency shifting will increase the dynamic range of the laser anemometer and will also increase the angle within which the velocity can be measured.

It is possible to resolve the sign of the velocity without a frequency shift by using a scheme based on the detection of quadrature signals. Two sets of interference fringes 90^0 out of phase are superposed. This can be done by inserting a so-called quarter wave plate in one of the beams. A quarter wave plate may convert linearly polarized light to circularly polarised light. A phase difference of 90^0 between two orthogonal polarization states is introduced. By having two detectors that detect orthogonal polarization components two quadrature signals are generated. A phase difference of $+90^0$ is seen for one velocity direction and -90^0 for the opposite direction. The problem with the method is that it requires polarization conservation light scattering. This is not always possible. Mathematically we may consider the two quadrature components as the real and imaginary parts of a complex signal, respectively. The Fourier transform of such a signal is single sidebanded as illustrated in Fig. 16c. Now the sign of the velocity determines whether the Doppler peak is located at a positive or a negative frequency.

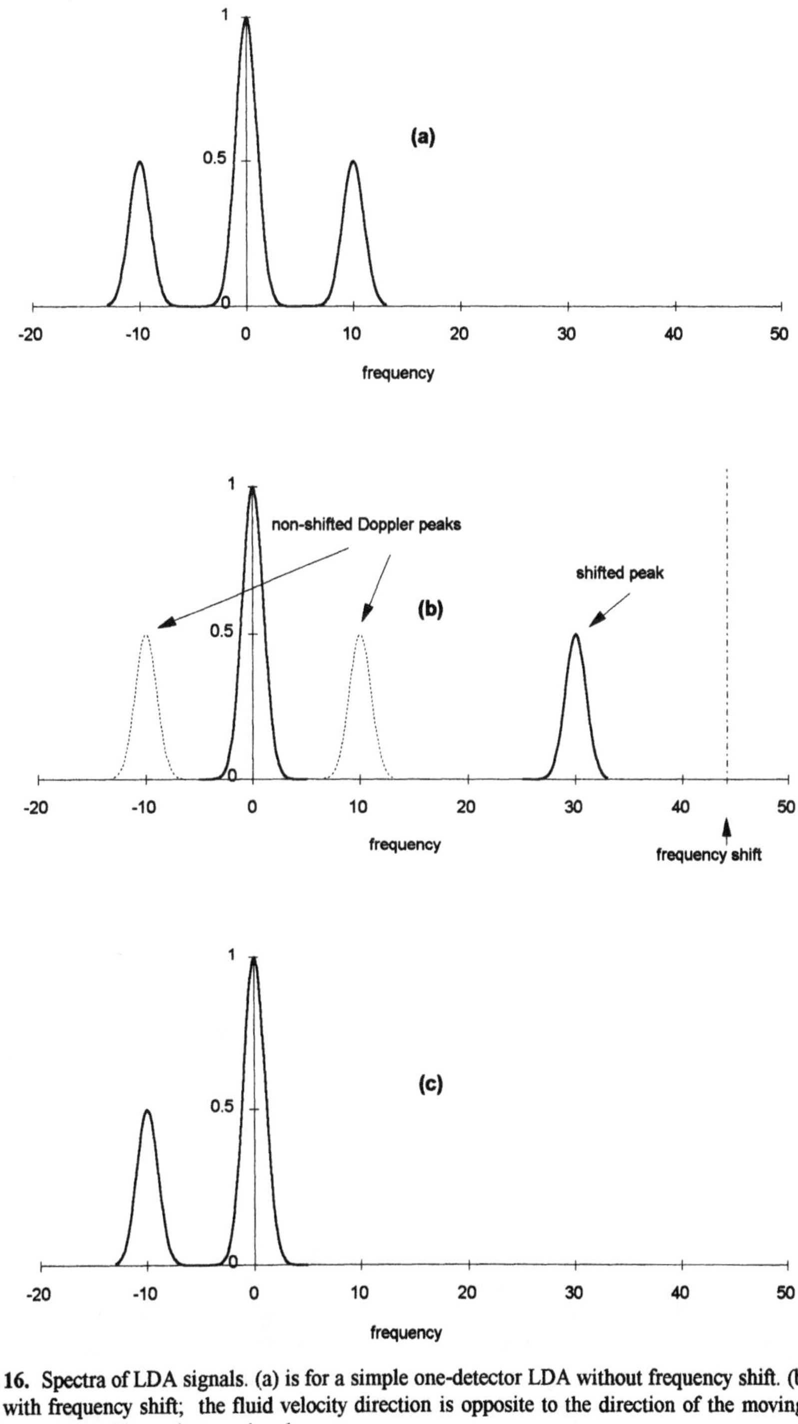

Figure 16. Spectra of LDA signals. (a) is for a simple one-detector LDA without frequency shift. (b) is for a system with frequency shift; the fluid velocity direction is opposite to the direction of the moving fringes. (c) is for a system with quadrature signals.

The LTA with two detectors can easily resolve the sign of the velocity. However, the smaller acceptance angle may still be a problem, and a velocity that is zero cannot be measured. In principle zero-velocity can never be measured because it would require an infinite amount of time; however, in an LDA with frequency shift a signal is observed even from a nonmoving particle in the measuring volume. This is not the case for an LTA. Fig. 17 shows the correlation functions of an LTA with one detector (a) and two detectors (b).

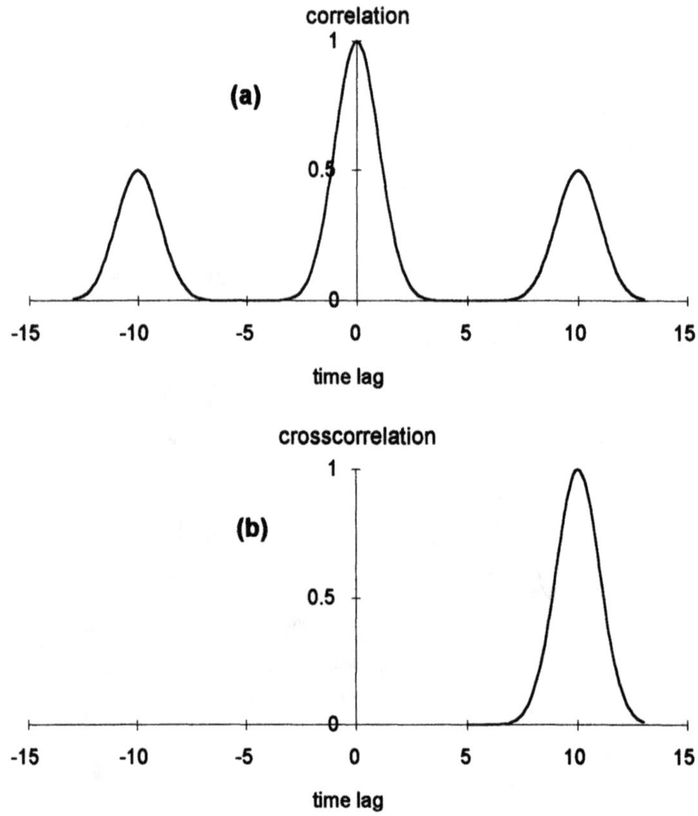

Figure 17. Correlation functions for an LTA. (a) is for a single-detector system and (b) is for a two-detector system.

SIGNAL PROCESSING

In this section we shall consider methods for signal processing for time resolved measurements of velocities. It is assumed that the averaging time is shorter than the coherence time of the velocity fluctuations (i.e. the Kolmogorov time scale or inner scale) and that the extension of the measuring volume is smaller than the spatial coherence length of the velocity (Kolmogorov length scale). We are not primarily concerned with methods for inferring velocity statistics on the basis of measured data (this is normally done in a general purpose computer), but rather with how to obtain these data on the basis of a given optical configuration. This type of processing is generally done by a dedicated signal processor. It is beyond the scope of the presentation given here to do a detailed derivation of the different processing schemes. We shall formulate the tasks of the signal processors, briefly outline a synthesis method, discuss the concept of *robust* processing, and go through some processing schemes.

Single Burst Detection. The procedure for estimating the localised velocity and, possibly, other associated parameters can then be outlined as follows (Fig. 18):

(1) At any given time it is decided whether a particle is or is not present in the measuring volume (hypothesis testing).

(2) If it is determined that a particle is present, velocity, residence time, and time of arrival are estimated. (For specific applications other parameters may also be relevant, e.g. phase, amplitude, and modulation index.)

(3) A validation is performed in order to ensure that the measurement fulfils certain quality requirements, e.g. that the estimated accuracy is adequately low.

(4) Relevant parameters for the fluid flow are then estimated on the basis of the measurements performed on the bursts of the individual particles.

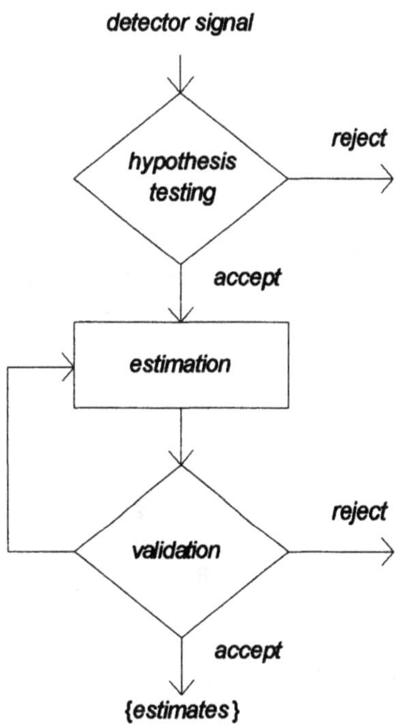

Figure 18. Detection, validation and estimation process. The way the different tasks are actually performed and how they are combined are different in different processors.

Task (3) may be incorporated into task (1) or task (2). Here we shall concentrate on tasks (2) and (1) (in that order). It is assumed that the required accuracy is so high that the hypothesis testing problem becomes trivial (in some very special cases this may not be true). It is noted that the expected signal from the passage of a single particle is deterministic, i.e. if the trajectory and the speed of the particle are given, the only stochastic element is the random arrival of photons. (This is in contrast to the many-particle case, where the random distribution of the particles implies that the signal is stochastic even in the case of no photon or electron noise.)

Maximum Likelihood Estimation, Lower Limit to the Measuring Uncertainty; Robust Estimation

Maximum likelihood estimation (MLE) is widely used for the development of algorithms for parameter estimation based on a set of data affected by noise. It has been shown that no unbiased estimator (i.e. an estimator that averaged over many trials will approach the true value without a bias error) can provide a lower variance of the estimate than an MLE. A lower bound for the variance can be evaluated from the reciprocal value of the so-called *Fisher number*. The Fisher number can often be calculated and then used for comparison of the performance of an actual processing algorithm. Evaluating the performance of an arbitrary processor is in general a far more demanding task than calculating the Fisher number.

An MLE will often imply that the observed signal is to be correlated with the expected signal if the noise power is independent of the signal dynamics (additive). The parameter of the expected signal that gives the largest correlation is the *estimate*. Thus, for the LDA

$$\int r(t)s(t-t_0,\omega_0,\omega_s,\phi)\mathrm{d}t \tag{26}$$

is to be maximised. This defines the estimated parameters

$$\{\hat{t}_0,\hat{\omega}_0,\hat{\omega}_s,\hat{\phi}\}.$$

$s(t)$ is given by Eq. (15). If the background is absent, then the signal should be correlated with the logarithm of the expected signal. However, such a procedure is not robust: if the assumption about no background appears not to be fulfilled, then the estimator will perform poorly, i.e. give a large variance of the estimate; whereas an estimator that is developed under the assumption of additive signal independent noise does also perform rather well - albeit not optimally - in the case of multiplicative noise.

Before we proceed with frequency estimation, let us consider the estimation of the temporal position of a pulse. This is the essential task in the LTA and is also done in the LDA to estimate the time of occurrence of a measurement. Let the expected pulse have a Gaussian shape. The scheme for the estimation is shown in Fig. 19: the measured signal, $r(t)$, is correlated with the expected signal by a filtering (which is actually a convolution; this is not essential here). If the maximum of the correlation exceeds a certain threshold, the presence of a pulse is established. Simultaneously, the signal is correlated with the derivative of the pulse giving an output that is the derivative of the correlation of the upper channel. The zero crossing of the correlation derivative defines the estimated temporal position of the pulse.

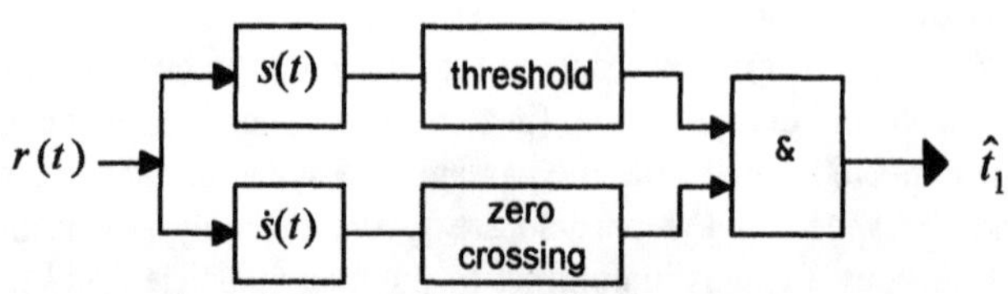

Figure 19. Scheme for estimating the temporal position of a pulse.

Now, the acceptance of a measurement should be conditioned by requiring that the estimated quantity has a variance below a certain specified level. For the pulse position estimation we may require that the variance of the estimated pulse position is much smaller than the square of the pulse width. Let us assume that the background dominates, i.e. that $a<<b$ (Eq. 18). It can then be shown that

$$\frac{a^2}{b\omega_s} > 1; \tag{27}$$

the term may be interpreted as the noise power (note that it is the noise power in the signal bandwidth - not the instrument bandwidth, which is often used when specifying signal-to-noise ratios). The noise power may be measured when no particles are in the measuring volume. If the particle concentration is low, this simply means that the threshold has to be larger than the r.m.s. of the detector output.

A similar threshold for the zero background case can be derived. However, the noise power is more difficult to estimate.

For the LTA the velocity estimation is a matter of estimating the time-of-flight between the two illuminated volumes. This implies the temporal estimation of two pulses (as shown in Fig.19) and taking the difference. A threshold for acceptance may be also be applied here. If the relative r.m.s. of the estimate is to be below a certain value μ and the ratio of beam radius to beam spacing is κ, then the threshold is given by

$$\frac{a^2}{noise\ power} \geq 8\left(\frac{\kappa}{\mu}\right)^2; \ \kappa = 1/(\omega_S \tau_0) \tag{28}$$

assuming that the background noise is dominating.

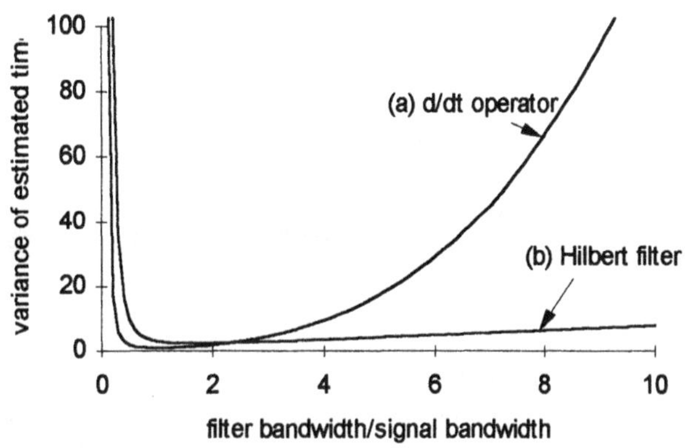

Figure 20. Variance of the estimated position of a pulse as a function of the relative filter bandwidth ω_f/ω_S. The estimate is given by the zero crossing of the filtered signal after the conversion of an even to an uneven pulse. This is done by (a) an d/d*t* operator and (b) by a Hilbert filter. The reference is the variance of a maximum likelihood estimator (minimum variance).

The bandwidth of the signal ω_s is needed for the estimation process, but the bandwidth is in general not known prior to a measurement. Two approaches are possible to solve the problem. One is the obvious procedure: to estimate the signal bandwidth. However, this is a rather demanding task and may not even be necessary. The other procedure is to use a value that is acceptable for a worst case estimate and then modify the estimation procedure so that it becomes more ***robust*** which in the present case means less sensitive to the value of ω_s. The maximum likelihood procedure implies that we find the maximum of the expression given in Eq. 26 by, e.g., finding the value of the argument for which the derivative is zero. It turns out that if the zero crossing of the Hilbert transform is used rather than the zero crossing of the derivative, then an unbiased estimator is obtained that is in general less sensitive to applying the correct signal bandwidth[13]; this is illustrated in Fig. 20. (Note that the Hilbert transform of an even function is an uneven function, and taking the Hilbert transform does not change the bandwidth; the bandwidth is increased by taking the derivative.)

An MLE for the Doppler frequency of an LDA implies that the signal is to be filtered in a filter bank as illustrated in Fig. 21. The filters are supposed to have a bandwidth equal to the bandwidth of the signal itself. The maximum in ω-space will then define the estimated Doppler frequency. We may again investigate the impact of the actual filter bandwidth and also the effect of the finite number of channels and find again that taking the Hilbert transform in ω-space provides a more robust estimation than taking the derivative.

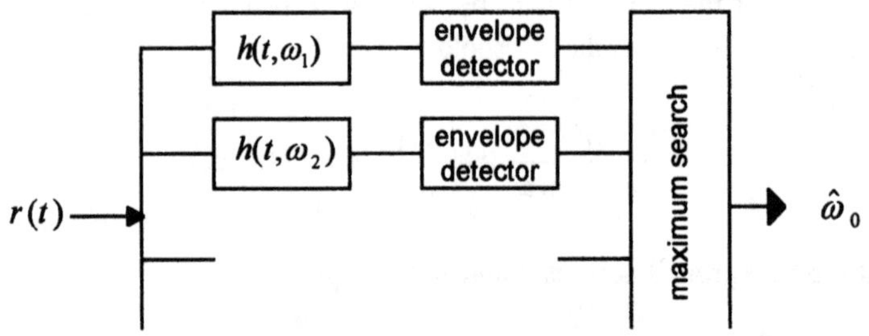

Figure 21. Scheme for estimating the Doppler frequency. The filters are bandpass filters with centre frequencies $\omega_1, \omega_2, \ldots$

A lower limit to the variance is given by

$$\frac{\text{var}\{\hat{\omega}_0\}}{\omega_0^{\;2}} = 8\frac{b\omega_s}{(a\gamma)^2}\kappa \tag{29}$$

in the background dominated case. Note that κ is the ratio of signal bandwidth to Doppler frequency. As in the LTA case we may also establish validation criteria based on the signal-to-noise ratio. Let the specified relative lower limit to the uncertainty be $\mu \ll 1$. For $b \gg a$ we find that

$$\frac{(\gamma a)^2}{\langle noise\ power\rangle} > \left(\frac{\kappa}{\mu}\right)^2. \tag{30}$$

And for $b \ll a$

$$\frac{f(\gamma)a^2}{\langle noise\ power\rangle} > \left(\frac{\kappa}{\mu}\right)^2; f(\gamma) < \gamma. \tag{31}$$

Instead of actually using a filter bank a scheme outlined in Fig. 22 may be applied utilising the FFT procedure. The measured spectrum may be fitted to the expected spectrum. The most important part of the fitting must obviously be at and right around the spectral peak. Furthermore, most practical schemes do only allow for the estimation of a finite number of spectral values. It has thus been proposed to apply a parabolic fitting based on just three estimates of the spectrum. Let us consider this method and compare it with the previous concepts.

Let three neighbouring values, $\hat{S}_1$, $\hat{S}_2$, and $\hat{S}_3$ be selected. The spacing between the values is ω_f. These values are to be fitted to a parabola given by

$$S_r(\omega) = \alpha(\omega - \omega_0)^2 + \beta \, . \tag{32}$$

This gives that the estimated frequency is given by

$$\hat{\omega}_0 = \omega_2 - \frac{\omega_f}{2} \frac{\hat{S}_1 - \hat{S}_3}{2\hat{S}_2 - \hat{S}_1 - \hat{S}_3} \, . \tag{33}$$

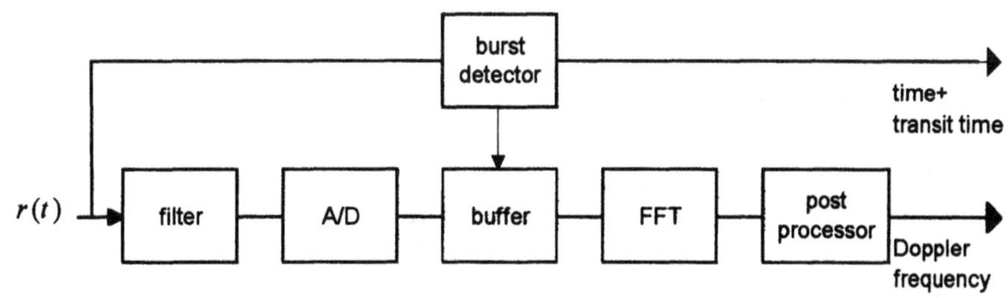

Figure 22. Scheme for FFT Doppler burst processor. The burst detector has several purposes: it shall identify the bursts, estimate the time of occurrence and the residence time (reciprocal signal bandwidth), and it shall initiate the FFT.

Simple analytic expressions have not been found for neither the bias error nor the variance. However, the estimator defined by Eq. (33) is in general biased as shown in Fig. 23. The plot shows the bias error normalised with the signal bandwidth versus the difference between the true Doppler frequency and the frequency at which the maximum spectral estimate is found normalised with the spacing between the spectral samples, i.e. .$(\omega_0 - \omega_2)/\omega_f$ It is noted that the maximum error increases with increasing spacing between the selected frequency components. A moderate noise level ($b\omega_f \approx 0.2\, a$) appears to reduce the bias error.

Simulations show that for signal-to-noise ratios above 10 dB referred to the signal bandwidth (0 dB if the reference is to the full input bandwidth - as often specified by manufacturers) the performance is essentially as for an MLE (i.e. no unbiased estimator can provide a lower uncertainty). Below that level the performance depends on the possible interpolation scheme that is applied. A procedure known as *zero-filling*[14] is often applied. The principle is here to enlarge the record length to, e.g., twice its original length with zeros. This does not provide additional information, but provides more closely spaced spectral estimates. The same result could also be obtained by interpolation on the spectrum, but zero-filling may be computationally faster. Inspecting Eq.(33) it also appears very plausible that

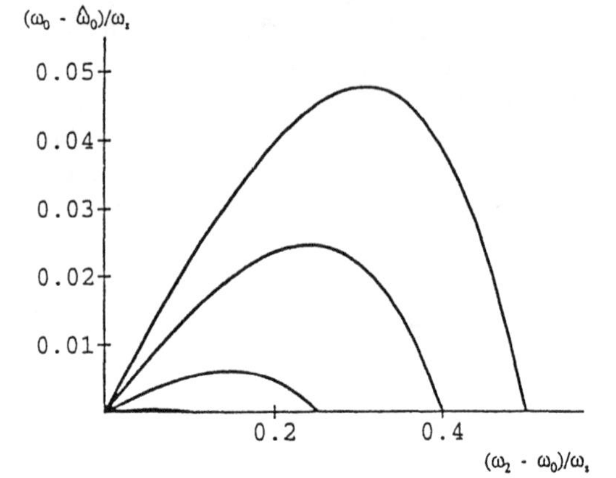

Figure 23. Normalized bias of the estimated frequency versus the relative difference between the true Doppler frequency and the frequency of the largest estimated spectral value. The parameter is the ratio of the spacing between the estimated spectral components to the signal bandwidth. The largest bias is observed when $\omega_f/\omega_s = 1$. The succeeding values are 0.8, 0.5, and 0.2.

the variance of the estimated spectral values must be lower than the square of the expected value itself; otherwise, the probability of getting zero in the denominator cannot be ignored.

Measurements performed on a dedicated Doppler burst processor with parabolic fitting give bias errors very close to the values obtained by simulation (the difference is about 10% - within experimental error). The uncertainty of the dedicated processor is generally larger than the values obtained by simulation (50-100%). It is noted that a parabolic fit to the logarithm of the spectrum gives no error in the noise-free case assuming a Gaussian spectrum. But with an FFT processor the temporal signal is often truncated giving a non-Gaussian spectrum. The sensitivity to noise appears to be much larger when fitting to the logarithm rather than to the spectrum itself. Several investigations have been performed to evaluate various fitting schemes . The basic result is that the closer the fitting curve is to the expected spectral values, the better is the estimator. A Gaussian fit appears to be much more robust both in terms of sensitivity to truncation because of the finite record length, and in terms of spacing between the spectral samples. However, this procedure is also computationally more demanding than the parabolic fit procedure.

Counters; Estimating the Doppler Frequency by Counting Zero Crossings

The so-called counter is in widespread use as LDA processor. The specific design may vary; however, most commercial instruments are based on the concept shown in Fig. 24. In addition to the scheme shown in Fig. 24 a validation circuit will often be incorporated. The scheme does not appear to be compatible with optimum estimation procedures. However, it has proven to be a very accurate and flexible instrument in sparsely seeded flows and good signal-to-noise ratios (especially if the user is very qualified). Thus it may be reasonable to investigate the performance and compare it with the procedures based on spectral analysis. A rigorous analysis is rather complicated. We shall apply some heuristic arguments. This appears adequate for establishing essential features of the performance and for comparing with other schemes.

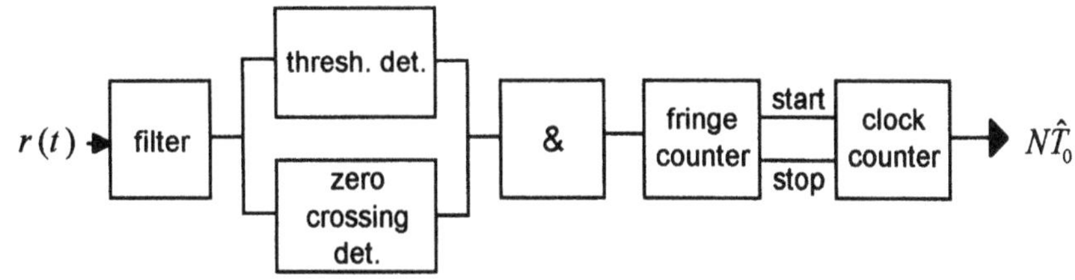

Figure 24. Basic scheme for an LDA counter without validation.

The frequency is estimated by measuring the times between a fixed number of zero crossings. The reciprocal value gives the frequency estimate. The number of detected zero crossings is usually much lower than the number of oscillations in a burst. The input filter may cause a bias if the Doppler frequency and the centre frequency of the filter are different - which they generally are. The centre frequency of the filtered signal is affected by the filter. The bias is zero if the Doppler frequency is identical to the centre frequency of the filter. This is also true for non-Gaussian filters if they are symmetric around ω_0. Also if the filter bandwidth ω_s is much larger than the signal bandwidth, the bias is negligible. The conditions are illustrated in Fig. 25. This filter-bias problem is in principle common to all processors. However, for a counterprocessor it is much more important to have a narrow prefilter bandwidth in order to reduce the uncertainty caused by noise than with other processors. And the bias error increases with decreasing prefilter bandwidth. Noise does also affect the bias error. With dominating additive noise the centre frequency of the signal approaches the centre frequency of the filter. If the filter bandwidth is much larger than the signal bandwidth, it is the slope of the filter around the Doppler frequency that is of importance. We may then approximate the bias error by

$$\langle \hat{\omega}_0 - \omega_0 \rangle = \omega_s \frac{\mathrm{d}/\mathrm{d}\omega \{ |H(\omega)|^2 \} \big|_{\omega=\omega_0}}{|H(\omega)|^2}. \tag{34}$$

Generally, the filter bandwidth, ω_f, is much larger than the signal bandwidth, ω_s, and is often also larger than the Doppler frequency. This implies that the noise will most likely cause a number of zero crossings around each zero crossing of the expected signal. The standard procedure for overcoming this problem is to introduce hysteresis in the zero crossing detector. The r.m.s. of the noise must then be smaller than the hysteresis interval. The hysteresis does introduce a shift in the temporal position of the zero crossings but it does not give any bias of the mean spacing between zero crossings of a narrowband signal. Noise in a spectral region below the Doppler frequency causes an uncertainty in the positions of the level crossings. Noise within a spectral region higher than the Doppler frequency causes a bias in the temporal positions of the level crossings.

Two sources of uncertainty are identified: quantization because of the finite clock frequency, and uncertainty of the zero crossings because of noise. The relative quantization uncertainty is given by

$$\frac{\mathrm{var}\{\hat{\omega}_0\}}{\hat{\omega}_0^{\,2}} = \frac{1}{12}\left(\frac{1}{N} \frac{\omega_0}{\omega_{clock}} \right)^2, \tag{35}$$

where N is the number of counts and the clock frequency. This uncertainty can be made arbitrarily small by increasing the clock frequency (or by interpolation if the phase of the clock is adequately stable and known).

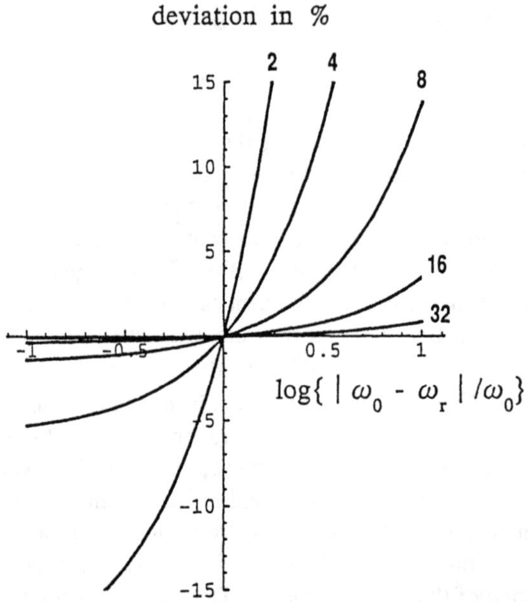

Figure 25. The bias error for a deterministic burst as a function of the relative difference between the centre frequency of the filter and the Doppler frequency. The parameter is given by the ratio of the filter bandwidth to the signal bandwidth.

The uncertainty caused by noise can be approximated by the following expression

$$\operatorname{var}\{\hat{\omega}_0\} = \frac{1}{2}\frac{b\omega_f}{a^2}\frac{^2}{N}. \tag{36}$$

Equation (36) is valid if $b >> a$ and the counting is performed over an interval shorter than the width of the burst.

For the case where the intrinsic signal noise dominates ($b << a$) we get that

$$\operatorname{var}\{\hat{\omega}_0\} = \frac{1}{2a}\frac{\omega_f^{\,2}}{(2\omega_f+\omega_s)^2}\frac{\omega_{0^2}}{N}. \tag{37}$$

As already mentioned, it is only noise with a coherence time longer than the time between the crossing of the upper and the lower levels of the level detector that causes an uncertainty in the spacing between the level crossings. It can be argued that the effective noise bandwidth of a counterprocessor is given by

$$\omega_n = \omega_0(1/\chi - 1(\pi N)) \tag{38}$$

which is in general larger than the Doppler frequency. The term χ is the ratio of the hysteresis interval to the expected amplitude, a. The noise bandwidth has to be used for ω_f in the equations for the uncertainty of the estimated frequency, Eqs. (36) and (37).

The quantization uncertainty increases linearly with the Doppler frequency, whereas the uncertainty caused by photon or electronic noise increases with the square of the frequency. The cross-over frequency can be evaluated on the basis of Eqs. (35), (36), and (37). This can be illustrated by the following example:

Laser power:	0.1 *W*
Beam diameter:	0.2 *mm*
Angle between beams:	6°
Receiver solid angle:	2.5×10^{-3}
Scattering cross-section:	10^{-12} m^2
Overall quantum efficiency:	0.01
N:	8
ω_{clock}	1 GHz

This yields that below 1.5 *kHz the quantization uncertainty is caused by the finite clock frequency smaller than the uncertainty caused by noise. It is assumed that the background is zero. The calculated cross-over frequency is thus a lower limit.*

Several commercial counters incorporate a validation scheme. Two methods are often applied. One is the so-called multilevel sequencing which implies that the signal has to pass different levels (often only two) in the same sequence that would be observed with a pure sine wave. The other method is based on the concepts proposed by Asher[15]. The basic idea is here to use two counters but counting up to different numbers, e.g. 5 and 8. If the frequency estimates derived from the two counters deviate with more than a preset amount, the measurement is rejected. The two methods may be applied jointly or only the multilevel sequencing is used. To the knowledge of the author a rigorous theoretical analysis has not been published. However, the following can be noted:

(1) The validations have a negligible effect if the hysteresis interval is much larger than the r.m.s. of the noise.
(2) If the r.m.s. noise is comparable with the hysteresis interval, then the validation does have an effect: the number of "outliers" (i.e. measurements far from the mean perceived to be erroneous) is reduced.
(3) A quantitative estimate of the change in performance cannot in general be given.

Covariance Processing for Frequency and Phase

The joint estimation of frequency and phase is done in order to obtain both the velocity and size of a spherical particle. Light is detected in two different directions. The expected frequency of the two signals is the same, but the phases are not. The phase difference yields information about the size. A so-called covariance processor was developed for this application[16]. The expected signals are of the same form as Eq. (15). The phases (times of arrival) are of course in general different. The amplitudes and the backgrounds may also be different. In radar systems the ratio of the real to imaginary parts of the complex covariance has been demonstrated as an estimator of frequency. The same concept may be used for phase estimation. The procedure is as follows: the signals are high-pass filtered to eliminate the background and the pedestal, so that only the oscillating part remains. For one of the signals a 90^0 phase shifted version is generated (mathematically this can be done by applying the Hilbert transform). For frequency measurements both the phase shifted and the nonshifted signals are multiplied by a delayed version of the same signal and integrated over the period of the burst (Fig. 26). We then get that

$$C_i = C_0(\tau)\sin(\omega_0\tau)$$
$$C_r = C_0(\tau)\cos(\omega_0\tau) \tag{39}$$

which can be considered the imaginary and real parts of the complex covariance. $C_0(\tau)$ is the covariance for the envelope. The ratio of the imaginary to real parts of the complex covariance is

$$\tan(\omega_0\tau) \tag{40}$$

from which the frequency can be obtained assuming that τ is known. Obviously, $|\omega_0\tau|$ must be smaller than π to avoid any ambiguity.

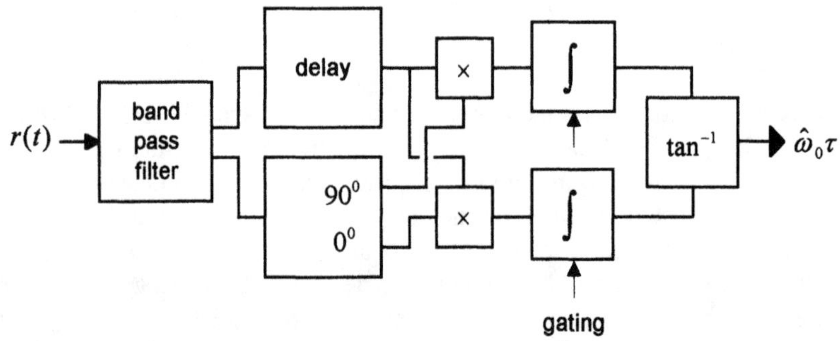

Figure 26. Scheme for measuring the frequency of a narrowband signal by the covariance method.

The phase difference may be determined by a slight modification of the scheme for measuring frequency: both the phase shifted and nonshifted signals originating from detector one are multiplied by the signal from detector two and integrated. The two terms that emerge are

$$C_s = C_0(\tau)\sin(\omega_0\tau + \Delta\phi),$$
$$C_c = C_0(\tau)\cos(\omega_0\tau + \Delta\phi). \tag{41}$$

The phase difference can then be obtained by evaluating the covariance functions at zero time lag, i.e.

$$\Delta\phi = \tan^{-1}\{C_s(0)/C_c(0)\}. \tag{42}$$

The processing is the same as illustrated in Fig. 26 except for the fact that two different input signals are applied and the delay is zero.

In general, three regions can be identified for the output noise; it may be given by (1) the product of the input noise and the input signal, (2) a product of input noise terms, or (3) the output noise may exceed the expected value of the output. The relations are illustrated in Fig. 27. Region (1) is preferred for actual measurements.

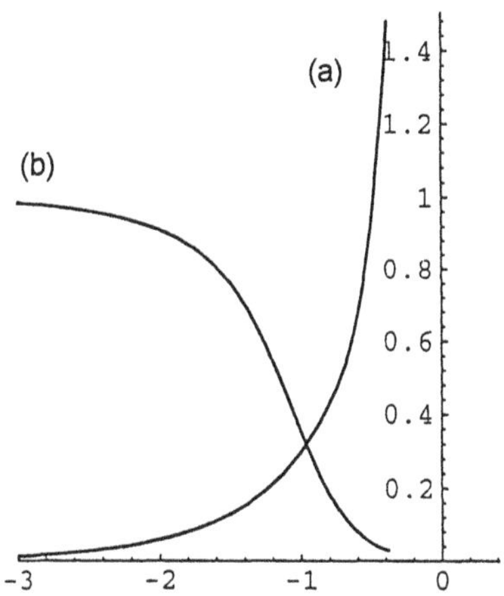

Figure 27. Log-log plot of the estimated phase versus the signal-to-noise ratio ($b>>a$) showing the regions where (1) the output noise is given by the signal multiplied by the input noise, (2) the output noise is given by the product of input noise terms, and (3) the output noise is larger than the maximum mean of the output. The estimated frequency exhibits the same generic dependence on the signal-to-noise ratio.

Successful application of the presented schemes implies that a good burst detector is incorporated. This is especially true if the signal-to-noise ratio is below one. A novel scheme proposed by K. Andersen from Dantec Measurement Technology for detecting the beginning and the end of a burst is applied in connection with covariance processing. The basic layout of the scheme is shown in Fig. 28. It is seen that the input to the logic block is the square of the product of the envelopes. Thus, detecting the levels at which a properly selected threshold is passed yields estimates of the beginning and the end of a burst, unaffected by the actual values of the frequency and phase. In the actual implementation a three-level validation is applied. Also, the delayed signal is clipped so the input to the threshold detector is proportional to the amplitude and not the square of the amplitude.

Proper operation of the burst detector essentially implies that the noise power within the bandwidth of the lowpass filters has to be smaller than the signal amplitude. It is noted that a slightly better burst detector performance could be obtained by using the signals from two detectors - rather than one. However, this would complicate the setup procedure; and it would not be possible to operate with one detector only. The practical implementation and experimental verification of this processing scheme are described in [17].

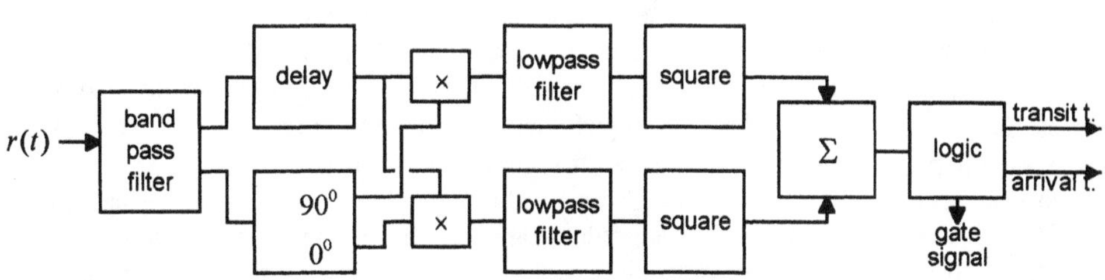

Figure 28. Burst detection scheme for the covariance processor.

Trackers

In the present context a tracker is a signal processing device that provides a running update of the velocity estimate. In order to obtain a replica of the temporal evolution of the velocity it is of course necessary that the coherence time - or persistence time - of the signal is smaller than any important time scale of the velocity fluctuations. It is also necessary that the response time of the tracker is smaller than the coherence time of the velocity fluctuations. (For a turbulent flow: the Kolmogorov or inner scale.) It is assumed that these requirements are fulfilled. The trackers considered here are assumed to be single-channel devices (as opposed to, e.g., a multichannel spectrum analyser). The state of the channel (e.g. its centre frequency) is continuously - or stepwise - updated. This is done on the basis of running estimates of the difference between the state of the tracker and the signal, respectively. The general structure of a tracker is shown in Fig. 29. The signal is compared with a reference signal in order to estimate the difference of the relevant parameters (e.g. frequency or time lag) of the signal and the reference, respectively. This difference is added to the previous state estimate and a new state estimate is obtained and used to update the reference signal.

Normally a tracker is considered as a processor that must operate on a continuous or quasi-continuous signal. However, this is not mandatory. It is only necessary that the mean time between bursts is much smaller than the coherence time of the velocity fluctuations. As an example of a tracker operating under such conditions let us consider a *coincidence tracker*. The fact that the mean time between particles giving rise to a measurement is smaller than the coherence time implies that the probability of a small difference between two consecutive measurements is larger than the probability of a large difference between the two measurements. We shall consider trackers both for the LDA and the LTA.

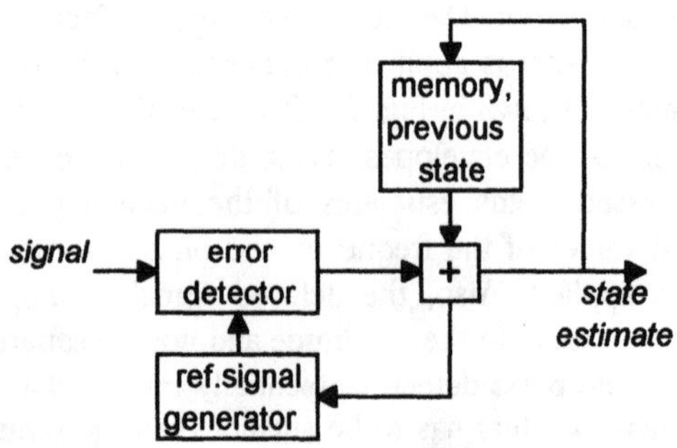

Figure 29. Basic scheme for a tracking processor.

The Coincidence Tracker. The coincidence tracker is a processor for the LTA operating in a sparsely seeded flow. Let us consider the general layout of the processor (Fig. 30). The filters and the discriminator perform the estimation of the temporal positions of the incoming pulses - as illustrated in Fig. 17. The *e*-filter (even) is closely matched to the expected pulse. The *u*-filter (uneven) is the Hilbert transform of the *e*-filter. This in order to get a more robust performance - as discussed in relation to Figs. 19 and 20. If the temporal difference between two consecutive measurements of the time-of-flight is smaller than what corresponds to the part of the shift register covered by the sloped part of the weight summation network, a full update of the shift register occurs. This lock range relative to the whole shift register is set to a value given by the anticipated turbulence intensity. Now, outside this range there should be a gradual tapering off of the weights if the previous analysis is to be followed. However, it is not considered very likely that measurements outside the lock range are caused by the type of noise considered so far. Based on experimental observations such occurrence was found to be caused either by electromagnetic interference or by a sudden change in the state of the wind velocity - excluding upstarts. Therefore, the persistence was considered the most reliable indication of the reliability of out-of-range measurements. (The processor is described in somewhat greater detail in[18].)

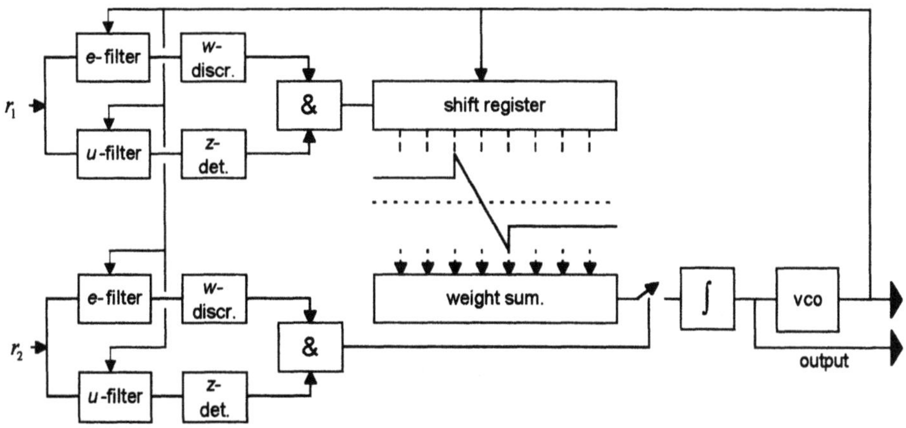

Figure 30. The coincidence tracker. The temporal scale of the tapped delay line filters (e and u) is controlled by the VCO. "z" is a zero crossing detector and "w" a window discriminator. When both these are activated, a pulse is generated by the &-gate. This occurs at the estimated temporal position of an incoming pulse of proper amplitude. The travelling time of the pulse to the middle of the shift register is updated each time a pulse is detected in both channels to match the time-of-flight. However, this only occurs if the difference between two consecutive measurements is within a preset lock range. Otherwise, a number of measurements are necessary in order to bring the tracker into a locked state.

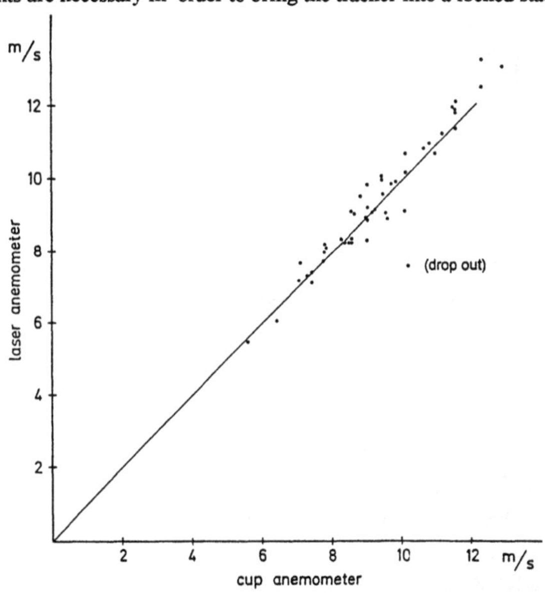

Figure 31. Scatter diagram of measurements obtained with an LTA and a cup anemometer, respectively. The cup anemometer was mounted on a mast at an altitude of 27 meters. The measuring volume of the LTA was positioned about 10 cm in front of the cup. The correlation coefficient between the two types of measurements is 0.97. Each measurement represents a 10 sec average performed at 30 sec intervals for a 2× 13 min uninterrupted interval. One measurement was declared "out-of-range" by the tracker.

Figure 31 shows a scatter diagram of measurements taken with a long-range LTA instrument with a coincidence tracker and with a cup anemometer. As can be seen, quite good agreement was observed. The two instruments do respond quite differently to fluctuating velocities. The cup anemometer exhibits a nonlinear response when exposed to a fluctuating velocity causing a positive bias of the mean velocity. The bias error of a laser anemometer operating with a sample and hold processor (as in the present case) goes to zero when the mean time between measurements becomes much larger than the coherence time of the velocity fluctuations and the tracker only operates within its lock range. In the other extreme with a very low measurement rate the normal arrival rate bias occurs.

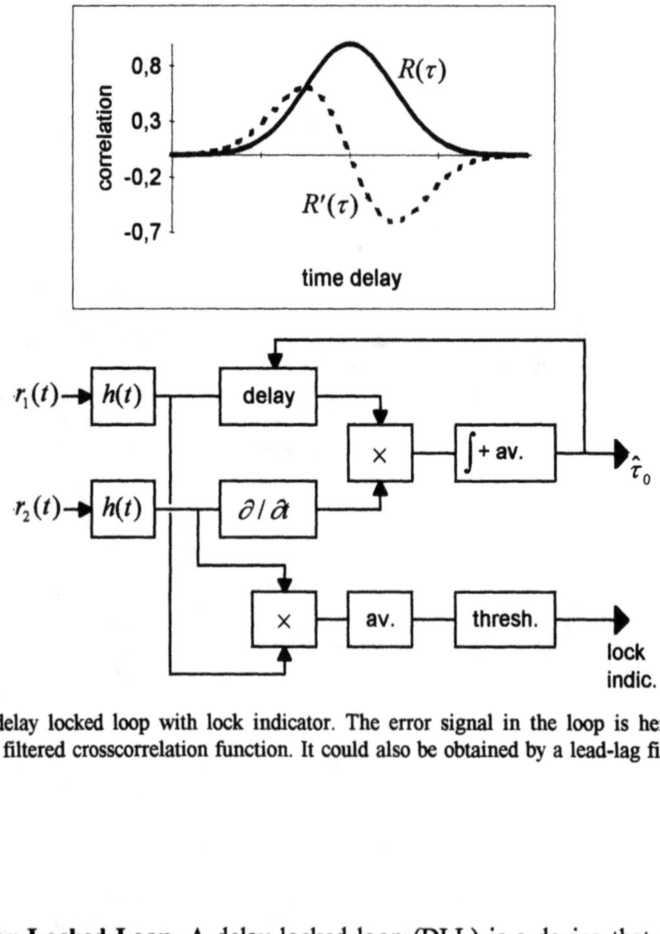

Figure 32. A delay locked loop with lock indicator. The error signal in the loop is here given by the derivative of the filtered crosscorrelation function. It could also be obtained by a lead-lag filter or a Hilbert filter.

The Delay Locked Loop. A delay locked loop (DLL) is a device that can track the temporal displacement between two signals. The basic idea is that the derivative of the crosscorrelation is used as an error signal in a feedback loop (Fig. 32). It can be shown that it may actually be an advantage to use the hard clipped signal (+1 when the dynamic signal is positive and -1 when it is negative) instead of the real signals[19]. A loop of this type has to be designed so that the steady-state error is zero and also so that it is unconditionally stable.

The Phase Locked Loop. The phase locked loop (PLL) is well established for demodulation and synchronisation. However, its use for tracking the centre frequency of narrowband random signals is less common although it has been applied to laser Doppler anemometry for a number of years. We shall briefly show how a PLL works and emphasise its main advantages and disadvantages.

Let the signal be given by

$$r(t) = A(t)\cos(\omega_0 t + \phi(t)) + n(t), \tag{43}$$

and let us assume that the averaging time is shorter than the coherence time of the signal. An MLE of ω_0 will then imply that $\hat{\omega}_0$ is given by the value for which the correlation between the observed signal and the expected signal is largest. This implies that the correlation with the expected quadrature signal must be zero. This is exactly what a PLL tries to do.

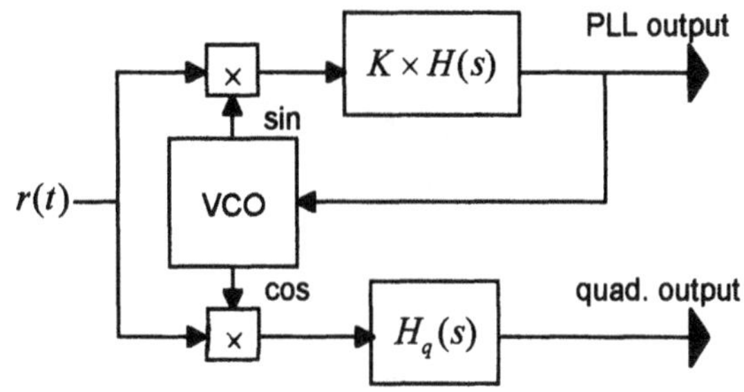

Figure 33. Basic scheme for the phase locked loop. The processor is generally also fitted with a quadrature detector working as a lock indicator.

Consider the correlation function

$$R_T(\tau) = \langle r(t) \times \sin(\omega_r(t+\tau)) \rangle_T. \tag{44}$$

For an infinite averaging time the correlation is zero irrespective of the averaging time. This reflects the fact that the spectral power of the signal (and the noise) is distributed - no power is concentrated at a single frequency. However, if $\omega_r = \omega_0$, the noise is absent, and the averaging time is shorter than the coherence time of the signal, then Eq. (44) yields a value proportional to the phase difference between the signal and the reference. This quantity is used as an error signal controlling the phase of the reference signal. Figure 33 shows the basic layout of such a phase locked loop. A quadrature output may be applied for monitoring the operational status of the tracker. It gives an estimate of the correlation between the input signal and the reference signal. With proper thresholding it can be used as a lock indicator.

Figure 34 shows a Doppler signal and the derivative of the phase of the signal as measured with a PLL in the case of a constant velocity and a high particle concentration. Note the correlation between the signal amplitude and the fluctuations of the phase derivative. The displacement between the spikes of the phase derivative and the minima of the signal envelope is caused by the delay of the loop.

The performance of the PLL operating on laser Doppler anemometer signals has been investigated by computer simulation and real experiments. The essential results are that with a response time shorter than the coherence time of the signal the loop serves as an excellent processor: the output has the statistics of the derivative of the phase as predicted by Rice[20] averaged over a time given by the loop response time, and very good correlation between the signal and the quadrature reference was observed. This is not a trivial result. A rigorous analysis of a PLL even in the case of a deterministic signal with additive noise is rather complicated and becomes even more difficult in the present case because there is a finite nonzero probability of an arbitrarily large value of the derivative of the phase. Any loop must have a finite response time. Thus the loop must occasionally lose track.

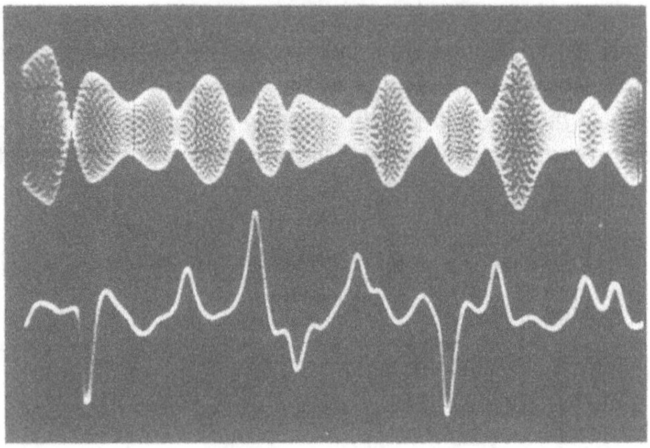

Figure 34. A simulated Doppler signal (only the envelope can be identified) and the output of a PLL with a response time corresponding to five oscillations of the Doppler signal.

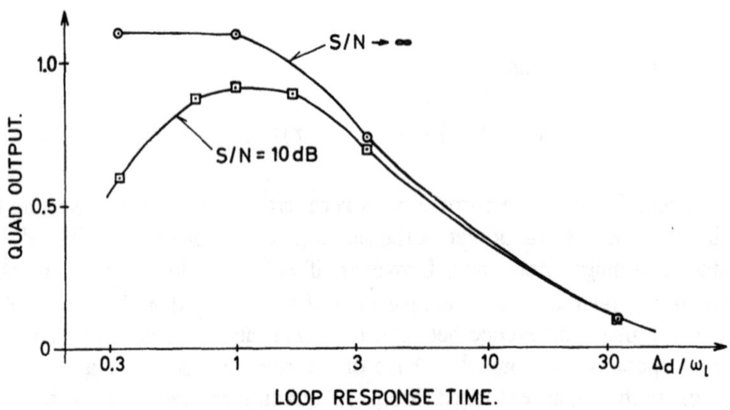

Figure 35. Quadrature output of a PLL versus the normalised response time (loop bandwidth/signal bandwidth). The signal-to-noise ratio is here defined as the peak of the Doppler power spectrum divided by the noise floor.

Figure 35 shows the normalised quadrature output of the tracker versus the loop response time for a constant mean signal level but at different noise levels. It is noted that there is in general an optimum response time that is close to the coherence time of the signal - a little larger for low signal-to-noise ratios and smaller for good signal-to-noise ratios. In the case of no noise the optimum response time is zero!

Figure 36 shows the output of a PLL in relation to the true turbulence both without and with electronic noise. In the first case the deviation between the input and the output is exclusively caused by the phase fluctuations associated with the random distribution of the particles. In the second case the fluctuations are partly caused by the fact that the tracker cannot distinguish between phase fluctuations of the signal and those caused by electronic noise.

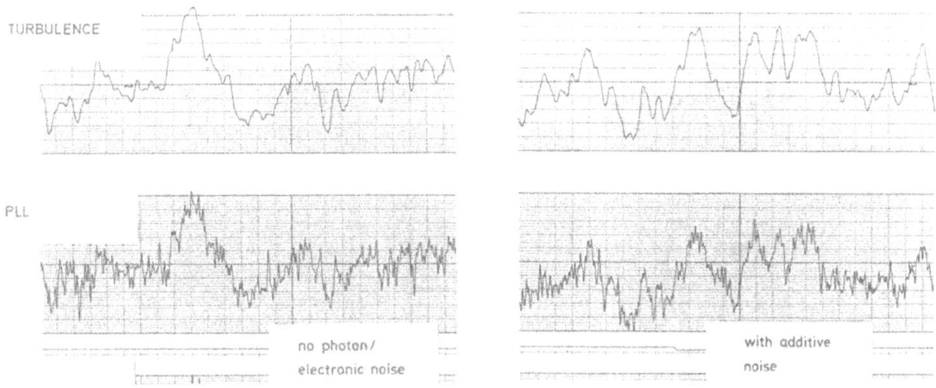

Figure 36. Examples of input and PLL output in the case of (a) no electronic/photon noise and (b) with noise.

The amplitude conditioned sampling described in the section on *Statistics of the Detector Signals* is to a certain extent intrinsic to the performance of the PLL: since the loop gain goes to zero when the instantaneous signal amplitude vanishes and the large phase gradients occur when the amplitude is small, the loop cannot follow the very fast phase changes. A rigorous implementation of the concept can be performed by gating the output with a threshold signal obtained from the quadrature output.

The Frequency Locked Loop. The phase locked loop required a loop response time shorter than the coherence time of the signal. This essentially gives a lower limit to the signal-to-noise ratio of around one (noise bandwidth = signal bandwidth). Because of the limited phase coherence it is not possible to increase the averaging time of the PLL itself. However, by using the power spectrum as the basis for the synthesis of a tracker rather than the signal itself a loop can be synthesised which allows averaging times larger than the coherence time of the signal. The principle is that now we will correlate the measured spectrum with the expected spectrum and find the maximum correlation by taking the derivative with respect to the centre frequency of the reference spectrum. This will generate an error signal that can be used in a control loop. (The principle is the same as that used for the DLL - Fig. 32 - except for the fact that the parameter to be estimated is now a frequency rather than a time delay.) Such a processor is denoted a frequency locked loop (FLL). Thus, the estimated Doppler frequency is found as the value of ω_r for which

$$\int \hat{S}(\omega)\frac{\partial S(\omega;\omega_r)}{\partial \omega_0} = 0. \quad (45)$$

It can be shown that this can be implemented by a scheme shown in Fig. 37a. However, the scheme shown in Fig. 37a requires a complex signal (i.e. also the quadrature component). This is generally not available, but can be generated by using an oscillator with quadrature outputs. Figure 37b shows a loop based on this concept for estimating incremental frequency changes. Normalisation of the error signal - as shown in Fig. 37a - is not performed in the loop. It is also here an advantage that the loop speed depends on the signal level. A limiter in the input avoids excessive amplitudes and will also have the effect that poor signal-to-noise ratios give a lower loop gain and thus a longer averaging time. We note that for the FLL the averaging time can in principle have any value in relation to the coherence time of the signal. Thus a poor signal to-noise-ratio can always be compensated by a longer response time (as opposed to the PLL) if the coherence time of the velocity fluctuations is adequately long.

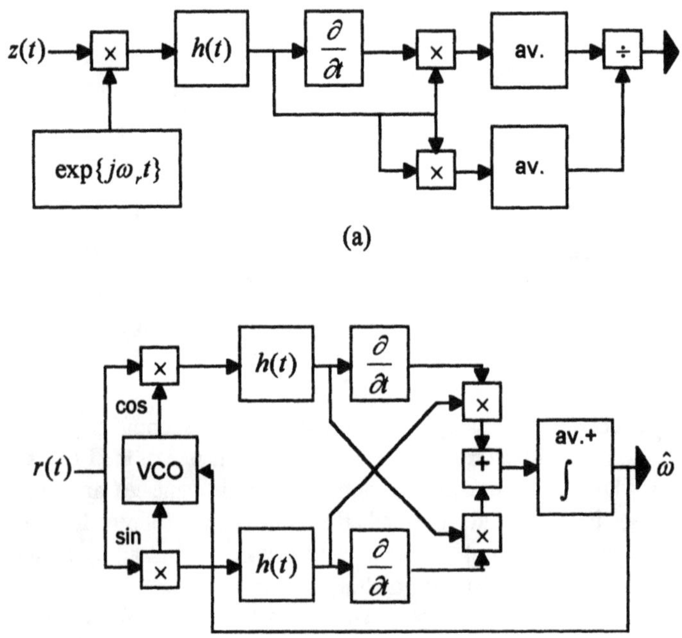

Figure 37. Scheme for estimating the frequency offset from the power spectrum. Note that the input signal is complex (a). This is necessary for a single sideband signal. (b) shows a possible scheme for a real signal with a loop response time that depends on the amplitude of the signal.

An FLL for laser anemometry was proposed and implemented by Wilmshurst et al.[21]. The basis for their design was somewhat different from the material presented here. They focused on a very fast response and not on good performance in the case of a low signal-to-noise ratio and developed the processor directly on the signal statistics.

The basic features of the different processors discussed here are presented in the table on signal processors.

Table 1. Summary of the features of the different processors that are discussed. The figures are indicative. It is noted that the signal-to-noise ratio is with reference to the processor bandwidth - not the signal bandwidth - and that additive noise is assumed.

PROCESSORS FOR LDA SIGNALS

	Complexity/ cost	Dynamic range	Threshold for operation	Slew rate/ range	Validation	Catch range + time	Particle conc.
Counter	low	high	high (S/N>10 dB)	∞	poor	n.a.	low
Spectrum Analyzer	high	1:10	S/N>-10 dB	∞	very good	n.a.	arbitrary
Correlator	high	1:10	S/N>-10 dB	∞	very good	n.a.	arbitrary
Covariance	high/ medium	1:10	S/N>0 dB	∞	very good good	n.a.	low (arbitrary)
Phase Locked Loop	low	1:10	S/N>0 dB	$1/\omega_f$	good	(dynamic range)× 10/max.freq.	high
Frequency Locked Loop	high	1:100	S/N>-10 dB	$1/\omega_f$	very good	full range	medium - high

PROCESSORS FOR LTA SIGNALS

	Complexity/ cost	Dynamic range	Threshold for operation	Slew rate/ range	Validation	Catch range + time	Particle conc.
Timer/ counter	low	high	S/N>-10 dB (depends on pulse det.)	∞	good (potentially)	n.a.	low
Correlator	high	1:10	S/N>-10 dB	∞	very good (with proper postproc.)	n.a.	high (unless gated)
Coincidence tracker	medium	1:10	S/N>-10 dB	> mean time between particles	good	full range	low
Delay Locked Loop	low	1:100	S/N>-10 dB	$1/\omega_f$	very good	full range	medium - high

COMMENTS AND CONCLUSION

The most important consideration in connection with the use of laser anemometers is related to *scales*: characteristic spatial and temporal dimensions that define both the measuring setup and the object of the measurement. Let us summarise the essential scales and parameters.

First for the flow itself:

- velocity range,
- spatial and temporal scales of the velocity fluctuations,
- particle concentration,
- particle scattering cross-sections,
- dimensions of the fluid confinement,
- solid angle(s) giving access to the measuring region.

Then for the measuring instrument:

- dimensions of the measuring volume,
- emitted photon rate (laser power),
- overall efficiency of the system (how many photons are lost),
- bandwidth and upper frequency of the signal processor.

In many cases several of these quantities are not known - that may be the very reason for doing a measurement. Properly implemented validation and acceptance criteria may take care of most problems related to signal quality. Inspecting the burst on an oscilloscope will also give a good indication of noise levels and coherence time of the velocity fluctuations.

In the practical use of laser anemometers it is often the alignment of the optical system that gives most problems. The actual alignment procedure varies for the different makes and types. The most important and trivial things to check are proper crossing of the beams in an LDA, and proper imaging of the measuring volume on the pinhole in front of the detector.

The hot-wire anemometer was *the* instrument to be used for velocity measurements in fluid flows with high spatial and temporal resolution. The instrument is still in widespread use. The laser anemometer provided solutions to several of the problems of the hot-wire: the perturbation of the flow, the nonlinear response, and resolving the sign of the velocity. Many measurements have been performed with laser anemometers that would be impossible with hot-wire anemometers. However, laser anemometers presented new sets of problems as discussed here. It is generally assumed that the light does not perturb the flow. However, the light pressure may play a role in some cases with very small particles and very high laser power. More often perturbation of the beam propagation caused by, e.g., thermal gradients is encountered. This can be the case in combustion systems or in long-range systems for atmospheric measurements. It is here important to notice that refractive index gradients far from the measuring volume are much more serious than gradients in the measuring volume.

Particle image velocimetry (PIV) systems have acquired a considerable amount of interest in recent years because simultaneous measurements are made in a plane with many "interrogation spots". However, for quantitative measurements giving the temporal evolution of the velocity laser anemometers are superior.

The laser anemometer is still a relatively expensive tool. This state of affairs will most likely continue as long as the instruments are based on expensive and fragile gas lasers in combination with high power fibre optics. The electronic processing in laser anemometers is rather demanding and can only to a limited extent utilise developments in mass produced electronics. In the future we may see more of semiconductor-based systems, possibly using processors based on general purpose digital signal processing integrated circuits on special PC-boards.

REFERENCES

1. Y.Yeh and H. Cummins, Localized fluid flow measurements with a He-Ne laser spectrometer, *Appl. Phys. Lett.* **4**, 176-178 (1964)
2. We can mention the series of Lisbon conferences *Proceedings of the n'th Symposium on the Application of Laser Anemometry to Fluid Mechanics* (now up to the 6th), the series *Laser Anemometry: Advances and Applications* and in connection with photon correlation *Photon Correlation Techniques and Applications.*
3. For an introduction to Fourier Optics see J.W. Goodman, *Introduction to Fourier Optics* (McGraw Hill, New York 1968) or G.O. Reynolds, J.B. DeVelis, G.B. Parrent, Jr., and B.J. Thompson, *The Physical Optics Notebook: Tutorials in Fourier Optics* (SPIE/Am. Inst. Phys., Bellingham/New York 1968).
4. M.J. Rudd, "A new theoretical model for the laser Dopplermeter", J. Phys. E.: Sci. Instrum. **2**, 55-58 (1969).
5. L. Lading, "A Fourier optical model for the laser Doppler velocimeter", Opto-electronics **4**, 385-398 (1972).
6. L. Lading, J. Adin Mann, Jr., and R.V. Edwards, "Analysis of a surface scattering spectrometer", J. Opt. Soc. Am. **A 6**, 1692-1700 (1989).
7. Note that the terminology applied here is not necessarily identical to that used in connection with communication and radar systems. *Homodyne* refers to the fact that scattered light is mixed with scattered light and not with a reference beam and does not here refer to whether frequency off-set is applied.
8. See, e.g., M.J. Post and E. Cupp, "Optimizing a pulsed Doppler lidar", Appl. Opt. **29**, 4145-4158 (1990), and M.J. Kavaya, S.W. Henderson, J.R. Magee, C.P. Hale, and R.M. Huffaker, "Remote wind profiling with a solid-state Nd:YAG coherent lidar system", Opt. Lett. **14**, 776-778 (1989).
9. H.T. Yura, S.G. Hanson, and T.P. Grum. "Specle: statistics and interferometric decorrelation effects in complex ABCD optical systems", J. Opt. Soc. Am. **10**, 316-323 (1992).
10. S. Hanson, "Broadening of the measured frequency spectrum in differential laser anemometers due to interference plane gradients", J. Phys. D: Appl. Phys. **6**, 164-171 (1973).
12. R.V. Edwards (Ed.), "Report on the special panel on statistical particle bias problems in laser anemometry", J. Fluids Eng. **109**, 89-93 (1987).
13. W.K. George and J.L. Lumley, "The laser Doppler velocimeter and its application to the measurement of turbulence", J. Fluid Mech. **60**, 312-362 (1973). L. Lading and R.V. Edwards, "The effect of measurement volume on laser Doppler anemometer measurements as measured on simulated signals", in *Proceedings of the LDA-Symposium* (LDA-Symposium, Copenhagen, 1975).
14. L. Lading, "Estimating time and time-lag in time-of-flight velocimetry", Appl. Opt. **22**, 3637 (1983); L. Lading and K. Andersen, "Estimating frequency and phase for velocity and size measurements," in *Laser Anemometry*, J.T. Turner, ed. (STI/Springer, Oxford, 1990), pp. 161-168.
15. L.A. Rabiner and B. Gold, *Theory and Application of Digital Signal Processing* (Prentice-Hall, New Jersey 1975).
16. A. Asher, *Laser Doppler System Development* (Tech. Inf. Ser. 72CRD295, 1972, General Electric Corporate Research & Development Dist., P.O.Box 43, Schenectady, N.Y. 12301).
17. L. Lading and K. Andersen, *Estimating Frequency and Phase for Velocity and Size Measurements* (in: *Laser Anemometry*, J.T. Turner, Ed., STI/Springer, Oxford 1990).
18. L. Lading and K. Andersen, "A covariance processor for velocity and size measurements", in "*Proceedings of 4th Symposium on Applied Laser Anemometry to Fluid Mechanics* (4th Symp. on Appl. Laser Anem. to Fluid Mech., Lisbon, 1988). L. Lading and K. Andersen, *Estimating Frequency and Phase for Velocity and Size Measurements* (in: *Laser Anemometry,* J.T. Turner, Ed., STI/Springer, Oxford 1990). L. Lading and K. Andersen, *Burst Detection in a Phase/Frequency Processor*, (Laser Anemometry; Advances and Applications, ASME, New York 1991), pp. 53-62.
19. L. Lading, "Remote measurement of wind velocity with the time-of-flight laser anemometer", in *Proceedings of Long and Short Range Optical Velocity Measurements,* Ed. H.J. Pfeifer R17/80, (Saint-Louis, France, 1980). C. Fog, "The coincidence tracker. The electronic equipment for a time-of-flight wind speed measurement system", J. Phys. E.: Sci. Instrum. **15**, 1184 (1982).
20. L. Lading and R.V. Edwards, "Laser velocimeters: lower limits to uncertainty", J. Appl. Opt. **32**, 3855-3866 (1993).
21. S. O. Rice, "Statistical analysis of a sine wave plus random noise", Bell. Syst. Tech. J. **27**, 109-157 (1948). (Rice's analysis of random noise provided the basic material for investigating the phase noise encountered in the many-particle case of laser anemometry.)
22. T.H. Wilmshurst and J.E. Rizzo, "An autodyne tracker for laser Doppler anemometry", J. Phys E.: Sci. Instr. **7**, 924-930 (1974).

NEW OPTOELECTRONIC TECHNOLOGIES FOR LASER ANEMOMETERS

Julian D. C. Jones

Department of Physics
Heriot-Watt University
Riccarton
Edinburgh EH14 4AS
United Kingdom

INTRODUCTION

Over the last ten years or so, new optoelectronic technologies have enhanced the design options for laser anemometers, thus extending their versatility. Most obvious is optical fibre beam delivery, now included in many commercial products. Solid state sources and detectors, such as diode lasers and avalanche photodiode detectors, are used increasingly in special applications. Other devices, such as guided wave modulators and holographically formed beam conditioning optics, are beginning to find their place.

New optoelectronic technologies can and have been applied to a very wide range of optical measurement techniques. However, in most techniques similar design considerations apply, and it will be helpful in this chapter to discuss the application of new technologies principally in terms of a single representative example instrument: the Doppler difference anemometer. However, applications in a range of other techniques will also be considered.

The Model Solid State Laser Anemometer

Figure 1 is an idealised representation of a Doppler difference anemometer constructed using classical optical components. Typically, a gas laser source used. The beam is amplitude divided by a system of bulk optic prisms to produce two beams which are conditioned to intersect at their waists, thus forming the measurement volume. Particles traversing the measurement volume scatter light from each beam, producing a different Doppler frequency shift in each case. The scattered light signals are collected, and interfere on the face of a detector, where they produce a signal intensity modulated with a frequency proportional to the component of the velocity of the particle in the plane of the beams and normal to the optical axis. Figure 1 shows a system where light is collected from the measurement volume in full backscatter; typically, the photodetector is a photomultipler tube. In order to distinguish the sign of the velocity component, it is usual to introduce an optical frequency shift between the two beams, generally in the MHz range. The frequency shift may be produced by various different types of bulk optic components.

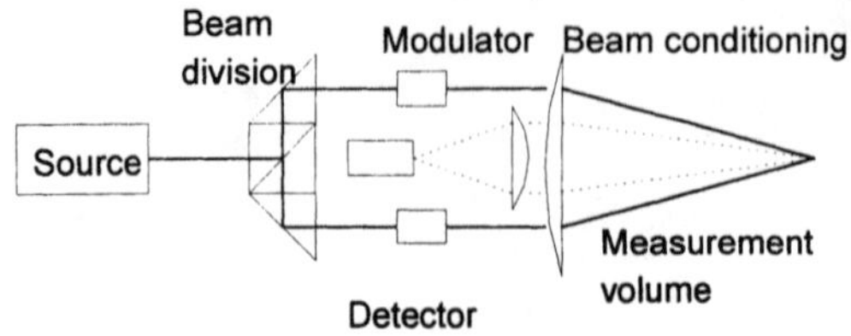

Figure 1. The model classical laser anemometer. Solid lines show the transmitted optical beams, and the broken lines the light scattered from the measurement volume.

Figure 2 shows an idealised *solid state* laser anemometer, embodying the various new optoelectronic technologies which are the subject of this chapter. For example, a diode laser is used rather than a gas laser as the optical source. Optical power is delivered by fibre optics, perhaps using diffractive optics rather than conventional lenses to condition the beam. The beam is modulated by fibre or integrated-optic devices, rather than e.g. with a bulk-optic Bragg cell. The photomultiplier is replaced with an avalanche photodiode.

OPTICAL FIBRES

Optical fibres were developed as a transmission medium for telecommunications, where the required properties are different from those needed in anemometry[1]. However, here the performance of practical fibres will be discussed from the viewpoint of their suitability for laser anemometry.

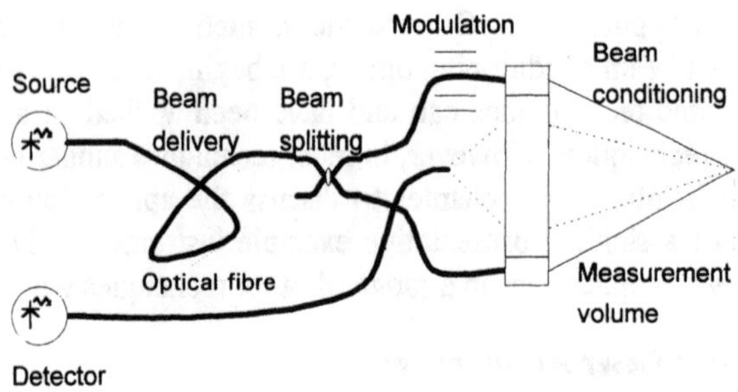

Figure 2. The model solid-state laser anemometer.

The Two Dimensional Dielectric Slab Waveguide

The waveguiding properties of an optical fibre are more easily understood by first considering the two dimensional dielectric slab waveguide[2], figure 3 . It comprises three planar layers of dielectric materials of infinite lateral extent, which are perfectly transparent and differ only in their refractive indices. The central layer is the *core*, refractive index n_1; the outer layers have refractive index n_2, where $n_1 > n_2$. Hence, a ray propagating in the core of the guide, incident on the core-cladding interface at greater than the critical angle, will experience total internal reflection, and thus be guided.

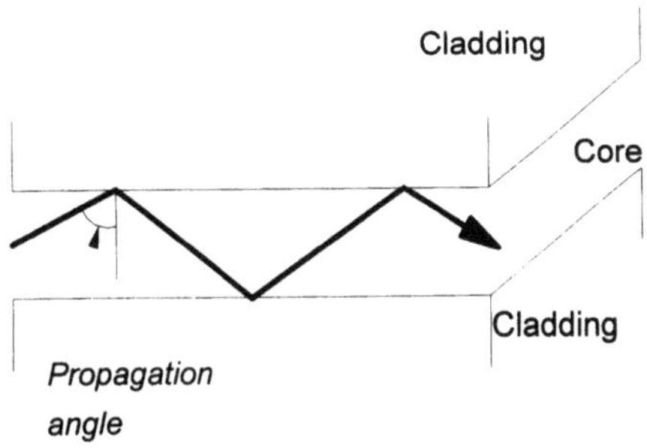

Figure 3. The two-dimensional slab waveguide.

Only particular ray directions yield stable electric field distributions, or *modes*, and the ray directions are hence quantised. For a stable electric field distribution the optical path length associated with a round-trip across the guide is an integral number of wavelengths[3], corresponding to an optical phase change of 2π radians so that

$$\cos\theta_m = (m\pi + \psi)\lambda / 2\pi d n_1 \tag{1}$$

where θ_m is the angle of incidence at the core-cladding interface, m is an integer, ψ is the phase change on reflection, λ is the wavelength of light, and d is the thickness of the core region. For example, consider two rays with equal angles of incidence, but propagating in opposite directions across the guide. The plane wavefronts associated with each ray hence intersect at an angle, yielding interference fringes.

The variable m is the *mode number*. It can take a minimum value of unity, corresponding to the *fundamental mode*, and a maximum value corresponding to the requirement that the angle of incidence must exceed the critical angle for total internal reflection; hence the maximum value of m, which corresponds to the number of guided modes is

$$m(\max) = \frac{2dn_1}{\lambda}[1-(\frac{n_2}{n_1})^2]^{1/2} - \frac{\psi}{\pi} = \frac{(V-\psi)}{\pi} \tag{2}$$

where

$$V = \frac{2\pi d n_1}{\lambda}(n_1^2 - n_2^2)^{1/2} \tag{3}$$

and is called the *normalised film thickness*. The mode number corresponds to the number of maxima in the interference pattern.

Practical Optical Fibres

Fibre materials. Several materials have been used for the fabrication of optical fibres. The first example used a stream of water as the core, and the surrounding air as the cladding[4]! Practical fibres at present use either polymers or glasses to form both the core and the cladding[5], although some low quality fibres use a polymer core, and rely on the air to provide a cladding. However, polymer cores present substantially higher attenuation than

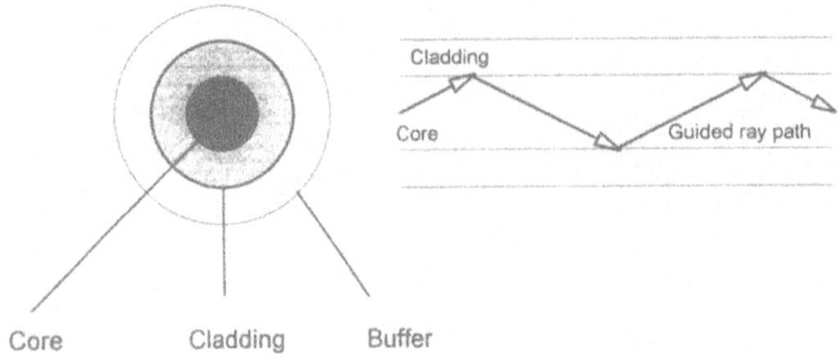

Figure 4. Cross sections of a step-index multimode optical fibre.

glasses. Amongst the glasses, pure fused silica yields the lowest attenuation over the visible and near infra-red and ultra-violet spectral regions, and is hence predominant in fibre fabrication.

The core has a refractive index higher than that of the cladding. This can be achieved by manufacturing them from different materials, of which the most common example is the use of a polymer cladding with a silica core – the so-called *plastic clad silica* (PCS) fibre. Much more usual, however, is to form both core and cladding from fused silica, but to incorporate dopants into either or both of the components in order to modify the refractive index. For example, germania (GeO_2) is often used to increase the refractive index of the core, so that a pure silica cladding can be used.

Optical fibres are nominally cylindrical optical waveguides, of coaxial structure. They have a core and a cladding region, each made from a transparent dielectric; an example is shown in figure 4.

Fibre fabrication techniques. The first-developed fibre fabrication techniques involved modifications of traditional liquid phase glass making methods[6], albeit with exceptional attention to the purity of the materials. From the molten phase, a *preform* is fabricated – essentially a cylinder of glass, containing the core and cladding regions. The preform is mounted vertically within a furnace, and from its base a fibre is drawn. In an alternative technique, the liquid core and cladding regions are injected downwards through vertical concentric nozzles, and a fibre is once again drawn.

For high quality optical fibres, typical of those used in telecommunications, and for all fibres of the single mode type (described below), vapour phase fabrication techniques[7] are used. Perhaps the most widespread example is the modified chemical vapour phase deposition technique. The starting point is a fused silica cylinder, destined to become the outer part of the cladding region of the preform. The starting materials are admitted to the inside of the cylinder as vapours mixed with oxygen; for example $SiCl_4$ is a suitable starting material for silica (SiO_2). The reaction vessel is raised to a sufficiently high temperature to oxidise the starting materials, hence depositing a silica 'soot' on the walls of the vessel. Many layers are deposited, with the flow rates of the starting materials adjusted to control the refractive index of each layer. Once the deposition process is complete, the layers are sintered, the preform is collapsed, and the fibre is drawn.

Step index multimode optical fibres. Perhaps the simplest of all fibre types is the step index multimode shown in figures 4 and 5. Step index denotes that there is a discontinuous change in refractive index across the core-cladding interface. A very wide range of core diameters of such fibres are available commercially, from a few tens of μm up to about 1 mm. However, in each case, the core diameter is large in comparison with the wavelength of light (typically 0.5 μm), so by analogy with the two dimensional guide and equation (2), many modes may be guided – hence the expression 'multimode'.

It is helpful to express the core size in a normalised form in terms of the 'V - number', defined as

$$V = \frac{2\pi a}{\lambda}(n_1^2 - n_2^2)^{1/2} \tag{4}$$

where a is the core diameter. It is useful to note that the requirement that the angle of incidence at the core-cladding interface exceeds the critical angle leads to

$$NA = (n_1^2 - n_2^2)^{1/2} = \sin\alpha(\max) = \sin[\frac{\pi}{2} - \theta(\max)] \tag{5}$$

where NA is the numerical aperture of the fibre and α (max) is the maximum angle which a guided ray can make with the optical axis of the fibre. Hence, when light enters the end face of the fibre from a surrounding medium of refractive index n_o, where the end face is normal to the optical axis, the maximum angle which a ray can enter the fibre and be guided is given by

$$\beta = \sin^{-1}(n_o \sin\alpha) \tag{6}$$

where β is called the acceptance angle.

For the two dimensional guide, it was easily seen that the number of guided modes scaled with V; for the three dimensional fibre, the derivation is more complex although it is intuitive that it should scale with V^2. In fact, for $V >> 1$, it may be shown[8] that

$$N \approx \frac{1}{2}V^2 \tag{7}$$

where N is the number of guided modes. Figure 5 implies that all mode paths intersect the fibre axis; however, only those which are *meridional* do so. Other *skew* mode paths are angular helixes, with vertices at the core-cladding interface[9].

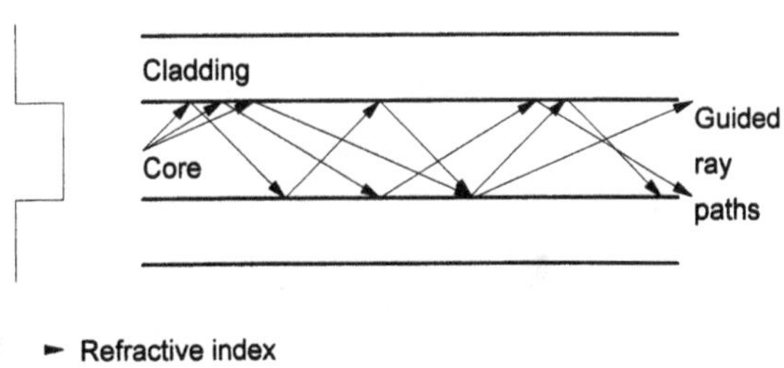

Figure 5. Refractive index distribution and cross section of a step-index multimode optical fibre.

The optical path lengths of the various modes differ. The time taken for an optical pulse to propagate along the fibre becomes dependent on the identity of the mode in which it propagates and the emerging signal is temporally broadened. This inter-modal dispersion[10] prompted the development of more sophisticated types of fibre for high-bandwidth telecommunications.

Graded index multimode optical fibre. A cross section of a graded index fibre is shown in figure 6. It is similar to the step index fibre, except that the refractive index is a function of radial distance from the fibre axis, with a maximum on the axis. The refractive index of the cladding remains uniform, and the index is continuous across the core-cladding interface. The effect of the varying refractive index is that the ray paths are curved by continual refraction. The numerical aperture of the fibre is a function of the radial position, so that the acceptance angle is a maximum on the axis, and zero at the core-cladding interface. Many practical graded index optical fibres have a refractive index distribution of quadratic form, such that

$$n_1(r) - n_2 = [n_1(0) - n_2](\frac{r}{a})^2 \tag{8}$$

where $n_1(r)$ is the core refractive index at a distance r from the axis. Hence light from an axial point source will be focused successively as it propagates. The distance between foci is the *fibre beat length*; in this sense, the fibre acts like a lens.

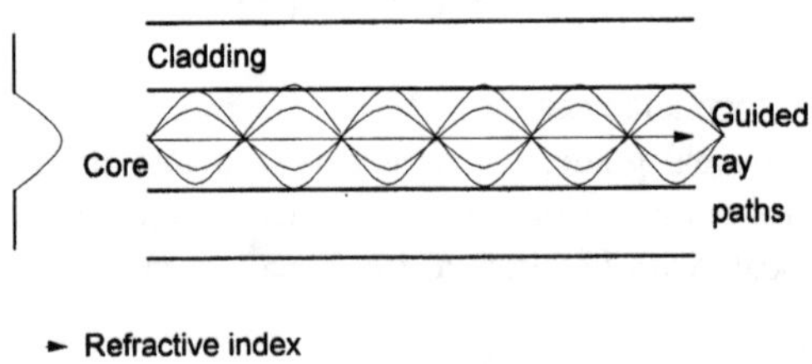

Figure 6. Refractive index distribution and cross section of a graded index multimode optical fibre.

The graded index shows considerably less inter-modal dispersion than the step index type[11]. Although the higher order modes travel greater physical distances, they do so in regions further from the core where the refractive index is lower, leading to a higher phase velocity.

Single mode optical fibres. From equation (2) it may be inferred that in an optical fibre as the core diameter is reduced, successively fewer modes propagate. Eventually, when the condition $V<2.405$ is satisfied, only the lowest order mode can propagate[12], and the fibre is said to be *single mode*. The V-number is, from equation (4), a function of wavelength. For a fused silica fibre with a difference of 1% between the core and cladding refractive indices and a core diameter of 3 μm, single mode operation will be achieved for wavelengths more than 0.85 μm. For such small dimensions, ray optics approximations are invalid, and the waveguiding properties must be deduced from electromagnetic theory[13]. There is no inter-modal dispersion and residual dispersion arises from the dependence of the refractive index and waveguiding properties on the wavelength of the light, but is relatively small[14]. Hence, single mode fibres are used to the exclusion of other types in high-bandwidth telecommunications, manufactured by vapour phase techniques.

Beam Quality in Fibre Delivery

Multimode fibres. Applying equation (7) for a step index fibre of core diameter 50 μm and a numerical aperture of 0.3 at a wavelength of 532 nm indicates that approximately 4,000 modes are guided. Because there is no correlation between the phase of the wavefronts for each of the guided modes, light emerging from such a fibre can be considered using a geometrical optics approximation. A measure of beam quality is the product of the core size and the numerical aperture, as shown in figure 7. However the near and far field distributions of the irradiance depend on the properties of the input laser beam, the launch optics, the fibre properties, and the presence of power coupling between the modes of the fibre (which is exacerbated by bending)[15,16]. The near field irradiance profile has approximately the same diameter of the fibre core, and the far field profile has approximately the same angular extent as the numerical aperture. The small-scale irradiance distribution in the far field is far from uniform: interference between the guided modes produces a strong objective speckle pattern. Even the slightest perturbations of the fibre produce phase changes between the modes, hence modulating the speckle[17]. The beam quality at the output of a multimode fibre is low, and thus unsuitable for focusing to a small spot, as would be required in the transmission optics of a laser anemometer[18].

Single mode fibres. The beam quality at the output of a single mode fibre is extremely high. The near field profile at the exit of the fibre is approximately gaussian, and the emerging light can be considered as a gaussian beam waist[19]. The waist diameter, defined at the $1/e^2$ irradiance points, is approximately equal to the fibre mode field diameter, which is slightly larger than the fibre core diameter – typically 4 to 6 μm in the 500 to 850 nm wavelength range. The numerical aperture of a single mode fibre is typically around 0.1.

Birefringence and polarisation

An ideal cylindrically symmetric single mode fibre offers no birefringence to the guided beam, and thus the state of polarisation of the guided beam is preserved. Practical fibres

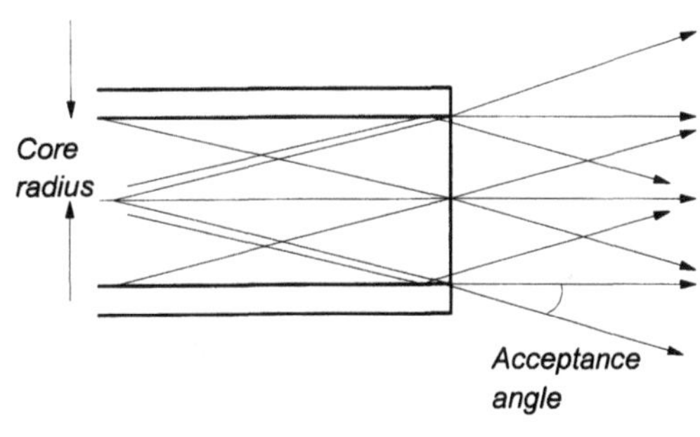

Figure 7. Cross section showing rays emerging from a step-index multimode optical fibre.

exhibit intrinsic and extrinsic birefringence[20]. Intrinsic birefringence is created in the manufacturing process. The geometry of the core is not perfectly circular, and residual anisotropic stresses are produced in the fibre. Even if the fibre had no significant intrinsic birefringence, it would remain subject to extrinsic birefringence. When the fibre is deployed, it is inevitably bent and twisted, thus producing anisotropic stresses. Birefringence causes the state of polarisation to evolve as the guided beam propagates. Many applications demand that a beam of known state of polarisation is delivered, for example to maximise visibility in interferometry.

Special fibres have been developed with very high intrinsic linear birefringence, induced either by producing an elliptical core geometry, or an approximately elliptical stress distribution. One successful design, shown in figure 8, is based on creating thermal stress in the fibre[21]. The cladding contains elements with a different coefficient of thermal expansion than their surroundings, so that when the fibre cools following manufacture, considerable anisotropic thermal stress is created across the core, thus leading to birefringence.

Such fibres are capable of preserving polarisation under special conditions. The intrinsic birefringence is arranged to be very much greater than the extrinsic birefringence. Hence, the axes of birefringence are well defined throughout the fibre. Thus, linearly polarised light of azimuth aligned with one of the birefringence axes of the fibre will propagate in a single polarisation *eigenmode* of the fibre, and its linear state of polarisation will thus be preserved.

In general, the input beam will couple to both fibre eigenmodes. Consider the coupling of a linearly polarised beam into the fibre, where the azimuth of the plane of polarisation is at an angle of 45° to the axes of birefringence. The beam thus couples equally to the eigenmodes. Because of the birefringence, the two states have unequal phase velocity, and a phase difference, or *modal retardance* accumulates between the eigenmodes given by

$$\Delta\phi = \phi_f - \phi_s = \frac{2\pi l}{\lambda}(n_f - n_s) \tag{9}$$

where ϕ_f and ϕ_s are the phases of the *fast* and *slow* eigenmodes, n_f and n_s are the corresponding refractive indices, and l is the distance travelled in the fibre. The corresponding evolution of the state of polarisation is shown in figure 9. The period of polarisation state evolution is called the *polarisation beat length*; high intrinsic birefringence corresponds to a short polarisation beat length, given by

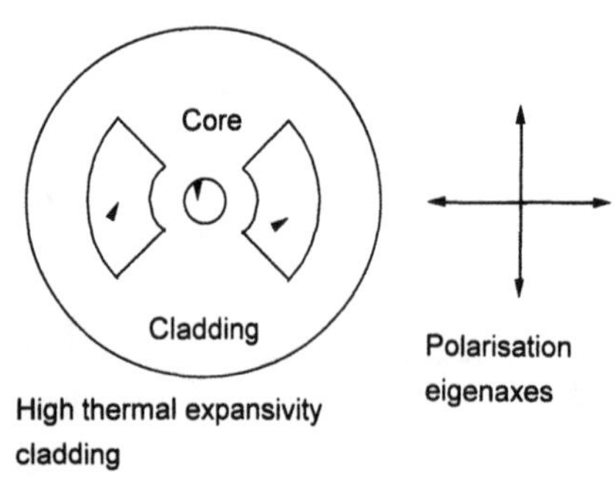

Figure 8. Cross section of a highly birefringent fibre of the thermally stressed type, and the corresponding directions of the polarisation eigenaxes.

$$L_b = \lambda / (n_f - n_s) \tag{10}$$

For practical highly birefringent fibres, the beat length is typically in the range 1 to 3 mm. For a nominally cylindrically symmetric single mode fibre, deployed in a typical arrangement, a polarisation beat length of a few m would be expected.

In multimode optical fibre, the birefringence is different for each mode, hence causing a loss of polarisation coherence between the modes. Hence, the output beam is not polarised.

Power Handling

Photorefractive effects. The small core diameters of single mode optical fibres leads to high power densities. Photorefractive effects set a limit to the power transmission of CW beams of wavelengths in the blue-green region of the spectrum, or shorter[22]. The high power density leads to the formation of absorbing *colour centres* in GeO_2 doped fibres. The effect becomes significant at power densities exceeding approximately 8 GWm^{-2} at a wavelength of 488 nm, corresponding to a power level of about 100 mW in 4 μm core diameter fibres. The effect manifests itself by an increase in the optical absorption.

Special fibres have been developed to provide an increased power handling capacity. The simplest technique involves using a lower dopant concentration, which also allows a larger core diameter to be used, leading to reduced power density. Even more effective is to fabricate the fibre without any core doping, and to use an index-reducing cladding dopant. In this way, practical CW power levels of at least 5 W at 488 nm are achievable.

Optical damage thresholds. High instantaneous pulsed power densities may exceed the optical damage threshold of the fibre material, dependent on wavelength, pulse duration and power density. The damage threshold of fused silica at a wavelength of 532 nm is 1.4 $kJcm^{-2}$ for a pulse of duration 15 ns and 17.5 Jcm^{-2} for 0.7 ns[23].

Damage effects are not relevant for point laser velocimeters using single mode

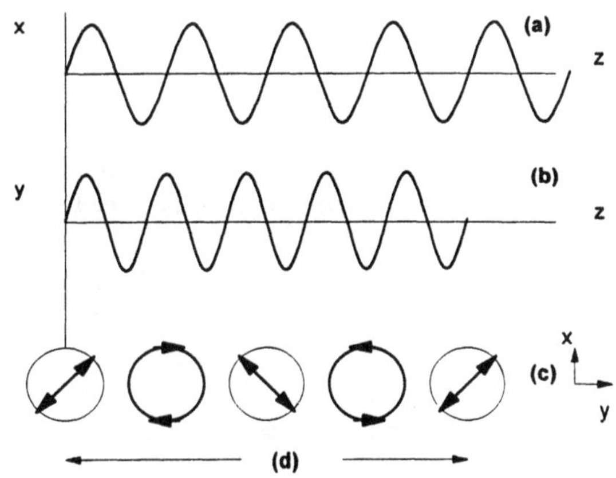

Figure 9. Evolution of the state of polarisation of the guided beam in a birefringent optical fibre, when a linearly polarised beam is coupled into the fibre with an azimuth of $\pi/4$ with respect to the eigenaxes, which are in the x and y directions, with the fibre axis in the z direction; (a) and (b) are the electric field strengths for each of the eigenmodes, (c) shows the resultant state of polarisation, and (d) is the polarisation beat length.

transmission fibres. They are important for pulsed laser applications such as particle image velocimetry (PIV). The damage threshold in a fibre is less than that in the corresponding bulk material, and is reduced by effects such as self-focusing in the fibre and imperfections at the core-cladding interface. For example, for a 200 μm core diameter fused silica fibre used to deliver 10 ns pulses at 532 nm, the damage threshold is approximately 1.5 mJ[24].

Environmental Effects

The phase of a fibre-guided beam is modulated by changes in the physical length and the effective refractive index of the core. In single mode optical fibres, this is the basis of interferometric optical fibre sensors[25]. Unwanted phase modulation can be a difficulty: for example, in a Doppler difference anemometer, phase changes between the transmitted beams lead to shifts in the interference fringes within the measurement volume.

In highly birefringent optical fibres, environmental effects produce a differential change in phase between the polarisation eigenmodes, with corresponding variations in the state of polarisation of the guided beam – unless only a single mode is excited. Table 1 shows typical values for the phase sensitivity in practical fibres[26].

Table 1. Phase sensitivity of the guided beam in a birefringent optical fibre. The data refer to a fibre of the thermally stressed type, with a beat length of 2 mm, measured at a wavelength of 633 nm, in a sample length of 0.1 m.

Stimulus	Phase sensitivity	Differential phase sensitivity	Units
Temperature	100	5	rad K^{-1}
Strain	6.5×10^6	6.5×10^4	rad m^{-1}
Acoustic pressure	5×10^{-4}	5×10^{-6}	rad Pa^{-1}

OPTICAL FIBRE COMPONENTS

Directional Couplers

The directional coupler is the basic optical fibre component for the division or recombination of guided beams; it is analogous to the bulk-optic beamsplitter[27].

The 2 × 2 directional coupler. The simplest coupler is the 2 × 2, which has two inputs and two outputs, shown in figure 10. Within the interaction region, the fibre cores are brought very close together, so that the guided modes couple. There are two common types, according to their fabrication techniques: fused and polished.

In the fused type[28], two fibres are brought into contact, and heated until they fuse. They are then drawn axially to produce a fused biconical taper. Light travelling along an input fibre experiences a steadily decreasing core size, and is forced further into the cladding. In the interaction region, the cores are vanishingly small, and waveguiding occurs at the cladding-air interface. The mode of the input fibre couples to the composite modes of the interaction region, shown as the lowest order symmetric and anti-symmetric modes in figure 11. Dispersion causes a phase difference to accumulate between the modes as they propagate, thus altering the power distribution across the interaction region. Hence the power entering each of the two cores at the output of the coupler is given by[29]

$$P_1 = \frac{1}{2} P_o (1 + \cos \Delta kl)$$
$$P_2 = \frac{1}{2} P_o (1 - \cos \Delta kl) \tag{11}$$

where P_o is the input power, Δk is the difference between the propagation constants for the symmetric and anti-symmetric modes, and l is the length of the interaction region. This elementary analysis assumes no losses in the coupler. The *coupling ratio* is defined by

$$K = \frac{P_2}{P_1 + P_2} \tag{12}$$

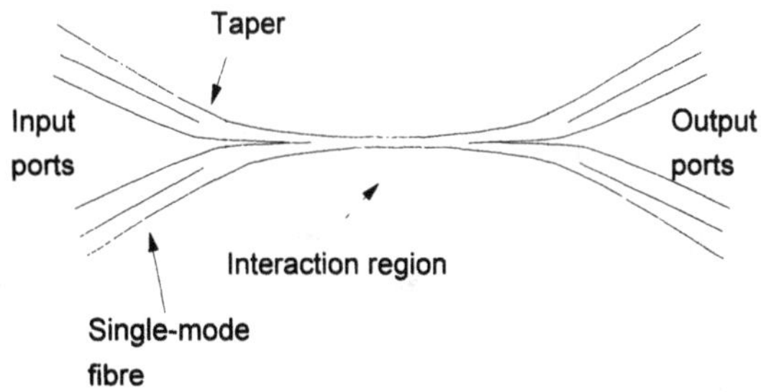

Figure 10. Directional coupler of the fused type.

An optical phase difference between the outputs is produced as a result of the coupling process, of $\pi/2$ radians in an ideal device. The coupling ratio is defined during the manufacture of the device, and any value between 0 and 1 can be specified. Fused directional couplers are now highly developed, and commercial devices with insertion losses of less than 0.1 dB are readily available.

Polished couplers. The other main type of directional coupler is produced by polishing the cladding material away to allow closer access to the core[30]. The fibre is embedded in a curved groove in a fused silica block. The fibre and block are simultaneously polished to within a few μm of the core. A second similar ***half block*** is produced, and when the two blocks are placed in contact and adjusted, then power is coupled from one fibre to another. Through mechanical alignment, the coupling ratio is adjustable. Losses even less than in the fused devices are possible, although manufacture is much more complex.

3 × 3 directional couplers. In a symmetrical coupler with three input and three output ports, the optical phase difference between the outputs is $2\pi/3$ rads, in an ideal device. Outputs of such a form are ideal for demodulating interferometer signals using phase-stepping algorithms, without requiring an active modulator[31]. 3 × 3 couplers are made using the fusion technique. Couplers with larger port counts are feasible, although for very large numbers of inputs and outputs, integrated optical fabrication techniques are preferred.

Chromatic properties of couplers. The coupling ratio is wavelength dependent, through the chromatic dispersion of the propagation constants of the modes in the interaction region; i.e., in equation (11), $\Delta k = \Delta k(\lambda)$. The wavelength sensitivity can be enhanced by making a long interaction length, so that $\Delta\ kl >> 2\pi$. It is hence possible to construct a coupler for which $K(\lambda_{,})$ is zero, and $K(\lambda_{,})$ is unity. Such couplers can thus be used for wavelength-division multiplexing[32]. In other applications, a coupling ratio constant with wavelength is required. *Wavelength-flattened* couplers are made with short interaction lengths, and the waveguides are made deliberately dissimilar[33].

Polarisation properties of couplers. The interaction region of a coupler is not symmetrical, and is hence birefringent. The coupling ratio is thus slightly sensitive to state of polarisation, and the state of polarisation is not preserved. Directional couplers made using highly birefringent fibre, generally by using the polishing technique[34] can be designed to preserve polarisation, or to behave like polarising beamsplitters.

Fibre Polarisers

Several in-fibre devices exist to polarise the guided beam. One design exploits a surface plasmon resonance, and is similar to one half of a polished directional coupler, in which a metal film, perhaps with a dielectric overlay, is deposited on the polished part of the fibre[35]. The TM mode of the guided beam excites a lossy plasmon resonance, whilst the TE mode continues to propagate without significant loss. A simpler design exploits bend loss in highly birefringent fibre[36]. The polarisation eigenmode in such fibres have unequal mode field diameters. The mode with the greater diameter is more susceptible to bend loss. Hence, a coiled birefringent fibre acts as a polariser, but with some inevitable insertion loss.

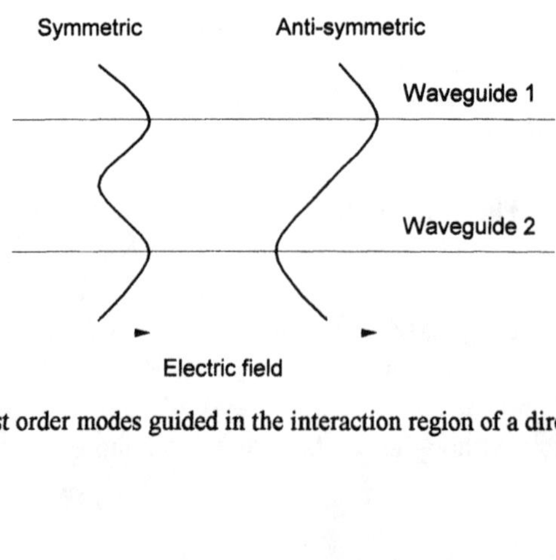

Figure 11. Lowest order modes guided in the interaction region of a directional coupler.

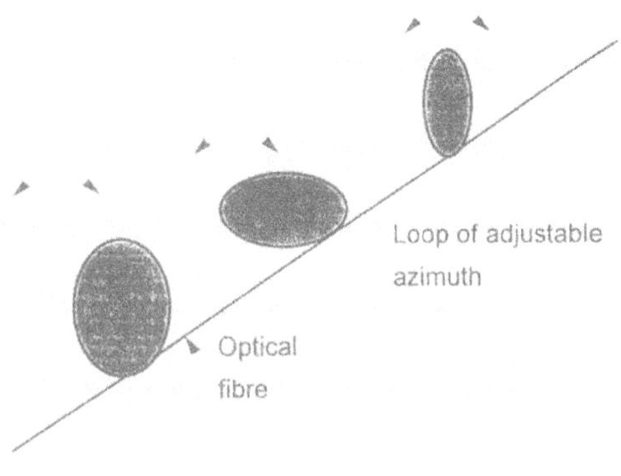

Figure 12. Fibre-optic polarisation controller.

Polarisation Controllers

In many situations, it is necessary to be able to control the state of polarisation of a fibre-guided beam – for example, it is often desirable to avoid the complexities inherent in the use of highly birefringent fibre, but still produce a linear state of polarisation. Polarisation controllers are made using the fibre equivalent of bulk-optic linear retarders or wave plates. A waveplate is easily synthesised in single mode fibre of nominal cylindrical symmetry, through the use of bend-induced birefringence[37]. When the fibre is bent into an arc, birefringence is induced with axes in, and perpendicular to, the plane of the bend. The birefringence induced depends on the dimensions of the fibre and the bend, and the optomechanical properties of the fibre. For example, for a fused silica fibre at a wavelength of 633 nm, the refractive index difference between the eigenmodes is approximately

$$\Delta n = c(\frac{r}{R})^2 \tag{13}$$

where $c \approx 0.133$.

A polarisation controller can be formed from a set of linear retarders, as shown in figure 12. For full birefringence control, three retarders are required, either as three quarter-wave plates, or two quarter-wave and one half-wave.

Fibre Optic Modulators

Piezo-electric phase modulators. The most common form of fibre phase modulator consists of a piezo-electric cylinder about which part of the fibre is wrapped[38]. The dimensions of the cylinder are controlled by the applied electrical potential, thus straining the fibre and producing a phase modulation in the guided beam. For a modulator formed by coiling 100 turns of fibre about a 50 mm diameter cylinder made from PZT-5H, a phase modulation of about 4.5 rad V^{-1} is produced. Frequency response is limited by mechanical resonances in the cylinder, and for the device described, the first resonant frequency is about 30 kHz.

Phase modulation is useful in path length control in an interferometer, or for phase stepping but cannot provide a true frequency shift. Various pseudo-frequency shifting techniques are available: in serrodyne (sawtooth ramp) phase modulation, the linearly rising part of the ramp corresponds to a constant rate of change of phase, and hence to a frequency shift[39]. The frequency shift is not single sideband, and harmonics result from the flyback. Furthermore, a rapid flyback excites mechanical resonances. Conversely, it is convenient to modulate the phase sinusoidally, and large modulation amplitudes at high frequencies are achieved by deliberately exciting at a mechanical resonance. However, signal processing from sinusoidal phase modulation is much more difficult than from a simple frequency shift, although various synthetic heterodyne techniques are available[40].

Surface acoustic wave frequency shifters. A true frequency shifter is based on the use of a piezoelectric element to produce travelling surface acoustic waves, and hence to induce frequency modulation of a beam guided in a highly birefringent optical fibre. A schematic of such a device is shown in figure 13[41]. At the input to the device, light is coupled into only one of the fibre eigenmodes. The polarisation axes of the fibre are aligned at 45° to the surface of the piezoelectric element. Hence, the surface acoustic waves produce stresses and strains in the fibre which produce local rotation of the polarisation axes, and couple power from the first eigenmode to the second (initially unpopulated) one. Because the mode coupling is produced by a moving disturbance, a frequency shift results. For significant total power transfer to occur, it is necessary that the power contributions coupled into the second mode at each point are in phase, thus imposing a phase matching condition between the propagation constants for the first and second modes,

$$k_f - k_s = k_{a,eff} \tag{14}$$

where $k_{a,eff}$ is the effective propagation constant of the surface acoustic wave. Thus, phase matching is achieved by aligning the fibre at an appropriate angle to the acoustic wavefronts, thus adjusting the effective acoustic propagation constant. Modulators of this type have been used to produce frequency shifts in the tens of MHz range with typical sideband suppressions of 20 dB.

Flexure wave modulators. Mode coupling by a moving disturbance in highly birefringent fibres can also be achieved using flexure waves[42]. A prototype device is shown in figure 14. A piezoelectric actuator, via an acoustic impedance matching horn, is used to excite travelling flexure waves in a birefringent fibre. The axis of the exciting vibration is arranged at 45° to the polarisation axes of the fibre. Similar to the surface acoustic wave modulator, a single eigenmode is populated at the input, and power is coupled to the second mode and frequency shifted by the travelling disturbance. An advantage over the surface acoustic wave device is that lower frequency shifts can be produced: 0.8 MHz in reference[41]. No geometrical adjustment can be made to ensure phase matching, and only frequency adjustment of the actuator is feasible.

Alternative versions of the technique have been based on coupling the two lowest order spatial modes, the LP_{01} and LP_{11}, rather than the two polarisation eigenmodes, using a fibre whose core diameter is just large enough to guide both modes, rather than being perfectly single mode, and which is not highly birefringent[43].Torsional waves[44] have also been used for mode coupling.

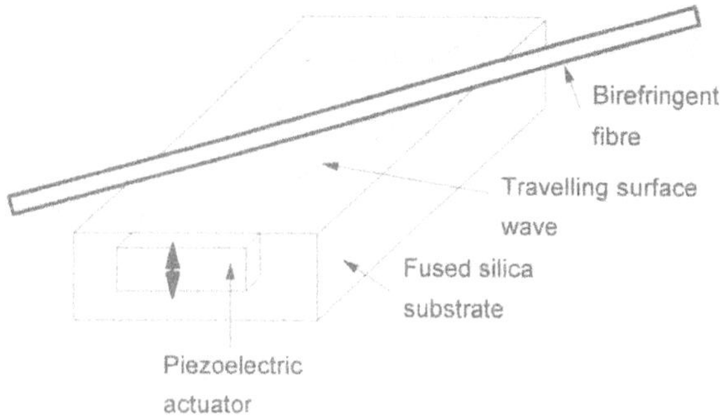

Figure 13. Fibre optic frequency shifter of the surface acoustic wave type.

Integrated-Optic Modulators

Integrated-optic devices are based on waveguides formed by doping a bulk material. For example, one type of guide is produced by using a titanium dopant to locally increase the refractive index in a $LiNbO_3$ substrate[45]. $LiNbO_3$ exhibits a strong electro-optic effect. Hence, through the application of an electric field across the guide, phase modulation is produced. Therefore, in some respects the $LiNbO_3$ modulator is analogous to the piezoelectric fibre optic phase modulator. However, electro-optic modulation can be very much more rapid than mechanical modulation. The bandwidth of integrated-optic modulators can thus be very high – extending even into the GHz range.

Integrated-optic devices can be produced using techniques similar to those employed

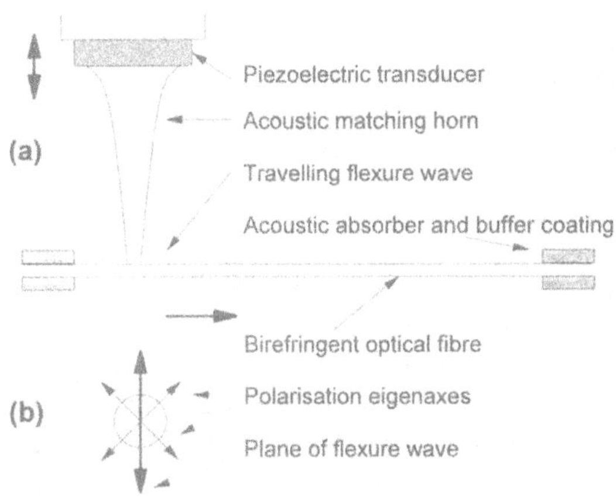

Figure 14. Fibre-optic frequency shifter based on coupling flexure waves into a birefringent fibre: (a) schematic of the modulator; (b) flexure wave polarisation and fibre polarisation eigenaxes.

for integrated electronics. Diffusion of the dopant into the substrate creates waveguides with a refractive index raised in comparison with the surrounding un-doped substrate. The guides thus produced are single mode, with widths of 5 to 10 μm, extending into the substrate over depths of a few μm.

The integrated-optic modulator is a phase modulator; frequency shifting is achievable through serrodyne phase modulation. Phase modulation can also be converted to rapid intensity modulation by forming an integrated-optic Mach Zehnder interferometer, with phase modulation in one arm[46].

A disadvantage with integrated-optic modulators is the difficulty of incorporating them into fibre optic systems: extremely accurate alignment is required. Nevertheless, many commercial products are available, generally supplied with single mode fibres already attached to form pigtailed devices. Almost all commercial products are intended for telecommunications applications, and are hence available for use at wavelengths of 1.3 μm and 1.55 μm.

DIODE LASERS

Operating Principles

Diode lasers are a development of light emitting diodes[47]. A typical device structure is shown in figure 15. A light emitting diode (LED) is a junction between *p* and *n* doped semiconducting materials. Hence, when a forward electrical bias is applied across the junction, an injection current flows, thus exciting electrons from the conduction to the valence band. Electrons subsequently return to the valence band, recombining with a hole, with a release of energy. In order for radiative recombination to dominate, a *direct bandgap* semiconductor material must be used. Hence, except under special circumstances, silicon is unsuitable, and GaAs is most commonly used, often alloyed with Al, or In and P for longer wavelength devices. The material composition determines the gain spectrum, and hence the wavelength of operation.

The end faces of the diode chip are cleaved to provide reflecting surfaces. The refractive index of GaAs is high, so that the Fresnel reflection coefficient is sufficient to allow the chip to form an optically resonant cavity. Through a combination of heavy doping and a strong forward electrical bias, a population inversion may be formed between the conduction and valence bands, thus permitting optical gain. In combination with the Fabry Perot cavity produced by the Fresnel reflections, laser action may take place.

Operating Characteristics

Power output. The dependence of optical output power on injection current is shown in figure 16. At low injection currents, cavity losses exceed optical gain, and lasing does not occur; the device behaves as an LED. Above the *threshold current*, optical gain exceeds the cavity losses, and the optical power output rises rapidly with current. The probability of non-radiative recombination rises with temperature, hence the threshold current is higher at elevated temperatures. In order to achieve the high current and optical power densities required to exceed threshold, without excessive total current, device structures are designed to constrain the dimensions of the active region in both dimensions. One practical design is shown in figure 15[48]. For confinement in the vertical direction, the active layer is sandwiched between two layers of a different semiconductor alloy of lower refractive index, thus forming a waveguide. Such use of different materials creates a *heterostructure*. In the

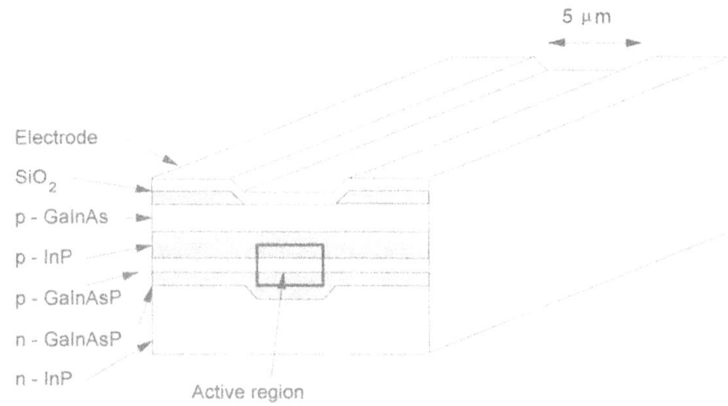

Figure 15. Diode laser of the channelled substrate heterostructure type.

device shown in the figure, the substrate and electrode are channelled, thus constraining the current flow to the centre of the structure.

Low-cost diode lasers, of the type used in compact disc players – for example – have maximum power outputs of only a few mW. However, commercial devices are readily available with powers of up to 150 mW, and devices of steadily increasing power are appearing. Diode lasers may also be fabricated in *arrays*, where many devices are combined into a single package. One and two dimensional arrays are available. Although power levels of many W are feasible, it is difficult to combine the outputs of the many facets into a single beam. The chief use of diode arrays is to pump other solid-state laser media. For example, the diode pumped Nd:YAG laser is commercially well developed.

Spectrum. GaAs lasers, with the addition of suitable dopants, are available with operating wavelengths in the range from about 0.67 to 1.65 μm. Rather shorter wavelengths are achieved using *quantum well* structures to perturb the energy levels of the bulk materials[49]. Very much shorter wavelengths are feasible using compounds from groups two and six of the periodic table: blue diode lasers based on ZnSe have been demonstrated.

At low injection currents below threshold, the spectrum is essentially that of an LED, and without structure covers a range of a few nm. Just above threshold, the longitudinal mode structure becomes apparent, and the output comprises discrete lines. The cavity length of a diode laser is short, and hence the mode spacing is wide – typically 0.5 nm. In a device with sufficiently uniform material composition throughout the active region, line narrowing by mode competition takes place as the injection current is increased, so that at the normal operating power the output is essentially in a single mode, giving a typical coherence length of a few m. Lower cost diode lasers have slightly non-uniform active regions, and hence operate on several longitudinal modes simultaneously, even at the highest injection currents.

Devices of higher coherence length are produced by controlled cavity feedback. In *distributed feedback* DFB lasers, periodic refractive index variations are created in the cavity to act as a Bragg grating, and hence provide a spectrally narrow reflector[50]. The highest coherence length devices are of the *external cavity* type, in which the cavity is defined by the rear laser mirror and an external mirror[51].

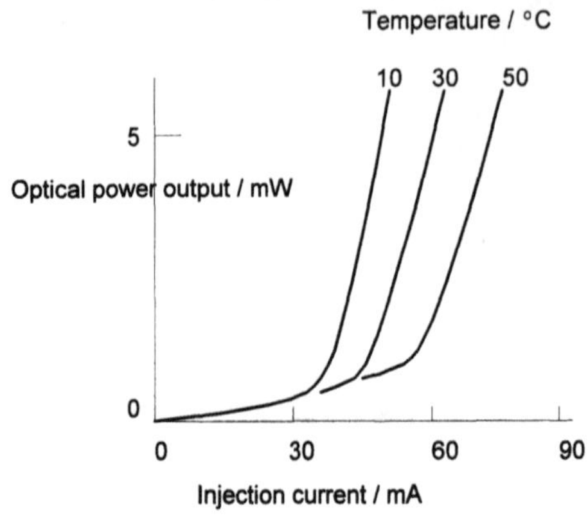

Figure 16. Optical power output of a typical diode laser, as a function of injection current and temperature.

Beam characteristics and coupling to fibres. The output facet of a diode laser is rectangular, with typical dimensions of 1 × 5 μm. The output beam is thus both strongly divergent and elliptical. The output is linearly polarised, with a typical extinction ratio of about 30, and where the polarisation axes are aligned with those of the ellipse. The beam may also be slightly astigmatic. For higher power devices, ellipticity and astigmatism are usually greater. To couple the beam into a single mode fibre, it is first collimated and then focused into the fibre core. If spherical collimating and focusing optics are used, and the ellipticity is left uncorrected, then a maximum coupling efficiency of about 30% from a low power device may be expected. The beam can be circularised by using anamorphic optics, typically by a prism pair, or crossed cylindrical lenses. Diode laser optics are commercially available. It is also possible to purchase modules comprising a diode laser permanently coupled to a fibre via suitable launch optics. In telecommunications applications, fibres are usually *pigtailed* to the laser: the fibre is cemented to the output facet of the laser without intermediate optics. Some beam transformation is possible by tapering the end of the fibre, or by polishing it into a lens shape[52], but coupling efficiency remains low.

Laser Diode Modulation

Temperature modulation. Changing the temperature of the active region produces two effects[53]. Firstly, the shape of the gain curve is temperature dependent. For example, for a typical GaAs diode laser with a nominal wavelength of 850 nm, the wavelength corresponding to the peak of the gain curve increases by approximately 60 pm K^{-1}. Secondly, the optical path length of the cavity changes, through thermal expansion and the temperature dependence of the refractive index. Hence, the wavelength of a particular laser mode increases accordingly, by typically 220 pm K^{-1} for the example of the 850 nm laser. In a single mode device, only the mode closest to the gain peak will lase. Hence, the effect of temperature tuning the laser is to produce regions of continuous wavelength tuning, corresponding to the wavelength shift of a single mode, separated by discontinuous wavelength shifts corresponding to a switch in laser operation to an adjacent mode. The continuous tuning range is approximately equal to the cavity mode spacing – 2.0 nm for a typical 300 μm cavity length. At higher temperatures, the injection current is increased, and the power output falls.

Current modulation. Changes in the injection current produce wavelength modulation through changes in temperature and charge carrier density[54]. The charge carrier density affects the refractive index of the active region, and hence the cavity length. At low modulation frequencies, the thermal effect is dominant, and in a typical device will produce an optical frequency shift of about 3 GHz mA^{-1}. At frequencies above about 100 kHz, the charge carrier effect is dominant, and the modulation rate falls to about 0.3 GHz mA^{-1}. The frequency response is set by the thermal time constants of the diode, but can be linearised by electronic equalisation[55]. Current-induced frequency modulation of diode lasers has been exploited to generate a frequency shift in laser anemometry, by the pseudo-heterodyne technique[56].

Optical Feedback and Isolation

Diode lasers are susceptible to the effects of both incoherent and coherent optical feedback into the cavity from any other part of the optical system. The absorption of the incoherently fed-back light causes a rise in temperature, and the generation of charge carriers within the active region, which has been used for the *optothermal* modulation of diode lasers[57].

The high gain of the diode laser, and the relatively low reflectivity of the internal cavity mirrors, renders them especially susceptible to coherent feedback. Even low levels of optical feedback can cause the laser to oscillate between various internal and external cavity modes, thus severely reducing the effective coherence length, and causing an increase in noise. Conversely, in the presence of a large and stable external reflection, it is possible to stabilise the operation of the laser to a single narrow external cavity mode.

Coherent feedback to diode lasers has been exploited usefully in Doppler shift frequency measurement by *self mixing*[58]. The injection of frequency-shifted light into the cavity is equivalent to feedback of constantly changing phase, with concomitant modulation of the laser power at the Doppler frequency.

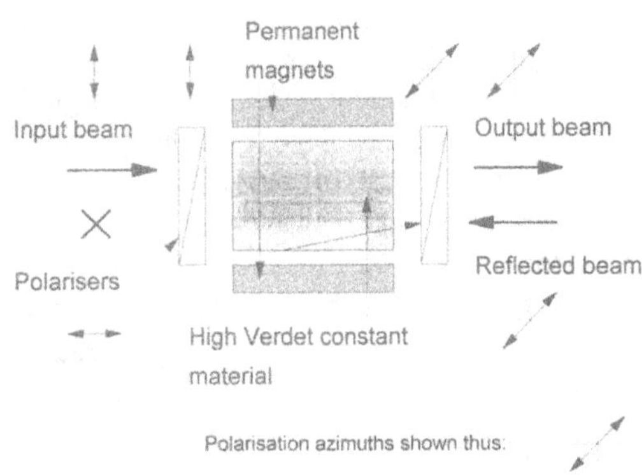

Figure 17. A Faraday isolator.

Considerable feedback is caused by reflection at the input face of a fibre optic system, reduced by polishing the end face of the fibre at an angle. But because of other reflections, such as the output face of the fibre, and it is usually necessary to use an optical isolation. The most effective optical isolator is of the Faraday type, and an example is shown in figure 17. The Faraday effect is the rotation produced in the plane of polarisation of an optical beam when it propagates in a dielectric material under the influence of a magnetic field parallel to the optical axis. The rotation is proportional to the field strength and to the *Verdet constant* of the material. The laser light enters via a polariser, aligned with the azimuth of the input beam. It passes through a material of high Verdet constant, surrounded by a permanent magnet. A rotation of 45° in the plane of polarisation is produced, and the light exits through a suitably aligned polariser. Fed-back light re-enters the isolator through the polariser. It is reversed in direction in comparison with the outgoing beam, and hence experiences a magnetic field of the opposite sense, thus producing an opposite 45° rotation. The fed-back beam therefore becomes linearly polarised with an azimuth orthogonal to that of the exit polariser, and is thus extinguished. Faraday isolators are commercially available, specified for the appropriate operating wavelength with extinction ratios of 40 to 60 dB.

PHOTODIODES

Semiconductor materials may be used for the detection of light[59]. The absorption of a photon may cause an electron to be excited from the valence to the conduction band, producing a measurable electrical signal by the change in conductivity produced. Such semiconductor devices are normally configured as p-n junctions – *photodiodes*. The sensitive region is the depletion layer between the p and n materials, where a potential gradient exists.

A typical device structure is shown in figure 18. Photons enter the device via a window with an anti-reflection coating. The top layer of the device, e.g., the n-doped region, is made thin as photons absorbed there will not contribute to the photocurrent. The depletion layer is made wide to maximise photon absorption. The depletion layer is widened by reverse bias. A small *dark* current flows in the absence of light, due to thermally-created charge carriers, increasing with reverse bias. The maximum reverse bias is set by the limit of dielectric breakdown. Hence, many practical photodiodes achieve a thick depletion layer by interposing a very lightly doped, or undoped (*intrinsic*) layer in the junction, leading to a diode of the p-i-n type, as shown in figure 18.

The proportion of photons absorbed in the depletion layer, and giving rise to detectable photelectrons, is called the *quantum efficiency*, and depends on the wavelength of the light. At short wavelengths, the absorption length is short, and a significant proportion of the photons are absorbed in the top layer of the device. At longer wavelengths, the absorption length increases, until photons pass through the depletion layer undetected. At even longer wavelengths, the photon energy is less that the band gap of the semiconductor, and photons cannot therefore excite photoelectrons. The quantum efficiency is hence dependent on the materials used in the photodiode. Silicon has a band gap corresponding to a maximum wavelength of about 1.1 μm. For longer wavelengths, germanium can be used, or alloys such as InGaAsP. A disadvantage of smaller band gap materials is that the probability of thermal carrier generation is increased, leading to higher dark currents.

For a given quantum efficiency, the electron generation rate is proportional to the photon arrival rate and hence incident power, with output current proportional to electron generation rate. *Responsivity*, is the current output per unit input optical power, 0.5 A W^{-1} for a typical silicon photodiode at 850 nm. The electrical bandwidth of a photodiode is set by its junction capacitance, and by the finite time taken for charge carriers to cross the depletion region. The ultimate bandwidth is typically in the GHz range.

Avalanche Photodiodes

For the detection of weak optical signals, it is possible to achieve internal gain in *avalanche photodiodes*[60] by using very high reverse bias. Charge carriers produced by photon absorption are accelerated by the field, and cause collisional ionisation to produce secondary charge carriers, which are in turn accelerated. Hence, a single photon can produce an arbitrary number of charge carriers.

The current gain is a sensitive function of the electric field in the depletion layer and to produce useful avalanche gain, specially designed photodiodes must be used. Even so, the gain is a sensitive function of the reverse bias and temperature, and the applied voltage must be stabilised. For a silicon avalanche photodiode, gains of 100 to 200 are typical with a reverse bias of 300 V.

Photodetector Noise

The fundamental noise limit is set by the quantised nature of the charge carriers, and the consequent statistical variation in their numbers – described by *shot noise*. The statistics are Poissonian, and hence the shot noise current scales with the square root of the photocurrent and dark current, given by

$$\langle i_n^2 \rangle = 2eB(I_p + I_d) \tag{15}$$

where e is the electronic charge, B the electrical bandwidth and I_p and I_d are the photo- and dark currents respectively.

The number of secondary charge carriers produced by collisional ionisation of a primary carrier is determined by a statistical distribution, giving noise modelled by multiplying the shot noise by an excess noise factor, $F(M)$, an increasing function of the

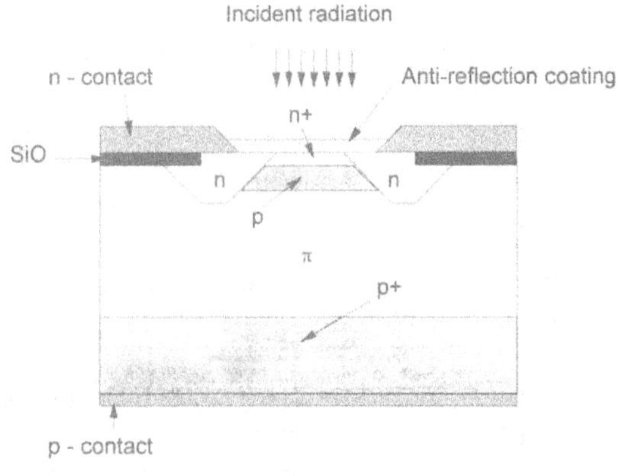

Figure 18. Structure of a typical avalanche photodiode – p, n: p-doped and n-doped semiconductor; p+, n+: heavy doping; π: intrinsic or lightly doped.

current gain M. Hence,

$$\langle i_n^2 \rangle = 2eB(I_p + I_d)F(M)M^2 \tag{16}$$

Additional noise arises in the external amplifying circuit: thermal noise arises in the load resistor, and additional noise is caused by the active components, represented by multiplying the thermal noise by a factor F_n. Hence the mean square thermal noise current is given by

$$\langle i_t^2 \rangle = 4kTBF_n / R_L \tag{17}$$

where k is Botzmann's constant, T is the absolute temperature, B is the electrical bandwidth and R_L is the load resistance.

From equations (15) and (16) the overall signal to noise ratio is found to be[61]

$$S/N = \frac{I_p^2}{2eB(I_p + I_d)F(M) + 4kTBF_n / M^2 R_L} \tag{18}$$

At high optical powers, the shot noise contribution is dominant. Hence, because $F(M)>1$, a simple photodiode without internal gain will give better noise performance. At low powers, thermal noise dominates, and hence by using the internal gain of the avalanche photodiode ($M^2 >> 1$), the signal to noise ratio is improved. However, if M^2 is increased too far, then $F(M)$ increases more rapidly, and the signal to noise ratio deteriorates. Hence, there exists an optimum value of the current gain, dependent on the device properties.

APPLICATIONS

Reference Beam Anemometers

The first laser anemometer based on fibre optics was of the reference beam type[62], and is shown schematically in figure 19; it is analogous to the Fizeau interferometer. Light from the source is guided by the fibre to the measurement volume. At the distal face of the fibre, a small proportion of the optical power is Fresnel reflected, and returns towards the source to form the reference beam. The remainder of the light expands from the fibre as the object beam. A fraction of the light scattered by particles entrained within the fluid flow is recaptured by the fibre, and returns to the detector, where it mixes with the reference beam, to generate an optical beat frequency equal to the Doppler shift induced by the moving scattering particles.

The measurement volume is ill defined, and a significant fraction of the return signal emanates from the region close to the distal end of the fibre, where the flow is perturbed. One refinement using lenses at the distal face of the fibre to produce a focused measurement volume[63]. Another requires the use of a low-coherence source: an interferometer is used at the detector to balance the path difference between the reference beam and scattered light from the measurement volume[64]. Interference is observable only for those parts of the measurement volume for which the path is balanced.

Fibre interferometers of the reference beam type have been developed extensively for the non-contact measurement of the motion, and specifically the vibration, of solid surfaces. Various instruments based on Michelson[65] or Mach Zehnder configurations, and using homodyne and heterodyne signal processing, have been developed[66].

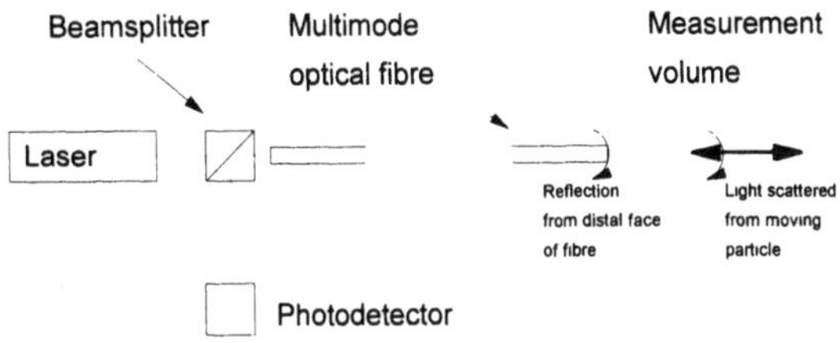

Figure 19. Simple fibre-optic laser anemometer of the reference-beam type.

An example is shown in figure 20, designed for the purpose of detecting acoustic emission generated during metal cutting operations, for tool wear monitoring[67]. The acoustic emission is detected via the surface acoustic waves which it generates on the workpiece, seen as out-of-plane displacements with amplitudes in the nm range, and at frequencies in the 0.1 to 1.0 MHz range. This example is chosen for discussion because it leads to a fibre interferometer design which embodies may of the features previously discussed in this chapter.

Light from a diode laser source is coupled into the single mode optical fibre Michelson interferometer, via a Faraday isolator. The beam is divided into the signal and reference arms by a directional coupler. The signal arm terminates in a lens which conditions the beam to form a waist on the target. Light scattered at the target is recaptured by the probe lens, and guided back towards the detectors. The reference fibre terminates in a reflector, which returns the reference beam to the detectors, where it interferes with the signal beam to give an output at detector 1 of the form

$$I_1 = \kappa_1[1+V_1\cos(\phi_s+\phi_d-\phi_r)] \tag{19}$$

where ϕ_s, ϕ_d and ϕ_r are optical phase terms corresponding to the signal, drift and noise, and the optical path length of the reference arm, respectively; V_1 is the visibility of interference, and κ_1 is a constant. The vibration produces a phase modulation

$$\phi_s = \frac{4\pi\Delta z}{\lambda}\cos\omega_a t \tag{20}$$

where Δz and ω_a are the amplitude and frequency of the out-of-plane vibration.

To maximise the visibility the recombining beams must be of equivalent states of polarisation. A difficulty is that optical fibres are birefringent, and that the birefringence is environmentally sensitive. The interferometer thus incorporates fibre polarisation controllers. Part of the signal beam fibre serves as a downlead, connecting the probe to the remainder of the interferometer. The downlead is subject to dynamic bending, and it is thus not practicable to compensate for the induced birefringence using the polarisation controller. Hence, the downlead is constructed using highly birefringent fibre. The polarisation controllers are adjusted such that only a single polarisation eigenstate is excited. Hence, the state of polarisation of the guided beam is preserved[68].

The Michelson interferometer requires a reflector in the reference arm Many techniques exist for producing reflective fibre ends, through the use of optical coatings, or attachment of bulk optic mirrors. However, an interesting alternative shown in figure 20 is to use a *fibre loop reflector*, which is an optical fibre Sagnac interferometer[69]. It comprises a directional coupler, where the two outputs are joined to give a continuous loop. Hence, light propagating in the two possible directions around the loop recombines and interferes. In an ideal and unperturbed Sagnac interferometer, the recombining beams are in phase, and the loop acts as a perfect reflector. When the loop is birefringent, then the reflectivity is controlled through the birefringence. Thus, by incorporating a fibre birefringence controller in the loop, a reflector of continuously adjustable reflectivity is formed. It is useful to be able to adjust the intensity of the reference beam in order to maximise the visibility of the interference.

In the Michelson interferometer, one of the outputs is directed towards the source. Nevertheless, it is useful to be able to access that complementary output, which has the form

$$I_2 = \kappa_2[1 - V_2 \cos(\phi_s + \phi_d - \phi_r)] \tag{21}$$

Hence, electronically balancing and subtracting the two outputs yields a signal

$$I_\delta = \kappa_\delta \cos(\phi_s + \phi_d - \phi_r) \tag{22}$$

where κ_δ is a constant. The signal-to-noise ratio is thus improved, through removal of the DC term. The complementary output is recovered from the rejected output of the polarising beamsplitter, used as the polariser in the Faraday isolator. The interferometer polarisation controllers are adjusted so that the beam returning to the Faraday isolator is orthogonally polarised with respect to the input beam.

The detection of acoustic emission involves monitoring very small amplitude, high frequency vibrations, in the presence of much larger amplitude, low frequency ambient vibrations, represented in equation (19) by the term ϕ_d. The ambient vibrations are compensated by a homodyne processing technique[70]. The differentially combined interferometer outputs (equation 20) are amplified and fed back to a phase modulator of the piezoelectric type, in the reference arm. The effect of the feedback loop is to maintain the differentially combined outputs at zero. The bandwidth of the feedback loop is designed to

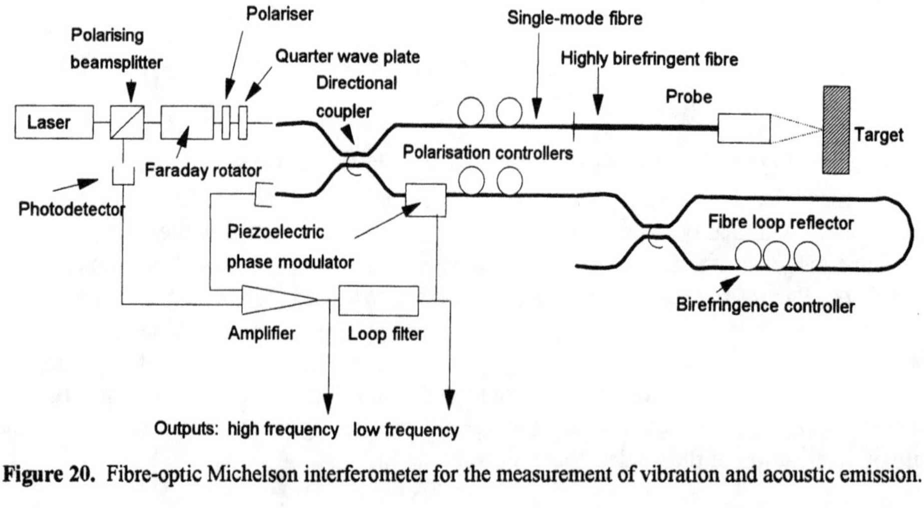

Figure 20. Fibre-optic Michelson interferometer for the measurement of vibration and acoustic emission.

be sufficient to compensate the ϕ_d, but the signal term ϕ_s is beyond the servo bandwidth; hence

$$\phi_d - \phi_r = \frac{\pi}{2} + N_i \pi \qquad (23)$$

where N_i is an integer. Thus, when the servo loop is closed, equation (21) becomes

$$I_\delta = \pm \kappa_\delta \sin \phi_s \propto |\phi_s| \qquad (24)$$

Where the small-angle approximation is valid, given that the amplitude of the out-of-plane vibration $\Delta z << \lambda / 2$. In practical applications a displacement amplitude resolution of 0.05 nm in the full-bandwidth frequency range from 0.1 to 1.0 MHz, equivalent to a resolution of 0.053 pm $Hz^{-1/2}$ was achieved.

Doppler Difference Anemometers

Doppler difference anemometers may be classified into the groups shown in figure 21[71].

Fibre links. In fibre linked designs, the optical fibres are used to separate the source and detector from a separate transceiver comprising optics for dividing the input beam into the two transmitted beams which are combined to form the measurement volume. The transceiver also contains the frequency shifter, where required for directional information. The transmitting fibre is single mode, to preserve the spatial coherence of the guided beam.

The transceiver contains optics to collect the light scattered from the measurement volume. Spatial coherence is not required in the return signal, and multimode fibre is suitable. It is only necessary that the dispersion in the return fibre is small enough to accommodate the bandwidth of the Doppler signal. Step-index fibre is often preferred because its acceptance angle is uniform over the core cross section Graded-index fibre has less intermodal dispersion, but the acceptance angle varies over the core area: maximum on axis, and zero at the core-cladding interface.

Fibre probes. The optical fibres are used to separate a compact and passive probe from the active elements of the system. The active elements comprise the optical source, means for dividing the beam into the two input beams, and the frequency shifting element, where required. The two transmitted beams are guided separately to the probe, normally by using two separate optical fibres. The fibres are single mode, to preserve the spatial coherence of the beam. The probe contains optics to condition the fibre outputs to produce beams crossing at their waists, hence forming the measurement volume. For high visibility interference in the measurement volume, the recombining beams must have similar states of polarisation. This is achieved either through polarisation control in transmitting fibres of nominally circular core geometry, or by using highly birefringent fibres.

The two transmitting fibres experience different environments, so that the relative phase of the two transmitting beams is environmentally modulated[72]. Temperature effects are generally too slow to be of significance, but vibration and acoustic perturbations may cause motion of the fringes within the measurement volume with frequencies comparable to the Doppler shift. The most serious effect is that of vibration, which induces strain in the fibres, and hence phase modulation. The perturbation is generally of an oscillatory nature, and thus has the effect of broadening the observed Doppler spectrum, equivalent to an increase in the apparent measured turbulence of the flow.

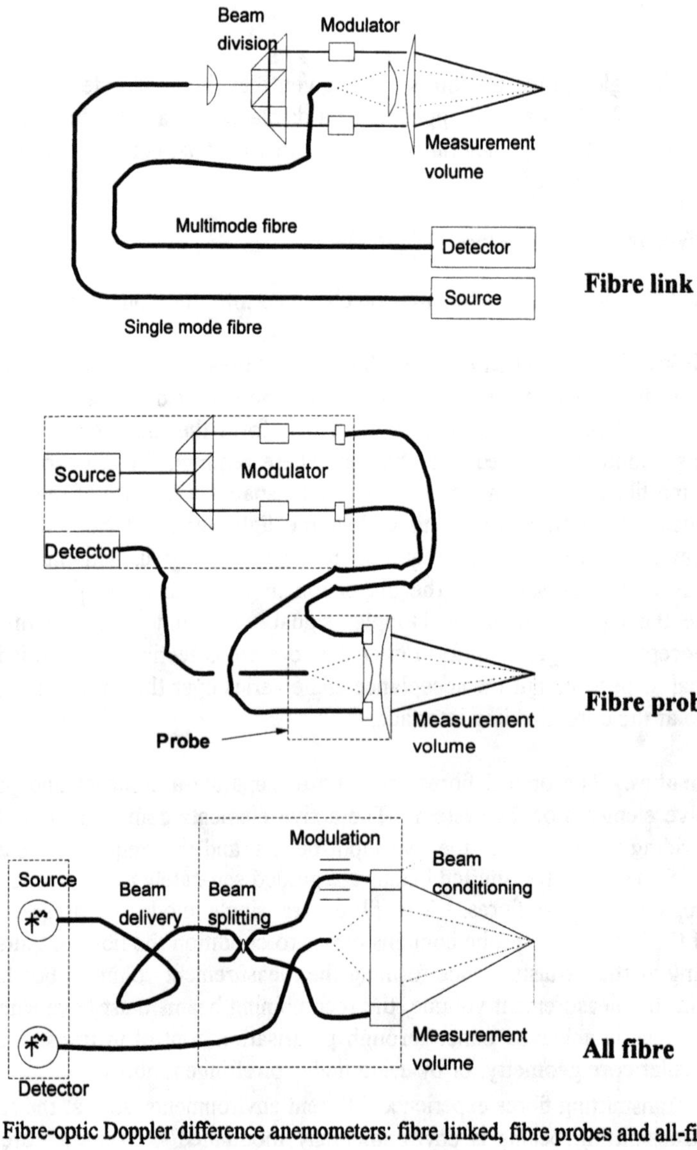

Figure 21. Fibre-optic Doppler difference anemometers: fibre linked, fibre probes and all-fibre designs.

One technique for reducing the relative phase modulation is to transmit the two beams as the orthogonal polarisation eigenmodes in a *single* highly birefringent fibre[73]. However, as may be seen from table 1, a residual phase modulation exists.

All-fibre. All-fibre systems are similar to those of the probe type, except that the probe contains fibre optic elements for the functions of beam division and frequency shifting. The earliest reported design made use of a directional coupler for beam division, and a piezoelectric phase modulator for frequency shifting[74]. More recently, integrated optic modulators have been used for frequency shifting. Other alternatives include deriving the two input beams from two different laser sources, of known frequency difference, using stabilised diode lasers[75].

Other Interferometers

The flexibility of fibre optic systems has allowed consideration of more complex interferometer designs for laser anemometry. One example is in the development of interferometers as frequency discriminators for the direct measurement of the Doppler shift. In bulk optics, Fabry Perot interferometers have been used in this way, although their limited frequency resolution restricts the technique to very high velocities. It is feasible to construct very large path length Fabry Perot interferometers using optical fibre. However, such frequency-discriminating interferometers are equally sensitive to source frequency fluctuations as to the Doppler shift, and are of limited practical value.

An alternative example is the Sagnac, illustrated in figure 22[76]. The single mode fibre loop is interrupted near to one end, and a probe is interposed. The probe takes light from the loop, reflects it from the target (a moving particle in the measurement volume, or a solid surface), and returns it to the loop to continue in the same direction. Both the clockwise and anti-clockwise propagating beams are reflected from the target, and they interfere at the detector.

Because the two beams are incident on the target at different times, then if the target is moving, a phase shift is produced between the recombining beams, given by

$$\phi_D = 2(2\pi/\lambda)\bar{v}\Delta t \tag{25}$$

where $\bar{v}$ is the mean target velocity during the loop delay Δt, and assuming normal incidence and reflection from the target. Because the interferometer is common path, it is (to first order) insensitive to environmental phase perturbations (shown as ϕ_d in equation 18, for example) and source frequency fluctuations.

The output from the interferometer is similar to that of any other two beam interferometer, as equation (19), with ϕ_D replacing ϕ_s. Hence, for maximum sensitivity it is necessary to bias the interferometer at one of its quadrature points, defined by equation (23). The fibre Sagnac interferometer can be phase biased through control of the fibre birefringence. Hence, a polarisation controller is included in the loop.

The fibre Sagnac interferometer has been exploited for high-frequency vibration measurement. In one application, the noise floor limited velocity resolution was 50 nm s^{-1} $Hz^{-1/2}$.

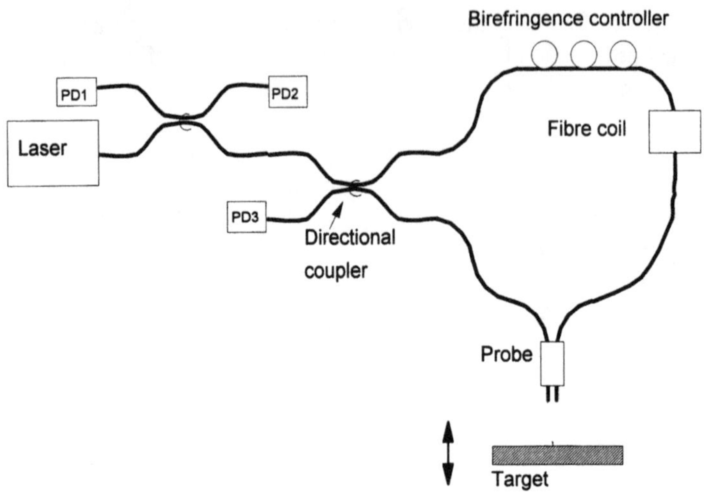

Figure 22. Doppler velocimeter based on the fibre-optic Sagnac interferometer.

Transit Anemometers

Fibre-based systems for transit anemometry have followed a similar route to those for Doppler difference systems, with fibre-linked and fibre probe systems realised. Single mode fibres provide effective spatial filtering of the transmitted beam, thus allowing the production of very small light spots in the measurement volume[77].

Particle Image Velocimeters

Fibre illuminators for particle image velocimetry would be useful where optical access to the flow is poor. Such illuminators must transmit high power pulses, with high beam quality. Typically, pulsed lasers such as Q-switched and frequency doubled Nd:YAG lasers operating a wavelength of 532 nm are used. Similar lasers are used for other flow diagnostics such as laser induced fluorescence. The high pulsed powers produce energy densities well beyond the damage limits of single mode fibres, and it is necessary to use large core multimode lasers. Practical measurements have shown the feasibility of transmitting ns pulses of energy 1 - 2 mJ through fused silica step-index fibres of core diameter 200 μm at wavelengths of 532 nm, but large core diameter fibres reduce beam quality and mode-mode interference in their far field causes speckle. The small-scale intensity fluctuations can be suppressed by ensuring that the fibre is long enough that the inter-modal dispersion exceeds the coherence length of the laser source[78].

CONCLUSIONS

This chapter has reviewed areas of new optoelectronic technologies relevant to laser anemometry. The most significant new source has been the diode laser. Power levels for single-devices remain relatively low, typically up to 150 mW. Even at these powers, diode lasers represent a suitable replacement for helium neon gas lasers. A disadvantage of diode lasers is their susceptibility to optical feedback, often necessitating the use of a Faraday isolator, which can be significantly more costly than the diode itself. Nevertheless, an advantage of the diode laser, which can be compelling, is the ease with which it can be modulated, both in power and wavelength, via its injection current or temperature. Available power levels are steadily increasing. However, an important application of the diode laser is in pumping Nd:YAG lasers. For pumping purposes, only low beam quality is required, and is readily supplied by diode arrays. Already, compact diode-pumped Nd:YAG lasers are available with frequency doubled outputs at 532 nm, power levels competitive with air-cooled argon ion lasers, and with excellent beam quality, coherence length and stability. Higher power devices are nearing the market, with power levels similar to large water-cooled argon ion lasers. Q-switched versions giving short, high energy pulses are also available.

Undoubtedly, the technology which has had the greatest impact on the optical systems used for laser anemometry in recent year is fibre optics. All major manufacturers now supply fibre optic beam delivery systems without compromising performance. At higher power levels non-linear and photorefractive effects limit the power transmittable through single-mode fibres to the order of 100 mW, or up to a few W when special fibres are used. For high power pulsed laser techniques, such as particle image velocimetry, the power densities are far too high for use with small-core single-mode fibres. Multimode fibres must be used instead, with attendant degradation in beam quality, and a continuing ultimate limitation on the power levels that can be used.

Attention must be paid to the effect of environmental perturbations of the fibre. Phase modulation may lead, for example, to a broadening of the apparent velocity spectrum. Polarisation changes lead to a loss of visibility in the interference in the measurement volume, but can be avoided by using special fibres.

Most commercial systems use fibre only for beam delivery. Nevertheless, at the research level, many systems have been reported which exploit guided wave components, such as fibre optic directional couplers, for beam division. A practical impediment to their widespread use is that directional couplers made from highly birefringent fibres are expensive. Alternative techniques for beam splitting involve the use of integrated-optic devices, where couplers with large numbers of ports can be fabricated with relative ease. A practical difficulty is the means of interfacing an integrated-optical component with an optical fibre. For free-space beam division, a technique offering far greater control than is achievable by conventional bulk optics is offered by diffractive optics, or holographic optical elements. Diffractive optics offers the realisation of optical components, effectively holograms, capable of combining the functions of several optical components simultaneously, such as the focusing and splitting of optical wavefronts in a single device.

Many laser anemometers require modulators to provide a frequency shift to give directional information. A variety of fibre modulators are available, although they tend to be slow. Rapid integrated-optic modulators are available, but require a fibre interface and are expensive. Techniques based on modulation of diode laser sources are becoming established.

The integration of new optoelectronic technologies into laser anemometry is a continuing process. Fibre technology is mature, and fibre delivery in laser anemometry is mature. The integration of fibres into other optical techniques, such as particle image velocimetry, is beginning. Newer optical sources, such as diode lasers and diode-pumped solid state lasers, are finding an increasing rôle in laser anemometry. Increasingly, solid-state optics, such as integrated-optic devices are being exploited. Already it is possible to envisage the realisation of wholly solid-state anemometers, using diode laser sources, integrated optic modulators, fibre beam delivery, and diffractive optics beam conditioning.

It is through such developments that the technique of laser anemometry is finding wider usage, and is reaching beyond the research laboratory into a broader range of engineering applications.

REFERENCES

1. S E Miller and A G Chynoweth (eds.) 'Optical Fiber Telecommunications' (Academic Press, New York, 1979)
2. D L Lee 'Electromagnetic Principles of Integrated Optics' (John Wiley, New York, 1986)
3 P K Tien and R Ulrich 'Theory of prism-film coupler and thin-film light guides' J Opt Soc Am 60 (1970) 1325 - 1337
4. J Tyndall, Royal Institution of Great Britain Proceedings 6 (1870) 189
5. S R Nagel 'Fiber materials and fabrication methods' in 'Optical Fiber Telecommunications II' (eds. S E Miller and I P Kaminow), (Academic Press, New York, 1988) Chapter 4
6. K Koizumi, Y Ikeda, I Kitano, M Furukawa and T Sumitomo 'New light focusing fibres made by a continuous process' Appl Opt 13 (1974) 255
7. J B MacChesney, P B O'Connor and H M Presby 'A new technique for preparation of low-loss and graded-index optical fibers' Proc IEEE 62 (1974) 1280
8. D Gloge 'Weakly guiding fibers' Appl Opt 10 (1971) 2252
9. J D Love and A W Snyder 'Optical fiber eigenvalue equation; plane wave derivation' Appl Opt 15 (1976) 2121
10. I P Kaminow, D Marcuse and H M Presby 'Multimode fiber bandwidth: theory and practice' Proc IEEE 68 (1980) 1209
11. J E Midwinter 'Optical Fibers for Transmission' (John Wiley, New York, 1979)
12. K Okamoto and T Okoshi 'Analysis of wave propagation in optical fibers having core with α-power refractive index distribution and uniform cladding' IEEE Trans Microwave Theory Tech MTT-24 (1976) 416
13. M J Adams 'An Introduction to Optical Waveguides' (John Wiley, Chichester, 1981)
14. W A Gambling, H Matsumura and C M Ragdale 'Mode dispersion, material dispersion and profile dispersion in graded index single-mode fibres' IEE J Microwaves, Optics and Acoustics 3 (1979) 239
15. D Su, A A P Boechat and J D C Jones 'Beam Delivery by large core fibres: effect of launching conditions on near field output profile' Appl Opt 31 (1992) 5816
16. A A P Boechat, D Su and J D C Jones 'Effect of launching conditions on output near field profile in large core graded index fibres' Appl Opt 32 (1993) 291
17. B Culshaw 'Fibre Optic Sensors and Signal Processing' (Peter Peregrinus, 1983)
18. D A Jackson and J D C Jones, 'Extrinsic fibre optic sensors for remote measurement: Part One', Optics and Laser Tech 18 (1986) 243
19. D Marcuse 'Loss analysis of single-mode fiber splices' Bell Syst Tech J 56 (1977) 703
20. I P Kaminow 'Polarisation in optical fibers' IEEE J Quantum Electron QE-17 (1981) 15
21. D N Payne, A J Barlow and J Ramskov-Hansen 'Development of low and high birefringence optical fibers' IEEE J Quantum Electron 18 (1982) 477
22. G Meltz, W M Morey and W H Glenn 'Formation of Bragg gratings in optical fiber by a transverse holographic method' Opt Letts 14 (1989) 823
23. W L Smith, J H Bechtel and N Bloembergen 'Dielectric-breakdown threshold and nonlinear-refractive-index measurements with picosecnd laser pulses' Phys Rev B12 (1975) 706
24. S W Allison 'Pulsed laser damage to optical fibers' Appl Opt 24 (1985) 3140
25. D A Jackson and J D C Jones, 'Fibre optic sensors', Optica Acta 12 (1986) 1469
26. J D C Jones, P Akhavan Leilabady and D A Jackson, 'Monomode fibre optic sensors: optical processing schemes for recovery of phase and polarisation state information', Int. Journal of Optical Sensors 1 (1986) 123
27. R A Bergh, G Kotler and H J Shaw 'Single mode fibre optic directional coupler' Electron Letts 16 (1980) 260
28. B S Kawasaki, K O Hill and R G Lamont 'Biconical taper single-mode fibre coupler' Appl Opt 6 (1981) 327
29. M T Feit and J D Fleck 'Propagating beam theory of optical fiber cross-coupling' J Opt Soc Am 71 (1981) 1361

30. O Parriaux, S Gidon and A A Kuznetsov 'Distributed coupling on polished single mode optical fibers' Appl Opt 20 (1981) 2420
31. K P Koo, A B Tveten and A Dandridge 'Passive stabilisation scheme for fiber interferometers using 3 × 3 directional couplers' Appl Phys Letts 41 (1982) 616
32. M Digonnet and H J Shaw 'Wavelength multiplexing in single-mode fiber couplers' Appl Opt 22 (1983) 484
33. D B Mortimore 'Wavelength flattened fused couplers' Electron Letts 21 (1985) 742
34. I Yokohama, K Okamoto and J Noda 'Fibre optic polarising beamsplitter employing birefringent fibre coupler' Electron Letts 21 (1985) 415
35. W Johnstone, G Stewart, B Culshaw and T Hart 'Fibre optic polarisers and polarising couplers' Electron Letts 24 (1988) 866
36. M P Varnham, D N Payne, A J Barlow and E J Tarbox 'Coiled birefringent fibre polarisers' Opt Letts 9 (1984) 306
37. H C Lefevre 'Single mode fibre fractional wave devices and polarisation controllers' Electron Letts 16 (1980) 778
38. G Martini 'Analysis of a single-mode optical fibre piezo-ceramic phase modulator' Opt Quant Electron 19 (1987) 179
39. R K Y Chan, J D C Jones and D A Jackson 'A compact all-optical fibre Doppler difference laser velocimeter' Optica Act 32 (1985) 241
40. J H Cole, B A Danver and J A Bucaro 'Synthetic heterodyne interferometric demodulation' IEEE J Quantum Electron QE18 (1982) 694
41. W P Risk, G S Kino and H J Shaw 'Fiber-optic frequency shifter using an acoustic wave incident at an oblique angle' Opt Letts 11 (1986) 115
42. C N Pannell, R P Tatam, J D C Jones and D .A Jackson, 'Optical fibre frequency shifter using linearly birefringent monomode fibre' Electron Letts 23 (1987) 847
43. B Y Kim, J N Blake, H E Engan and H J Shaw 'All fiber-optic acousto-optic frequency shifter' Opt Letts 11 (1986) 389
44. H E Engan, B Y Kim, J N Blake and H J Shaw 'Propagation and optical interaction of guided acoustic waves in two-mode optical fibres' J Lightwave Tech 6 (1988) 428
45. W J Tomlinson and S K Korotky 'Integrated optics: basic concepts and techniques' in 'Optical Fiber Telecommunications II' (eds. S E Miller and I P Kaminow), (Academic Press, New York, 1988) Chapter 9
46. R C Alferness 'Waveguide electro-optic modulators' IEEE Trans Microwave Theory and Techniques MTT-30 (1982) 1121
47. H Kressel and J K Butler 'Semiconductor Lasers and Heterojunction LEDs' (Academic Press, Orlando, 1978)
48. H C Casey Jr and M B Pannish 'Heterostructure Lasers' (Academic Press, Orlando, 1978)
49. U Koren 'Wavelength division multiplexing light source with integrated quantum well tunable lasers and optical amplifiers' Appl Phys Letts 54 (1989) 2056
50. M Okai 'Corrugation-pitch-modulated MQW-DFB laser with narrow spectral linewidth (170 kHz)' Photonic Technol Letts 2 (1990) 529
51. C A Park 'Single-mode behaviour of a multimode 1.55 μm laser with a fibre grating external cavity' Electron Letts 22 (1986) 1132
52. H M Presby, N Amitay, R Scotti and A F Benner 'Laser-to-fiber coupling with optical fibre up-tapers' J Lightwave Tech 7 (1989) 274
53. P Gill 'Laser interferometry for precision engineering metrology' in 'Optical Methods in Engineering Metrology' ed. D C Williams (Chapman and Hall, London, 1993)
54. A Dandridge and L Goldberg 'Current induced frequency modulation in diode lasers' Electron Letts 18 (1982) 302
55. D Anderson, S R Kidd, P G Sinha, J S Barton and J D C Jones 'Scheme for extending the bandwidth of injection-current-induced laser diode optical frequency modulation' J Mod Opt 38 (1991) 2459
56. J D C Jones, M Corke A D Kersey and D A Jackson 'A miniature solid-state directional laser Doppler velocimeter' Electron Letts 18 (1982) 1081
57. D Anderson and J D C Jones 'Optothermal frequency and power modulation of laser diodes' J Mod Opt 39 (1992) 1837
58. H W Jentink, F F M de Mul, H E Suichies, J G Arnoudse and J Greve 'Small laser Doppler velocimeter based on the self-mixing effect in a diode laser' Appl Opt 27 (1988) 4475

59. S R Forrest 'Optical detectors for lightwave communication' in 'Optical Fiber Telecommunications II' (eds. S E Miller and I P Kaminow), (Academic Press, New York, 1988) Chapter 14
60. R P Webb, R J McIntyre and J Conradi 'Properties of avalanche photodiodes' RCA Rev 35 (1974) 235
61. T P Lee and T Li 'Photodetectors' in 'Optical Fiber Telecommunications' (eds. S E Miller and A G Chnoweth), (Academic Press, New York, 1979) Chapter 18
62. R B Dyott 'The fibre-optic Doppler anemometer' Microwaves, Optics and Acoustics 2 (1978) 13
63. K Kyuma 'Laser Doppler velocimeter with a novel optical fibre probe' Appl Opt 20 (1981) 2424
64. K Weir, W J O Boyle, A W Palmer, K T V Grattan and B T Meggitt 'The measurement of vibration using a fibre probe and a Michelson interferometer' in 'Sensors: Technology, Systems and Applications' (ed. K T V Grattan), (Adam Hilger, Bristol, 1991) 269
65. J Valera, D Harvey and J D C Jones 'Automatic heterodyning in speckle pattern interferometry using laser velocimetry' Opt Eng 31 (1992) 1646
66. A C Lewin, A D Kersey and D A Jackson 'Optical fibre interferometer incorporating an air path' J Phys E Sci Instrum 18 (1985) 604
67. R McBride, J S Barton, W K D Borthwick and J D C Jones 'Fibre Optic interferometry for acoustic emission serving in machine tool wear monitoring' Meas Sci Technol 4 (1993) 1122
68. Duncan P Hand, Tom Carolan, James S Barton and Julian D C Jones 'Extrinsic fibre Michelson interferometric sensor with antiphase outputs and polarization insensitive downlead' Opt Comm 97 (1993) 295
69. R. McBride and J.D.C. Jones 'A passive phase recovery technique for Sagnac interferometers based on controlled loop birefringence' J Mod Opt 39 (1992) 1309
70. D A Jackson, R Priest, A Dandridge and A B Tveten 'Elimination of drift in a single-mode optical-fibre interferometer using a piezo-electrically stretched coiled fibre' Appl Opt 19 (1980) 2926
71. D A Jackson and J D C Jones 'Extrinsic fibre optic sensors for remote measurement: Part Two', Optics and Laser Tech. 18 (1986) 299
72. C N Pannell and J D C Jones 'Acoustically induced phase noise in optical fibre interferometers' Meas Sci Technol (1994)
73. J Knuhtsen, E Olldag and P Buchave 'Fibre optic laser Doppler anemometer with Bragg frequency shift utilising polarisation-preserving single-mode fibre' J Phys E: Sci Instrum 15 (1982) 1188
74. J D C Jones, R K Y Chan and D A Jackson, 'Design of fibre optic systems for Doppler difference laser velocimetry' in 'Laser Anemometry in Fluid Mechanics' (Ladoan, Lisbon, 1984), pp. 69-83.
75 H Muller and D Dopheide 'Direction sensitive laser Doppler anemometer using the frequency shift of two stabilised laser diodes' Proc SPIE 2052 (1993) 323
76. D Harvey, R McBride and J D C Jones `Fibre optic Sagnac interferometer based velocimeter' Meas Sci Technol 3 (1992) 1077
77. C N Pannell, J H Midgley, J D C Jones and D A Jackson 'Fibre optic transit velocimetry using laser diode sources' Electron Letts 24 (1988) 24
78. R D Morgan, D J Anderson, J D C Jones, W J Easson and C A Greated 'Design of fibre optic beam delivery system for particle image velocimetry' SPIE Proc 2052 (1993) 675

NEW TECHNOLOGIES FOR LASER ANEMOMETERS

V. Strunck, H. Wang, H. Többen,
R. Kramer, H. Müller, V. Arndt, D. Dopheide

Physikalisch-Technische Bundesanstalt
Department for Fluid Mechanics
Bundesallee 100
D-38116 Braunschweig

ABSTRACT

An introduction of modern techniques in laser Doppler anemometry is presented. Laser diodes and Nd:YAGs are today's light sources for miniaturized LDAs. Several investigations have been performed in PTB as to measure flow profiles without scanning the sensor, to acquire the instantenous velocity vector using only one source, to sense the direction of flow by passive frequency shifting using Brillouin scattering , or, active by matching two Nd:YAG lasers or stabilized diode lasers. To increase the performance of detection, detailed selection criteria of photodiodes have been developed and used. Advanced techniques concerning the optics, the electro-optics, the electronics and the data processing will be presented.

INTRODUCTION

The application of diode lasers to laser anemometry not only allows the development of miniaturized optical flow sensors by substituting huge gas lasers but also enhances the possibilities of new techniques to realize directional sensitive and multiple velocity component LDA systems. All arrangements in the present paper deal with the fringe type anemometer thus using the heterodyning effect of light from an individual scatterer in the measuring volume formed by the cross section of two laser beams. At least one of the two light waves is scattered and Doppler shifted by the scatterer. The second may be a light wave of fixed frequency serving as reference outlined in the first part, or it may be another Doppler shifted wave such that the difference of the Doppler shifts enables the measurement of velocity described in the following parts.

Standard techniques use the light wave originating from one laser source to obtain one LDA component. In contrast, multiplexing one diode laser for multiple velocity component purposes is described. These optical multiplexed LDA signal components at a single receiver can be demultiplexed by hardware electrically then using several standard acquisition systems or by software when using a single acquisition system.

The benefits from measuring more than one component can be the accession of instantaneous velocity vectors or the sense of the scatterer crossing the different fringe systems, if two components are parallel but slightly shifted from each other in order to

Optical Diagnostics for Flow Processes
Edited by L. Lading *et al.*, Plenum Press, New York, 1994

obtain a quadrature signal for direction discrimination. One component systems including directional sensitivity need a frequency shift between the crossing beams. Moving gratings in form of acousto-optical cells which employ Bragg diffraction have been applicated successfully in frequency shifted systems, an alternative method with the same result is using self induced acoustical waves in fibres, called Brillouin effect.

Frequency shifted fringe systems are also obtainable from two laser sources with the advantage of twice the energy density in the measuring volume. The lasers for such a set-up have to have extremely narrow bandwidths compared to the Doppler information. YAG ringlasers will work for this purpose as outlined.

The problem of the large bandwidth of diode laser most notably has been overcome by a mixing technique and quadrature demodulation.

All frequency shifting techniques in this paper allow a wide range of setting the shift frequency, they are more comfortable and require less or no power compared to using Bragg cells. The two laser source systems also do not need costly beam splitters.

Progress has been done also on the most important receiving part, the detector. Whereas the common GaAlAs diode laser frequency fits well to the sensitivity of Si detectors, the quantum efficiency in the frequency region of 1 μm like that of Nd:YAG lasers is only 5%. Recently selected as well as newly available detectors allow a good performance without increasing the power of the sources. The efforts having been done to the selection of better photo receivers conclude the overview on activities in PTB.

For years Laser Doppler anemometry has been one of the most popular diagnostic tool in fluid research and a variety of systems exists. This paper promotes the usage of modern semiconductor and optical devices for state of the art sensors not have been possible up to date.

SPATIAL SENSOR BASED ON REFERENCE SCATTER METHOD

As other methods common LDA techniques rely on measuring the velocity in a small probe volume and then varying its position in the test section. The spatial resolution of such velocity profile is limited in principal by the size of the probe volume (say 1 mm).

The enhanced reference scatter method and its variations improve the spatial resolution by at least one order of magnitude and enables to measure a small scale velocity profile while leaving the sensor fixed.[1]

Such a set-up is made possible by smaller intensity fluctuations and longer coherence lengths of diode lasers compared to gas lasers and by pin diodes that can operate with a higher optical power compared to photo multipliers. Noiseless lasers are necessary because the noise power in the reference beam must be smaller than the power of the Doppler shifted radiation. Traditional techniques overcome this problem by keeping the reference beam power as low as 1% of the illuminating beam, but the presented system, due to better laser diodes, allows to use equal power in both crossing beams. Thus, the set-up in figure 1 looks equivalent to a forward scatter dual beam set-up, but instead the detector PD1 is located inside beam 1. The measuring volume is not oriented along the optical axis of both beams but along the axis of the illuminating beam 2.

Introducing a second detector PD2 in the previous illuminating beam 2 and calling it now reference beam, this dual reference scatter LDA recognizes not only two Doppler signals but also the time of flight of a scatterer from one beam to the next. Note, there are several ways to measure time of flight, first the Doppler signals (bursts) and second the extinction signals from the beams. Also a one detector configuration is possible. The travelling time depends on the speed of the scatterer and on the distance of the two beams crossed.

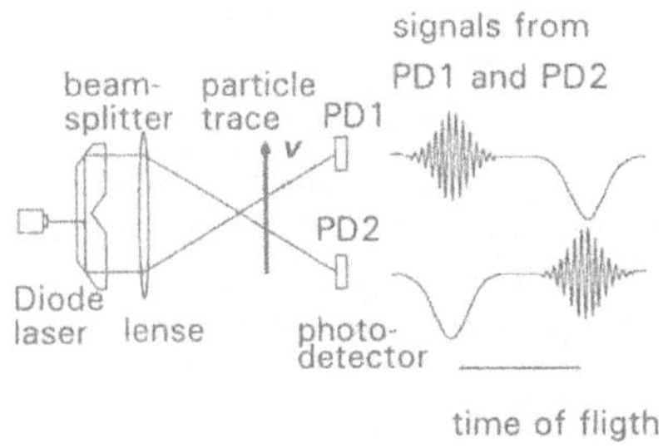

Figure 1. Set-up of a crossbeam reference scatter LDA including scatter signals received at PD1 and PD2.

Time lag thus is a measure of the on-axis distance from the cross point of both beams. If a scatterer traverses the cross point, both detector signals are identical. Signal pairs outside the cross point exhibit a time lag not in the phase of the Doppler frequency but in the envelopes of their signals. Speaking in terms of the fringe model, both (virtual) fringe systems coincide, but the geometrical detection volumes do not. This gives the opportunity to measure the time lag of Doppler signals either using the envelopes (Hilbert transform) or cross correlating the modulation. In the latter case the time resolution reduces to the time of flight from one fringe to the next, but the spatial resolution is already increased typically by 10.

Another advantage knowing the passage through the measuring volume is the possibility to calibrate the distance of the fringes along the optical axis and to take into account the typical spreading of the fringe system in this direction.

Limitations to extend the geometric detection volumes are given by the 'aperture coherence criterion'.[2] The angle between the scattered radiation and the reference beam has to be smaller than the ratio of wavelength to aperture diameter of the receiving device. To increase the size of the measuring volume, the aperture of the pin diode used is small. Another approach is to increase the distance between the measuring volume and the detector. Both methods show an enlargement of measuring volume exceeding the pure cross section. Figure 2 shows the application of the spatial sensor to a free jet. A complete profile has been measured with a resolution of 25 μm without moving the sensor.

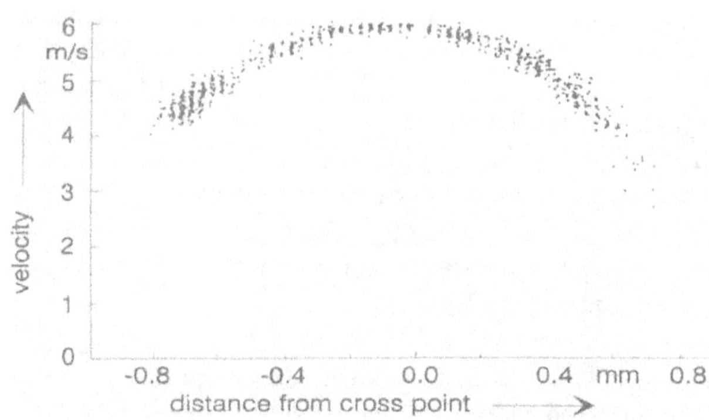

Figure 2. Profile of a free jet from a 3 mm pipe.

Instead of using two components from both beams, one can interchange one component with the (backscattering) component of the standard dual beam set-up. Then restrictions are that only signals out of the cross section itself can be used to measure the passage and the spatial resolution will be half of the set-up before. The method of measuring the passage distance to the cross point of both beams can also be applicated to 3-D flows and extended to multiple component measurements. But to get signal pairs for each component, the area of geometric measuring volume is limited to the cross section of the beams.

MULTIPLE COMPONENT MEASUREMENT USING ONE LASER DIODE

The reduction of components and enhancement of signal-to-noise ratio are main advantages of a high frequency pulsed multiple component LDA.[3,4,5] Two or three high frequency pulsed diode lasers will be required if two or three velocity components are measured with a single acquisition chain. A new technique has been verified, which allows to take advantage of the peak power enhancement by the pulsing technique and to use simply one high frequency pulsed diode laser for measuring all components of flow velocity.[6] This can be accomplished by dividing the output beam of a HF pulsed diode laser into several beams and time multiplexing the beams by different path lengths. Another approach is to use cw-laser and an optical switch, the Mach-Zehnder modulator.

The basics of multiplexed measuring volumes are different fringe systems that are transient for a short time interval, short compared to the duration of the passage of a scatterer. Each of the sequential pulses belongs to a different velocity component to be measured. The pulsed information is converted to current by a single receiving detector with a bandwidth fast enough to reconstruct the pulses. Now two schemes can be applied to sort the different components:

- *coherent sampling* - A transient recorder synchronizes with the pulsing frequency. The recorder has to have a shorter aperture time than the pulse duration in order to sample the peak of the pulse.[3] The different components are sorted after acquisition by address decoding in a computer. To take no regard at special triggering, a pulse pause has to be introduced into the pulse chain to identify the sequence of pulsed information.[5,7]

- *coherent demultiplexing* - A fast electrical switch synchronized with the pulse frequency demultiplexes the different components to different electrical channels. Classical acquisition using counters or burst analyzers is made possible.[6]

As shown in figure 3, the PTB′s two dimensional HF pulsed LDA is modified to consist of only one diode laser .[4,7] A 240 MHz oscillator serves as an internal time base for the whole system. This base frequency is divided into three 80 MHz pulse trains by a three stage shift register. One of these electrical pulse trains is applied to the diode laser driver to trigger the diode laser, giving the first laser pulse train at 80 MHz. By using the beam splitter 2 and mirrors, a second laser pulse train with a delayed phase to the first one is obtained. These two laser pulse trains, which are subsequently split by beam splitters 1 and 3 respectively and focussed into the measuring volume, result in two orthogonal sequentially pulsed fringe systems. The phase difference between the first and the second laser pulse train can be optimized by moving mirrors 1 and 2.

The receiver of the system is a single fast Si avalanche photodiode (APD) connected to a broad band amplifier. The combination of the AlGaAs diode laser with a Si APD has the highest quantum efficiency and allows the best signal to noise ratio (SNR) to be obtained.[8]

The received pulse signal is sampled coherently by a transient recorder with a synchronized sampling frequency at 240 MHz. Based on the coherent sampling technique according to Dopheide et al.,[4,7] demultiplexing of the electrical pulse signal into pulses corresponding to laser pulse trains 1 and 2 and the pulse pause is done by reordering the transient recorder's memory via software program.

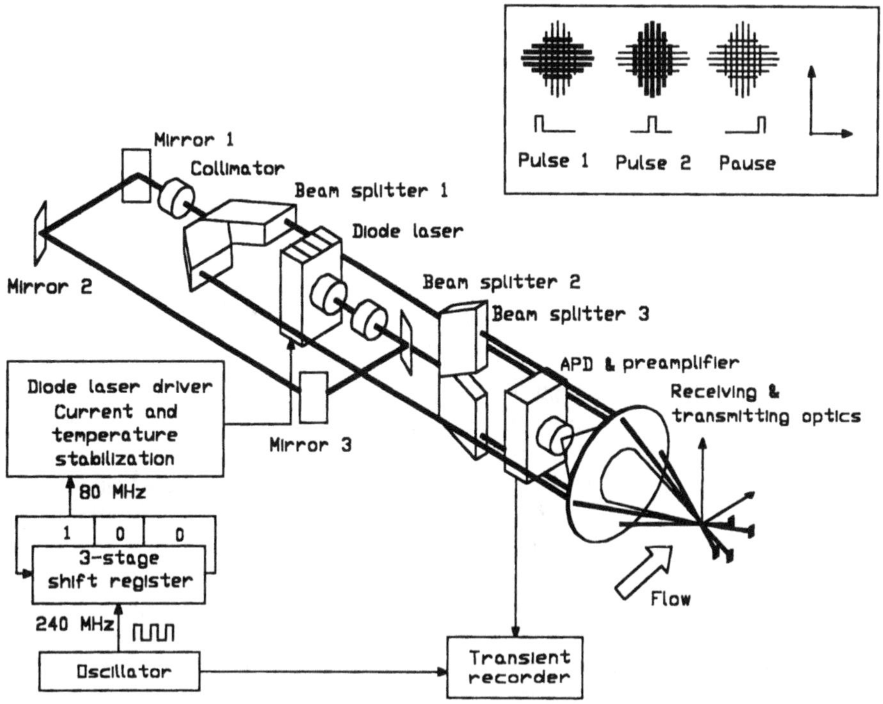

Figure 3. Set-up of the HF pulsed two dimensional LDA with only one diode laser and "coherent sampling". The delay is produced by moving mirrors, so that two 80 MHz pulse trains and one 80 MHz pulse pause are observed in the measuring volume sequentially. The inset in the right hand corner shows the cross sections of successive fringe patterns in the measuring volume.

The output signal of the HF pulsed two dimensional LDA is shown in figure 4.The periodic structures can be seen in the mixture of the two Doppler signals relevant to the two sequential laser pulse trains. By assigning every third sample to one velocity channel, three channels with the Doppler signals can be obtained. Two channels mainly reconstruct the Doppler signals corresponding to two flow velocity components. The third channel is essentially a mixture of both signals with a smaller modulation indicating the crosstalk of the two information channels into the pulse pause. Because crosstalk is small, the point distribution of the pause part is separated in the plot from the signal mixture. Crosstalks effects to some extend can be rejected by software.

Other practical realization of the pulse LDA outlined is to use fibres instead of mirrors to produce different path lengths and to use fibre couplers for beam splitting, and pigtailed diode laser would reduce the effort on adjustments of the optical parts.

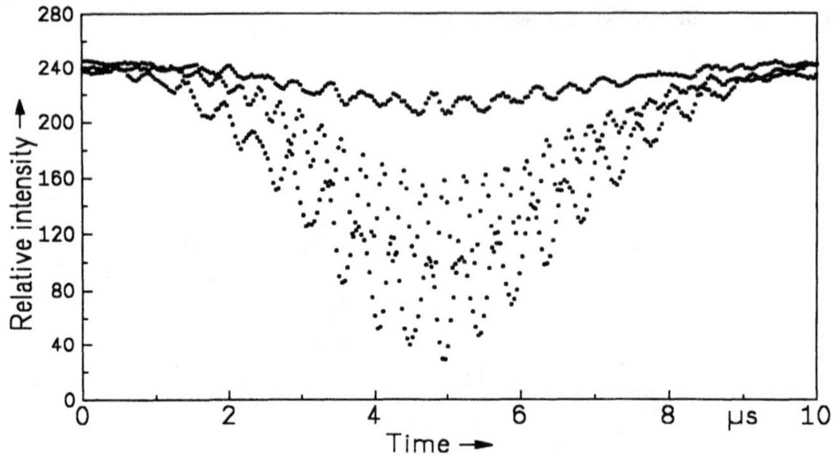

Figure 4. Signal received by the HF pulsed two dimensional LDA using "coherent sampling".

However, LDAs based on pulsed diode lasers suffer from maintaining single mode emissions, because short injection current pulses can produce several modes of light frequency, so the diodes have to be selected carefully.

New ways using integrated optical devices offer possibilities to use a single diode laser without pulsing the diode and to get the same type of pulsed multiple velocity component LDA. Applying an electrical modulation to the centre electrode of a Mach-Zehnder interferometer, the incoming beam is modulated in accordance.[9] Hybrid electro-optical Mach-Zehnder modulators allow to switch the incoming beam between two outputs. The advantages are small losses, switching frequencies up to 5 GHz and single mode radiation in the pulsed beams. Because each output of the Mach-Zehnder multiplexer serves like a single pulsed diode laser, no time multiplexing using different path lengths is necessary.

As shown, integrated optical devices and fibres allow a highly miniaturised design of flow sensors for multiple velocity component LDA. If more than one LDA component is available in a system, directional sensitivity can be obtained by using two LDA components for one velocity component with directional sensitivity. Instead of placing two fringe systems orthogonally, they have to be paralleled. They should not coincide exactly but shifted in direction of the plane vector of fringes by an amount of a quarter of the fringe spacing. Traversing scatterer now produce two Doppler signals with a phase difference indicating the direction of the flow. How to process these signal pairs will be explained later on. In the following, other methods of evaluating the sense of scatterer through the measuring volume will be exhibited.

BRILLOUIN FREQUENCY SHIFT LDA

In laser Doppler anemometry a frequency shift technique is used to determine magnitude and sign of fluid velocities. This technique needs two coherent light beams with slightly different frequencies. For the generation of the frequency shift, additional components are usually employed, such as Bragg cells, rotating phase gratings as well as opto-electronical components for phase modulation techniques. To realize an adjustment insensitive fibre optical LDA system for directional velocity measurements stimulated

Brillouin scattering (SBS) occurring in a single mode fibre can be used as frequency shift mechanism.

SBS can be described as a coupled three wave interaction involving the incident pump wave, the generated acoustic wave, and the scattered Brillouin (Stoke) light wave. In the classical picture the pump creates a pressure wave in the fibre due to electrostriction, and the resultant variation in density changes the optical susceptibility. Thus, the pump induced index grating scatters the pump light in the backward direction through Bragg diffraction[10]. The scattered light, called a Brillouin wave, is downshifted in frequency by an amount equal to the frequency of the acoustic wave.[11] Applying typical communication single mode fibres at 1.3 μm wavelength, the shift frequency is about 13 GHz as shown in figure 5.

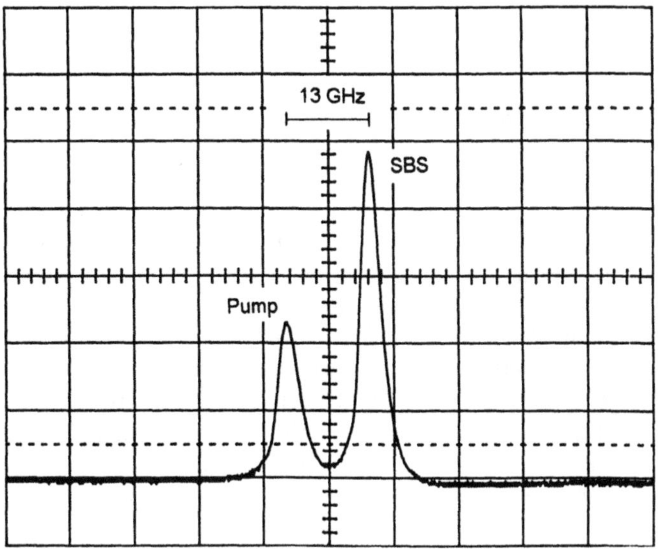

Figure 5. Typical Fabry-Pérot frequency spectrum of pump and Brillouin light.

For a frequency shift in the region of LDA applications, two Brillouin waves must be used. This can be accomplished by using two fibres with a core refractive index different up to 0,01 or by heating one of the fibres. Figure 6 shows the set-up of a Brillouin shift LDA. A pigtailed 1.32 μm Nd:YAG laser with a non-planar ring oscillator design, 5 kHz line widths and 150 mW fibre output power (P_L) was used as pump source. Via fused fibre couplers the pump power was launched into fibre 1 and fibre 2 (P_{L1} and P_{L2}). Again, via the fibre couplers half of the stimulated Brillouin waves (P'_{B1} from P_{B1} and P'_{B2} from P_{B2}) was focussed into the measuring volume by pigtailed gradient index (GRIN) lenses. Both fibres had a length of 6 km. At least 1/10 of the input power can be used in the measuring volume.

The shift frequency in the set-up outlined was at once 257 MHz. First experiences with the system show its applicability. Because the shift frequency is fluctuating about a few MHz with a bandwidth of about 100 kHz, it must be measured synchronously with the Doppler signal. Future set-ups will have a lower shift frequency to simplify electronic processing of the signals and the stability of shift frequency will be increased.

In due course an electronic processing scheme will be denoted that is able to handle easily the signals obtained by Brillouin and similar LDAs. The following applications will deal with systems using two different laser sources for each component.

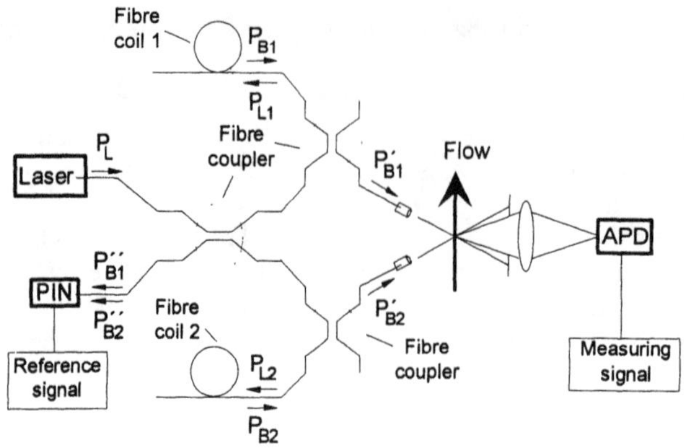

Figure 6. Set-up of Brillouin shift LDA.

FREQUENCY SHIFT LDA USING TWO Nd:YAG RINGLASERS

The new technique uses two tuneable lasers for the two beams of a LDA, which intersect in the measuring volume .[12,13] The difference of the emission frequencies of the lasers is identical to the shift frequency of a shift LDA. The shift frequency can be varied by tuning the emission frequency of one laser or two lasers. The small line width of Nd:YAG ringlasers allows electronic signal processing with conventional techniques like transient recorders. In order to eliminate the unknown shift frequency, i.e. beat frequency, the frequency also must be measured simultaneously with the signal. Only one signal chain is necessary if the beat frequency is stabilized actively by phase locked loop techniques. At a wavelength of 1064 µm an output power 300 mW for both lasers is typically. Due to the high output power the system is well suited for high velocity flows or small scatter particle detection when SNR gets poor.

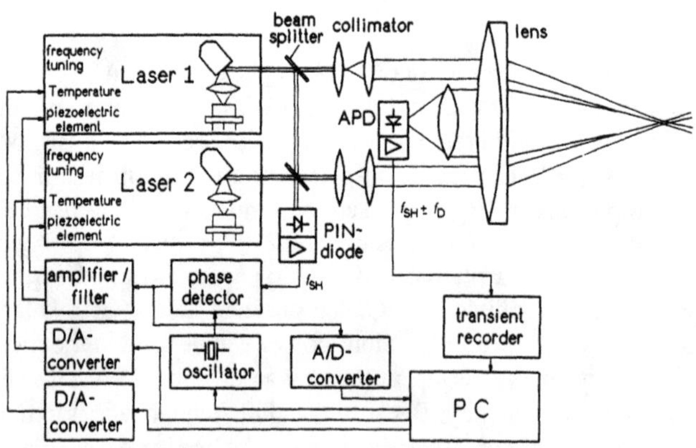

Figure 7. Two Nd:YAG LDA with PLL stabilized shift frequency.

Figure 7 illustrates the set-up and figure 8 measured signals. The output beams are expanded with two collimators to achieve a beam diameter of roughly 100 microns in the measuring volume. The distance of the beams is 65 mm and the focal length of the front lens amount to 310 mm. The light scattered by scatter particles in the measuring volume was detected in backward direction by a Si avalanche photodiode (APD) with IR enhanced characteristic which has a high sensitivity at the wavelength of 1064 nm. With two beamsplitters small fractions of both laser beams are superposed at a PIN diode to detect the beat frequency of the lasers.[14]

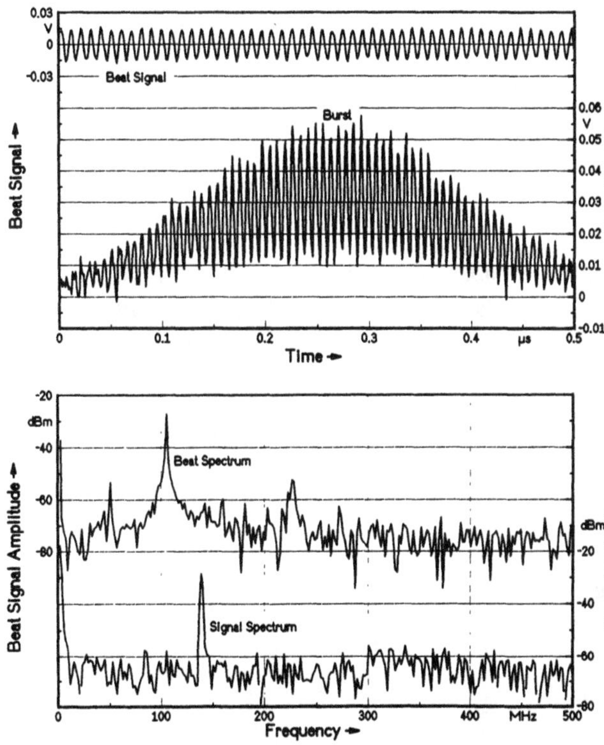

Figure 8. Beat signal and burst signal in the time domain measured in an air jet. The particle traverses in opposite sense to the fringe movement. Frequency distributions of the upper signals show the sense of fringe movement and the quality of the system.

In order to eliminate a beat frequency drift on the evaluation of the Doppler frequency a two channel transient recorder can be used. The transient recorder is triggered by burst signals from the receiving APD and digitizes the burst signal and the beat signal simultaneously. Using FFT procedures for both signals it is possible to calculate the signed Doppler frequency exactly. The difference of the actual beat frequency to the desired shift frequency is evaluated to generate an error signal that is fed back by D/A-converters to the temperature controllers of the lasers. In figure 7 the active piezoelectric control with a local oscillator and a phase locked loop is used instead. Then only one signal channel is necessary.

The shift frequency has been set up to 110 MHz and the fringe distance is about 6 μm. In a nozzle jet a single scatter particle signal and the beat signal have been measured and plotted in figure 8. The velocity of the scatter particle was 200 m/s and the flow direction was contrary to the movement of the fringes (-600 m/s) to simulate a high speed flow at Mach 2,4. The transient recorder has to be fast for this purpose, because of the short duration (500 ns) of the scatter particle inside the measuring volume. The excellent SNR exceeds the noise floor at more than 30 dB in the plot.

The lasers can be continuously tuned to shift frequencies from 0 Hz to 10 GHz without mode hopping. Shift frequencies that can be adapted to our acquisition system work up to 800 MHz.

It has to be denoted that no special effort has to be done to diminish the influence of the bandwidth of beat frequency, because it is as low as 15 kHz.

Another technique that does not take care on small radiation bandwidths but on subtracting their influence is using two laser diodes with a large radiation bandwidth. This technique has also been shown successfully when adapted to the Brillouin LDA.

FREQUENCY SHIFT LDA WITH TWO STABILIZED DIODE LASER

The new technique doubles the intensity in the measuring volume and saves components for beam splitting and frequency shifting by applying two frequency stabilized monomode diode lasers. In this case the frequency shift is given directly by the optical frequency difference of the two laser diode beams which are focussed into the measuring volume outlined in figure 9.[14]

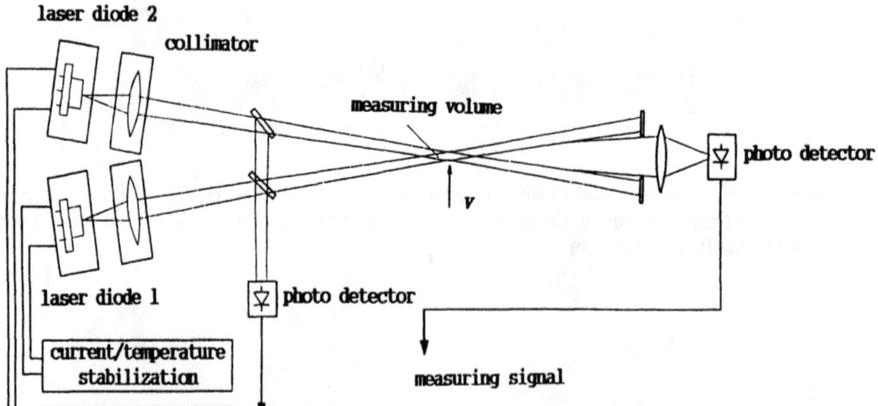

Figure 9. Principle of a frequency shift LDA set-up using of two stabilized monomode lasers.

In opposite to conventional techniques the stability of the resulting shift frequency will depend on the stabilization of the emission frequencies and the line width will be determined by the emission line widths of the laser diodes. Considering line widths of about 20 MHz up to 50 MHz by applying conventional monomode diode lasers, the line widths of the resulting shift frequency as well as the occurring shift frequency fluctuations caused by drift effects will be in the 100 MHz range.

The generation of a frequency shift by using the optical frequency difference of two stabilized monomode laser diodes in principle allows the realization of simple directional LDA set-ups (see figure 9). Therefore, two laser diodes with almost equal emission frequencies are required.

The selection of appropriate monomode laser diodes requires the measurement of their spectral characteristics depending on the current and the temperature of the laser diodes.

In order to eliminate the shift frequency fluctuations it is necessary to generate a reference signal. By interfering a fractional part of both diode laser beams the resulting beat frequency ω_{sh} is disturbed by the same frequency fluctuations $\Delta\omega$ as the burst frequency $\omega_D+\omega_{sh}$. Mixing the measuring signal with the reference signal eliminates the shift frequency and all frequency and bandwidth disturbances caused by instabilities of the laser diodes.[15] Using Euler's formula, it is

$$\left(e^{j(\omega_{sh}+\Delta\omega)t} \mp e^{-j(\omega_{sh}+\Delta\omega)t}\right)\cdot\left(e^{j(\omega_D+\omega_{sh}+\Delta\omega)t} + e^{-j(\omega_D+\omega_{sh}+\Delta\omega)t}\right) =$$

$$= \left(e^{j\omega_D t} \mp e^{-j\omega_D t}\right) + \left(e^{j(\omega_D+2\omega_{sh}+2\Delta\omega)t} \mp e^{-j(\omega_D+2\omega_{sh}+2\Delta\omega)t}\right)$$

whereas the second term on the right hand side can be neglected after low pass filtering. The minus sign on the leftt hand side can be realized by phase shifting the reference signal at 90° and using a second mixer unit. We can further write the down mixed components as $\sin(\omega_D\, t)$ and $\cos(\omega_D\cdot t)$. When the Doppler shift frequency changes sign, corresponding to a change of sense of the flow direction, only the component $\sin(\omega_D\cdot t)$ changes sign. Now comparison of the sign of phase difference of the obtained signal pair allows to detect the direction of the scatter particle through the measuring volume, and shift frequencies and disturbances have been rejected. A typical signal pair is shown in figure 10.

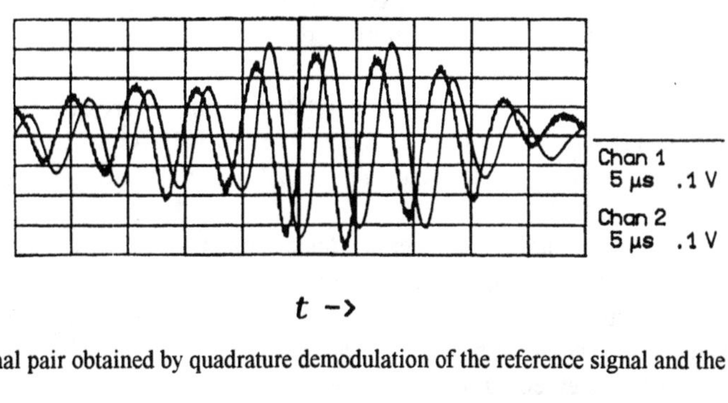

Figure 10. Signal pair obtained by quadrature demodulation of the reference signal and the receiver signal.

There are different ways to process the signal pair:
Rewriting the terms $(e^{j\omega_D t} \mp e^{-j\omega_D t})$ as $(e^{j\omega_D t} + e^{-j\omega_D t}) + j(e^{j\omega_D t} - e^{-j\omega_D t})$ allows to calculate a phase angle $\varphi(t) = \omega_D t$ and the angle velocity $\dot{\varphi}(t) = \omega_D$. The velocity v of the scatterer then can be calculated via $v(t) = \frac{\omega_D(t) \cdot d}{2\pi}$ when taking d as fringe distance. Note the possibility to measure the velocity and the sign as a function of time. A data reduction system can be specially designed for this operation, or a two channel transient recorder and specialized software using the proposed algoritm can do this task.

A second opportunity is to mix the signal pair to desired frequencies again. This is realized by a local oscillator at frequency ω_B, of course with a second 90° shifted output. Adding both frequency mixer outputs

$$(e^{j\omega_D t} + e^{-j\omega_D t}) \cdot (e^{j\omega_B t} + e^{-j\omega_B t}) = (e^{j(\omega_B+\omega_D)t} + e^{-j(\omega_B+\omega_D)t}) + (e^{j(\omega_B-\omega_D)t} + e^{-j(\omega_B-\omega_D)t})$$

and

$$(e^{j\omega_D t} - e^{-j\omega_D t}) \cdot (e^{j\omega_B t} - e^{-j\omega_B t}) = (e^{j(\omega_B+\omega_D)t} + e^{-j(\omega_B+\omega_D)t}) - (e^{j(\omega_B-\omega_D)t} + e^{-j(\omega_B-\omega_D)t})$$

cancels difference terms and gives the desired frequency $\omega_B + \omega_D$. This frequency corresponds to classical Bragg shifted frequency of standard LDAs and is acquirable with one transient recorder channel or counter type processor.

It is to denote that these signal pair procedures are feasible for all direction sensitive sensors creating a quadrature modulation signal.

A variety of new developments of miniaturized LDA systems have been outlined using GaAlAs diode lasers, Nd:YAG lasers and Si photodiodes. Special attention should now be given to the receiving stage of these systems.

SELECTION OF SUITABLE PHOTODETECTORS FOR LDA

Advantages of diode lasers over gas lasers are not only their small size, low energy consumption and high reliability but also from the physical point of view, the fact that the spectral sensitivity of Si photodiodes shows a maximum at a wavelength of about 830 nm which is the emission wavelength at which GaAlAs laser diodes emit maximum power. In this region, the quantum efficiency of Si photodiodes can attain 90% and maintain a very low self noise.

In contrast, Nd:YAG lasers emit at wavelengths of 1 μm, where quantum efficiency of regular Si photodiodes is reduced to only 5 - 10 %. Ge and InGaAs achieve their maximum responsivity at higher wavelengths. In figure 11 typical materials and their responsivity are shown.

For the selection of suitable receivers, comparison measurements have been done. Several commercially available photodetectors were tested in both Nd:YAG laser and diode laser Doppler anemometer (LDA), and the attainable SNR and its dependence on laser power in the LDA probe volume was recorded in a turbulent free jet.[16,17]

Preamplifiers are implemented in some detectors (EG&G's C 30919, C 30986), others have been optimized with commercially available amplifiers to compact receiving modules in a way described by Dopheide and Durst[18] and Dopheide et al. (1988).[8]

Various receiver models have been analyzed. Typical SNR curves of photodiodes start with an approximately linear rise as presented by Durst and Heiber[19] and Dopheide and Durst.[20] SNR is decreased at high laser power due to overload phenomena in the photodiode or the amplifier. A simple sensitivity comparison is given by the position of the slopes.

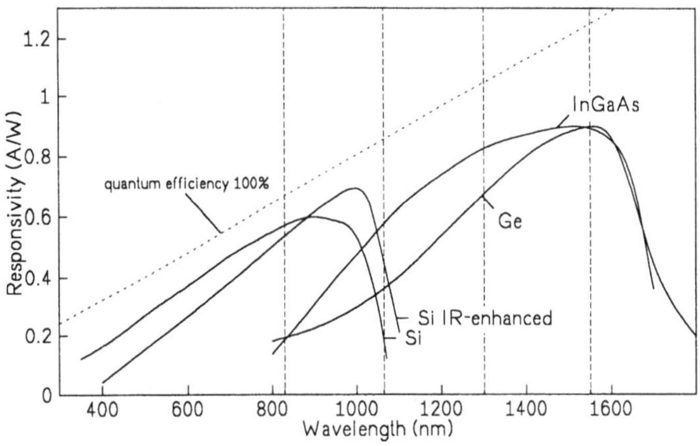

Figure 11. Responsivity of available semiconductor materials for photodetectors (see EG&G catalogues). The dashed line indicates the maximum quantum efficiency of 100%.

Figure 12 shows a selection of the best detectors of different types of material. The minimum power necessary for detection, the maximum attainable SNR and the efficiency can be read from the diagram.

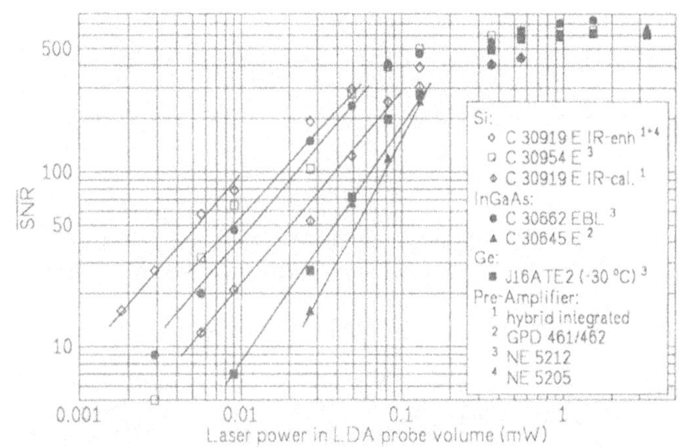

Figure 12. Attainable averaged SNR of Doppler signals as a function of applied laser power in LDA probe volume in a forward scattering Nd:YAG laser LDA ($\lambda = 1064$ nm, $v \sim 90$ m/s, $f_D \sim 11$ MHz, bandpass filtering of signals: 5-20 MHz). Only the results of the best receiving modules of each semiconductor family are taken into account.

- Si-IR-enhanced types (C 30919 E IR enh., C 30954 E) provide the best results in competition. There is an advantage to regular Si APD at the low laser power.
- Sensitive InGaAs receivers are also advantageous. Because of their lower gain compared to Si APDs, a low noise preamplifier should be used with a spectral noise current density comparable to the photodiode's noise current density.
- The new Ge-APDs do not achieve as good efficiencies as Si or InGaAs receivers do, even when cooling down to -30 °C for reduction of dark current noise.
- Normal Si-APDs offer a low sensitivity at 1064 nm due to a small quantum efficiency.

Commercially available InGaAs-APDs often have small chip areas and, therefore, high cut-off frequencies compared to Si-APDs. Hence they are suitable especially for high signal frequencies like in LDA frequency shift systems. But the spreading in quality is quite large.

CONCLUSION

Considerable improvements in the LDA system performance have been evidenced in PTB. In particular a set of new dual beam systems with different properties have been developed and tested succesfully. Special attention was paid to miniaturization for having smaller sensors, improvements in SNRs, more velocity components, directional sensivity and appropiate detectors at less expense. Mechanical alignment problems have been simplified by physics, optical devices like fibres, or electrical devices like mixers.

All of the LDA configurations currently in operation as well as considered in future designs are supposed to perform multiple velocity component measurements in turbulent flows. The different techniques allow to

- acquire spatial velocity information,
- use a pulsed diode for multiple velocity component measurements,
- profit from Brillouin waves for directional sensivity,
- lock (two) Nd:Yag lasers for directional velocity information,
- use two stabilized diode laser for frequency shift applications.

Mixing the applicated techniques will create additional designs, but the three aspects of

- acquiring more (spatial) velocity information,
- multiple velocity component measurements, and
- directional sensivity

will remain. Investigation in direction of the topics exhibited will be the subjects of future developments.

REFERENCES

1. V. Strunck, E.G. Grosche, and D. Dopheide, New laser Doppler sensors for spatial velocity information, in: "Proc. 5th Int. Congress on Instrumentation. in Aerospace Facilities," IEEE Pub. 93CH3199-7, 36:1-5, St. Louis, France (1993).
2. L.E. Drain, Coherent and noncoherent methods in Doppler optical beat velocity measurement, *J. Phys. D: Appl. Phys.* 42:837 (1972).
3. D. Dopheide, H.J. Pfeifer, G. Taux, and M. Faber, Multi-component laser Doppler anemometers with high-frequency pulsed diode lasers, in: "Proc. of Intern. Conference on Fluid Dynamic Measurement and its Applications, (A-1)," Beijing, China (1989).
4. D. Dopheide, V. Strunck, and H.J. Pfeifer, Miniaturized multi-component laser Doppler anemometers using high-frequency pulsed diode lasers and new electronic signal acquisition systems, *Experiments in Fluids* 9: 309-316 (1990).

5. V. Strunck, D. Dopheide, and M. Rinker, Pulsed laser Doppler anemometer for multicomponent applications, in: "Sixth Int. Symp. on Appl. of Lasertechniques to Fluid Mechanics, (11.2)," Lisbon, Portugal (1992).
6. H. Wang, D. Dopheide, H. Müller, and V. Strunck, Optimized signal separation in a high-frequency pulsed diode laser Doppler anemometer for multi-component measurements. in: "Tagungsband 2. Workshop Lasermethoden in der Strömungsmeßtechnik, Aktueller Stand und neue Anwendungen, (23)," Shaker Verlag, Aachen, Germany (1993).
7. D. Dopheide, M. Rinker, and V. Strunck, High-frequency pulsed laser diode application in multi-component laser Doppler anemometry. *Optics and Lasers in Engineering* 18: 135-145 (1993).
8. D. Dopheide, M. Faber, G. Reim, and G. Taux, Laser and avalanche diodes for velocity measurement by laser Doppler anemometry, *Experiments in Fluids* 6: 289-297 (1988).
9. M.A. Powell, An ally for high speed and the long haul: Integrated optical modulator applications in communication systems, *in: Photonics Spectra, Special CLEO issue,* 102-108 (1993).
10. G.P. Agrawal."Stimulated Brillouin Scattering` in Nonlinear Fiber Optics," Academic Press, Boston-London-Toronto (1989).
11. D. Cotter, Stimulated Brillouin scattering in monomode optical fiber, *Journal of Optical Communications* Vol. 4, 1:10-19 (1983).
12. R. Kramer, H. Müller, V. Arndt, D. Dopheide, and V. Strunck, Neue Aspekte des Einsatzes von diodengepumpten Nd:YAG-Lasern und selektierten Photodioden in der Laser-Doppler-Anemometrie. *Laser Magazin* 5-6: 16-24 (1993).
13. H. Müller, R. Kramer, and D. Dopheide, Neue Verfahren zur Erzeugung einer Frequenzshift in der Doppler-Anemometrie mit abstimmbaren Festkörperlasern, *in: PTB Mitteilungen* 103, 1:3 - 17 (1993).
14. H. Müller, and D. Dopheide, Direction sensitive Laser Doppler Velocimeter using the optical frequency shift of two stabilized laser diodes, in: "Laser Anemometry Advances and Applications," SPIE Vol. 2052: 323-330, (1993).
15. J. Czarske, F. Hock, and H. Müller, Quadrature demodulation - a new LDA-burstsignal frequency estimator, in: "Proc. 5th International Conference of Laser Anemometry - Advances and Applications, (79-86)," Konigshof, Veldhoven, Netherlands (1993).
16. V. Arndt; H. Müller; D. Dopheide, Vergleichsmessungen zur Selektion von Photoempfängern für die Optimierung von Nd:YAG-LDA-Systemen, in: "Tagungsband 2. Workshop Lasermethoden in der Strömungsmeßtechnik, Aktueller Stand und neue Anwendungen, (41)," Shaker Verlag, Aachen (1993).
17. D. Dopheide, and M. Faber, 1990: Einsatz von Diodenlasern und Photodioden in der Strömungsmeßtechnik, in: "Lasermethoden in der Strömungsmeßtechnik, 13-69)," B. Ruck., AT-Fachverlag, Stuttgart (1990)
18. D. Dopheide, and F. Durst, Einfluß der Axialmoden von CN-Lasern auf Laser-Doppler-Messungen bei hohen Frequenzen. Teil 2: Experimental verification of the extended theory of laser-Doppler-anemometry, *Technisches Messen* 49: 69-77 (1982).
19. F. Durst, and K.F. Heiber, Signal-Rausch-Verhältnisse von Laser-Doppler-Systemen, *Optica Acta* 24: 43-67 (1977).
20. D. Dopheide, and F. Durst, High frequency laser Doppler measurements using multiaxial mode lasers. *Appl. Opt*, 20, pp 1557 (1981).

PHASE DOPPLER ANEMOMETRY AND ITS APPLICATION TO LIQUID FUEL SPRAY COMBUSTION

Graham Wigley

Flow Measurement Consulting Service
A-8020 Graz
Austria

INTRODUCTION

In conventional laser Doppler anemometry (LDA) knowledge of the exact size of the seeding particles or droplets is non-essential. This is provided that all the seeds are small enough to follow the highest turbulent fluctuations occurring in the flow under investigation yet, large enough to provide adequate scattered light intensities while maintaining good signal modulation. However, many flows, particularly of importance in industrial processes, are classed as two- or multi-phase flows. These contain, mix or transport particulates, bubbles or droplets whose size and density cause significant differences to be observed in their behaviour from that of the continuous phase. Small light particulates would follow the small scale structures in the continuous phase while large heavy particulates may not even be representative of the large scale flow structures.

If LDA is to be applied unambiguously to two or multi-phase flows some form of phase and/or particulate size discrimination must be applied simultaneously with the instantaneous velocity measurement. However, the particle size, shape and refractive index determine the scattered light intensity and its angular distribution so the required information is intrinsic to the LDA signal and only has to be decoded.

This Chapter presents discussion on light scattering by particles and droplets and is followed by a description of the application of an optical dropsizing technique to liquid fuel spray combustion. The theoretical part covers the background to the physical properties of how light waves or rays interact with particles and droplets and evaluates what characteristics of the scattered light can be used to determine a particle or droplet's velocity and size. The practical part is concerned with how the phase Doppler (PDA) anemometry technique is applied and, in particular, considers a study of transient high pressure and temperature liquid fuel sprays. Specific instrumentation and optical scattering geometries are presented and their performance evaluated. Comments are also made on data reduction and presentation.

Optical Diagnostics for Flow Processes
Edited by L. Lading *et al.*, Plenum Press, New York, 1994

LIGHT SCATTERING ANALYSES

In the following general analysis the term, scatterer will be used to indicate particle or droplet. The problem to be solved is how to relate the scatterer's properties of shape, size and refractive index to the angular distribution of the scattered light intensity in terms of, the light wavelength, amplitude, phase and polarization. The appropriate light scattering analysis is determined by the size of the scatterer, d, relative to the light wavelength, λ. The controlling parameter is the size parameter, α, where

$$\alpha = \pi d / \lambda . \tag{1}$$

Rayleigh scattering occurs when $\alpha << 1$ i.e. scatterers smaller than 0.05 μm for a visible wavelength of 0.5 μm. The angular distribution of parallel polarized scattered light is a dipole and has only two lobes with maxima in the forward scatter direction, $\theta = 0$ °, and one in the back scatter direction, $\theta = 180°$. For perpendicular polarized light the distribution is spherical. The scattered light levels are low and increase approximately as d^6. In an LDA/PDA experiment these scatterers would exhibit a very high flow fidelity but would be barely detectable.

Lorenz-Mie theory describes the light scattering processes taking place for $\alpha \sim 1$ i.e. for the scattering size class 0.05 to 5 μm. However, the latter does not represent an upper limit for the Lorenz-Mie theory but rather the lower limit for which the much simpler, and easier to apply, geometric optics theory can be used reliably. Lorenz-Mie and geometric optics are the only analyses applicable to LDA/PDA applications. They describe a complex angular distribution of the scattered light with a dominant forward scatter lobe and side and backscatter lobes increasing in number proportional with the increase in the size of the scatterer. The total scattered light intensity increases with the square of the size of the scatterer, d^2.

Lorenz-Mie Theory

This describes the solution to Maxwell's wave equation for the electro-magnetic scattering by a sphere. Although most recognition is given to the German, Mie[1] it was the independent work of the Dane, Lorenz[2,3] and who derived the practical formulae in common use today. The solution involves the expansion of the electro-magnetic waves inside and outside the sphere in terms of spherical harmonics with matching boundary conditions at the surface of the sphere. Good general descriptions of the mathematics can be found in van de Hulst[4] and Kerker[5] and, specifically for LDA/PDA applications, by Negus and Drain[6].

At a large distance, r, from the sphere, the farfield solution of the electric field, E, of the wave scattered at an angle θ is represented by

$$E = \frac{-i}{kr} E_o \exp[i(wt - kr)]\, S(\theta) \tag{2}$$

where E_o and ω are the amplitude and angular frequency of the incident wave, k is the wave number, $\frac{2\pi}{\lambda}$, and $S(\theta)$ is an Amplitude Function. For an incident light wave, plane polarized with the electric vector perpendicular to the plane of scattering then,

$$S(q) = S_1(q) = \sum_{n=1}^{n=\infty} \frac{2n+1}{n(n+1)} \left[a_n \Pi_n(\cos q) + b_n t_n(\cos q) \right] \tag{3}$$

while for the incident light wave polarized parallel to the plane of scattering

$$S(q) = S_2(q) = \sum_{n=1}^{n=\infty} \frac{2n+1}{n(n+1)} \left[b_n \Pi_n(\cos q) + a_n t_n(\cos q) \right] \tag{4}$$

The terms a_n and b_n are the Scattering Coefficients which are expressed as Ricatti-Bessel functions of order n and with arguments x and y given by

$$x = pm_2 d l_o \text{ and } y = pm_1 d l_o \tag{5}$$

where m_1 and m_2 are the refractive indices of the scatterer and medium respectively and λ_0 the wavelength in vacuum. The terms π_n (cosθ) and τn (cosθ) are the Angular Functions and are expressed as Legendre functions of order n.

Computer programs for calculating the Scattering Coefficients and Angular Functions generally reduce the Ricatti-Bessel functions and Legendre functions to series and recurrence relationships. Particular effort has been put into the development of stable recursion schemes, necessary for the solutions to converge, without the accumulation of numeric round-off errors. This is specifically appropriate to large scatterers since the number of terms, n, required for convergence is equivalent to the size parameter, α.

One of the earliest programs available for calculating the electro-magnetic radiation scattered by a sphere was by Dave[7]. Later programs, with the emphasis on particle sizing or optimized solution algorithms, have been published by Bohren and Huffman[8], Grehan and Gouesbet[9] and Wiscombe[10]. Designers of particle sizing instrumentation do have their own computer programs, albeit based on those just mentioned, but the only company to market a code is Invent, it is called "Streu", german for scatter. On a more practical note these computer programs can be run on a PC but an upper limit of $\alpha < 500$ or $d < 100$ µm has to be accepted.

Geometric Optics Theory

Analogous to the Amplitude Functions derived from the Lorenz-Mie theory Geometric Amplitude Functions can also be derived from the classical laws governing the amplitude and direction of light rays with allowances being made for the state of the incident polarization and emerging phase. The geometric approximation to describing the scattered light pattern from a large dielectric scatterer assumes a superposition of the three principal light rays emerging from the scatterer after reflection and first and second refraction as well as the higher order refracted rays from multiple internal reflections. Van de Hulst[4] showed that for a large dielectric spherical scatterer with a refractive index sufficiently different from the surrounding medium, i.e. $2\alpha(m-1) \gg 1$, that in the asymptotic limit of $\alpha \rightarrow \infty$, or $\alpha \gg \lambda$, the geometric and Lorenz-Mie Amplitude Functions were equal.

A scattering diagram in geometric optics is shown in Figure 1. All the rays lie in a single plane, the scattering plane, which includes the sphere's centre. The directional dependence of the scattered rays is given by the following three laws and is independent of the sphere's diameter.

Law of Reflection:	Angles of incidence, θ_i, and reflection are equal	
Law of Refraction:	$m_1 \sin\theta_i = m_2 \sin\theta_r$	(6)
Emerging ray direction:	$\theta_p = 2(P\tau_r - \tau_i)$	(7)

The angles of incidence, reflection and refraction are with respect to the normal to the surface, but for the emerging ray direction they are defined with respect to the tangent to the surface. The refractive index, m, is complex and is defined by

$$m = n - ik \tag{8}$$

where n is the real part and k, the imaginary part, which determines the light absorption within the scatterer. In equation (7), P is the interfacial integer, this equation is essential to relate the angle of incidence, θ_i, to the scattering angle, θ_p.

The partition of energy at each interface between the reflected and both the external and internal refracted rays is given by the Fresnel coefficients r_1 and r_2. These are derived from the wave equation but under highly simplified conditions. The direction of scatter can be

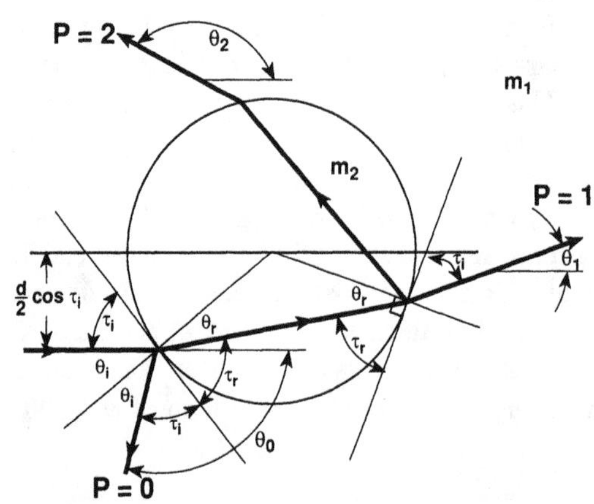

Figure 1. Ray trace of geometric scattering

neglected since it is given by geometric optics and the incident ray is only considered as two polarized components, parallel and perpendicular to the scattering plane. Conservation of energy at the interface for the incident, reflected and refracted components then leads to a significant simplification of Maxwell's equations. The amplitudes of the polarized components for the reflected ray are given below. The subscripts 1 and 2 refer to the incident ray polarized perpendicular and parallel to the scattering plane.

$$r_1 = -\frac{\sin(\theta_i - \theta_r)}{\sin(\theta_i + \theta_r)} = \frac{m_1 \cos\theta_i - m_2 \cos\theta_r}{m_1 \cos\theta_i + m_2 \cos\theta_r} \tag{9}$$

$$r_2 = \frac{\tan(\theta_i - \theta_r)}{\sin(\theta_i + \theta_r)} = \frac{m_2 \cos\theta_i - m_1 \cos\theta_r}{m_2 \cos\theta_i + m_1 \cos\theta_r} \tag{10}$$

The amplitudes of the polarized components for the refracted ray are simply, t_1=1-r_1 and t_2 = 1-r_2 i.e.

$$t_1 = \frac{2m_1 \cos qi}{m_1 \cos qi + m_2 \cos qr} \tag{11}$$

and

$$t_2 = \frac{2m_1 \cos qi}{m_2 \cos qi + m_1 \cos qr} \tag{12}$$

From these relatively simple trigonometric relationships it is easy to see that certain special properties can be identified. For $m_2 > m_1$, and ($\theta i+\theta r$) = $\pi/2$, Snell's law, equation (6), reduces to

$$\tan qi = \frac{m_2}{m_1} \tag{13}$$

and the Fresnel coefficient, r_2, equation (10), for the light component polarized parallel to the scattering plane, reduces to zero. This angle of incidence is Brewster's, or the polarizing, angle, θ_B, and indicates that the reflected light can have no component parallel to the scattering plane, i.e. the plane of incidence, for angles of incidence greater than θ_B. When $m_2 < m_1$, i.e. light travelling from a more dense to a less dense medium then Snell's law, equation (6) gives the upper bound for which a refracted ray, P=1, can exit since

$$\sin qr = \frac{m_1}{m_2} \sin qi > 1 \tag{14}$$

Above this critical angle, θ_c, total internal reflection of the light occurs. These limiting cases are applicable to any surface geometry unlike the Rainbow angle,θ_{RB}, which is specifically for spherical droplets where the rays after total internal reflection at P=1 exit the droplet at P=2 to produce a peak in the external intensity distribution at the angle

$$\cos^2 qr = \left[\frac{m^2-1}{3}\right] \tag{15}$$

In the traditional approach of geometric optics the Amplitude Functions are defined as

$$S_j^P(a,m,qi) = \sqrt{i_j^P(a,m,qi)}\, \exp\left[i\, s_p(a,m,qi)\right] \tag{16}$$

where the subscript j=1, 2 for the perpendicular and parallel polarization components respectively. The phase dependence, σ_p (α, m, θ_i), is given by van de Hulst[4] as

$$\frac{\pi}{2} + 2\alpha \cos\theta i \tag{17}$$

for the reflected ray and

$$\frac{3\pi}{2} + 2\alpha(\cos\theta i - m\cos\theta r) \tag{18}$$

for the refracted ray.

Glantschnig and Chen[11] discuss how practical closed form Amplitude Functions can be derived by recasting these as functions of the scattering angle, θ, and noting that over 90 % of the incident light is forward scattered and over 99 % of this is contained in the rays emerging from the first two interfaces i.e. those drawn in Figure 1.

The Amplitude Functions for the rays reflected from the first interface are presented as,

$$S_1^{(1)}(\alpha,m,\theta)=\alpha\frac{\sin\frac{\theta}{2}-\sqrt{m^2-\cos^2\frac{\theta}{2}}}{\sin\frac{\theta}{2}+\sqrt{m^2-\cos^2\frac{\theta}{2}}}\mathrm{x}\frac{1}{2}\exp\left[j\left(\frac{\pi}{2}+2\alpha\sin\frac{\theta}{2}\right)\right] \tag{19a}$$

$$S_2^{(1)}(\alpha,m,\theta)=\alpha\frac{m^2\sin\frac{\theta}{2}-\sqrt{m^2-\cos^2\frac{\theta}{2}}}{m^2\sin\frac{\theta}{2}+\sqrt{m^2-\cos^2\frac{\theta}{2}}}\mathrm{x}\frac{1}{2}\exp\left[j\left(\frac{\pi}{2}+2\alpha\sin\frac{\theta}{2}\right)\right] \tag{19b}$$

For the second interface, the Amplitude Functions for the first refracted ray are,

$$S_1^{(2)}(a,m,q)=a\left[1-\left(\frac{1+m^2-2m\cos\frac{q}{2}}{1-m^2}\right)^2\right]x\sqrt{\frac{m^2\sin\frac{q}{2}\left(m\cos\frac{q}{2}-1\right)\left(m-\cos\frac{q}{2}\right)}{2\sin q\left(1+m^2-2m\cos\frac{q}{2}\right)^2}}x\exp\left[j\left(\frac{3p}{2}-2a\sqrt{1+m^2-2m\cos\frac{q}{2}}\right)\right] \tag{20a}$$

$$S_2^{(2)}(a,m,q)=a\left[1-\left(\frac{(1+m^2)\cos\frac{q}{2}-2m}{(m^2-1)\cos\frac{q}{2}}\right)^2\right]x\sqrt{\frac{m^2\sin\frac{q}{2}\left(m\cos\frac{q}{2}-1\right)\left(m-\cos\frac{q}{2}\right)}{2\sin q\left(1+m^2-2m\cos\frac{q}{2}\right)^2}}x\exp\left[j\left(\frac{3p}{2}-2a\sqrt{1+m^2-2m\cos\frac{q}{2}}\right)\right] \tag{20b}$$

There are limitations to the accuracy of these relationships for light rays incident at grazing incidence, θ_i = 90°, and for scattering angles greater than 60°. No simple closed form relationship exists for the Amplitude Functions for light emerging at the third interface, P = 2, i.e. for second order refracted light.

Comparisons of the Mie and geometric optics calculations for the scattered light intensity distributions as functions of the incident polarization and the scattering angle are shown in Figure 2. The incident light beam was plane, monochromatic and polarized while the scatterer was a water droplet, 10μm diameter, and the receiver was a perfect point detector. Other beam characteristics can be found in Table 1. The intensity distributions for: (a) reflection, (b) 1st and (c) 2nd order refraction are shown independently. The Brewster angle, with parallel polarization, and the critical and rainbow angles are readily distinguished at scattering angles of 73.8°, 82.4° and 137.4° respectively. The high frequency fluctuations on the Mie curve are indicative of the large number of scattering lobes which are a result of interference between the individual scattering mechanisms.

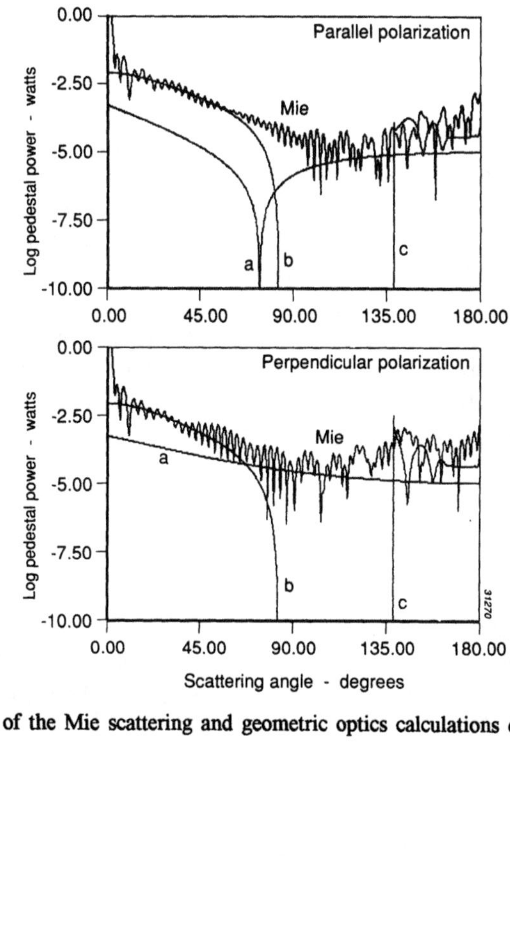

Figure 2. Comparison of the Mie scattering and geometric optics calculations of the total scattered light intensity distribution.

SIZING METHODS

Intensity Methods

Glantschnig and Chen[11] considered a simple light scattering experiment with a light beam of intensity, Io, incident on a scatterer of diameter, d, and collection of the scattered light at an angle, θ, by a detector of finite aperture, Δθ. The scattered light intensity, I, is

$$I(d,m,q) = I_o K(m,q) d^2 \tag{21}$$

where K is a proportionality constant based on the sum of the Amplitude Functions. However, there is a lower limit to the size that can be measured given by

$$\alpha \Delta\theta \geq 4\pi \tag{22}$$

In general, the more simple a measuring instrument is, the more difficult it is to extract accurate information in complex practical problems. Essential to this technique are accurate measurements of the incident and scattered light intensities. However, these are determined by light transmission between the source and scatterer, and scatterer and detector and are therefore very much dependent on the concentration of scatterers. Furthermore, for laser light sources the intensity distribution across the beam is Gaussian, so the trajectory across the beam has to be known. Some of these effects can be reduced by comparing, or ratioing, the light scattered at two different scattering angles. However, in the context of this Chapter, where velocity and size measurements are required, the intensity method is not applicable since the velocity of the scatterer can only be estimated from its time of flight through the incident light beam which, again, highlights the unknown trajectory problem.

Visibility Method

Negus and Drain[6] considered a more complex optical geometry with two incident light beams crossing to create a fringe pattern, as in LDA, and derived an estimate of the scatterer's size from the fluctuation intensity of the scattered light as it traversed the fringe pattern. The fluctuation frequency does, of course, give a very good estimate of the scatterer's velocity so a brief discussion of this sizing method is presented. The scattered intensity is a modification of the equation for one incident beam

$$I = I_o K d^2 \left[1 + \cos 2p \frac{Q}{l} \cdot V \right] \tag{23}$$

where Θ is the angle between the incident beams and V, the visibility is

$$V = \left| 2 \frac{J_1 p \frac{d}{df}}{p \frac{d}{df}} \right| \tag{24}$$

This visibility function is the fluctuation intensity where J_1 is the Bessel function of the first order and d/df is the ratio of the scatterer's diameter to the fringe spacing in the beam crossover. The general form of this function for forward scattered light is shown in Figure 3 together with a typical scattered light signal. The form of this curve obviously limits the accurate sizing range from 0.2 df $\leq$ d $\leq$ 0.8 df. A large size range can only be covered at the expense of a large fringe spacing and therefore low velocity and spatial resolutions.

The visibility function is very sensitive to the use of off-axis apertures and aperture stops and Negus and Drain[6] showed that for a light detector placed at a forward scattering angle of $\theta = 9°$ and with a receiving aperture of $\Delta\theta = 10°$ the sizing range could be dramatically extended from approximately 0.2 df $\leq$ d $\leq$ 6df but with an upper limit of $\Theta \approx 3°$ for the beam crossover angle.

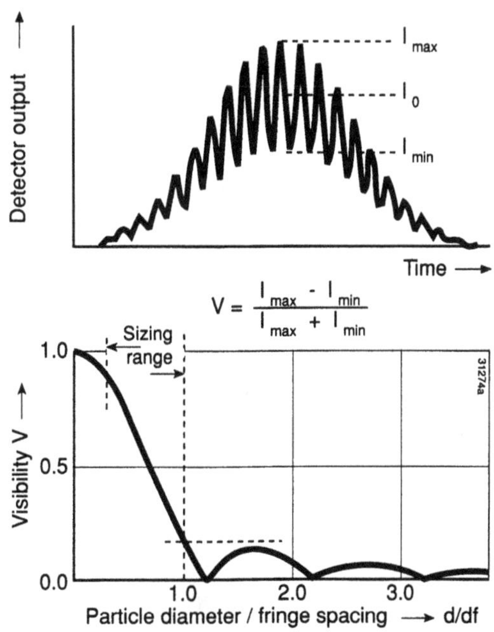

Figure 3. Relationship of fluctuating signal intensity to the visibility function.

It is worth while here to consider the optical geometry of the crossover and the velocity size range and resolution. The fringe spacing, df, is

$$df = \frac{\lambda}{2\sin(\theta/2)} = 12.1\ \mu\text{m} \tag{25}$$

with $\lambda = 0{\cdot}6328\ \mu$m, so a sizing range of 2.4 to 72 μm could be expected. For a reasonable estimate of velocity a minimum of 10 fringes must be considered i.e. a crossover volume diameter, D, of 0.12 mm and, since the crossover volume length is given by, D divided by $\tan\Theta/2$, results in a crossover volume length of 4.62 mm. A scatterer with a velocity of 1ms^{-1} would produce, a fluctuation frequency of only 80 kHz.

Coupled with the relatively low size range, spatial and velocity resolutions this technique also suffers from any degradation in the modulation intensity of the fringes themselves. These can be caused by imperfect optical components, windows and a high concentration of scatterers.

Phase Method

The third optical sizing method to be considered goes a long way to improving the short comings of the previous methods but at the expense of another increase in the optical and analytical complexity. This method again considers the fluctuation intensity in the scattered light caused by a scatterer traversing two crossed laser beams but examines how a phase difference between two scattered signals can be generated that is proportional to the scatterers size. This method is called phase Doppler anemometry, PDA, and is in essence the general form of LDA. Some of the earliest works describing this technique are given by Bachalo and Houser[12], Bauckhage et.al.[13] and Saffman et.al.[14] while more up to date works are given by Naqwi and Durst[15,16].

The term phase, so far, has been used to describe the time relationship of harmonic wave fronts but from now it will also be used to indicate the temporal relationship between two scattered light signals. A simple geometric fringe model can help describe how the phase between two fluctuating signals can be generated and used to determine size. In Figure 4 a transparent droplet is shown at two different locations during its traverse through the fringe pattern generated by two crossed laser beams. The droplet is simply considered as a spherical lens that images the fringe pattern onto a plane in the far field. The ratio of the fringe spacing in the far field to the fringe spacing in the beam crossover is the magnification of the spherical lens which is a direct function of its radius of curvature. To measure the far field fringe spacing two detectors are required. In Figure 4 we see how, as the droplet moves down through the fringe pattern, the far field image also moves down past the detectors, detector 2 would obviously receive its signal earlier than detector 1. In this schematic signal 2 leads signal 1 by three fringes i.e. a phase of $3x2\pi$.

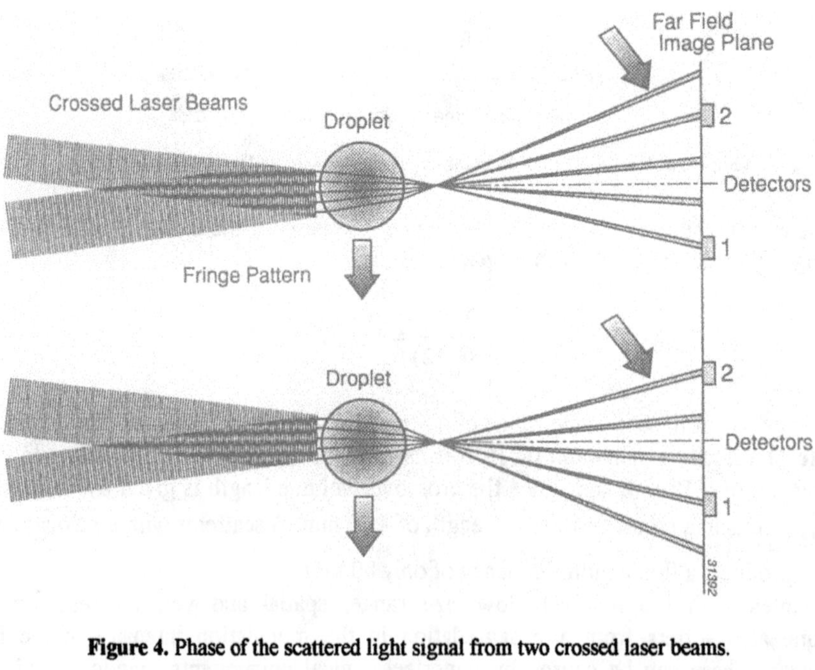

Figure 4. Phase of the scattered light signal from two crossed laser beams.

The fringe pattern is in, and parallel to, the scattering plane while the detectors are perpendicular to it. The far field pattern is seen as being constructed from rays of interface integer P=1 i.e. only from first order refracted light. However, this fringe pattern, in the near forward scattering angle, i.e. $\theta < 20°$, would be complicated due to interference with light diffracted around the droplet. For a simple linear relationship to exist between signal phase differences and dropsize scattering angles have to be chosen where one scattering mechanism is dominant. The region in which diffraction can be expected to exist has just been described and it is relatively simple, by referring to Figure 2 and using the special properties discussed in the geometric optical theory, to identify scattering angles where reflection, 1st and 2nd order refraction are dominant.

The phase shift is calculated from the path length difference existing between the two incident rays as they are reflected from or refracted through the scatterer. In practice these are actually calculated from the path length difference between each incident ray with respect to a hypothetical central reference beam. A straight forward and general approach to deriving the phase shift due to path length differences is given by Sellens[17].Three scattering diagrams are shown in Figure 5, representative of, (a) a liquid spray in a gas, (b), bubbles of gas in a liquid and, (c), solid spherical particles in either medium.

The rays Rn lie in a single plane that includes the sphere centre and the reference rays, Xn. It can be seen that in geometric optics the direction of the scattered rays is independent of the dropsize diameter, d. Determination of the total scattering process must be considered for four paths, i.e. the paths between the two input beams from the transmission optics and the pair of detectors in the receiver optical system.

Sellens[17] derives two different relationships to express the angle of incidence, θ_i, with respect to the scattering angle, θ. For refraction

$$\theta_i = \frac{\pi}{2} - \tan^{-1}\left(\frac{\left|\frac{1}{m} - \cos\frac{\theta}{2}\right|}{Sin\,\frac{\theta}{2}}\right) \tag{26}$$

where $m = \frac{m_2}{m_1}$ i.e. the ratio of the dispersed phase to that of the continuous phase and is valid for both $m < 1$ and $m > 1$ and, for reflection

$$\theta_i = \frac{\pi}{2} - \frac{\theta}{2} \tag{27}$$

The path length differences between the incident ray and the reference ray are given for refraction as,

$$\Delta L = (\Delta R_o + \Delta R_4 - d)m_1 + R_2 m_2 \tag{28}$$

and, for reflection as

$$\Delta L = (\Delta R_o + \Delta R_1 - d)m_1 \tag{29}$$

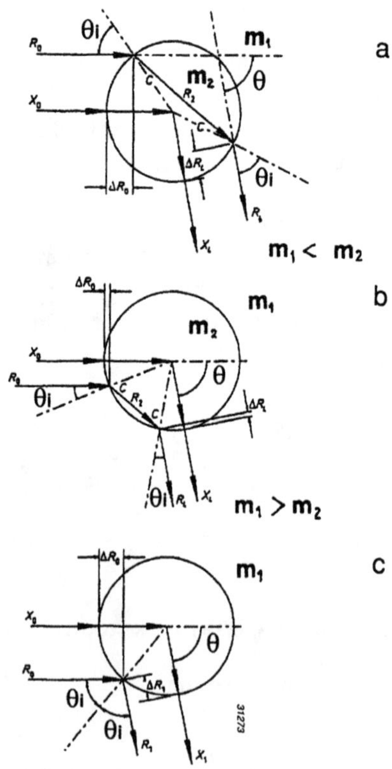

Figure 5. Geometric scattering diagrams for, (a), liquid droplet in a gas, (b), gas bubble in a liquid and, (c), solid particle in a gas or liquid. Reproduced from Sellens [17].

For these to be useful they must be expressed in terms of the light scattering geometry and the diameter, d, of the scatterer. Both path length differences can be reduced to

$$\Delta L = \beta d \tag{30}$$

where the geometric factor, β, for refraction is

$$\beta = m_2\left(1 - \frac{Sin^2\theta i}{m^2}\right)^{1/2} - m_1 \cos\theta i \tag{31}$$

and for reflection

$$\beta = -m_1 \cos\left(\frac{\pi - \theta}{2}\right) \tag{32}$$

The simple optical geometry shown in Figure 4 has 2 transmission rays and 2 detectors so there are four scattering angles to be considered of the form θ_{ij} where i is the incident ray number and j the detector number. These are fixed by the geometry of the optical system. The geometric factors β_{ij} are then determined for refraction and reflection from equations (26), (31) and (32) respectively.

The relationship between the phase difference between detectors 1 and 2, $\Delta\Phi_{12}$, is then given by

$$\Delta\Phi_{12} = \frac{2\pi d}{\lambda}(\beta_{12} - \beta_{11} + \beta_{21} - \beta_{22}) \tag{33}$$

The general optical geometry for PDA is shown in Figure 6 with two incident beams in the xz plane, crossing at an angle Θ, and two detectors positioned symmetrically about the scattering plane, yz, with elevation angles of ±ψ, at a scattering angle θ.

The phase differences Φ_{12} for this optical geometry for reflection and refraction are given by Saffman et al.[14] as

$$\Phi_{12} = \left(\frac{\pi d m_1}{\lambda}\right)2\sqrt{2}\left\{\left[1 + Sin\frac{\Theta}{2}\cdot Sin\theta \cdot Sin\psi - \cos\frac{\Theta}{2}\cdot\cos\theta\right]^{1/2} - \left[1 - Sin\frac{\Theta}{2}\cdot Sin\theta\cdot Sin\psi - \cos\frac{\Theta}{2}\cdot\cos\theta\right]^{1/2}\right\} \tag{34}$$

and

$$\Phi_{12} = \left(\frac{\pi d m_1}{\lambda}\right)4\left[\left\{1 + m^2 - \sqrt{2}m\left[1 + Sin\frac{\Theta}{2}\cdot Sin\theta\cdot Sin\psi + \cos\frac{\Theta}{2}\cdot\cos\theta\right]^{1/2}\right\}^{1/2} - \left\{1 + m^2 - \sqrt{2}m\left[1 - Sin\frac{\Theta}{2}\cdot Sin\theta\cdot Sin\psi + \cos\frac{\Theta}{2}\cdot\cos\theta\right]^{1/2}\right\}^{1/2}\right] \tag{35}$$

There is no closed form solution existing for the geometric factor for scattered rays emerging at P = 2 i.e. second order refraction, the result must be determined numerically through an iterative process.

Practical Aspects of a Phase Doppler Anemometer

The material covered so for has discussed the physical principles behind the PDA method but has given little practical description of the instrumentation, sizing range and resolution. This section discusses these with respect to specific hardware configurations and for three different types of scattering representative of reflection, 1st and 2nd order refraction.

The transmission optical system geometry is shown in Figure 6 and has the specification as given in the Table below.

Table 1. Specification of transmitter optical system.

Light wavelength, λ	0.514 μm
Beam diameter	5 mm
Beam separation	50 mm
Lens focal length	450 mm
Angle between beams, Θ	6.36°
Fringe spacing, df	4.64μm
Measurement volume diameter, D	60.00μm
Light power, per beam	250 mwatts

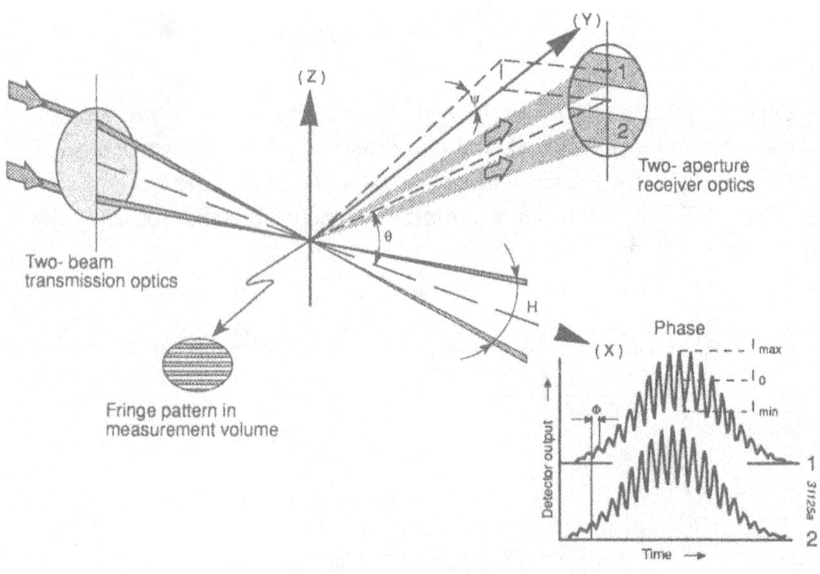

Figure 6. Optical geometry for phase Doppler anemometry.

The receiver optical system in Figure 7 is shown as a single lens divided into two identical apertures symmetrically positioned, either side of the lens centre which is in the xy scattering plane. The design practicalities of a receiver optical system merits some comment here. The best form for a lens is circular so one would expect the highest resolution to be obtained with two independent circular lens systems. However, apart from the high degree of positional accuracy that would be required, the shape of the scattered fringe pattern at the receiver lens must also be considered. In general, the receiver will also see a fringe pattern parallel to the xy scattering plane so, making the receiver from a single lens with slit like apertures, allows very easy symmetrical positioning possible while the greater aperture length allows more light to be collected than would be by two circular apertures.

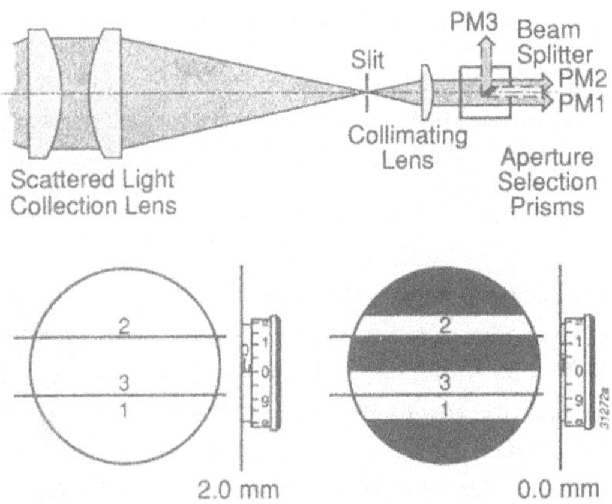

Figure 7. Receiver optical system with adjustable apertures.

To partition the receiver lens, apertures have been used as well as split lenses in a common housing. The alternative to partitioning the received light at the lens is to use a small beam splitter after spatially filtering and re-collimating the scattered light. There are two types of beam splitter in common use, a stack of prisms with different orientations, to direct the received light to the relevant photo-detectors or, a beam splitting cube with a central reflective coating on the internal hypotenuse. The latter is shown in Figure 7, it is a schematic of the DANTEC 57x10 PDA receiving optics. One advantage of this unit is that the aperture receiving areas can be varied using micrometer controlled knife edges. The range of variation in the apertures is shown in Figure 7, from which the approximate aperture sizes and centres can be obtained by scaling the lens diameter to 78 mm.

Before proceeding to establish typical phase/dropsize relationships for specific scattering geometries it should be noted here that the angles between the incident light beams, Θ, and the receiving aperture centres, 2ψ, are limited by the lens diameters so are generally small. The equations, (34) and (35) describing the phase relationship can therefore be simplified since the sine function can be replaced by the angle directly and the cosine functions put to unity. The phase shift therefore increases proportionately with increasing Θ or ψ.

To examine the phase/dropsize relationship or phase factor, as a function of scattering angle, θ, we now consider three specific geometries corresponding to reflection, 1st and 2nd order refraction for the case of water droplets in air. The refractive indices for air, m_1, and water, m_2, are 1.000 and 1.333 respectively.

An examination of the light intensity distributions in Figure 2 shows that excellent agreement between the Mie and geometric optics calculations occurs according to the following Table.

Table 2. Choice of scattering angle

Scattering Mechanism	Scattering Angle	Polarization
(a) Reflection	100°	perpendicular
(b) 1st Refraction	70°	parallel
(c) 2nd Refraction	146°	parallel

The phase/dropsize relationships were calculated, for each of the above geometries, for aperture 2, with respect to the lens centre, for the maximum and minimum aperture settings, over the dropsize range 0 to 50 μm. The results from these calculations are shown in Figure 8 with the best fit phase factors, in degrees/μm, summarized in Table 3, as for apertures 1 and 2, i.e. double the value given by the plots. The dropsize increment was 0·25 μm i.e. 200 iterations were performed and the calculation time for each Mie curve was 70 minutes but only 10 minutes for the geometric optics calculation on a Compaq 486 33MHz PC. In principle the latter could be significantly reduced since 200 iterations are not needed for a linear relationship.

Table 3. Phase factors for the different scattering geometries

Scattering Mechanism	Reflection	1st Refraction	2nd Refraction	
Max. aperture	3·60	5·73	8·60	°/μm
Min. aperture	2·43	3·99	5·95	°/μm

A comparison of the plots for the two different cases, Figure 8, shows how the larger aperture always gives the steeper slope and therefore the greater resolution. It would also produce a larger signal amplitude but, for a fixed upper limit on the phase difference, would produce a lower sizing range than if the smaller aperture was used.

In comparing the plots for the different scattering mechanisms it is seen that for reflection a negative relationship is produced while refraction produces positive slopes i.e. the translation of the farfield fringe pattern for reflection is in the opposite direction. Furthermore, use of reflection always brings a lower size resolution, but a greater size range and a phase factor independent of the refractive index of the scatterer. One drawback to the use of reflection appears in the poor agreement between the Mie and geometric optics calculations for the small size ranges. By far the best agreement between the Mie and geometric optics calculations is for 1st order refraction which extends down to the smallest size classes i.e. below the expected limit for validity of the assumption. The phase factor indicates a medium resolution and size range when compared with the other scattering mechanisms. The use of second order refraction indicates the highest size resolution and therefore the lowest size range but, unfortunately, a poor agreement between the Mie and geometric optics plots exists. Use of the linear geometric optics phase factor would overestimate the size of drops below ≈ 25μm but underestimate the size for drops over 25μm diameter.

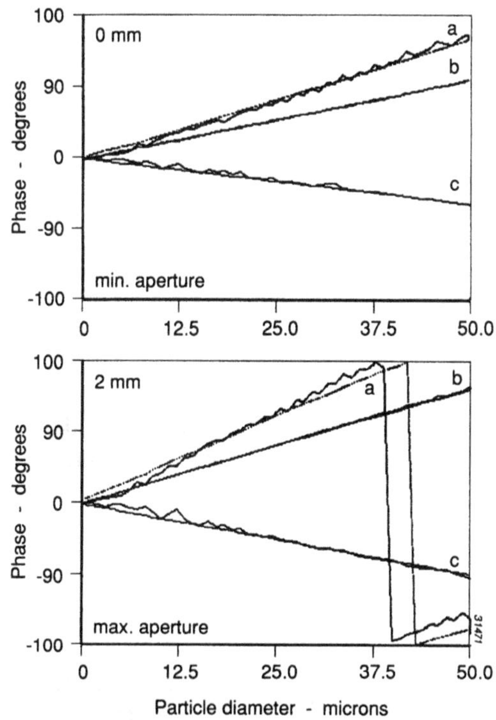

Figure 8. Phase/dropsize relationships for (a) reflection (b) 1st and (c) 2nd order refraction.

These plots, Figure 8, indicate that for large dropsize ranges, and particularly where a high size resolution exists, the phase difference between apertures 1 and 2, Φ_{12}, can exceed 2π. To increase the drop sizing range and avoid the 2π ambiguity the third aperture, shown in Figure 7, closer to the lens centre is used. This provides two extra low resolution estimates of the phase difference, Φ_{13} and Φ_{23}, which when compared with Φ_{12} allows the dropsize to be determined unambiguously.

There are, however, other limitations to the maximum dropsize range that can be measured. As the dropsize increases, the fringe spacing imaged onto the receiver aperture decreases and this must be limited before one complete fringe, or a phase of 2π, exists across the aperture itself as this would reduce the signal visibility to zero. A maximum of 260° represents a good compromise between signal visibility and dropsize range. The photo-detectors, amplifiers and filters have a particularly stringent requirement to meet; they must have a linear response over a wide frequency and amplitude range or they would introduce their own phase shifts onto the signals. The major problem lies with the dynamic range of the signal amplitude which increases as the square of the dropsize. Maximum dropsize ranges of 40 to 1, an amplitude range of 1,600 to 1, are practical upper limits. So although the scattering geometry and the use of three aperture receivers can indicate a wide dropsize range the experimental range is really determined by the gain settings in the signal detection and amplification chain.

So far only the limitations inherent in the measurement technique have been discussed but the physical processes of the fluid flows must also be taken into account. Surface tension keeps droplets and bubbles spherical but as the size increases or its velocity relative to the continuous phase then the droplet will deform and ultimately break-up. It is the Weber number, We, the ratio of the momentum force on the droplet to the surface tension forces, which determines the droplet stability

$$We = \frac{\rho u^2 d}{\sigma} \tag{36}$$

where ρ is the density of the air, u is the droplet velocity d its diameter and σ the surface tension of the liquid. This dimensionless ratio must be less than 1 for surface tension to dominate and the droplet to remain spherical.

The theory underlying this PDA size measurement technique is only applicable to spherical scatterers, however, the fundamental measurement is the local radius of curvature and for three-aperture receiver systems the departure from sphericity can be deduced since

$$\Phi_{12} + \Phi_{23} + \Phi_{31} = 0 \tag{37}$$

In general, non-spherical droplets are rejected, by software in the data reduction stage, to obtain dropsize distributions and mean quantities.

Despite the above limitations of the PDA measurement technique it represents a major advance in optical diagnostics for velocity and size measurements. As such it has virtually rendered redundant the Intensity and Visibility techniques although both of these can still be considered appropriate for low cost monitoring applications.

APPLICATION TO LIQUID FUEL SPRAY COMBUSTION

For liquid fuel to burn efficiently a combustible air-fuel mixture must be formed. The liquid is fed through an atomizer to generate a fuel spray which dramatically increases the surface area of the fuel and promotes both intensive mixing with the surrounding air and the formation of fuel vapour. The mechanics and hydraulics of the atomizer determine the liquid flow through it and these can be used to control the nature and turbulent properties of the emerging liquid. Once the liquid is outside the atomizer aerodynamic forces are mainly responsible for the disintegration of the liquid into droplets. It is the interaction between the internal and external processes that determine the fuel spray's external characteristics of shape and penetration and the detailed internal properties of droplet number density, velocity and size distributions as a function of space and time, Lefebvre[18].

Knowledge of the droplet size, absolute velocity and slip velocity relative to the surrounding air provides information on droplet trajectories, Reynolds numbers and drag coefficients and would allow the calculation of the liquid fuel flux. The vaporization of individual droplets can also be obtained, it is a function of the initial size, slip velocity, heat transfer to the droplet surface and the vapour pressure of the environment surrounding the droplet. In this way the local air-fuel ratio and its distribution can be estimated and this governs combustion ignition, flame stability, combustion efficiency, pollutant and soot formation as well as combustor life-time, Chigier[19].

Optical sizing diagnostics for velocity and dropsize play an essential role not only in spray atomizer design but also in combustion technology. However, although optical diagnostics are non-intrusive their performance can be severely limited if the ambient conditions are not suitable. In liquid fuel spray combustion the ambient gas conditions into which the fuel is sprayed can vary widely in temperature and pressure. In furnaces steam heated heavy fuel oil is sprayed directly into the high temperature flames and combustion products while in gas turbines, kerosene fuel is first sprayed into, and mixed by, highly turbulent and swirling airstreams. The two types of internal combustion engines, Diesel and spark ignition, also exhibit wide variations in ambient conditions. In the former Diesel fuel is injected at high pressure into the combustion chamber at critical or super critical temperatures and pressures while in spark ignition engines gasoline is sprayed into the relatively cool, low pressure inlet manifold. The introduction of the liquid fuel spray can also modify the ambient conditions. The injected fuel will lower the temperature of the charge due to quenching and the supply of latent heat for liquid vaporisation. Effective atomization, yielding high droplet number densities, could lead to vapour rich air-fuel mixtures where saturated vapour conditions could inhibit further vaporisation and lead to an increase in the ambient gas density.

In the application of optical diagnostics to liquid fuel spray combustion the effects of temperature and pressure must be taken into account, for these affect the fuel, air and vapour densities and consequently their refractive indices. Whereas these variables govern the accuracy of measurement there are other parameters that determine whether any measurement is feasible or not. These determine the transmission of the light rays through the air-fuel mixture and combustion products. The spray and flame zones must be optically thin i.e. the input laser beams must cross, there must be only one droplet present in the measurement volume and the scattered light must reach the receiver with a sufficient intensity. Dense sprays and soot attenuate the laser beams and scattered light and therefore limit the spatial measurement range. Liquid fuel i.e. jets, ligaments, films or globules and strong temperature and density gradients can refract the incident laser light to prevent direct transmission and the formation of a measurement volume.

For an accurate measurement of dropsize, PDA requires knowledge of the refractive index of the droplet and the surrounding gases. These can rarely be measured or deduced with sufficient accuracy so the problem reduces to considering how the optical scattering properties of a fuel droplet changes with the maximum possible variation in refractive index.

Data is required for the refractive index of fuels at room, or atomizer, temperature and at the boiling and critical points for atmospheric and high pressure combustion systems respectively. Practical liquid fuels are a mixture of many hydro-carbons and aromatics so the density and refractive index, $m = n - ik$, changes according to fuel composition. Hydro-carbons and aromatics have real refractive indices, n, between 1·4 to 1·5 and 1·5 to 1·6 respectively at room temperature. The variation of the imaginary refractive index, k, which is determined by the light absorption in the liquid, depends mainly on the aromatic fractional content. These range from $10^{-5} \leq k \leq 10^{-2}$ for light and heavy fuel oils respectively for light wavelengths of approximately 0.5μm i.e. blue-green. The real refractive indices for some common fuels and fuel components, are given in Table 4.

Table 4. Refractive indices for fuels and liquids.

n	Liquid	Temperature
1·22	Ethanol (critical temperature)	(516 K)
1·27	(Brewster's angle = critical angle)	
1·33	Water	(273K)
1·36	Ethanol	(273K)
1·39	Heptane and Dodecane	(273K)
1·45	Diesel	(273K)
1·50	Benzene	(273K)

Data on the variation of refractive index with temperature is rare, however, fuel density as a function of temperature is more readily available and can be used, in conjunction with the Eykman relationship, equation (37), to estimate the refractive index at the elevated temperature. Eykman gives the variation of the liquids real refractive index, n, with density, ρ, as

$$\left(\frac{n^2 - 1}{n + 0\cdot 4}\right) = \rho \cdot \text{constant} \tag{38}$$

This equation was derived empirically, specifically for hydro-carbons and actually performs better than the more common theoretical relationship derived by Lorentz-Lorenz.

There are two practical notes to make here, the first is that all heavy fuel oils are multi-component, with each component having different physical properties and volatilities e.g. the boiling points of the components of Diesel fuel range from 425 to 695 K. The second note is that commercial fuel oils may contain dyes, for identification purposes, and therefore exhibit very high absorption and fluorescence and modify drastically the angular distribution of the refracted light intensities.

From Table 4 above, limits can be established for the maximum variation in refractive index for hydro-carbon fuels as a function of temperature i.e. 1·50 down to 1·22. In determining the general light scattering geometry of a PDA system this range of refractive index must be considered. The angular dependence of the scattering mechanisms of reflection, 1st and 2nd order refraction are easily determined by plotting the variation of Brewster's angle, the critical angle and the rainbow angle respectively, as in Figure 9. As a fuel droplet rises in temperature and its refractive index falls, Figure 9 shows that Brewster's angle increases gradually while the critical and rainbow angles decrease. From an optical point of view Brewster's angle and the critical angle are equal for a refractive index of n = 1·27, at a scattering angle of 76°, while Brewster's angle and the rainbow angle are equal for a refractive index of n = 1·1, at the scattering angle 80°. It must be recognised then that in any PDA scattering geometry, apart from near forward and back scatter, that the dominant mode of light scatter can change as the refractive index decreases.

To investigate the choice of light scattering geometries further an analysis will be performed for the PDA configuration given for the transmitter in Table 1 and for the receiver, with fully open apertures, in Figure 7. The scattering angle is measured with respect to the receiving lens centre. This is an F/4 lens i.e. a focal length of 310 mm and aperture of 78 mm, so the light collected is integrated over an angular width of ± 7°. The laser light polarization is parallel to the fringe pattern i.e. in the xy plane of Figure 6. The droplet size is 5μm diameter and refractive indices of n = 1·45 and 1·22 will be taken as representative of cold and hot droplets respectively.

The angular variation of scattered light intensities and phase difference, Φ_2, relative the lens centre, are presented for (a) reflection, (b) 1st and (c) 2nd order refraction and Mie scattering as a function of scattering angle in Figures 10 and 11 respectively. The scattered intensities are presented for the full range of scattering, 0 to 180°, while the phase is only presented from 15° to 165°. Good agreement between the geometric scattering curves and

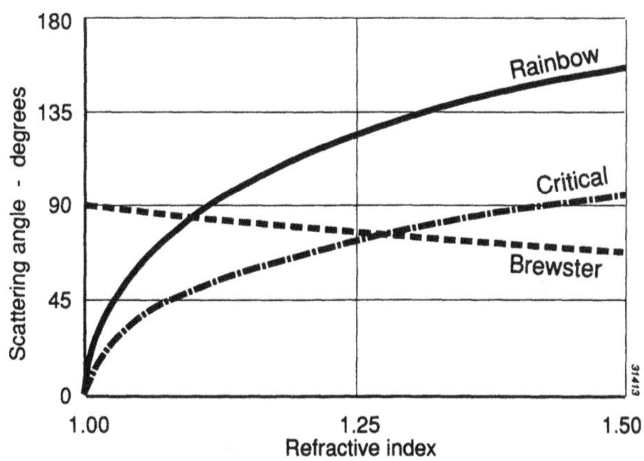

Figure 9. Variation of the Brewster, critical and rainbow angles with refractive index.

the Mie curves can only be found for the 1st order refracted light. In particular the phase shows excellent agreement for the cold drops for scattering angles from 30 to 80° but for hot drops from only 50° to 70°. The intensity distribution for 1st order refracted light shows that for the latter case the scattered intensity has decreased dramatically by 70° scattering angle.

The use of reflected light for drop sizing in fuel spray combustion would appear the logical choice since the phase factor is independent of the refractive index of the scatterer. However, as first shown in Figure 8 and displayed here, in Figures 10 and 11, relatively poor agreement between the geometric and Mie phase factors exist for small dropsizes. To an even greater extent the same is true for 2nd order refraction, but the major difficulty here is that the optimum scattering angle for 2nd order refraction, nominally between the rainbow angle +5° and +10°, is seen to demonstrate a very large excursion for the considered minimum and maximum refractive indices.

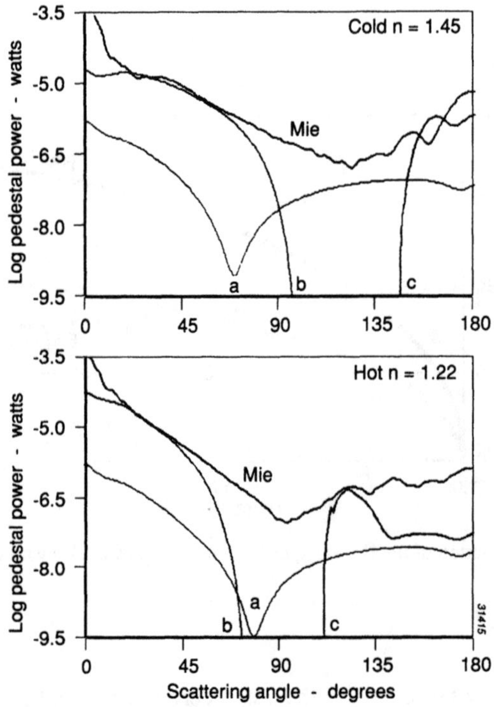

Figure 10. Variation of scattered light intensity as a function of scattering angle.

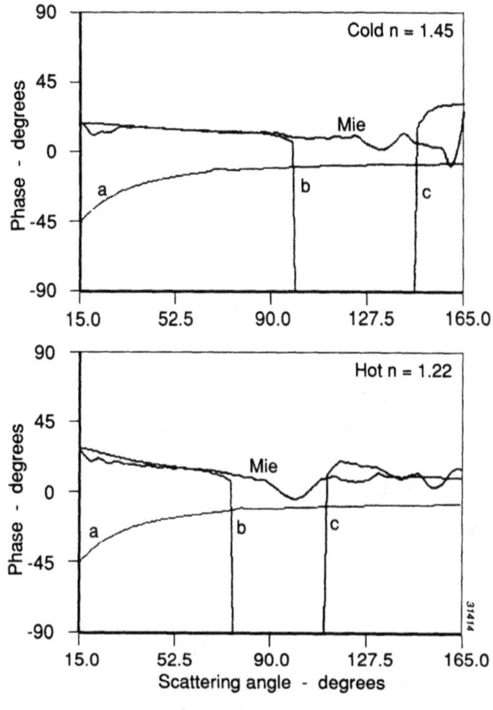

Figure 11. Variation of signal phase as a function of scattering angle.

This analysis supports the work described by Pitcher and Wigley[20] where the use of small receiver aperture geometries was evaluated. To complement this earlier work the phase-refractive index relationship for Mie and 1st order refraction scattering for fixed drop sizes 5, 10, 20, 30 and 40 μm are compared, over the range n = 1.22 to 1.60, at scattering angles of 30° and 70° for the fully open aperture, in Figure 12. The general trend at the 70° scattering angle is for flat profiles with excellent agreement between the Mie and geometric calculations except for refractive index indices less than 1.30. For the 30° scattering angle good agreement between the Mie and geometric calculations are seen for drop sizes of 20μm and greater but the profiles show a near linear decrease in phase with increasing refractive index.

The poor comparison below n = 1.30 for the 70° scattering angle can be expected since the Brewster and critical angles are then comparable. However, it is the Mie curve that determines the true performance of any scattering geometry so any refractive index greater than 1.30 can be chosen as a refractive index representative for all cases. For the 30° scattering geometry no single refractive index can be identified. This is illustrated in Figure 13, for the 30° and 70° scattering angles respectively, where the phase is plotted as a function of dropsize for the Mie analysis for the following fixed values of refractive index; 1.22, 1.27, 1.33, 1.39, 1.45 and 1.50 i.e. those listed in Table 4. A phase factor for 70° scattering angle can be estimated from Figure 13 which is practically insensitive to variations in refractive index. With a phase factor of 5.7 °/μm a precision of ±2% can be obtained over the refractive index range of 1.25 to 1.5 for drop sizes up to 50μm. The situation is very different for the 30° scattering angle where the slope of the plots increases proportionately with decreasing refractive index. Furthermore, the Mie plots show non linearities, when compared with the linear geometric plots, for dropsizes less than 12.5μm.

There are two questions to be considered here: (1), why is the high scattering angle insensitive to refractive index change even though the dominant scattering mode is 1st refraction? and (2), can these different results be used to provide other fluid property data? The insensitivity of the phase factor to refractive index changes can be seen from considering equations (16) and (17). The angle of incidence related to 70° scattering angle is 83° i.e. close to grazing incidence and therefore these rays would have the minimum path length inside the droplet. For fuel spray combustion this means that the temperature of the outer shell of the droplet determines the light scatter rather than the droplet's core temperature.

The PDA technique can be extended by using two receiver systems one arranged at a high scattering and the other set at a low scattering angle. The ratio of phases can be used to estimate the droplet's refractive index. Knowledge of the refractive index has two potential applications in: (1), fuel spray combustion, and (2), multi-phase flows. In the former Eykman's relationship, equation (37), can be used to calculate the density of the droplet which, inturn, can be used to estimate the droplet temperature. For multi-phase flows the refractive index allows the phases to be discriminated to allow independent fluid dynamic properties to be evaluated in liquid/liquid and liquid/particle flow systems, Naqwi and Durst[16].

The final section in this chapter considers what can be measured, the data reduction and data presentation. The PDA technique is a single particle counting technique and as such provides the velocity and size measurement for each droplet. Droplet size distributions can be calculated e.g. Rosin-Rammler, but for most calculations involving flow and transfer processes it is much more convenient to work with mean diameters. Different mean diameters can be derived from the individual dropsizes to correlate with the physical process being investigated. Both Lefebvre[18] and Sowa[21] give a good discussion on representative

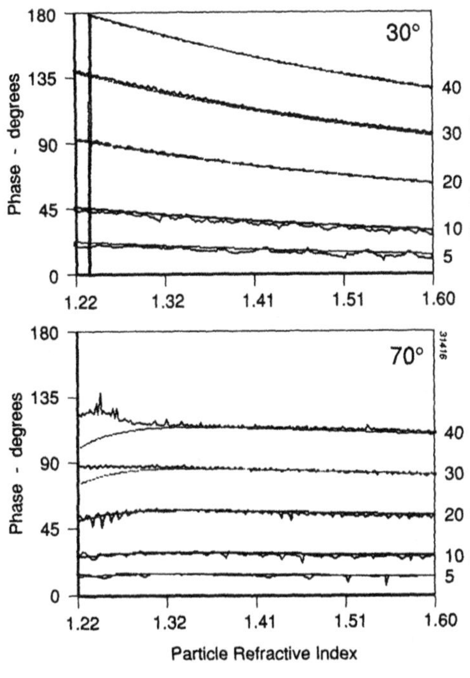

Figure 12. Phase/refractive index relationships for the 30° and 70° scattering geometries as a function of dropsize.

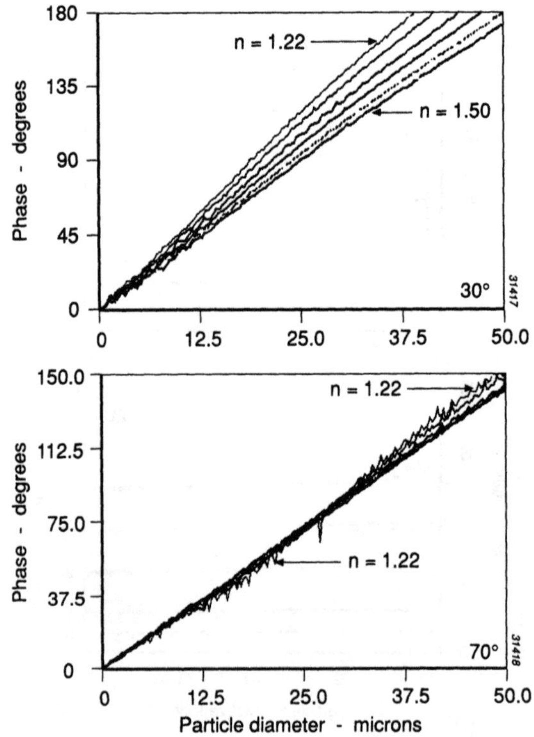

Figure 13. Phase/dropsize relationships for the Mie analyses of the 30° and 70° scattering geometries as a function of refractive index.

mean droplet diameters and their interpretation. Mean diameters are calculated from the following relationship and are summarized in Table 5.

$$D_{ab}^{(a-b)} = \frac{\sum N_i d_i^a}{\sum N_i d_i^b} \tag{39}$$

where d_i = representative diameter of class i.
N_i = number count of droplet in class i.
and a, b are integers.

Table 5. Representative mean droplet diameters

a	b	Name of mean diameter	Symbol	Expression	Application
1	0	Length	D_{10}	$\frac{\Sigma N_i di}{\Sigma N_i}$	Comparison
2	0	Surface area	D_{20}^2	$\left(\frac{\Sigma N_i d_i^2}{\Sigma N_i}\right)$	Surface area
3	0	Volume	D_{30}^3	$\left(\frac{\Sigma N_i d_i^3}{\Sigma N_i}\right)$	Volume
2	1	Surface area-length	D_{21}	$\frac{\Sigma N_i d_i^2}{\Sigma N_i d_i}$	Absorption
3	1	Volume-length	D_{31}^2	$\left(\frac{\Sigma N_i d_i^3}{\Sigma N_i d_i}\right)$	Evaporation, molecular diffusion
3	2	Sauter (SMD)	D_{32}	$\frac{\Sigma N_i d_i^3}{\Sigma N_i d_i^2}$	Mass transfer, droplet vapourization
4	3	De Brouckere or Herdan	D_{43}	$\frac{\Sigma N_i d_i^4}{\Sigma N_i d_i^3}$	Combustion equilibrium
		Average spray surface area		$\frac{\pi \Sigma N_i d_i^2}{\Sigma N_i}$	
		Average spray volume		$\frac{\pi \Sigma N_i d_i^3}{6 \Sigma N_i}$	Volume Flux

In applying these to PDA data several important aspects should be considered. Where large dropsize distributions are encountered the mean velocity can be misleading since different drop sizes can take up different velocities i.e. both speed and direction. The velocity-size correlation determines the usefulness of a single mean velocity. Table 5 indicates that the volume flux may be calculated but this is a temporal distribution. The estimation of the volume flux from a photograph would yield a spatial distribution. Discussion on the calculation of the volume and mass flux can be found by Bachalo et al. in Chigier[19] and Saffman[22]. Since this calculation contains d_i^3 it is very sensitive to the larger droplets found in the tail of the dropsize distribution and for sparse data sets they can totally dominate and bias the results.

This is certainly the case for high pressure, transient Diesel fuel sprays. The greatest fraction of the droplets have D_{10} diameters $\leq 10\mu m$ but occasionally on the spray tip and especially in the spray tail, at the end of injection, some large droplets are produced which, if area or volume weighted means were used to calculate a mean dropsize, would seriously obscure the true picture. The Diesel spray represents a severe challenge to optical velocity and drop sizing techniques. The sprays are optically thick so light transmission through or into the spray is very low but the spray, close to the injector, can be very poorly atomized with liquid jet and ligaments dominating the flow processes. The transient, high pressure nature of the spray also brings experimental difficulties with very low duty cycles for data arrival and very high velocities. Nevertheless, measurements are possible and one result is shown in Figure 14. It is one time frame of velocity and dropsize data taken from a computer animation of the spatial and temporal development of the spray injecting into quiescent flow conditions inside a Diesel engine. The fuel line-pressure diagram shows the spray behaviour at the time that peak pressure, 386 bar, is reached.

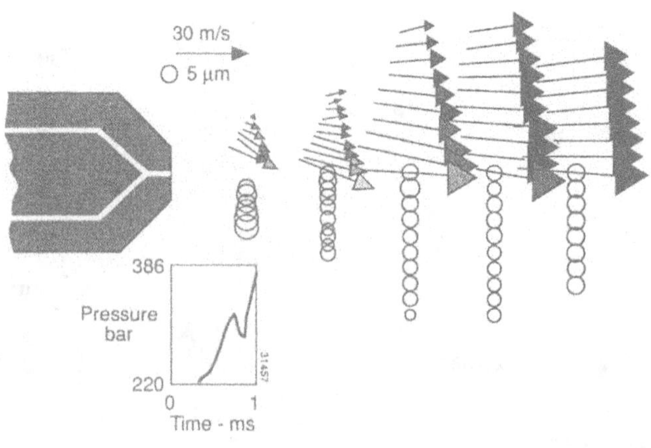

Figure 14. Spatial distributions of mean droplet trajectories and dropsizes in a Diesel spray.

The axial and radial velocities and dropsize were measured simultaneously for each droplet detected. Upto 15,000 velocity/size samples were recorded at each measurement point during, typically, 1,000 injections. To record the time history of the spray the time between droplets was measured, relative to a trigger pulse, with a resolution of 1 μs. A high spatial resolution is also required with increments in the radial measurement position of 0.1 mm for the 5, 10 and 25 mm axial locations and 0.2 mm for the 15 and 20 mm axial locations. The data has been time averaged in bin intervals of 40 μs. The display in Figure 14 shows the droplet trajectory as vectors and the diameters, D_{10}, as circles, mirrored about the spray axis.

The volume flux is impossible to quantify under these injection conditions close to the nozzle but some indication of the spatial variation in droplet number density can be estimated from the droplet number per time bin divided by the number of injections. This droplet number density is represented by the grey scale in Figure 14. On the spray periphery i.e. in the optical thin zones then the spatial variation is quantitatively correct, however, elsewhere in the spray low densities indicate the presence of either dense concentrations of droplets or liquid only. The latter can be seen occurring on the spray centre line close to the nozzle where no data is presented at the 5 and 10 mm axial measurement planes. The radial growth of the spray tip is readily seen as well the entertainment of droplets into the spray just downstream from the nozzle. Perhaps the most interesting feature of the spray is that the mean droplet velocity increases with distance downstream. This is purely an indication of the exchange of momentum between the penetrating spray and the original quiescent air flow in the combustion chamber.

SUMMARY

This chapter has reviewed the essential physical and analytical processes necessary to quantify a light scatterer's size from the scattered light intensity distribution. The sizing methods of Intensity and Visibility were described since their signal characteristics are contained in the scattered light signals utilized by the phase Doppler anemometry PDA technique for size and velocity measurements. Limitations of the PDA technique have been described but from an applications point of view it is a robust technique and likely to find a considerable future, not only in research institutes, but in solving industry's flow process problems.

The references cited have been chosen to give the reader a broad overview not only of optical diagnostic methods for size and velocity measurements but also there applications to spray atomization and combustion.

REFERENCES

1. Mie, G., Beiträge zur optik über medien, speziell kolloidealer metallösungen, *Ann. Physik. 25:377* (1908).
2. Lorenz, L., *Videnskab. Selskab. Skifter*. 6 (1890).
3. Lorenz, L., "Oeuvres Scientifigues". Copenhagen, Denmark. 1:405 (1898).
4. van de Hulst, H.C., "Light Scattering by Small Particles", *J. Wiley and Sons,* New York (1957).
5. Kerker, M., "The Scattering of Light and other Electromagnetic Radiation", *Academic Press,* New York (1969).
6. Negus, C.R. and Drain, L.E., Mie calculations of the scattered light from a spherical particle traversing a fringe pattern produced by two intersecting laser beams, *J.Phys.D.Appl.Phys.*, 15, 375-402 (1982).

7. Dave, J.V., Subroutines for computing the parameters of the electromagnetic radiation scattered by a sphere, Report 320 3237, IBM Scientific Center, Palo Alto, California, USA (1968).
8. Bohren, C.F. and Huffman, D.R., Absorption and scattering of light by small particles, *J.Wiley and Sons,* New York (1983).
9. Grehan, G. and Gouesbet, G., Mie theory calculations: New progress, with emphasis on particle sizing, *Applied Optics*, 18, 3489-3493 (1979).
10. Wiscombe, W.J., Improved Mie scattering algorithm, *Applied Optics,* 19, 1505-1509 (1980).
11. Glantschnig, W.J. and Chen, S.H., Light scattering from water droplets in the geometrical optics approximation, *Applied Optics,* 20, 2499-2509 (1981).
12. Bachalo, W.D. and Houser, M.J., Phase Doppler spray analyzer for simultaneous measurements of dropsize and velocity distributions, *Optical Engineering,* 23, 583-590 (1984).
13. Bauckhage, K., Floegel, H.-H., Fritsching, U., Hiller, R., The phase Doppler difference method, a new laser Doppler technique for simultaneous size and velocity measurements, part 2: Optical particle characteristics as a base for the new diagnostic technique, *Part.Part.Syst.Charact.*, 5, 66-71 (1988).
14. Saffman, M., Buchhave, P. and Tanger, H., Simultaneous measurement of size, concentration and velocity of spherical particles by a laser doppler method, laser anemometry in fluid mechanics - II (Editors: Adrian, Duras, Durst, Mishina and Whitelaw), Ladoan, Lisbon, pages 85-104 (1984).
15. Naqwi, A.A. and Durst, F., Light scattering applied to LDA and PDA measurements, Part 1: Theory and numerical treatments, *Part.Part.Syst.Charact.*, 8, 245-258 (1990).
16. Naqwi, A.A. and Durst, F., Light scattering applied to LDA and PDA measurements, Part 2: Computational results and their discussion, *Part.Part.Syst.Charact.,* 9, 66-80 (1992).
17. Sellens, R.W., A derivation of the phase Doppler measurement relations for an arbitrary geometry, *Experiments in Fluids*, 8, 165-168 (1989).
18. Lefebvre, A.H., "Atomization and Sprays", *Hemisphere Publishing Corp.* (1989).
19. Chigier, N., "Combustion measurements", *Hemisphere Publishing Corp.* (1991).
20. Pitcher, G., Wigley, G. and Saffman, M., Sensitivity of dropsize measurements by phase Doppler anemometry to refractive index changes in combusting fuel sprays, *5th Int. Symposium on Applications of Laser Techniques to Fluid Mechanics,* Lisbon, 1990. Editors Adrian, R.J. et al., Springer Verlag, 227-247 (1991).
21. Sowa, Interpreting mean drop diameters, *Atomization and Sprays,* 2.1 (1992).
22. Saffman, M., Automatic calibration of LDA measurement volume size, *Applied Optics,* 26. 13, 2592-2597 (1987).

TWO PHASE FLOW MEASUREMENTS

A. M. K. P. Taylor

Thermofluids Section
Department of Mechanical Engineering
Imperial College of Science, Technology & Medicine,
London SW7 2BX, U. K.

INTRODUCTION

The range of technical applications of two phase flows is indeed impressive: it extends from particulate flows in gases, economically important examples of which include atomized kerosene in gas turbine combustors, pulverised coal in utility boilers and the preparation of freeze dried foods, to bubbly flows in liquids, such as Argon stirring in secondary steel making, and cavitation. There has thus been an incentive to understand and hence control these industrial flows for several decades but the range of instruments at our disposal for this purpose was limited. However, the last decade has seen intensive development of many, predominantly laser-based, techniques which are now being applied to two phase flows to measure, primarily, the *size, velocity, flux and concentration* of the dispersed phase. While many applications will remain too hostile – or too optically disturbed – for measurement by these techniques, many new, exciting applications of the instruments have begun to be made and this trend will continue to increase for the foreseeable future.

The field of two-phase flow measurement, in the sense of *optical techniques* rather than in the sense of the physical insight obtained, has already been recently extensively reviewed, in the particular context of combustion applications, by Taylor [1,2], Jones [3] and Chigier [4]. In the preparation of this contribution, it has been stimulating to find that, in little more than two years, many important references have been added to the literature on two phase flow optical measurement techniques. The purposes of this contribution are therefore to review these recent additions, to provide an entry to these publications and, in addition, to include specifically references which are not related to combustion. To do justice to this new material in the space allowed for this contribution, the reader is thus directed to the first four references for an introduction to the basic principles of the techniques mentioned below and, indeed, this paper is intended to be read in conjunction with these earlier publications. The emphasis on recent work does not, however, mean that older, well-established instruments have been superseded. In many cases, recent work has involved defining more precisely the limitations of the newer instruments and this information deserves a wide audience. Finally, the literature referenced in this article is by no means exhaustive and many references have been omitted, arbitrarily, to limit its length. Where choices had to be made, I favoured the most recent publications.

The material below is grouped under three headings, namely the Phase Doppler velocimeter (PDV), Dispersed phase temperature measurement and Non-PDV measurements. These headings tend to emphasize the domination of the PDV in two phase flow measurements and this has arisen because of its ease of use, its good accuracy, competitive cost and commercial availability. The choice of headings may be questioned on the ground that these are an ungainly mixture of technique and measurement objective: the defence is that these headings reflect the scope of recent publications. The references have a slight bias to those published by the author and his co-workers, for reasons of familiarity only: I have tried to include as wide a cross-section of other workers' publications as well. Finally, it is as well to be explicit about the scope of this review: it is to review recent advances in optical measurement techniques which, with the one explicit exception of particle image velocimetry applied to a bubbly flow, are concerned primarily with the properties of the dispersed, rather than with the continuous, phase. This article does not review the well-established probe based techniques which are important for bubbly flows, nor the fluid mechanics insight gained from such measurements.

THE PHASE DOPPLER VELOCIMETER (PDV)

The PDV provides simultaneous measurements of single droplet[††] size and velocity with good temporal- and spatial-resolution. From these measurements, *derived* quantities can be produced, such as the size distribution (and hence to moments such as the arithmetic and Sauter mean diameter) and the *local* droplet number density and dispersed phase mass flux (or volume fraction for bubbles). This is so, provided that the variation of the cross-sectional area, and volume, of the optical probe as a function of the droplet diameter is known or, better, can be measured on-line.

The PDV is a well-established technique, in the sense that the basic theory of operation of the instrument has been understood for well over half a decade, that it is applied routinely by a many *non-specialist* researchers world-wide and that second generation commercial instruments are on the market. However, there are two aspects which are the subject of current research and these are that errors may sometimes arise in measurement of size and that there is little consensus on the accuracy with which measurements of number density and flux are possible. These two items are, arguably, of interest to a wide audience and are thus considered first. The PDV, and its close derivatives, are undergoing continuous development and evaluation and the third sub-heading, below, reviews recent investigations. This section is intentionally brief because many of these developments have yet to be corroborated by other researchers and hence the value of the developments has yet to be established.

[††] or *spherical* solid particle or gas bubble: in the remainder of this article, unless otherwise explicitly mentioned, the word droplet will be used to imply spherical bubbles and solids as well

Sizing Errors due to Trajectory Ambiguity Effects

It is straightforward to demonstrate that, when the diameter of a droplet is larger than about one-fifth [6] that of the probe of the PDV *and* the droplet is positioned at the edge of the probe on the side *farthest* away from the detectors, the light irradiance reflected by the droplet surface is comparable to that scattered by refraction. Even for an optimally adjusted optical system [7], the phenomenon occurs at a droplet:probe diameter ratio must be about 0.5. In many circumstances, it is usual to find the response curve (i.e. the relationship between the measured phase and the droplet diameter) assuming that the *opposite* is true: that is, that the reflected component is much weaker than the refracted. With trajectory ambiguity, droplet size will be measured erroneously and the magnitude of the error has been investigated by a number of authors, partly by calculation and partly by experiment, who show it is a strong function of the trajectory of the droplet through the probe volume. Sankar *et al.* [5] have shown, using the so-called geometrical optics approximation, that a 45 μm droplet placed in a test volume with 80 μm waist might result in an underestimation as large as 50 and 67% of the true diameter when collecting scattered light at 30° and 150°, respectively. Calculation [8] suggests that scattering angles around 100° result in smaller sizing error, although the laser power required to make this a viable measurement angle is large. Detailed "trajectory maps" of the measured phase error as a function of location within the probe volume have recently been provided [6] using the computationally intensive, but exact, theory termed Generalized Lorenz-Mie theory (GLMT). These maps provide useful insight into the magnitude of the problem and show, in particular, that isophase lines are perpendicular to the fringes. This fact has potentially important implications, as discussed later.

A number of remedies have been proposed to the trajectory ambiguity error, the most direct being to increase the diameter of the test volume: thus, in the numerical example quoted above, the error reduces to less than 10% if the probe diameter is 200 instead of 80 μm [5]. The disadvantage of doing this is that, to avoid multiple occupancy, the maximum number density of the dispersed phase that can be tolerated is reduced which, in many applications, is a great disadvantage. Other ways proposed include the use of an additional detector assembly [6], but it is difficult to align widely separated detectors and the complexity - and hence cost - of the associated electronics would be high. The use of analyzers to block light of inappropriate polarization has been mooted for a long time [9]. As an extension to this idea, it might be thought that collection of scattered light near the Brewster angle, where reflection is therefore at a minimum, would result in small errors: surprisingly, this appears not to be so [5]. Another elegant idea [6] is to analyze the Doppler signal using wavelet transforms which allow the detection of the phase *evolution* of the signal, the change of phase along a signal being one of the characteristics of trajectory ambiguity. The origin of trajectory ambiguity is the change in the relative *magnitudes* between the reflected and refracted light: it is argued [5] that trajectory ambiguity can be eliminated by changing the relative *phases* between these two components, which is achieved by increasing the intersection angle. Experiment and theory show that by increasing this angle up to 5.4°, the errors can be reduced to less than 5% for diameters above 27 μm. Although somewhat smaller angles can be used, depending on the droplet:probe diameter ratio, the disadvantage of this approach is that the maximum diameter which can be sized decreases. In the author's laboratory, reliance is placed on a phase ratio validation scheme using *two* pairs of photomultipliers [10] to distinguish between signals due to reflection and refraction. Haugen and von Benzon [11] recently reported a similar strategy and they also analyze the magnitude of the tolerance on, effectively, the acceptable phase ratio which will still identify trajectories giving rise to ambiguity: values around 5 to 20% are quoted. In addition, they note that the detector separation must be optimised to accommodate these values. It should be noted that PDV systems based on *one* pair of photodetectors only cannot use this important strategy.

The observation that the isophase lines are perpendicular to the fringes has led to an interesting proposal [12] for a PDV free from trajectory ambiguity effects. The use of a *planar* PDV layout, in which the detectors are in the plane of the incident laser beams from the transmission optics, as opposed to the orthodox PDV arrangement where the detectors are placed at angles of typically 30° to this plane, leads to the isophase lines being parallel to the fringes. Therefore, the iso-lines of erroneous phase are associated with approximately half of the probe volume *farthest* from the detectors. In addition, it has long been known [8] that this half of the probe *also* gives Doppler signal amplitudes which are about only one-tenth that of the other half. As a consequence, it is necessary to process only a part of the signal around the *peak* of the Doppler burst to obtain the phase accurately, which most signal processors do already. [It should be noted, for the discussion in the next sub-heading, that this result suggests that signals from up to *half* the nominal probe area are unusable.] In this way, signal processing is greatly simplified and the direction of motion of the droplet is immaterial, as - it seems - is the angle of the trajectory through the probe. Experiments with droplet diameters which ranged from 0.3 to 1.3 of the probe diameter confirmed that the errors were less than 10%. This newly developed system has a number of disadvantages at present. The transfer function of the system (i.e. degrees phase shift per μm) is a strong function of the beam intersection angle and the elevation angle of the detector. In addition, the transfer function is always smaller than that of an orthodox PDV layout. The calculated response curve for the planar layout does not have good linearity below about 25 μm and it may be necessary to introduce either a third photodetector or use signal amplitude validation to avoid the difficulties this creates.

Although the *potential* for individual errors due to trajectory ambiguity is large, these are certainly "worst case" analyses which pertain to the relatively rare event of particles passing through the edges of the probe volume, as also noted elsewhere [5]. Calculations of the appearance of a high-passed Doppler signal subject to trajectory ambiguity confirm [12] that the part due to reflection has about of one-tenth the amplitude of the refracted signal. The measurement of phase made by signal processors which are designed to provide excellent sensitivity and ability to detect signals with low amplitude and poor signal-to-noise ratio (SNR), e.g. those based on discrete Fourier transforms, may indeed be affected by these low amplitude contributions. However, for zero-crossing counters, examination[8] of the signature of the signals suggests that it is likely that the low-amplitude parts of the signal which carry the trajectory ambiguity effects fail to be be detected because these fail to exceed the amplitude threshold (which is a normal feature of these instruments) or are rejected by the "5/8" frequency validation circuitry or, perhaps most importantly, by the imposed *minimum* number of fringes in a signal [13]. The experience of the author with the use of zero-crossing counters over a number of years, particularly in flows with strong size-velocity correlations, suggests that phase measurement errors due to trajectory ambiguity should, in future, be examined taking into account the operation of the signal processor. This is particularly important in the light of new, attractive concepts such as real-time SNR-based burst detection [14] and burst-and-peak detection circuitry which purposely limit the operation of the phase measurement† to the parts of the signal with the highest SNR [15].

Errors in the Measurement of Number Density and Flux

A recent review [16] of the literature on the assessment of the uncertainty on the integrated measurement of number density and flux by the PDV suggests that it can be large, up to 100%, and is reported even by experienced workers (who can be relied on to set the voltage on the photomultiplier tube correctly and thus avoid the well-known sensitivity of PDV measurements to this setting)¶. It is perhaps for this reason that few papers emphasize either number density or flux measurements despite the fact that in much applied research it is frequently the flux of particles which is of greater interest than either velocity or size (although in simple flows such as an axisymmetric jet with monodisperse glass beads [17], for example, any errors should be systematic, rather than random, and hence it is justified to report the flux). For example, the flux [16] of liquid droplets in a swirl-stabilised kerosene-fired burner [18], or the concentration of quasi-monodisperse glass spheres within the recirculation zone formed downstream of a sudden step expansion as a function of the upstream pipe flow Reynolds number [19], are fundamental to the description of the flow and the most striking phenomena on visual observation. The accurate measurement of particle size is a basic requirement for accurate measurements of density and flux, so that the remarks made above about trajectory ambiguity are important here also. Further, it is understood that account must be taken of the fraction of signals which fail to satisfy validation criteria imposed by signal processing equipment or software. A particularly important validation criterion is that there be no multiple occupancy in the measuring volume of the PDV: this restriction is sometimes also referred to in the literature as the single particle constraint. The criterion comes into play particularly for the measurement of number density and flux in densely loaded sprays, such as those produced by a diesel injector [20]. A conceptual framework now exists [21] for estimating the likely effects on these measurements, based on Poisson statistics for the spatial distribution of droplets and for the droplet arrival rate. Thus it is possible to calculate the probability that a droplet, which enters the probe volume at an arbitrary instant in time, will be the measured by the PDV by requiring that it is also the *only* droplet in the volume for the duration of its residence time within the volume. This calculation has been used to simulate the results of measurements of flux and density made by the PDV with three important outcomes. First, the welcome conclusion from this study is that measurements made on the leading and trailing edges of the spray are likely to be accurate. Second, the expected result that - in the densest regions of the spray on the centreline and aft of the leading edge - no measurements *at all* are possible [20] is confirmed. Third, it is shown that

† Their use results in the centre of the Doppler burst being detected and hence improving, relative to a *fixed* trigger level, the SNR of the measured portion by about 10 dB although, somewhat perversely, the improvement *decreases* with decreasing particle size – which is opposite to the desired effect!

¶ It must be noted that others have reported agreement with independent measurement of the volume flow rate of dispersed phase which is within a few percent. Generally, however, the consensus is that the uncertainty is large.

although measurements between the leading and trailing edges of the spray *can* made by the PDV at off-axis locations which have the correct qualitative profile shape, the results can be in error by over 50%, reflecting the effects of the so-called Poisson efficiency. Presumably, this is similar to the validation rate recorded by most PDV instruments and implies that the correction of measurements by this number is imperative. In reading the remainder of this section, it should be borne in mind that this correction factor can be as high as 1.25, and sometimes is higher, and that this is frequently the largest identifiable contributor to inaccuracies.

Number density is a volume-based concept and it is necessary[22] to know the increase of the probe volume with droplet size (this arises because large droplets scatter more light than do small ones). Thus, for a particular diameter class, d_i, the number density $C(d_i)$ is

$$C(d_i) = \frac{\sum_{j=1}^{n(d_i)} \tau(d_i)_j}{T_s} \cdot \frac{1}{V(d_i)} \tag{1}$$

where $\tau(d_i)_j$ is the jth sample of residence time of this diameter, the summation is made over the number of measurements, $n(d_i)$, made over a sampling time of T_s and $V(d_i)$ is the effective volume of the optical probe for this diameter. The extension of this formula to the number density of all diameters is trivial. Note that early formulations for establishing number density based on the mean velocity of the *fluid* are fundamentally incorrect and appear to have been abandoned [23] in favour of the measurement [22] of the residence time of the droplets in the volume.

The flux of a particular size class, $G(d_i)$, requires careful definition and leads to algebraically fierce equations! The flux is here defined as that crossing an area defined by having unit normal perpendicular to the plane of the fringes of the PDV,

$$G(d_i) = \frac{\rho\pi}{6T_s} \frac{d_i^3}{A_\perp(d_i)} \sum_{j=1}^{n(d_i)} \frac{U_\perp(d_i)_j}{|U_\perp(d_i)_j|} \tag{2}$$

where ρ is the droplet density, $A_\perp(d_i)$ is the area of the probe volume for size class d_i with unit normal perpendicular to the plane of the fringes, $U_\perp(d_i)_j$ is j^{th} sample of the velocity component normal to the fringes for a droplet with diameter d_i. The fraction under the summation symbol has the value +1 or -1 and thus provides a positive or negative sign for each measurement, depending on the sign of the velocity component, so that if an equal number of droplets cross the probe in the positive as in the negative sense, the net flux is *zero*. It should be noted carefully that the *magnitude* of the velocity attached to each measurement is immaterial to magnitude of the droplet flux. The cross-sectional area is found from

$$A_\perp(d_i) = 2\, z_p \,.\, y(d_i)_{max} \tag{3}$$

where z_p is the known length of the image of the spatial filter incorporated in the receiving optics and $y(d_i)_{max}$ is the maximum width over which a measurable signal can be obtained and it has now been assumed that the flow is unidirectional and its direction coincides with the normal to the plane of the fringes. The width $y(d_i)_{\max}$ can be related [24] to $<x^2(d_i)>$, the average of the square of the distances over which droplets of diameter d_i are detected in the direction perpendicular to the plane of the fringes. For each trajectory, the so-called burst length, $x(d_i)$ (units: μm), is given by

$$x(d_i) = U_{\perp}(d_i)_j \, \tau(d_i)_j \tag{4}$$

The analysis has been extended[22] to include the effects of frequency shifting and the result is

$$y(d_i)_{\max} = \frac{3b_y}{2.\sqrt{2}} \sqrt{\left\{ \cos^2\left(\frac{\theta}{2}\right) \frac{<x^2(d_i)>}{b_y^2} - \left(\frac{N_o}{N_f}\right)^2 \frac{1}{n(d_i)} \sum_{j=1}^{n(d_i)} \left[\frac{|U_{\perp}(d_i)_j|}{(U_{shift} + U_{\perp}(d_i)_j)} \right]^2 \right\}} \tag{5}$$

where b_y is the Gaussian waist diameter at the intersection region, θ is the beam intersection angle, N_o and N_f are respectively the minimum number of zero crossings required by the signal processing electronics and the number of fringes within the Gaussian waist and U_{shift} is the equivalent velocity of the frequency shift. It should be noted that z_p, the length corresponding to the image of the spatial filter incorporated in the receiving optics is subject to ambiguity due to receiver slit blur circle [25]. The estimation of the probe cross-sectional area and droplet occurrence *at largest diameters* is particularly important for the accurate estimation of flux but, as mostly recently noted by Oldenburg and Ide [26], the sampling statistics of the PDV are inevitably poor for these diameters and hence so is the estimate of the probe volume. Zhu *et al.* [23] use the *boundary* of the scatter plot of the burst length *vs* d_i to estimate $y(d_i)_{\max}$, presumably to mitigate the effects of small sample size, although the effectiveness of this procedure, as opposed to the use of the average of the square of the burst length, has not been quantified. Hardalupas (personal communication, 1994) suggests that the best statistical accuracy may be obtained by fitting the frequently used theoretical solution for $A_{\perp}(d_i)$ at the point with the modal diameter.

In contrast, the accurate measurement of number density is dependent on the accurate detection of the *smallest diameters*, which generally have the worst SNR and therefore are the most likely to fail validation criteria. In some cases, it may be wise to estimate the number density, the droplet flux and mean diameters using explicit limits on the lower and upper diameters that will be considered: thus, for example, a Sauter mean diameter could be defined based on the measurements between, say, 5 and 80 μm. The lower limit avoids the difficulties associated with the uncertainty of measurement of the smallest diameters - which in many cases do not contribute to the phenomenon being investigated - and the upper limit serves to remove the uncertainty associated with the statistical inaccuracy due to small sample size at the largest droplets. It is usually not possible to justify the imposition of the upper limit from the physics of the flow, because the flux contained by these droplets is sometimes a large fraction of the total. However, the proposal *does* have the merit that the uncertainty attached to the quoted measurement would be improved and, particularly for those who attempt to evaluate CFD techniques, would clarify the basis on which the measurement is quoted. In addition, the following paragraphs demonstrate that many of the quantities required in the equations above are subject to uncertainty.

Recently, it has been noted [15, 27] that the measurement of the transit time, $\tau(d_i)j$, and the burst length, $x(d_i)$, by use of a fixed threshold level applied to the high passed Doppler signal, even if the arbitrary setting of the trigger level is ignored, may be subject to 90% error at an SNR of 8 dB. It is likely that a similar effect ("break up") also arises in regions of long transit time due to near-zero mean droplet velocity. Qiu and Sommerfeld [15] suggest that, for instruments which digitise the Doppler signals (i.e. *not* zero crossing counters), the logarithmic mean of the signal amplitude be used instead of the burst length to estimate the transit time because this quantity is relatively immune to the effects of SNR. Computer simulations suggest that this method results in an error of only 4% at an SNR as low as - 5 dB. Apart from being a computationally somewhat intensive method, errors also arise due to clipping of signals due to the finite resolution of the analogue-to-digital converters over large ranges of diameters. A pessimistic estimate of this error is, for a 50:1 ratio of diameters, about 14%.

The preceding paragraphs have not mentioned the difficulties encountered when the flow is *not* predominantly directed normally to the fringes. If the flow can be considered to be two dimensional in the plane normal to the the direction of propagation of the laser beam, then a two channel velocimeter can be used to extend the preceding analysis. If the flow has a large component in the direction along the laser beam which cannot be measured without a three-channel velocimeter then the above analysis fails completely[23]. Qiu and Sommerfeld [15] claim that the measurement of the droplet size-dependent probe cross-section by means of the logarithmic mean of the signal amplitude can be used in a *three*-dimensional flow using a *one*-channel PDV system, and they have recently extend the approach to estimate the *instantaneous three* dimensional particle velocity vector by a *one*-channel PDV system. The approach is claimed [14] to provide accuracies of the order of 5% of the volume flow rate in a liquid spray. It is worth noting that these measurements have been corrected to take account of the fact that the validation rate is less than 100% and this is frequently a larger correction factor than is the difference between calculation methods for probe area, *particularly* if a two-channel velocimeter is used. It should, however, be carefully noted that even when the the flow is predominantly unidirectional, the *direct* measurement of dispersion, i.e. *normal* to the main flow direction, is difficult for reasons connected to the cone of acceptance of the velocimeter[28].

Finally, we return to the findings of the previous sub-heading, which showed that it seems likely that up to about half of the nominal cross-sectional area of the optical probe must be disregarded because of the effects of trajectory ambiguity, at least for the largest diameters which also make the largest contribution to the measurement of droplet flux. No account - at least, no explicit account - has been taken of the loss of half of the cross-sectional area in any of the above analyses, although some preliminary calculations have shown the effect [8] for the size of the probe, and thus the good agreement for flux that has been obtained in the most recent experiments would not have been anticipated. This suggests that other factors are at work and that further research on this important topic is warranted.

Recent Developments in, and Evaluations of, the PDV and DCW

Development of the DCW. The dual cylindrical wave (DCW) system is a variant of the PDV for particle sizing and velocimetry. The instrument will not be described here, except to remark that the size is related to the *frequency* of the signal, rather than the phase, and that the potential advantages of the DCW over the PDV are that the signal processing is simpler and that the so-called "2π" ambiguity of the PDV - which arises when the phase difference between two signals exceeds 360° - may be obviated. However, the beam ellipticity is large (ratio of major to minor axes of about 1500) and hence the effects of non-uniform droplet illumination, which give rise *in extremis* to trajectory ambiguity in the PDV and thus are signals to be detected and rejected, are more important and less avoidable. Hence knowledge of the effects of non-uniform illumination is central to the development of the DCW and this characteristic makes GLMT (see above) the theoretical framework of choice for the analysis of the performance of the DCW [29]. Recent results for the collection of light at small angles to forward scatter have produced closer correspondence with experiment in terms of the relation between frequency and diameter than was possible with classical Lorenz-Mie theory.

PDV applied to small diameters. The lower limit of a PDV *adjusted to measure diameters up to the order* of 100 µm is of the order of a few µm due to oscillations at this size range in the otherwise closely linear response curve of the instrument [30], which implies that care needs to be exercised in the designation of phase measurements which are deemed to correspond to seeding particles. Recently, it has been demonstrated [31, 32] theoretically and experimentally that size resolution of the order of ±0.3 µm is possible *if* the optical arrangement of the PDV is modified to provide a *maximum* size range of about 10 µm. The modifications required are primarily to the beam crossing angle which must be of the order of 14°, as opposed to the usual few degrees in a PDV system for sizing maximum diameters an order of magnitude larger. The absolute limit of sizing by the PDV method has been investigated recently [7, 33] and is reported to be of the order of 300 nm currently, although 100 nm may be possible.

PDV applied to bubbles. There have been relatively few publications related to the sizing of bubbles and the measurement of void fraction by the PDV, even though the diameter range over which bubbles remain spherical, or nearly so, has many widely varied, though specialised, applications such as fermentation and food technology applications, the oxygenation or chlorination of liquids, cavitation, particularly in the marine context, and the nucleation of bubbles during boiling. Some of the earliest work [34] on the PDV was, however, performed in the pursuit of sizing of bubbles and the application of the PDV to the detection of the *concentration* of microbubbles in the context of cavitation nuclei for the control of water quality in cavitation tunnels has been found [35] to result in agreement to within 25% of the directly measured value for diameters over 80 µm. This figure falls to 50% as the bubble size decreases to 40 µm, although the result is reproducible which is an important attribute in this field. Recently, the direct comparison [36] between theoretical predictions of the response curve of the PDV and direct observation has been reported for bubbles in the diameter range between 10 and 1500 µm, yielding agreement to within 5%. A very brief, preliminary investigation of the effect of the Gaussian nature of the beams has recently been reported using GLMT [37], in the course of an investigation of the transient growth of bubbles in water and the coolant FC-77 from a heated surface. More work on this subject is certainly needed. Recently, Tassin and Nikitopoulos [38] have compared the performance of the PDV with direct video imaging to measure the size and velocity of bubbles from 0.25 up to about 1.1 mm diameter produced by a single injector in a quiescent environment and found agreement within 7%. Their investigation of the near injector field shows the unusual result that for bubbles below about 0.6 mm diameter, produced at high frequency and thus approaching the so-called "jet injection mode", the velocity *decelerates* with distance from the injector. The explanation given is that the detaching bubble "pushes" the preceding bubble to velocities higher than the terminal, to which the bubble stream must ultimately, however, decelerate.

The problem of sizing bubbles in turbid, or opaque, liquids, as are frequently encountered in biological applications which contain microorganisms of less than 1 μm size, can be partially alleviated by using light of long wavelength and the use of wavelengths as high as 1064 nm (from Nd:YAG lasers) have been suggested, although the advantage may be reduced due to the higher absorptivity of water at this wavelength [39]. As one example, experiments using concentration of commercial black ink up to 100% (wt/wt) and a 830 nm laser showed that bubbles up to 2000 μm diameter could be measured over depths up to 7.5 mm. The general result, as might have been expected, is that the broadening of the measured size distribution is linearly dependent on the mass concentration, diameter of inhomogeneity [40] and immersed incident path length of the inhomogeneous liquid, so that a concentration of latex spheres, 0.245 μm diameter, of 0.15% by mass resulted in a standard deviation of about 15 μm.

Evaluation of PDV. PDV measurements have begun to be made in flows in which other instruments have been the standard, or only, source of information for some years. The demonstration of some degree of correspondence between the PDV results and those obtained by these established techniques is necessary, particularly for users in industry. A good example is the measurement of the Sauter mean diameter of fuel sprays [41], for which the standard technique remains, largely, Fraunhofer diffraction. The comparison is not straightforward because the PDV provides point measurements and the Fraunhofer diffraction provides a line of sight "spatial" average. The way in which to convert PDV results into a suitable form for comparison, together with a critical assessment of the results, has been discussed [42]. Another example of evaluation of the PDV is in droplet sizing for icing cloud tunnels which have been described by Rudoff *et al.* [26, 32], it being found that agreement is good up to about 30 μm but deteriorates at larger diameters, mainly because of sampling statistics. Finally, there is difference in the optical and signal processing designs used by commercially available PDVs and it reassuring, for both experts and users, to know whether there is substantive difference in the results from these instruments. Domnick *et al.* [43] have made this comparison in a steady water spray and found agreement for arithmetic and Sauter mean diameters to within 10 and 4% (expressed as a standard deviation of seven operating conditions using three manufacturers' instruments) respectively. In contrast, there was large scatter in the integrated volume flux, the rms of the measurements obtained from the seven conditions being as poor as 28% as might be expected from the discussion above. Direct imaging of two locations in a spray has also been used to assess the measurement of mean diameters by a PDV[44]. After allowance for the conversion from temporal to spatial averages, the *worst* systematic errors between the two instruments were up to about 50 and 20% in the arithmetic and Sauter mean diameters respectively. These disappointingly large errors were attributed to the effects of multiple occupancy in the densest regions of the spray, as implied by Chin *et al.*[22].

The use of the PDV in backscatter is attractive in many applications because of the convenient optical access which this arrangement implies. Preliminary work on the calculation of the response curve in backscatter [45], which did not consider the integrating effects of finite apertures in the receiving optics, suggested that backscatter might result in non-monotonic response curves, because of the existence of two branches of the first internal reflection component of the scattered light. It should be noted that this work showed that the same result is obtained whether Lorenz-Mie theory or the so-called geometrical optics approximation is used. Subsequently [46], the incorporation of the integrating effects of the apertures suggested that this conclusion was wrong and that satisfactorily linear response curves *could* be obtained, at least up to about 50 μm, corresponding to a phase shift of 100°. However, recent work [47] has confirmed that the complete response curve *does* become strongly *non*-monotonic beyond these values, for the reasons suspected originally [45].

PDV applied to near-unity relative refractive index. The standard application of the PDV is for the measurement a dispersed phase with *relative* refractive index which is large, i.e. typically about 1.5 or its reciprocal. There are applications for the PDV, e.g. in the mixing and separation of liquids, where the *relative* refractive index is close to unity, i.e. of the order of 1.1 or its reciprocal, for which the standard theory used in geometrical optics is not applicable. The performance in these unusual situations has been examined [48] and it is suggested that the appropriate choice of scattering angle is, for refraction, 22° and parallel polarization or, for reflection, 45° and parallel polarization.

PDV applied to inhomogeneous materials. There arise, particularly in the chemical, pharmaceutical and food industries (e.g. condensed milk and solutions of instant coffee), process fluids which contain inhomogeneities (e.g. as colouring pigments, or pigments for protection against sunlight) that disturb the light scattering process of the PDV due to absorption and scattering. The magnitude of the effects depend on the size and concentration of the impurities. It should be noted that a medium must be treated as inhomogeneous when either the difference between the refractive indices of the constituents is large or when the constituents are large compared to the wavelength of light, in which case scattering by the fine particles introduces preferential forward scattering in competition with the refracted light. The disturbance can lead to useless sizing results, or at least to misinterpretation, and means to quantify and mitigate the effects of inhomogeneities have been studied [49, 40, 50]. The results show the extent to which the use of infra-red diode lasers can be advantageous (because the Mie scatter parameter of the inhomogeneities reduces, although the lower limit of measurable diameter may be raised as high as 15 μm, and because the absorption constant of some impurities in the infra-red is on many occasions reduced relative to the visible), as is the importance of perpendicular polarization of the incident light and also the use of the preferential collection of reflected light at approximately 90° side-scatter. However, in the latter case, the smallest droplets in the distribution may fail to be detected because the scattered intensity is low.

Guidance on effective PDV configurations can come from knowledge of the relative intensities of the reflected and refracted components which can be calculated using the concept of an "effective complex refractive index" (ecri): inhomogeneities cause attenuation of laser light due to scattering and to absorption which results in an imaginary part of the complex refractive index. The ecri concept is useful for predicting the relative intensities because, although the ecri cannot usually be *calculated* from information about the the size and concentration of the inhomogeneities, it can be *measured* in the laboratory using standard techniques. Experiments show that the net practical effect of the inhomogeneities is that the size distribution is broader than it is in reality. These references [49, 40, 50] have proposed a means for post-correction of the measurement to estimate the true distribution. The reader interested in this aspect is directed to the references quoted earlier for detailed evaluation of the effectiveness of the strategy.

Applications of the PDV to Sprays

There have been many PDV-based investigations of sprays in the contexts of basic studies [51] and applied work related either to gas turbines [52], utility burners (described in greater detail below) and diesel [20, 53, 54] and gasoline [41, 55] sprays and the associated spray-wall interaction [56, 57]. The preceding references are quoted to provide the interested reader with some of the most recent entries to the literature, although the list is very far from comprehensive. Here, some emphasis is placed on the publications from two groups on burner-related flows. PDV measurements (i.e size-resolved velocity and mean diameter characteristics) of atomized sprays have been reported in the inert [58, 59] and reacting [60] flows aft of bluff bodies, for the interaction of sprays with the large-scale eddies in the inert wake of a cylinder as a function of the free stream velocity [61] and inert swirling flow fields downstream of a vane swirler arrangement as a function of the swirl number [62, 63]. Detailed spray measurements (i.e size-resolved velocity, mean diameter and flux characteristics) have also been reported downstream of the exit diffuser ("quarl") in a swirl-stabilised flame of a small kerosene burner [18, 16] and *within* the quarl of another kerosene burner [64], the particular advantage of the latter two sets of experiments being that the upstream boundary conditions of the swirling air and of the atomized spray are available and thus make the data sets valuable for the evaluation of CFD codes. In the context of utility burners, it is important to be able to understand the findings and extrapolate these to full scale operation, which may involve several tens of megawatts (thermal) heat release. The approach adopted [18, 65, 64] has been to normalise typical time scales associated with the turbulence and with the mean axial and swirl (i.e. centrifuging) motion with the Stokesian time scales of droplets associated with up to three size classes in the flow. Finally, it should be noted that it is frequently useful to measure the correlation $\overline{u_p v_p}$, where the subscripts denote "particulate", as a function of diameter. This is usually done by using a two channel PDV [66], but recently it has been shown [67] that a *single* channel PDV can be used to measure this correlation using Hoagland's method but without having to move the receiving optics location. This has obvious advantages in terms of cost over a two channel system, although the accuracy is somewhat reduced.

MEASUREMENT OF THE TEMPERATURE OF THE DISPERSED PHASE

In this review, the measurement of the temperature of the dispersed phase will be discussed, briefly, in the context of two methods, namely Rainbow Refractometry (for droplets) and Two Colour Pyrometry (for combusting fuel droplets or particles), although the existence of other methods should be noted, such as fluorescence [68], intensity ratioing thermometry [69] and the use of using two receiving optical units at different mean scattering angles in an extended PDV arrangement [65]. In the case of Rainbow Refractometry, the temperature is inferred by shining laser light onto a droplet and measuring the primary rainbow angle which is approximately [70] linearly dependent on refractive index **and** quadratically dependent on droplet diameter below about 100 µm. Provided that these two quantities are measured, the liquid temperature can be found and so, in general, simultaneous particle sizing measurement is required. The experimental details for this simultaneous measurement have been described using, for sizing, either the high frequency "ripple" structure of light intensity in the vicinity of the rainbow [71, 72], or by combining with a PDV [70]. The object of measuring refractive index can be to examine transient droplet heat-up or to provide a correction to the response curve of the PDV, which is - through refractive index - sensitive to change in temperature. To do this, it has usually been assumed, for both the response curve of the PDV and for the location of the rainbow, that the refractive index is radially uniform during droplet heat up. The defence [65] of the assumption of radial uniformity, apart from the advantage of simplicity, is that at least for scattering *intensities* it is the refractive index near the droplet's surface that is the main region of importance. This assumption is, however, undoubtedly questionable and hence its implications have been recently examined using the conduction - limit (i.e. in the absence of convective currents within the droplet) which is also expected to be a worst case analysis for decane droplets initially at 300 K and transiently heated to 2000 K at 10 bar, typical of gas turbine operation. The refractive index changes, in this extreme case, from 1.41 and 1.28. For the PDV [73], forward and backscatter collection geometries were investigated and, even though the latter was chosen to be at an angle (160°) well clear of the rainbow angle, the error in backscatter is by far the worse, being up to 80%. For rainbow refractometry, it is found [74] that the monotonic dependence between rainbow location and temperature is, unfortunately, lost at these conditions. This subject is still one of intense research and, because it has long been known that, at 30° *forward* scatter angle, a 7% error in the refractive index results in a sizing error [75] of 10%, an estimate which coincides with that recently made by others [70, 73], a cost-effective recommendation, at least for the present, at this scatter angle is to account for the effect of temperature on the response curve of the PDV by noting this as an upper bound on the systematic error of the measurement. This crude statement should not be used to *correct* the results during transient heat conduction through a droplet, because the slope of the response curves of the PDV are *not* bounded by the corresponding values for the initial and final values of the refractive index (i.e. low and asymptotic approach to the high temperature states, respectively). It should be noted that, in any case, the variation of refractive index with temperature is seldom known accurately: if the dependence of density with temperature is known, however, the refractive index can be estimated using the Eykman equation. If there is sufficient laser power to do so, extensive analysis of a wide range of angles [65] has shown that use of scatter angles of about 70° results in low sensitivity to refractive index change. *Backscatter* angles are unlikely to be generally suitable because of the strong dependence of the position of the rainbow angle, a region which must be avoided in PDV, with refractive index and hence, by implication, with temperature. Even if the collection optics are arranged to avoid receiving light from the rainbow, a change in the refractive index from 1.4 to 1.33 can result [47] in sizing errors of the order of 100% for droplets of about 60 µm diameter. However, the number of local intensity maxima on the backscattered fringe pattern, along the bisector plane, is monotonically related to the droplet diameter and could provide [47] size information, albeit with a systematic 25% error due to variations of refractive index between 1.4 and 1.33.

The instruments of the preceding paragraph measure the liquid temperature during the heat-up of droplets. This paragraph is concerned with the next stage of the process, namely measuring the temperature at which a *single* droplet or particle *burns*, using two colour pyrometry. The instrument is based on collecting the *thermal* radiation which is *emitted* (i.e. not *scattered laser* light, as in all other instruments described above) by a hot material, which in practice is, for a burning droplet, the soot in the surrounding flame mantle, or, for combustion of pulverised coal, *either* the glowing char particle after complete devolatilisation *or* the combined emission from the soot mantle and glowing coal particle *during* devolatilisation of the coal. The temperature is inferred from the ratio of the intensity of light at two wavelengths and the principle is based on the common observation that the maximum value of the radiation intensity moves to shorter wavelengths with increasing temperature, quantitatively described by Wien's law. The design and optimisation of the instrument have been described over many years in the context of heated *inert* solids and for char *after* the volatiles have been completely released, including the need - in a *two* colour pyrometer - to assume that the emissivity is from a so-called grey body: a combined pyrometer and sizing velocimeter, not based on Doppler velocimetry, has been described[76]. Recently, a combined two colour pyrometer and PDV has been used [77] to measure droplet diameter, velocity and the temperature of the surrounding soot mantle in an atomized *kerosene* fired-, swirl-stabilised flame which, apart from providing the temperature of the burning mantle, quantifies the location at which droplets are burning and, hence, which part of the droplet flux remains inert. The extension of this instrument to the measurement of comparable quantities in pulverised coal combustion is complicated by, first, smaller emitted intensities as compared to droplet mantle combustion and, second, that - in general - the signal can derive *simultaneously* from emission of the burning volatiles *and* from the glowing particle. Fortunately, it has been shown [78] that the process is such that the light emitted during devolatilisation is predominantly from the burning volatiles and, only after release and combustion of all volatiles, is from the glowing char. Thus, the problem reduces to discrimination between the two possibilities and this is necessary because the response function of the pyrometer (i.e. between the temperature and ratio of emitted intensities at the two wavelengths) is different for soot and for a hot solid. Current work [78] is directed to establishing this discrimination criterion. Finally, once again in the context of pulverised coal combustion, it is worth noting that char is a mixture of substances including coal and fly ash and that it is desirable to monitor the progress of burn-out in a furnace. Card and Jones [79] have proposed a light scattering technique to discriminate between these two constituents which remains applicable during combustion and is successful in the range below 15 μm. Thus the transition from coal to fly ash can be followed continuously, enabling combustion rates to be studied.

NON-PDV MEASUREMENT OF TWO PHASE FLOW

PIV measurement. A brief overview of the *applications* of particle image velocimetry (PIV) will be considered here, although the reader interested in the *technique* is directed to other, specialist, references. Hassan *et al.* [80] have used PIV to measure the motion of the continuous (liquid) phase induced by a rising column of bubbles and this work demonstrates one way of removing the light "corona" reflected at the air/water interface, which otherwise swamps the light scattered by seed particles in the liquid phase in the vicinity of the bubbles. The resulting images provide information about the liquid flow pattern (streamlines, particle trace and vorticity plots) surrounding each bubble. Similar results, obtained in a slightly different way, have also been reported [81]. PIV has also been used to characterize particle dispersal in jets [82].

Measurement of non-spherical particle size. Despite the strong economic, technical and scientific incentives to measure flows with non-spherical particulates, the past decade has seen surprisingly little work reported on either the development of new, or application of existing, instruments. To some extent, this observation reflects the difficulty with which the measurement of size of non-spherical particles is made, as compared with the PDV for spherical particles. The instruments available for this purpose have been known, in most cases, since at least the late 1970s, are based on Fraunhofer diffraction and have already been reviewed [1,3,4]: suffice it here to say that the particles are illuminated by a laser beam and the maximum intensity of the predominantly *diffractively* scattered light is related to the particle size by earlier calibration. Thus, the sizing information is amplitude-, rather than frequency-, modulated and this implies the scrupulous maintenance of alignment, of the intensity of the laser beam and of detector sensitivity and the daily - at least - checking of calibration. This is made the more critical by the need to use a small laser Doppler velocimeter probe so as to place a so-called "pointer volume" within the larger Gaussian intensity distribution of the sizing beam to avoid trajectory ambiguity.

Although the end purpose of these instruments is the same, in broad terms, as that of the PDV, there are some important differences. First, these instruments must be insensitive to the refractive index, and particularly the imaginary part of the complex refractive index, of the material. This requirement is especially important to sizing coal particles in burners because the magnitude of the imaginary part is usually unknown and, in any case, combustion changes this property by a large amount. The degree to which this insensitivity exists has been assessed in the past by the use of Lorenz - Mie theory, thus implying that the particle was spherical and, in general, the insensitivity is acceptable. Second, the meaning of "accuracy and precision" of sizing provided by the instrument requires careful definition but the sizing tolerance required - at least for a simple instrument - is less than for the PDV. For example, little meaning can be assigned to the statement that the size of *rough* coal particle is, say, 60 ± 1 μm: but equally, a statement that the size is 60 ± 30 μm renders the measurement useless. The accuracy of sizing non-spherical particles has been reported recently [83, 84], by comparing the "optical size", i.e. the measurement of size inferred from the maximum amplitude using a calibration curve derived from an experiment with precision pinholes, with an "independent" size measurement. A preliminary study [83] defined the "independent" size from measurement of the projected area of the particle and established that the tolerance was of the order of ±5 μm. The main drawback to this experiment was that, because the procedure was so laborious, only a few tens of particles could be size in this way. Thus, later, an alternative "independent" size measurement was obtained [84] from the use of a rapidly accelerating flow in a nozzle, so that the measurement of the particle exit velocity from the nozzle could be directly related to its aerodynamic size, from a calculation of the particle motion in the - given - nozzle flow field (in a similar context, the deceleration of a particle-laden flow upstream of a cylinder has served a similar purpose [85], although surface splashing or bouncing can result in ambiguities with this approach [86]. Nevertheless, the usefulness of ideas such as these for checking the performance of instruments cannot be over-emphasized). This permitted large numbers of measurements to be made and the accuracy and precision of sizing were found to be +2, +7 and - 20 μm and ±2, ±7 and ±10 μm for nominal sizes of 15, 30 and 60 μm, the large loss of accuracy for the largest diameter likely to be due to photodetector saturation with increasing particle size. This instrument has been used for an extensive detailed investigation of coal particle trajectories in a 10 kW (thermal) model of a swirl-stabilised burner [87]. A parallel theoretical investigation [88] was conducted by modelling the coal as ellipses with a distribution of aspect ratios to assess whether the measured tolerances are those that might be expected. The agreement between theory and experiment was encouraging and suggests that much of the imprecision *is* due to gross aspects of irregularity, i,.e. aspect ratio. A simple bi-zonal aperture collection lens, together with one additional photodetector, was proposed to detect the particles with large aspect ratio and it is suggested that, by rejecting approximately half the measurements, it is possible to double the precision. A sounder means for characterizing the irregular particles than using the aspect ratio is through the probability distribution of radius and a correlation function at the surface [89]. At angles not too far from the forward scattering direction, and for irregularity heights within a limited range, irregular particles may be represented approximately as spheres with a size distribution equal to the probability distribution of radius. This implies that the error in size measurement can be estimated and, because each individual irregular particle behaves as though it is a size distribution of spheres, the effect is to broaden - artificially - the true size distribution by an amount of the order of the height of the irregularity.

The direct imaging of some two phase flows provides information on not only size, but also on other aspects such as velocity and flow visualisation, which supplements the spatially precise instruments highlighted above. In bubbly two phase flows, where non-spherical (initially, ellipsoidal) bubbles occur above about 1 mm diameter, the ability to measure even a "representative" average diameter may become difficult and certainly depends on the orientation of the bubble. For example, for bubbles about 1.1 mm diameter, early work suggested that the PDV size measurement was a good representation of the minor axis of the ellipsoidal bubble. More recent work, particularly for bubbles up to volume-averaged diameter of 2.7 mm, suggests that errors of up to 20% exist [38] and it is likely that, given the technical incentives to investigate bubbles of this size and larger, the *optical* technique of choice will be some kind of direct imaging [§]. Another excellent example of the value of direct imaging is provided by investigation of wall/droplet collision using an elegant technique [90] which permitted a high speed event, which would require a 50 000 frame/s camera, to be recorded by a standard video camera. Thus, for example, the work [90] not only confirms the surprising results reported earlier [56] which suggested that secondary atomization can result in *larger* droplets being ejected from the wall than were incident on it, but also shows *how* this comes about. Another recent example of direct imaging [91], once again concerning the secondary atomization of droplets, is an investigation into the wetting, transition and

---

[§] always bearing in mind that the optical disturbance to optical techniques, with even a few such "large" bubbles, is great and it seems likely that *physical* probe measurements will remain dominant [82]

non-wetting regimes associated with specific impingement conditions of droplet size, velocity, surface temperature and ambient pressure. Direct imaging permits information such as an average number of "splashed" droplets per incident droplet to be measured, as well as determining the coefficient of restitution.

A novel instrument [92] which combines some aspects of direct imaging with those of the laser Doppler velocimeter is the so-called Shadow Doppler anemometer which involves the imaging of the probe volume onto linear, or two dimensional, detector arrays. The instrument permits the size, shape and velocity of the particle to be found in the range from about 5 to several hundred µm, with the advantages over the Fraunhofer instruments with pointer volumes that the optical arrangement is robust, is insensitive to variations in laser power or detector gain and hence should be suitable for applications where the beams have to pass through windows and seems to be immune from trajectory ambiguity effects. The disadvantage is the need to detect out-of-focus images which occur, typically, when the particle is displaced about ±0.050 mm from the centre of the probe volume. The technique appears promising but requires much development effort.

CONCLUSIONS

1. The contribution has reviewed techniques for the measurement of the *size, velocity, flux and concentration* of the dispersed phase, primarily in the context of recent additions to the literature over the past two years in the context of reacting, as well as inert, flows.

2. The main developments have been classified under the headings of the Phase Doppler velocimeter, the measurement of the temperature of the dispersed phase and non-PDV measurement of two phase flow.

3. The dominance of the PDV has arisen because of its ease of use, its good accuracy, competitive cost and commercial availability. Although it is well-established instrument, the subjects of trajectory ambiguity and errors in the measurement of number density and flux require further research:

3.1 it is likely that the errors due to trajectory ambiguity will be found to be small, either because of the mode of operation of *existing* signal processing techniques rejects signals suffering from gross trajectory ambiguity or because of the application of measurement strategies as described in the article;

3.2 the errors in the measurement of density and flux need further research. Currently, the largest readily identifiable factors are the signal validation rate and the error in the measurement of particle residence time in the probe volume due to the effects of noise;

3.3 the range of applications of the PDV is currently dominated by fuel sprays. Recently, however, steps have been made towards the measurement - amongst others - of small gas bubbles and of inhomogeneous materials.

4.1 The temperature of heated droplets can be measured in a variety of ways, including rainbow refractometry;

4.2 the effects of changes in refractive index on the response curve of the PDV due to changes in liquid temperature are of the order of 10% and 80% at 30° and 160° scatter angles. A scatter angle of 70° results in the lowest sensitivity to changes in refractive index.

4.3 the temperature of burning droplets can be measured by two colour pyrometry: that of burning coal requires the use of a recently developed criterion to distinguish between a glowing coal particle and a particle surrounded by a burning soot cloud.

5. The non-PDV measurement of two phase flows has been characterized by PIV measurements and the measurement of non-spherical particle size. Of the latter, the Shadow Doppler velocimeter offers a novel route for this application which promises to be free of some of the disadvantages of existing Fraunhofer diffraction-based instruments.

ACKNOWLEDGEMENTS

I gratefully acknowledge the critical guidance of Prof. Jim Whitelaw on the development of two phase flow instrumentation over many years. This article has benefitted from many discussions with Dr Yannis Hardalupas and Prof Koichi Hishida and from the assistance of Messrs F Israel, H Morikita, N G Orfanoudakis and M Posylkin. The financial support for the work has come from, among other sponsors, the SERC Electro-Mechanical and Process Engineering Committees and the CEC DG XII, JOULE programmes.

REFERENCES

1. A M K P Taylor, Optically-based measurement techniques for dispersed two phase flow, *in*: "Combusting Flow Diagnostics", D F G Durão *et al.*, eds., Kluwer Academic Publishers, Dordrecht (1992).
2. M V Heitor, S H Stårner, A M K P Taylor, J H Whitelaw, Velocity, size and turbulent flux measurements by laser-Doppler velocimetry, *in*: "Instrumentation for Flows with Combustion", A M K P Taylor, ed., Academic Press, London (1993).
3. A R Jones, Light Scattering for particle characterization, *in*: "Instrumentation for Flows with Combustion", A M K P Taylor, ed., Academic Press, London (1993).
4. N Chigier (ed.). "Combustion Measurements", Hemisphere Publishing Corporation, Washington (1991).
5. S V Sankar, A S Inenaga, W D Bachalo, Trajectory dependent scattering in phase Doppler interferometry: minimizing and eliminating sizing errors, *in* "Sixth International Symposium on Applications of Laser Techniques to Fluid Mechanics", R J Adrian *et al.* eds., paper 12.2, Lisbon (1992)
6. G Gouesbet, Generalized Lorenz-Mie Theory and Applications, *in* "Third International Congress on Optical Particle Sizing", p. 393, M Maeda *et al.*, eds., Yokohama (1993).
7. A A Naqwi, Innovative Phase Doppler Systems and their application, *in* "Third International Congress on Optical Particle Sizing", p. 245, M Maeda *et al.*, eds., Yokohama (1993).
8. C H. Liu, The applications of laser- and phase-Doppler velocimetry, Ph D Thesis, University of London (1992).
9. M Saffman, The use of polarized light for optical particle sizing, *in* "Laser Anemometry in Fluid Mechanics III", pp 387 - 398, R. J. Adrian *et al.*, eds., LADOAN - Instituto Superior Técnico, Lisbon (1986).

10. Y Hardalupas, A M K P Taylor, Phase validation criteria of size measurements for the phase Doppler technique, accepted for publication, *Expts Fluids* (1993).
11. P Haugen, H-H von Benzon, Phase Doppler measurement of the diameter of large particles - considerations for optimization, *in* "Third International Congress on Optical Particle Sizing", p. 301, M Maeda *et al.*, eds., Yokohama (1993).
12. Y Aizu, F Durst, G Gréhan, F. Onofri, T-H Xu, PDA system without Gaussian beam defects, *in* "Third International Congress on Optical Particle Sizing", p. 461, M Maeda *et al.*, eds., Yokohama (1993).
13. Y Hardalupas, Description of the Fluids Section 'Model 2' Phase Doppler Counter. Report FS/90/29, Fluids Section, Department of Mechanical Engineering, Imperial College, London (1990).
14. J Domnick, F Durst, A Melling, H-H Qiu, M Sommerfeld, M Ziema, A new generation of phase-Doppler instruments for particle velocity, size and concentration measurements, *in* "Third International Congress on Optical Particle Sizing", p. 407, M Maeda *et al.*, eds., Yokohama (1993).
15. H-H Qiu, M Sommerfeld, A reliable method for determining the measurement volume size and particle mass fluxes using phase-Doppler anemometry, *Expts Fluids*, 13: 393-404 (1992).
16. Y Hardalupas, A M K P Taylor, J H Whitelaw, Mass flux, mass fraction and concentration of liquid fuel in a swirl-stabilised flame. Submitted to *Int J Multiphase Flow* (1994).
17. Y Hardalupas, A M K P Taylor, J H Whitelaw, Velocity and particle flux characteristics of turbulent particle-laden jets, *Proc. R. Soc. Lond. A* **426**: 31 - 78 (1989).
18. Y. Hardalupas, A M K P Taylor, J H Whitelaw, Velocity and size characteristics of liquid–fuelled flames stabilized by a swirl burner, *Proc. R. Soc. Lond. A* **428**, 129–155, (1990).
19. Y. Hardalupas, A M K P Taylor, J H Whitelaw, Particle dispersion in a vertical round sudden expansion flow, *Phil. Trans R Soc Lond A* **341**: 411 - 442, (1992).
20. Y Hardalupas, A. M. K. P. Taylor, J. H. Whitelaw, Characteristics of unsteady Diesel sprays, *Int. J. Multiphase Flow*, 18: 159 - 179 (1992)
21. W K Chin, K D Marx, C F Edwards, Effect of particle statistics on measurements of transient sprays by single-particle counters, paper 18.4.1, in "Sixth International Symposium on applications of laser techniques to fluid mechanics", R J Adrian *et al.*, eds., Lisbon (1992).
22. Y. Hardalupas, A M K P Taylor, On the measurement of Particle concentration near a stagnation point, *Expts in Fluids*, 8:113 - 118, (1989). See also Erratum, *Expts. Fluids*, 9:356 (1990).
23. J Y Zhu, R C Rudoff, E J Bachalo, W D Bachalo, Number density and mass flux measurements using the phase Doppler particle analyzer in reacting and non-reacting swirling flows, AIAA 93 - 0361, *31st Aerospace Sciences Meeting & Exhibit* (1993).
24. M Saffman, Automatic calibration of LDA measurement volume size, *Applied Optics*, 26: 2592 - 2597 (1987).
25. W D Bachalo, R C Rudoff, A Breña de la Rosa, Mass flux measurements of a high number density spray system using the phase Doppler particle analyzer, *AIAA 88-0236* (1988).
26. J R Oldenburg, R F Ide, Comparison of two droplet sizing systems in an icing wind tunnel, AIAA 93 - 0668, *31st Aerospace Sciences Meeting & Exhibit* (1993).
27. Y Hardalupas, M Posylkin, Evaluation of the phase-Doppler counter in noisy conditions, Internal Report TF/93/06 of the Thermofluids

Section, Department of Mechanical Engineering, Imperial College, London SW7 2BX (1992).

28. Y. Hardalupas, A M K P Taylor, J H Whitelaw, Fringe count limitations on the accuracy of velocity and mass flux in two-phase flows, *in* “Applications of laser techniques to fluid mechanics”, pp. 183 - 203, R J Adrian *et al.*, eds., Springer Verlag, Berlin (1991).
29. G Gréhan, K F Ren, G Gouesbet, A Naqwi, F Durst, Evaluation of a particle sizing technique based on laser sheets, *in* "Third International Congress on Optical Particle Sizing", p. 127, M Maeda *et al.*, eds., Yokohama (1993).
30. Y Hardalupas and A. M. K. P. Taylor, The identification of LDA seeding particles by the phase-Doppler technique, *Expts. in Fluids.* 6: 137 - 140 (1988).
31. S V Sankar, B J Weber, D Y Kamemoto, W D Bachalo, Sizing fine particles with the phase Doppler interferometric technique, *Applied Optics.* 30: 4914 -4920.
32. R C Rudoff, E J Bachalo, W D Bachalo, J R Oldenburg, Performance of the phase Doppler particle sizing analyzer icing cloud droplet sizing probe in the NASA Lewis icing research tunnel. *AIAA 92-0162, 30th Aerospace Sciences Meeting & Exhibit* (1992).
33. H-H Benzon, P Buchhave, The phase-Doppler method applied to very small particles, *in* "Third International Congress on Optical Particle Sizing", p. 261, M Maeda *et al.*, eds., Yokohama (1993).
34. M Saffman, P Buchhave, H Tanger, Simultaneous measurement of size, concentration and velocity of spherical particles by a laser Doppler method, *in* "Second International Symposium on the Applications of Laser Anemometry to Fluid Mechanics", paper 8.1, D F G Durão *et al.*, eds., Lisbon (1984).
35. H Tanger, E-A Weitendorf, Applicability tests for the phase Doppler anemometer for cavitation nuclei measurements, *Trans ASME, J Fluids Eng.,* 114:443 - 449 (1992).
36. A Breña de la Rosa, S V Sankar, B J Weber, G Wang, W D Bachalo, A theoretical and experimental study of the characterization of bubbles using light scattering interferometry, *Trans ASME, J Fluids Eng.* 113: 460 - 468 (1991).
37. T Nonn, Z Dagan, Characterizing bubble growth using phase Doppler techniques,*in* "Third International Congress on Optical Particle Sizing", p. 449, M Maeda *et al.*, eds., Yokohama (1993).
38. A L Tassin, D E Nikitopoulos, Non-intrusive measurements of bubble size and velocity. Submitted for publication, *Expts Fluids* (1993).
39. K Bauckhage, T Wriedt, T Meyer, Sizing bubbles in an opaque liquid using an infrared laser-diode PDA, *in* "Third International Congress on Optical Particle Sizing", p. 415, M Maeda *et al.*, eds., Yokohama (1993).
40. K Bauckhage, T Wriedt, Phase Doppler particle sizing using infrared laser diodes, *in* "Third International Congress on Optical Particle Sizing", p. 435, M Maeda *et al.*, eds., Yokohama (1993).
41. Hardalupas, Y , Taylor, A M K P , Whitelaw, J H , "Unsteady sprays by a pintle injector" *JSME International Journal*, series II, 33: 177 - 185 (1990).
42. E Cossali, Y Hardalupas, Comparison between laser diffraction and phase Doppler velocimeter techniques in high turbidity, small diameter sprays, *Expts Fluids* 13: 414 - 422 (1992).
43. J Domnick, V Dorfner, F Durst, M Yamashita, Performance evaluation of different phase-Doppler systems, *in* "Third International Congress

on Optical Particle Sizing", p. 425, M Maeda *et al.*, eds., Yokohama (1993).

44. D R Alexander, D F F Gutierrez, Spray characterisation of a NASA MOD-1 nozzle, *J Laser Applications*, 2: 49 - 54 (1990).
45. S. A. M. Al-Chalabi, Y. Hardalupas, A. R. Jones, A. M. K. P. Taylor, Calculation of calibration curves for the phase Doppler technique: comparison between Mie theory and geometrical optics, in G. Gouesbet and G. Gréhan (Eds.), *in* "Optical particle sizing: theory and practice", Plenum Press, New York, pp. 107 – 121 (1988).
46. S V Sankar, W D Bachalo, Response characteristics of the phase Doppler particle analyzer for sizing spherical particles larger than the light wavelength, *Appl. Optics* 30: 1487 (1991).
47. Y Hardalupas, C H Liu, Backscatter phase-Doppler anemometry for transparent non-absorbing spheres, *Expts Fluids* 14: 379 - 390 (1993).
48. P. Haugen, E I Hayes, PDA measurements of drop size at near unity relative refractive index, *in* "Third International Congress on Optical Particle Sizing", p. 285, M Maedà *et al.*, eds., Yokohama (1993).
49. U Manasse, T Wriedt, K Bauckhage, Optimization of PDA parameters for sizing of optically inhomogeneous liquid droplets, *in* "Third International Congress on Optical Particle Sizing", p. 453, M Maeda *et al.*, eds., Yokohama (1993).
50. U Manasse, T Wriedt, K Bauckhage, Reconstruction of real size distributions hidden in phase-Doppler anemometry results obtained from droplets of inhomogeneous liquids, *in* "Third International Congress on Optical Particle Sizing", p. 293, M Maeda *et al.*, eds., Yokohama (1993).
51. Y Levy, Laminar spray combustion, paper 34.3.1, *in* "Sixth International Symposium on applications of laser techniques to fluid mechanics", R J Adrian *et al.*, eds., Lisbon (1992).
52. V G McDonell, G S Samuelsen, Application of two component phase Doppler interferometry to the measurement of particle size, mass flux, and velocities in two phase flows, *in* 22nd Symp (Intl) on Combution, The Combustion Institute, Pittsburgh (1988).
53. G Pitcher, G Wigley, A study of the breakup and atomization of a combusting diesel fuel spray by phase Doppler anemometry, paper 25.1.1, *in* "Sixth International Symposium on applications of laser techniques to fluid mechanics", R J Adrian *et al.*, eds., Lisbon (1992).
54. C Arcoumanis, E Cossali, G Paál, J H Whitelaw, Transient characteristics of mutli-hole diesel sprays, *SAE paper 900480.*
55. F Vannobel, J B Dementhon, D Robart, Phase Doppler anemometry measurements on a gasoline spray inside the inlet port and downstream of the induction valve: steady and unsteady flow conditions, paper 29.2.1, *in* "Sixth International Symposium on applications of laser techniques to fluid mechanics", R J Adrian *et al.*, eds., Lisbon (1992).
56. Y Hardalupas, S Okamoto, A M K P Taylor, J H Whitelaw, Application of a phase Doppler anemometer to a spray impinging on a disc, *in* "Applications of laser techniques to fluid mechanics", Springer - Verlag. In the Press (1993).
57. C Arcoumanis, J-C Chang, Heat transfer between a heated plate and an impinging transient diesel spray, *Expts Fluids* 16: 105 - 119 (1993).
58. R C Rudoff, M J Houser, W D Bachalo, Experiments on spray interactions in the wake of a bluff body, *Trans ASME, J Eng Gas Turbines and Power* 110: 86 - 93 (1988).
59. A Breña de la Rosa, W D Bachalo, Particle diagnostics and turbulence

measurements in an isothermal spray under the influence of stabilizing bodies, paper 44, ICLASS - 91 (1991).

60. Y Hardalupas, C H Liu, J H Whitelaw, Experiments with disk stabilised kerosene-fuelled flames, submitted to *Comb Flame* (1993)
61. W D Bachalo, E J Bachalo, J Hanscom, S V Sankar, An investigation of spray interaction with large-scale eddies, *AIAA 93-0696, 31st Aerospace Sciences Meeting and Exhibit* (1993).
62. A Breña de la Rosa, G Wang, W D Bachalo, The effect of swirl on the velocity and turbulence fields of a liquid spray, *Trans ASME, J Eng Gas Turbines and Power* 114: 72 - 81 (1992).
63. A Breña de la Rosa, S V Sankar, G Wang, W D Bachalo, Particle diagnostics and turbulence measurements in a confined isothermal liquid spray, *Trans ASME, J Eng Gas Turbines and Power* 115: 499 - 506 (1993).
64. V D Milosavljevic, Natural gas, kerosene and pulverised fuel fired swirl burners, Ph D Thesis, University of London (1993).
65. G Pitcher, G Wigley, M Saffman, Sensitivity of dropsize measurements by phase Doppler anemometry to refractive index changes in combusting fuel sprays, *in* "Applications of laser techniques to fluid mechanics", R J Adrian *et al.*, eds, pp. 227 - 247, Springer Verlag, Berlin (1991).
66. V G McDonell, G S Samuelsen, Evolution of the two-phase flow in the near field of an air blast atomizer under reacting and non-reacting conditions, *in* "Applications of laser anemometry to fluid mechanics", R J Adrian *et al.*, eds., pp. 255 - 278, Springer Verlag, Berlin (1989).
67. Y Hardalupas, C H Liu, Size-discriminated cross-correlation measured by a single channel phase Doppler velocimeter, *in* "Sixth International Symposium on Applications of Laser Techniques to Fluid Mechanics", R J Adrian *et al.* eds., paper 18.2, Lisbon (1992).
68. A M Murray, L A Melton, Fluorescence methods for determination of temperature in fuel sprays, *Appl. Optics* 24: 2783 - 2787.
69. P Massoli, F Beretta, A D'Alessio, M Lazzaro, Temperature and size of single transparent droplets by light scattering in the forward and rainbow regions, *Applied Optics* 32: 3295-3301 (1993).
70. S V Sankar, K M Ibrahim, D H Buermann, M J Fidrich, W D Bachalo, An integrated Phase Doppler/Rainbow Refractometer system for simultaneous measurement of droplet size, velocity, and refractive index, *in* "Third International Congress on Optical Particle Sizing", p. 275, M Maeda *et al.*, eds., Yokohama (1993).
71. K. Anders, N Roth, A Frohn, Light scattering at the rainbow angle: information on size and refractive index, *in* "Third International Congress on Optical Particle Sizing", p. 237, M Maeda *et al.*, eds., Yokohama (1993).
72. N Roth, K Anders, A Frohn, Measurement of the rainbow position using a PSD-sensor, *in* "Third International Congress on Optical Particle Sizing", p. 183, M Maeda *et al.*, eds., Yokohama (1993).
73. M Schneider, E D Hirleman, Influence of refractive index gradients on size measurement of spherically-symmetric particles by phase-Doppler anemometry. *Applied Optics*. In the press.
74. M Schneider, E D Hirleman, H Saleheen, D Q Chowdhury, S C Hill, Rainbows and radially-inhomogeneous droplets, *in* "Third International Congress on Optical Particle Sizing", p. 323, M Maeda *et al.*, eds., Yokohama (1993).

75. I Hardalupas, Experiments with isothermal two phase flows, Ph D Thesis, University of London (1989).
76. D A Tichenor, E R Mitchell, R K Hencken, S Niksa, Simultaneous in situ measurements of size, temperature and velocity of particles in a combustion environment. *Twentieth Symposium (International) on Combustion,* The Combustion Institute, pp. 1213 - 1221 (1984).
77. F Israel, A M K P Taylor, J H Whitelaw, Simultaneous measurement of droplet velocity, size and flame mantle temperature measured by a PDA & two colour pyrometry in a piloted kerosene, swirl stabilised flame, paper 34.2.1, *in* "Sixth International Symposium on applications of laser techniques to fluid mechanics", R J Adrian *et al.*, eds., Lisbon (1992)
78. F Israel, Ph D Thesis in preparation (1994).
79. J B A Card, A R Jones, A light scattering study of coal combustion in a drop tube furnace, *in* "Third International Congress on Optical Particle Sizing", p. 159, M Maeda *et al.*, eds., Yokohama (1993).
80. Y A Hassan, T K Blanchat, C H Seeley, Jr., R E Canaan, Simultaneous velocity measurements of both components of a two phase flow using particle image velocimetry, *Int J Multiphase Flow*, 18: 371 - 395 (1992).
81. T Uemura, K Ohmi, F Yamaoto, Mixing flow in a cylindrical vessel agitated by a bubbling jet - application of image processing velocimetry, *in* "Applications of laser techniques to fluid mechanics", R J Adrian *et al.*, eds, pp. 521 - 536, Springer Verlag, Berlin (1991).
82. D R McCluskey, W J Easson, D H Glass, C A Greated, The characterisation of particle jet dispersal, *in* "Sixth International Symposium on Applications of Laser Techniques to Fluid Mechanics", R J Adrian *et al.* eds., paper 9.2.1, Lisbon (1992).
83. N G Orfanoudakis, A M K P Taylor, The effect of particle shape on the amplitude of scattered light for a sizing instrument, *Part Part Syst Characterisation* 9: 223 - 230 (1992).
84. N G Orfanoudakis, A M K P Taylor, Evaluation of a sizing anemometer and application to a small scale swirl stabilized coal burner, paper 5.3, *in* "Second International Conference on Combustion Technologies for a Clean Environment", M G Carvalho *et al.*, eds, Lisbon (1993).
85. R C Rudoff, D Y Kamemoto, W D Bachalo, Effects of turbulence and number density on the drag coefficients of droplets, *AIAA 90-0074*, 29th Aerospace Sciences Meeting (1990).
86. G S Jones, D Y Kamemoto, L R Gartrell, An investigation of the effects of seeding in laser velocimeter systems, *AIAA 90-0502*, 28th Aerospace Sciences Meeting (1990).
87. N G Orfanoudakis, Ph D Thesis in preparation (1994).
88. Y Hardalupas, H Morikita, A M K P Taylor, Amplitude measurement of size of non-spherical particles: tolerances and validation schemes evaluated by scalar diffraction theory, *in* "Third International Congress on Optical Particle Sizing", p. 485, M Maeda *et al.*, eds., Yokohama (1993).
89. S A M Al-Chalabi, A R Jones, Development of a mathematical model for light scattering by statistically irregular particles, *in* "Third International Congress on Optical Particle Sizing", p. 113, M Maeda *et al.*, eds., Yokohama (1993).
90. K Anders, N Roth, A Frohn, The velocity change of ethanol droplets during collision with a wall analysed by image processing, *Expts Fluids.* 15: 91 - 96 (1993).

91. J D Naber, P V Farrell, Hydrodynamics of droplet impingement on a heated surface, *SAE* paper 930919 (1993).
92. Y Hardalupas, K Hishida, M Maeda, H Morikita, A M K P Taylor, J H Whitelaw, A shadow Doppler technique for sizing particles of arbitrary shape, accepted for publication, *Applied Optics* (1994).

COLLECTIVE LIGHT SCATTERING : AN INTRODUCTION

Dominique M. Grésillon and Cyril Honoré

Laboratoire de Physique des Milieux Ionisés du CNRS
Ecole Polytechnique
F-91 128, Palaiseau
France

INTRODUCTION

Light scattering is one of the most common diagnostics of every day's life. In addition to the object to be looked at, it requires only two devices, a light source, and a detector, which are both provided by the sun and the eye. The sun light that is scattered by solid body outer surfaces gives the first information we have of our environment. Light can also be scattered by other matter's state than solid : "collective" light scattering refers to that kind of scattering that is obtained out of non-uniform transparent fluids. In that case, the scattered light can be thought of as an addition of electromagnetic waves that are scattered by each of the many individual particles constituting the transparent medium. The many atoms distribution (and motion into the gas volume) collectively shapes the scattered light intensity and frequency spectrum. Thus instead of a two-dimensional picture of an object outer surface, collective scattering provides a three-dimensional volume information. This is not so familiar : what is the three dimensional information provided by volume scattering, what does it mean, and how is it to be extracted and interpreted ? These questions will be dealt with in two steps : by first investigating the scattering from a single particle. And second, by building the "collective scattering" as the result from a collection of independant atoms with a non-uniform density. The relation between the collective scattered light signal and fluid structure and motion will finally be analyzed.

SINGLE PARTICLE SCATTERING

Since a gas is made of many individual atoms or molecules, it is useful to first investigate scattering by a single particle. Scattering occurs in two steps : an incident electromagnetic wave induces an internal electric charge distribution perturbation within an atom, at the incident wave frequency ; then a new electromagnetic wave is radiated by this non stationary electric charge. The frequency of this new "scattered" wave is near to the incident wave frequency.

Atom sizes ($\approx 10^{-10}$ m) are small compared to the usual electromagnetic wavelengths ($\approx 10^{-6}$ m), which themselves are small compared to the atom-to-observer distance ($\approx$ m). The "dipole approximation" is then appropriate. The atom electric charge perturbation is assumed to be a simple electric dipole, of electric dipole moment $\mathbf{p}$ (t). These ingredients are shown in Figure 1.

Optical Diagnostics for Flow Processes
Edited by L. Lading *et al.*, Plenum Press, New York, 1994

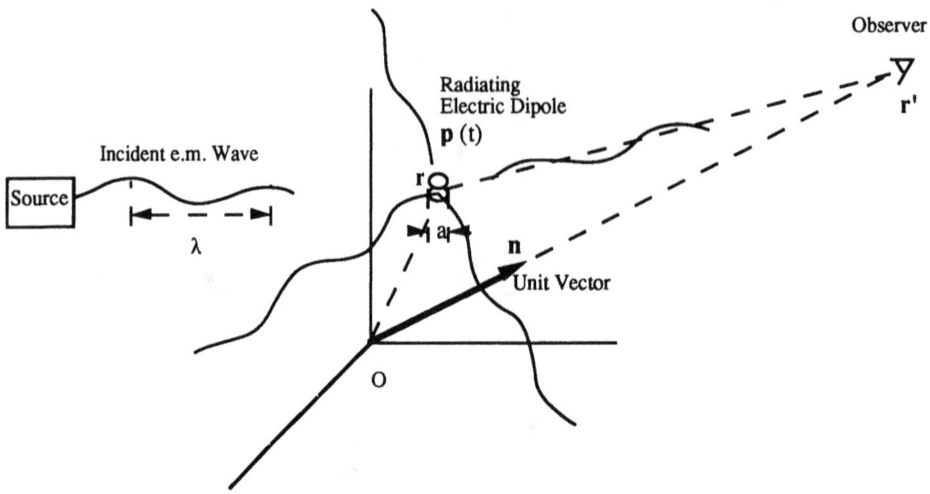

Figure 1. Scattering from a single particle. A light source (left side) emits an electromagnetic, monochromatic plane wave of wavelength λ . It perturbs the electron distribution of an atom of size a, situated at position **r**, inducing an oscillating electric dipole at the incident wave frequency. The dipole radiate a new electromagnetic wave, the "scattered" wave, at the same wavelength, in all space directions. An observer collects this scattered wave, propagationg in the unit vector direction **n** , at a large distance **r'** from the origin. Scales are ordered as $a \ll \lambda \ll r'$.

The electric field radiated at position **r'** from an unstationary electric dipole **p**(t) situated at position **r** is (see e.g. Jackson[1])

$$\mathbf{E}_R(\mathbf{r}',t) = \frac{\mu_0}{4\pi}\frac{1}{|\mathbf{r}'|}\mathbf{n}\wedge\left[\mathbf{n}\wedge\frac{\partial^2\mathbf{p}}{\partial t^2}\right]_{t=t^*} \tag{1}$$

where t* is the "retarded time", i.e. the observer's time minus the time it took for the electromagnetic wave to propagate from the atom to the observer position,

$$t^* = t - \frac{|\mathbf{r}'-\mathbf{r}|}{c} \tag{2}$$

where position **r** can be a function of time as the atom moves, provided this motion is nonrelativistic.

A useful expansion of t*, consistent with the dipole approximation, splits the distance in two terms depending on the atom and observer positions r and r' respectively,

$$t^* = t - \frac{|\mathbf{r}'|}{c} + \frac{\mathbf{n}\cdot\mathbf{r}}{c} \tag{3}$$

The field of the incident plane monochromatic electromagnetic wave, at the atom position, is

$$\mathbf{E}_i(\mathbf{r},t) = \mathbf{E}_0 e^{i(\mathbf{k}_i\cdot\mathbf{r}-\omega_i t)} \tag{4}$$

The inner atomic electric field is much larger than the wave electric field, so this later field only creates a weak perturbation to the atomic electron distribution. A displacement of the light electron negative charge takes place with respect to the heavy ion positive charge, at the rate of the incident field. This induces an electric dipole moment, oscillating at the same frequency as the incident wave,

$$\mathbf{p}(t) = \varepsilon_0 \alpha \mathbf{E}_i(\mathbf{r}, t) \tag{5}$$

where α is the atom "polarizability", a coefficient depending on the atom type.
Using this dipole moment into the radiated wave, Eq. (1) provides the scattered wave field,

$$\mathbf{E}_S(\mathbf{r}', t) = -\frac{k_i^2 \alpha}{4\pi r'} \mathbf{n} \wedge \left[\mathbf{n} \wedge \mathbf{E}_0\right] e^{i(k_i r' - \omega_i t)} \, e^{-i(k_i \mathbf{n} - \mathbf{k}_i).\mathbf{r}(t)} \tag{6}$$

where k_i is the incident wavenumber.
The scattered field is made of two elements :
-an elementary electromagnetic wave, spherically propagating from the origin to the observer

$$\mathbf{E}_{sph}(\mathbf{r}', t) = -\frac{k_i^2 \alpha}{4\pi r'} \mathbf{n} \wedge \left[\mathbf{n} \wedge \mathbf{E}_0\right] e^{i(k_i r' - \omega_i t)} \tag{7}$$

its phase and amplitude depend on the observer position, but not on the atom position. Its amplitude decreases as the inverse distance, and increases as the squared incident wavenumber (or frequency), inasmuch α is a constant. The electric field polarization is along the projection of the incident electric field onto a plane perpendicular to the observation direction **n**.

-and a phase modulating term (the last exponential in Eq.6), that only depends on the atom position.

$$e^{-i(k_i \mathbf{n} - \mathbf{k}_i).\mathbf{r}(t)} \tag{8}$$

The phase argument is a product of the atom position **r**, times the scattering "analyzing" wave-vector,

$$\mathbf{k} = k_i \mathbf{n} - \mathbf{k}_i \tag{9}$$

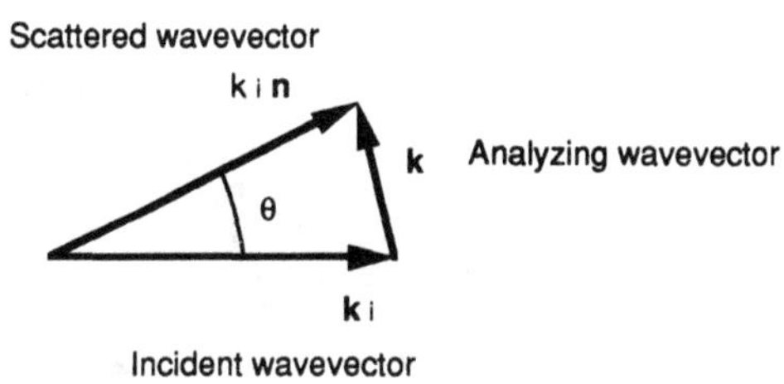

Figure 2. Construction of the "analyzing" scattering wavevector **k**, as the difference between the scattered electromagnetic wavevector (in the observer's direction) k_i **n**, and the incident electromagnetice wavevector $\mathbf{k}_i$. θ is the "scattering angle".

The scattering device is thus "printing" onto the observation volume a virtual parallel interference fringe system, of wavevector **k**, with which it is analyzing the atom position. This wavevector is shown in figure 2.

In Eq. 6 and 8, we made the provision of a moving atom : position r is a function of time.

The scattered wave frequency can then be shifted from the incident wave frequency. The Doppler frequency shift is the product of the particle velocity times the analyzing wavevector (Eq.9).

This looks very similar to seeded particle scattering used in Laser Doppler Anemometry. However, the fringe system here is only virtual, there is only one incident electromagnetic wave. And furthermore, the number of observed atoms in a real gas experiment (10^{18}) is obviously by far larger than any usable number of calibrated LDA particles.

SCATTERING FROM A GAS OF IDENTICAL PARTICLES

Gas is made of a large number of identical atoms or molecules. In the incident wave forward direction (where the analyzing k vector is zero), the scattered field from all these particles, causes a phase change to the propagating wave, independantly of particle velocities. This is the cause of the wave phase velocity and optical index modification from their vacuum values. The electromagnetic incident and scattered wavevectors to be taken into account in the scattering problem, will be those into the dielectric medium. We will further assume the effect of gas density non-uniformities on the optical index to be small enough for the phase front distortion to be neglected.

The total scattered field from a set of N identical observed molecules is then the sum of the individually scattered plane wave fields

$$\mathbf{E}_T(\mathbf{r}',t) = \mathbf{E}_{sph}(\mathbf{r}',t)\sum_{j}^{N} e^{-i\mathbf{k}.\mathbf{r}_j(t)} \tag{10}$$

where $\mathbf{r}_j(t)$ is the jth molecule position at time t.

This total wave frequency is modified form the incident wave frequency (at which $\mathbf{E}_{sph}$ oscillates), because of the combined effect of molecular motions. Two components can be recognized in these motions : a mean mass motion, at the fluid local velocity, and a microscopic Brownian motion wandering about the average local motion. These two contribute to $\mathbf{E}_T$ time variation.

Many informations on the components of the fluid motions can be obtained by recording the E-field complex amplitude. This recording is not possible with usual optical signal detectors, which are quadratic devices, detecting the radiated power (Poynting vector surface integrated), and furthermore effect a time average over a period of time that is much longer than the e.m. wave period. However, these defects can be overcome by the "heterodyne detection" technique, by which the detector is lighted not only with the scattered light, but also with a "local oscillator" signal, a secondary light beam taken from the source laser [2]. The quadratic and average detector properties are then made efficient in selecting the beat component between "scattered" and "local oscillator" beams, thus eliminating the light wave harmonic high frequency component. The beat component is precisely the complex scattered field amplitude.

It is thus interresting to analyze the complex field amplitude evolution in terms of the molecular and gas dynamics. This can be done by looking at the molecules trajectories. Each molecule interacts with its neighbours by binary collisions. Around a point of spatial coordinate r, define an elementary spherical volume of a few mean free path radius l_{mfp} (a few tenth of a micrometer). This is usually smaller than the analyzing wavelength, but this small sphere already contains hundreds thousands of molecules. This number, divided by the sphere volume, provides the local, instantaneous density n $(\mathbf{r}, t)$. Also a "local" and instantaneous average can be taken over these molecules velocities, $\mathbf{U}(t)$. At initial time, the spatial coordinate $\mathbf{r}$ is very close to the position average of the molecules situated in the neighbourood l_{mfp}.

As time goes on, the molecules are rapidly moving around r. However an elementary slow fluid motion can be uniquely defined, at any time. This is done by moving the mean position at the rate of the instant mean velocity, then taking an average over the molecules that are in the neighbourood l_{mfp} of the new position, and integrating step by step. Let $\mathbf{R}$ (t) be the fluid trajectory of that "point" of the fluid that was at position $\mathbf{r}$ at time t.

Now single out one of the molecule within that sphere at initial time. Its motion is random, but its time average (an average taken over several sub-nanoseconds collision times) is restrained to the spatial average local fluid velocity by colliding with the other particles. We then divide its trajectoy $\mathbf{r}_j$ (t) into the mean fluid trajectory $\mathbf{R}$ (t), and a microscopic random zero-mean component ρ (t) :

$$\mathbf{r}_j (t) = \mathbf{R} (t) + \rho_j (t) \tag{11}$$

We assumed the elementary volume to be small compared to the analyzing wavelength ($k\ l_{mfp} << 1$). In summing Eq.[10] over the inner molecules, the phase variation from one term to the other is small. The scattered field from each of these molecules can be taken as being equal, and the summation in Eq.[10] reduces to[3]

$$\mathbf{E}_T(\mathbf{r}',t) = \mathbf{E}_{sph}(\mathbf{r}',t) \int_V dr^3\, n(\mathbf{r},t)\, e^{-i\mathbf{k}.\mathbf{r}} \tag{12}$$

where V is the observed volume.
The integral in Eq.12 is a space Fourier transform of the gas number density :

$$n(\mathbf{k},t) = \int_V dr^3\, n(\mathbf{r},t)\, e^{-i\mathbf{k}.\mathbf{r}} \tag{13}$$

The scattered E-field complex amplitude is the space Fourier transform of the density field, at the scattering analyzing wavelength.

This simple result however leads to a desapointing conclusion for an equilibrium gas : its density is uniform, the scattered field complex amplitude is vanishingly small. However, this continuous fluid description left out the (small amplitude) space Fourier components due to each molecular point source. The sum of these fields averages to zero, which is what Eq.13 shows, but the sum of the squared fields does not. This is the so-called "incoherent" scattering out of uniform gas, that is the case of Rayleigh scattering devices.

For non-uniform fluid instead, the sum electric field in Eq.10 or 12, at any given time, is not zero, thus its squared value can be very large. This macroscopic cooperative effect is the source of the so called "collective scattering". Scattering experiments done on non stationnary gases, at sufficiently small analyzing k vectors, exhibit a scattered light intensity that is very much larger than would be expected from the incoherent (Rayleigh) scattering intensity. This is because practical flowing gases often have a large enough non-uniformity for the collective process to dominate the incoherent one. The collectively scattered light depends on large scale fluid motion characteristics, but as it can be shown[4], its frequency spectrum may still keep track of its molecular origin.

AN ILLUSTRATIVE EXPERIMENT

To give a feeling of how a scattering experiment is done, and what kind of signal is obtained, let us look at a defined experiment, with its scattering device, the flow being observed, and typical scattering signals.

Figure 3 shows the optical lay-out of a laser scattering device.

Gas scattered light intensity is usually small, with narrow frequency spectrum. It is thus necessary to use intense, coherent sources and sensitive detectors. These devices are available in the range of visible or infrared light. However, the laser beam wavevectors are not tuned to the wavevectors to be observed, since their wavelengths (in the micrometer range) are usually far smaller than the gas fluctuation scales (in the millimeter range). This is overcome by using long wavelength efficient laser (namely, infrared 10.6 μ wavelength, continuous wave CO_2 lasers delivering a few watts of e.m. power), and small scattering angle θ.

The scattering angle can be defined by reference to Figure 2. Since wavenumbers are proportionnal to the inverse wavelength, an analyzing wavelength of one millimeter can be obtained with an incident wavelength of 10.6 micrometers provided the scattering angle is reduced to 10.6 milliradians. This small angle requires well directed beams, for the scattered

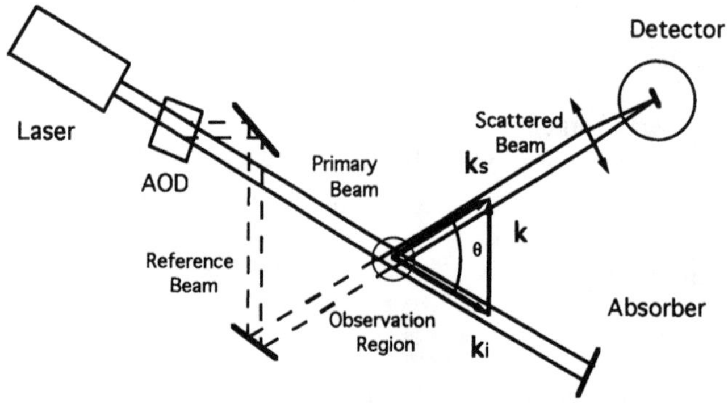

Figure 3. Laser scattering device lay-out. A continuous laser electromagnetic wave source (top left) lights the volume to be observed (center) as the "primary beam". The gas density spatial irregularities in the observation region scatter a small part of the incident "primary" light beam in many directions. A part of this scattered light is beamed towards a photodetector (upper right). A "reference beam" is formed from the primary beam, by partial diffraction in an acousto-optic deflector (AOD). Both "scattered" and "reference" beams are sent to the detector. The total light intensity is converted into an electric current that is recorded as a function of time. $\mathbf{k}_i$, $\mathbf{k}_s$, $\mathbf{k}$ are the incident, scattered, and analyzing wavevector, respectively.

beam in the forward direction to be separated from the incident one. This is obtained with "single mode" lasers : these provide gaussian radial light intensity beam profiles, with minimum natural divergence. The natural divergence of a 2 millimeter diameter beam is three times smaller the scattering angle implied for observing a 1 mm analyzing wavelength (whatever is the laser wavelength); this divergence is small enough for the scattering beam to rise out of the primary beam.

The scattered E-field is detected via a beat wave "heterodyne" technique. The detector is lighted not only with the scattered light, but also with a reference light beam. This reference beam also defines the detector field of view since, when the reference beam is properly focused on a uniformly sensitive detector, the reception antenna beam can be shown to be identical to the reference beam[3, 5].

The reference beam is taken out of the primary laser beam by partial diffraction out of an RF acoustic wave propagated in a solid state cristal, the "acousto optic deflector" (AOD). The reference beam frequency is that of the laser light, plus the frequency of the AOD acoustic wave RF signal. The detector performs a time average of the product of the scattered electric field, times the constant amplitude, pure sine wave reference electric field. The time average is done over a few nanoseconds, a time much longer than the wave period, but shorter than the characteristic fluid (and scattered E-field amplitude) evolution time. It extracts the low frequency beat wave component, at the AOD frequency, whose amplitude is proportionnal to the scattered electric field complex amplitude[5], Eq. 10.

The RF frequency AOD sine wave component is further eliminated in a second stage of demodulation, using two analogic mixers in two channels to obtain separatly the real and imaginary parts, as shown in Fig.4. These are simultaneously recorded as time sequences.

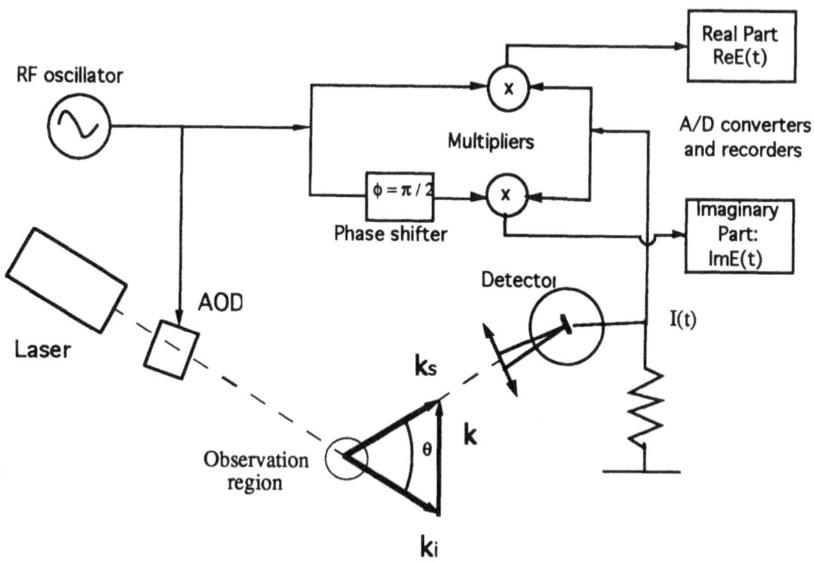

Figure 4. Electrical heterodyne detection circuit, to obtain and record the scattered E-field complex amplitude.

A non-stationnary flowing gas that is simply obtained, while it shows many of the turbulence "universal" features, is a free jet of compressed air blown out of a hole. A schematic is shown in Figure 5. The nozzle diameter is one millimeter, and pressure in the reservoir is three times the atmospheric pressure. The optical scattering device is observing 30 mm downstream the nozzle, in the so-called "developed turbulence" region.

An example of the signal obtained in this experiment, recorded as a function of time, is shown in Figure 6.

The analyzing wavevector is parallel to the mean flow velocity, its modulus corresponds to a wavelength of 0.7 millimeter. The observed region is a cylinder of 2.1 millimeters diameter and

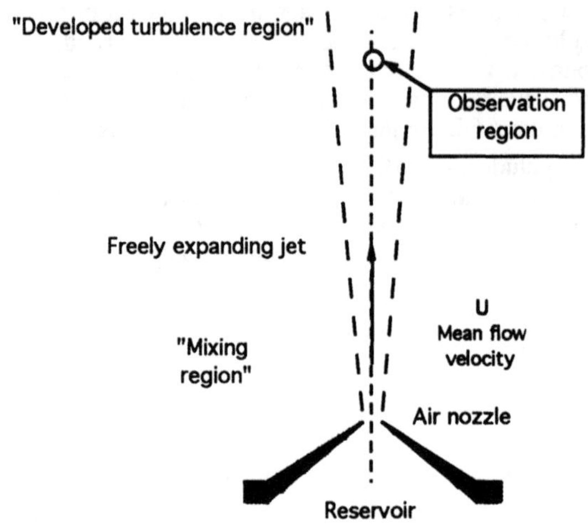

Figure 5 Free jet of air, blown out of a conical nozzle. The turbulent mixing of the compressed air with ambiant atmosphere is a typical exemple of an unstationnary, nonuniform dielectric.

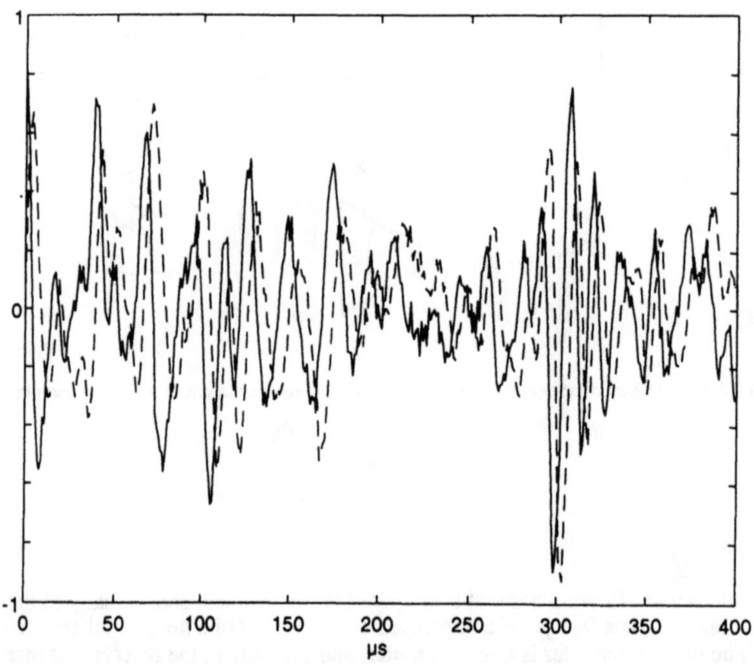

Figure 6. Time variation of the real (full line) and imaginary (dot-dashed line) parts of the scattered electric field observed in the observation region shown in Figure 5. The scattering analyzing wavelength is 0.7 millimeter. Time horizontal unit are microseconds.

approximatly 56 centimeters length. This length results from the small scattering angle ; it is larger than jet diameter. Two traces are shown : The full line is the real part of the scattered electric field (the molecular density space Fourier transform), and the dot-dashed line is the imaginary part. The random looking variations manifest density irregularities of various shapes passing by the observation region, and continuously renewed.

Various analysis can be performed on this signal, its amplitude and phase. The most common is the time Fourier transform and frequency spectrum, which will now be considered. We will later come back to look at an interresting phase property.

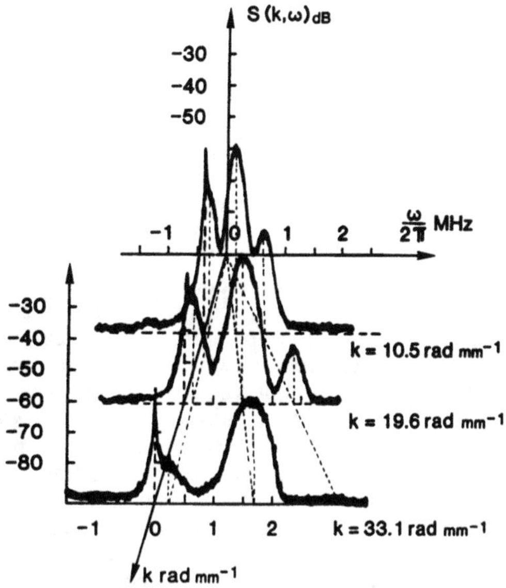

Figure 7. Three collective light scattering frequency spectra observed in a windtunnel supersonic wake experiment. Vertical axis is the frequency spectral density (logarithmic scale), horizontal axis is the frequency. The three spectra are obtained for different k-vector, and ordered along the outside directed linear k-axis. The k-vector is aligned with the mean flow. Three different frequency peaks are obtained at each k-value. Their phase velocity is independant of k. The largest amplitude peak is the (constant pressure) entropy wave, convected at the mean flow velocity, and the middle and smallest amplitude peak are the (constant specific entropy) slow and fast sound waves.

DENSITY FLUCTUATIONS WAVE ANALYSIS

Collective scattering naturally provides a space Fourier transform of the density field. It is a privileged observation means for mode analysis, and specifically for projecting the fluctuating field over plane wave normal modes. The projection is completed by time Fourier transforming the scattered E-field amplitude, either by numerical FFT, or by plugging the signal into an analogic frequency spectrum analyzer.

What can be expected ? For given optical set-up, a given k-vector component is selected. If the fluctuating density field is not too far from the stationary and uniform equilibrium, the linear perturbation properties can dominate. Given a k-vector, different waves may exist, each having a given frequency as defined by their characteristic dispersion relation. In a compressible gas, there are three different kinds of wave : vorticity, entropy, and sound modes[6]. Vorticity fluctuations should not be observed directly by collective light scattering since the vorticity modes are independant from the density. Entropy modes are stationnary in the fluid reference frame, non-propagating, and involve a constant pressure. While the sound mode is propagating in the fluid reference frame at the sound speed phase velocity, with a constant specific entropy.

This can be readily observed. Figure 7 shows a series of CLS signal frequency spectra obtained in an aerodynamics experiment, in a supersonic windtunnel[7]. CLS is performed on air flowing downstream, in the wake of a thin plate whose plane is parallel to the mean flow velociy. The flow Mach number is M = 1.1, and the local temperature (internal energy density) is 195 °K. Three different spectra are shown in Figure 7, each corresponding to a different wavevector and scattering angle. Frequencies are along the horizontal axis (linear scale), and the spectral density is along the vertical axis, in logarithmic (dB) scale. The three spectra are ordered along an oblique k-axis.

Three peaks at three different frequencies are observed in any one of these spectra. This suggests three different modes in the density fluctuation field. A large central peak is surrounded by a lesser amplitude peak on the low frequency side, and a still smaller amplitude peak on the high frequency side, at a symetric frequency position.

An indication of the respective dispersion relations is obtained by observing how the peak frequencies change as k is modified. This is shown as light full lines below the three family of peaks. The frequencies are seen to increase linearly with k. The rate of frequency increase is a measurement of the phase velocity. One finds 321, 26, and 616 meter per second, respectively. The mean velocity between slowest and fastest modes is equal to that of the middle one (321 m/s). The phase velociy shift between the middle and a side peak is 295 m/s. This is very near to the sound speed in an atmosphere at 195°T temperature. The two external peaks correspond to sound waves, convected in the mean flow reference frame : the slow and the fast sound waves, corresponding respectively to the low an high frequency peaks. The largest amplitude peak propagates at a phase velocity of 321 m/s, in the ratio of 1.09 with the sound speed. Its velocity is the mean flow velocity, and the ratio is the Mach number. This peak corresponds to the entropy wave, convected at constant pressure in the gas reference frame.

It is interresting to note how such a turbulent flow still retain its linear, first order perturbation density fluctuations characteristics. These observations show a remarquable, physical distinction between isobaric and isentropic density fluctuations. Well distinct frequency peaks on the CLS frequency spectrum are separated by a deep hole of at least 30 dB amplitude, and the entropy peak is the largest in amplitude, dominating the others by about 20 dB.

The entropy peak corresponds to force-free convected density fluctuations. Its frequency is the Doppler frequency occured in the sweeping of density irregularities in the CLS analyzing wavelength. If the convection velocity is not uniform, the Doppler frequency is broadened. The entropy peak frequency profile is a characteristics of turbulence. Its precise shape depends on the turbulence correlation scale length compared to the scattering analyzing wavelength : For long correlation length, the frequency profile provides the velocity probability distribution[8], while for short correlation length instead, the.frequency profile takes on a Lorentz shape, a characteristics of turbulent diffusion[9,10].

PHASE ROTATION AND INSTANT VELOCIMETRY

Since the main part of the scattered signal is coming from force free convected density irregularities, its time evolution is an information on the fluid dynamics in the observed volume. We already mentioned how these irregularities being swept at the mean flow velocity **U** along the scattering **k** vector, produced a line in the frequency spectrum at the Doppler frequency ω

$$\omega = \mathbf{k} \cdot \mathbf{U} \tag{14}$$

More interestingly, since the frequency spectrum is loosing a great deal of informations by taking a time weighted average, it is tempting to use the instant signal itself. The spatial mode analysis for instance is appropriate for both wave-wave coupling investigations[11], and for the observation of information transfer between different scales, as expected in the so-called Kolmogorov, fully developed turbulence phenomenology[12]. Here we deal with the simple question of comparing the rate at which the signal phase is changing, and the fluid instant velocity.

CLS Signal Instant Phase Rotation

The approximation of an infinitly large scattering volume will be made in this paragraph. The complex scattering signal s(t), obtained from the scattered field (Eq. 12) after heterodyne demodulation, is proportional to the density field Fourier transform $n(\mathbf{k}, t)$ (Eq. 13)

$$s(t) = A\, n(\mathbf{k}, t) \tag{15}$$

where A is a physical constant defining the scattering system sensitivity.
After an infinitesimal time increment Δt, $n(\mathbf{k}, t)$ becomes

$$n(\mathbf{k}, t+\Delta t) = \int_V dr^3\, n(\mathbf{r}, t+\Delta t)\, e^{-i\mathbf{k}.\mathbf{r}} \tag{16}$$

This can be related to the signal at time t, when use is made of particle conservation equation along the fluid trajectory. Be $[n(\mathbf{r}, t+\Delta t)\, dr^3]$ the number of molecules in the vicinity dr^3 around position $\mathbf{r}$ at time $(t+\Delta t)$. These position and volume element can be traced back to the original position $\mathbf{r}'$ and volume element dr'^3 the same molecules occupied at time t. Molecules number conservation implies

$$n(\mathbf{r}', t)\, dr'^3 = n(\mathbf{r}, t+\Delta t)\, dr^3 \tag{17}$$

Define $\Delta\mathbf{r}$ as the position increment, from $\mathbf{r}'$ to $\mathbf{r}$, occured in the fluid motion during the incremental time Δt,

$$\mathbf{r} = \mathbf{r}' + \Delta\mathbf{r} \tag{18}$$

Then

$$n(\mathbf{k}, t+\Delta t) = \int_V dr'^3\, n(\mathbf{r}', t)\, e^{-i\mathbf{k}.(\mathbf{r}'+\Delta\mathbf{r})} \tag{19}$$

For isobaric, force-free fluctuations, fluid motion is independant from density. Over the observed volume, $\Delta\mathbf{r}(\mathbf{r}')$ and $n(\mathbf{r}')$ distributions behave like the probability distribution of two random independant variables. Eq.19 volume integral can be split in two terms

$$n(\mathbf{k}, t+\Delta t) = \left[\frac{1}{V}\int_V dr'^3\, e^{-i\mathbf{k}.\Delta\mathbf{r}(\mathbf{r}',t)}\right] \bullet \int_V dr'^3\, n(\mathbf{r}', t)\, e^{-i\mathbf{k}.\mathbf{r}'} \tag{20}$$

The first term on the right hand side is a volume average which, in the limit of large volumes, is equivalent to an ensemble average, which usually is not a function of time. It can then be recognized as the "characteristic function" of the random variable $\Delta\mathbf{r}$,

$$\left[\frac{1}{V}\int_{v} dr'^{3}\, e^{-i\mathbf{k}.\Delta r(r',t)}\right]=\left\langle e^{-i\mathbf{k}.\Delta r}\right\rangle \tag{21}$$

The second term in Eq.20 r.h.s. is the signal at time t. Then Eq.20 relates the signal at time $(t + \Delta t)$ to its value at time t

$$s(t+\Delta t)=\langle e^{-i\mathbf{k}\,.\,\Delta r}\rangle . s(t) \tag{22}$$

Furthermore, for infinitesimal time increment Δt, the displacement $\Delta \mathbf{r}$ is

$$\Delta \mathbf{r}(\mathbf{r},t)=\mathbf{U}(\mathbf{r},t)\,\Delta t \tag{23}$$

where $\mathbf{U}(\mathbf{r},t)$ is the local, instant velocity. A moment expansion of the characteristic function is appropriate,

$$\left\langle e^{-i\mathbf{k}.\mathbf{U}\Delta t}\right\rangle \cong 1+i\,\mathbf{k}\,.\langle \mathbf{U}\rangle \Delta t \tag{24}$$

Where $\langle \mathbf{U} \rangle$ is the volume average velocity. From Eq. 22 and 24, the signal increment is

$$\Delta s = i\,\mathbf{k}\,.\langle \mathbf{U}\rangle \Delta t\ s(t) \tag{26}$$

and a differential relation is obtained

$$\frac{1}{s}\frac{\partial s}{\partial t}=i\,\mathbf{k}\,.\langle \mathbf{U}\rangle \tag{27}$$

But the logarithmic derivative of a complex function of time $Z(t)$ is made of two terms. The real part is the logarithmic derivative of its modulus $\rho(t)$; and the imaginary part is the derivative of the phase argument $\theta(t)$,

$$\frac{1}{Z}\frac{\partial Z}{\partial t}=\frac{1}{\rho}\frac{\partial \rho}{\partial t}+i\frac{\partial \theta}{\partial t} \tag{28}$$

Thus within our approximation of an infinitly large observed volume, Eq.27 shows the signal logarithmic derivative should be purely imaginary. In addition, according to Eq.28, the signal phase argument time derivative is the instant, volume average velocity along the scattering k-vector.

$$\frac{\partial \theta}{\partial t}=\mathbf{k}\,.\langle \mathbf{U}\rangle \tag{29}$$

Free Air Jet Volume Average Instant Velocity

This concept has been applied to the previously shown free air jet scattering experiment. The signal phase derivative is calculated from its real and imaginary parts $X(t)$ and $Y(t)$,

$$\frac{\partial \theta}{\partial t}=\frac{\frac{\partial Y}{\partial t}X-Y\frac{\partial X}{\partial t}}{X^{2}+Y^{2}} \tag{30}$$

Figure 8 is the phase derivative of the signal shown in Figure 6. The vertical units are given in terms of radians per second.

The short time spiky structures result from the noise enhancing time derivation procedure. More significant are the mean value (established around 250.10^3 rad/sec), and the slow oscillations (of about 100.10^3 rad/sec rms amplitude) on the 20 microsecond time scale. These can be interpreted as the (volume average) flow velocity time variation. Since the scattering wavelength was 0.7 mm, the corresponding velocities can be obtained from Eq.29 as 28 m/s (mean) and 11 m/s rms fluctuation.

As it was previously shown[8,9], another velocity measurement is obtained from the raw signal (real or imaginary part) frequency spectrum. In the condition of long correlation length,which are appropriate, the density fluctuation spectrum is the Doppler frequency shifted

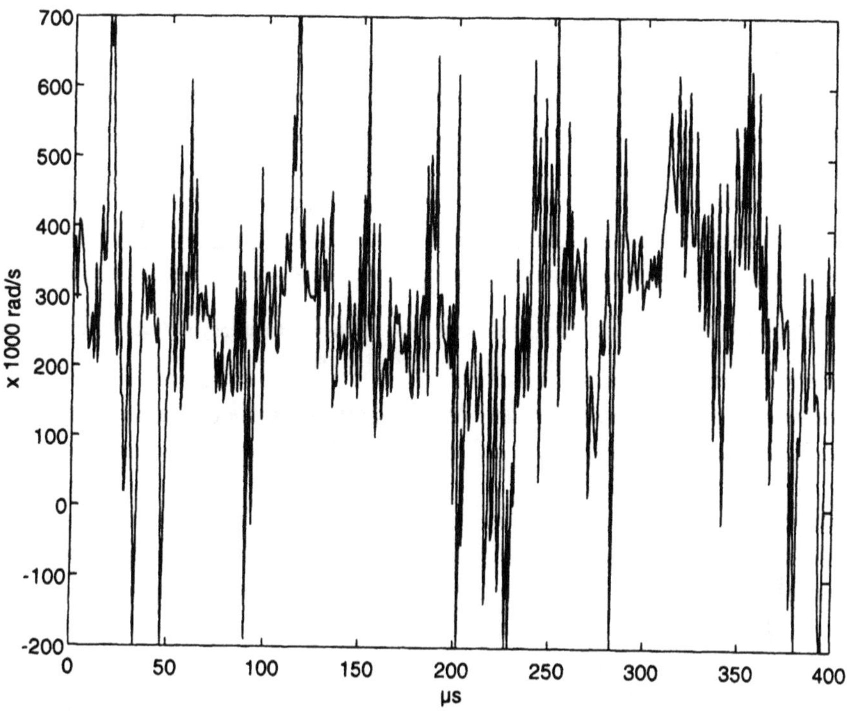

Figure 8. Free jet volume averaged velocity obtained from the time rate of change of the scattered signal complex phase (Eq.29 and 30), for the same signal and time sequence as in Fig. 6. The vertical units are shown as angular frequencies, in rad/sec (k is 10^3, and M is 10^6). Since the scattering wavevector was (2π / 0.7 mm), the velocity can be obtained from the ratio of the angular frequency to the scattering wavenumber.

fluid velocity probability distribution in the observed volume. This frequency spectrum is shown in Figure 9. A broad maximum is seen at (40 ± 30) KHz, i.e. a velocity of (28 ± 21) m/s. Away from this maximum, the spectral density decays rapidly, as expected from the gaussian shape of a probability distribution. The most likely velocity corresponds to the mean velocity obtained by the phase method, and the velocity width is larger, but of the same order as its phase evaluated rms value.

This is a very satisfactory result. The slight difference in the two velocity variation estimate can be understood from their different physical origins : the phase obtained velocities are likely to be more steady since they are volume averaged.

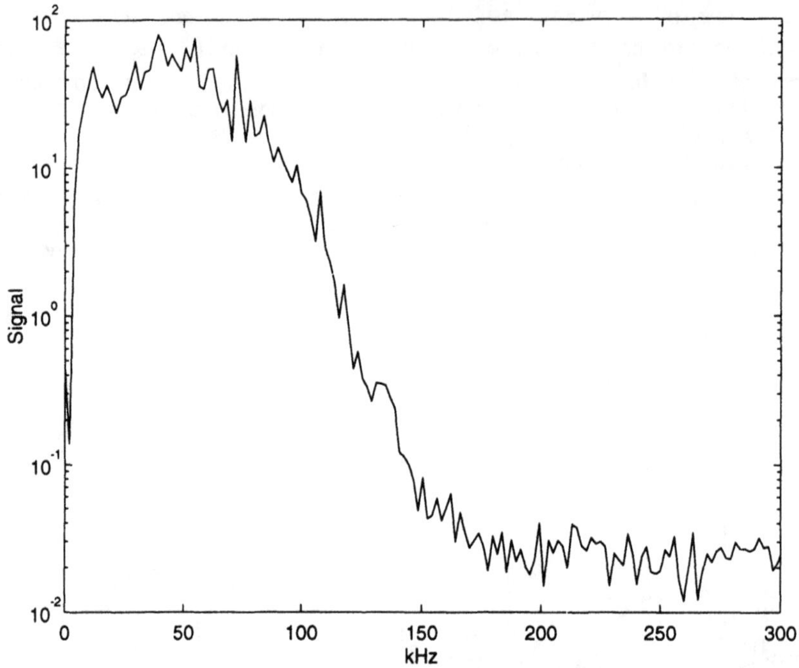

Figure 9. CLS signal frequency spectrum, corresponding to a (Doppler frequency shifted) velocity probability distribution (the scattering wavelength is 0.7 mm). Same data as in Fig.6.

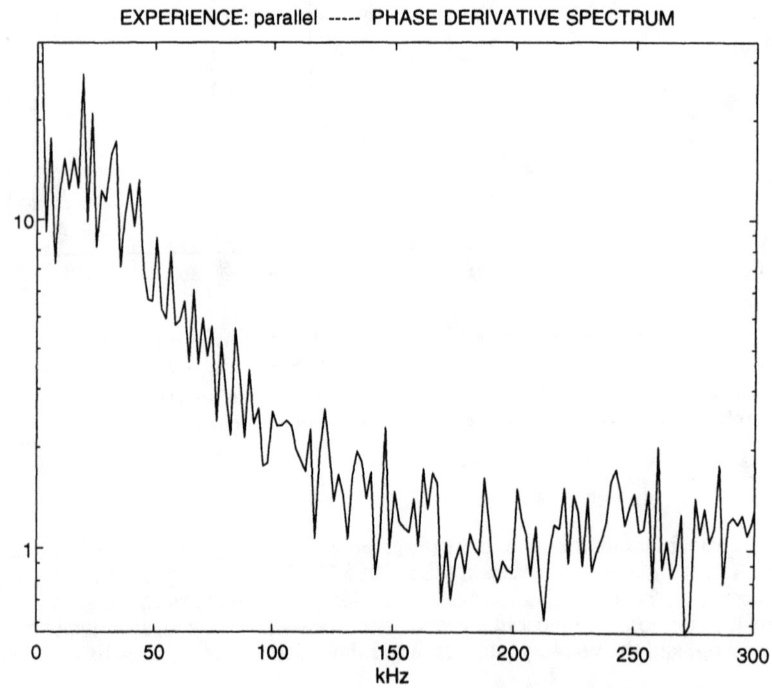

Figure 10. Frequency spectrum of Fig.8 instant velocity signal.

The phase-obtained velocity contains an additional information which is not in the signal Doppler frequency spectrum : it is the velocity time variation, which is not given by usual particle Doppler velocimetry.

This time information is thus worth looking at. Figure 10 is the frequency spectral density of this instant velocity signal obtained in Figure 8. The vertical units are logarithmic, the horizontal axis is the frequency axis, within the same interval as in Figure 9. The two spectra are different. Fig.10 spectrum shows a flat maximum at low frequencies, between zero and 30 KHz, then a slow decay down to the noise level around 150 KHz. This velocity fluctuation spectrum is narrower than Fig.9 density fluctuation spectrum, with a slower, positive second derivative decay. This shape may correspond to a Lorentz profile and a correspondingly exponential velocity time correlation.

These interpretations should not be too conclusive : a freely expanding jet is strongly non-uniform, the hypothesis of infinite volume observation of a stationary and uniform turbulent field is not appropriate. This being taken into account, it is surprising how well the phase-rate velocity measurements recover many expected results.

These measurements are illustrative of the many informations that can be extracted from a fluctuating transparent volume by simply looking at the real-time collective scattering signal.

Acknowledgments

All of these results were obtained with the collaboration of early students who became colleagues, and of well-established senior coworkers. It is a pleasure to remind and thank all of them for those many moments and feelings of common discoveries.

REFERENCES

1. J. D. Jackson, "Classical Electrodynamics" (Second Edition), John Wiley and Sons, New-York (1975).
2. E. Holzhauer and J.H. Massig, "An analysis of optical mixing in plasma scattering experiments" Plasma Physics 20, p.867 (1978).
3. D. Grésillon and B. Cabrit, "Collective Scattering and Fluid Velocity Fluctuations" , in "A Variety of Plasmas"(Proceed. of the Internat. Conf. on Plasma Physics, New-Delhi, 1989), A.Sen and P.K.Kaw, Editors (Pramana, Bengalore, India, 1991), p. 173 (1991)
4. B. Cabrit, "Diffusion collective de la lumière par un gaz turbulent : Dispersion moléculaire et turbulente", Doctoral Thesis, Université de Paris 6, (1992).
5. D. Grésillon, C. Stern, A. Hémon, A. Truc, T. Lehner, J. Olivain, A. Quémeneur, F. Gervais and Y. Lapierre, "Density fluctuation measurement by far infrared light scattering", Physica Scripta T2:2, p.459 (1982).
6. B-T. Chu and L. S. G. Kovasznay, "Non-linear interactions in a viscous heat-conducting compressible gas", J. Fluid Mech. 3, p.494 (1958).
7. D. Grésillon, G. Gémaux, B. Cabrit, and J. P. Bonnet, "Observation of supersonic turbulent wakes by Laser Fourier Densitometry (LFD)", in Eur. J. Mech., B/Fluids, 9, p. 415 (1990)
8. D. Grésillon, J.P. Bonnet, B. Cabrit and V. Frolov, "Collective light scattering and supersonic flow velocimetry" (in preparation) (1993).
9. D. Grésillon, B. Cabrit, J.P. Villain, C. Hanuise, A. Truc, C. Laviron, P. Hennequin, F. Gervais, A. Quemeneur, X. Garbet, J. Payan, and P. Devynck, "Collective Scattering of Electromagnetic Wave and Cross-B Plasma Diffusion" (Invited topical conference, 1992 Internat. Conf. on Plasma Physics, Innsbrück, July 1992); Plasma Physics and Controlled Fusion 34, p. 1985 (1992).
10. C. Hanuise, J.P. Villain, D. Grésillon, B. Cabrit, R.A. Greenwald and K.B. Baker, "Interpretation of HF radar ionospheric Doppler spectra by collective wave scattering theory" , Annales Geophysicae 11, p. 29 (1993).
11. D. Grésillon, and M. S. Mohamed-Benkadda, "Direct mode-mode coupling observation in the fluctuations of nonstationary transparent fluid", Physics of Fluids, 31, p. 1904 (1988).
12. D. Grésillon, B. Cabrit & N. Iwama, "Nonlinear signal analysis in a fluid turbulence : a comparison between bispectrum and mutual information." , Proceedings of the Miniworkshop on "Wave phenomena in Solar-Terrestrial Plasmas" (P. Malby and H. Pecseli, Ed.), Institute for Theoretical Astrophysics, University of Oslo, 14-16 June 1993, p. 165-179 (1993).

Whole Field Measurements

PARTICLE IMAGE VELOCIMETRY

Preben Buchhave

Physics Institute
The Technical University of Denmark
DK-2800 Lyngby, Denmark

INTRODUCTION

Particle image velocimetry (PIV), also called particle image displacement velocimetry (PIDV), is a method of whole flow field velocimetry, which is currently causing a lot of interest. This method relies on the recording of particle positions in a section of the flow at a given instant in time. The measurement region may be defined by a thin laser sheet (a 2-D cross section of the flow) or by illumination of an entire volume (a complete 3-D flow field). At any rate, PIV is a method, which allows the measurement of an entire velocity distribution in a whole flow field at a specified point in time. The method thus allows spatially resolved velocity measurements.

In this chapter we shall present some fundamental considerations regarding the recording and processing of PIV data. In particular we shall look at the selection of optical parameters for the illumination of the particles and the recording of the particle images. Also, we shall bring some elementary considerations regarding the optimal image processing method, and in

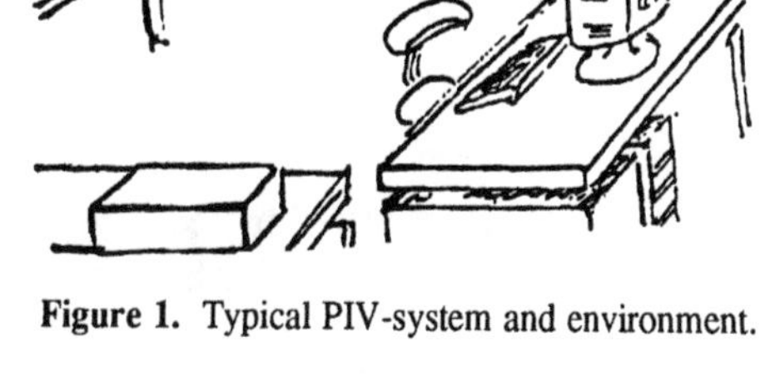

Figure 1. Typical PIV-system and environment.

Optical Diagnostics for Flow Processes
Edited by L. Lading *et al.*, Plenum Press, New York, 1994

particular take a look at optical methods or combinations of optical and numerical methods. More details regarding computer processing of PIV images will be presented in the article by Keane[1], and holographic and other, more advanced optical methods will be treated by Hinsch[2]. Recently reviews have been published on the state of the art of whole flow field velocimetry[3], optical processing of PIV-images[4] and particle image velocimetry[5].

Figure 1 shows a typical (maybe still a bit futuristic!) PIV flow measurement environment. The system consists of some optical components, which may include a pulsed or chopped laser, optics to form and direct the laser beam into the measurement region (shown as a compact optical head connected to the laser by a fiber optic cable) and some signal- and data processing equipment. We shall discuss these components in more detail, but in general the equipment consists of three main blocks: Means of recording the image, an image or signal processing unit, which extracts the velocity information from the image, and a data processing unit, which further processes the measured velocity information, controls the display and prepares the presentation of the final results. The data processor may also be utilized in an active control of the experiment. PIV-systems are under development at many research laboratories and some commercial systems or components are beginning to appear on the market. However, the PIV-method must still be categorized as an experimental method, and the user must be prepared to spend time to experiment with the recording process as well as with the image processing.

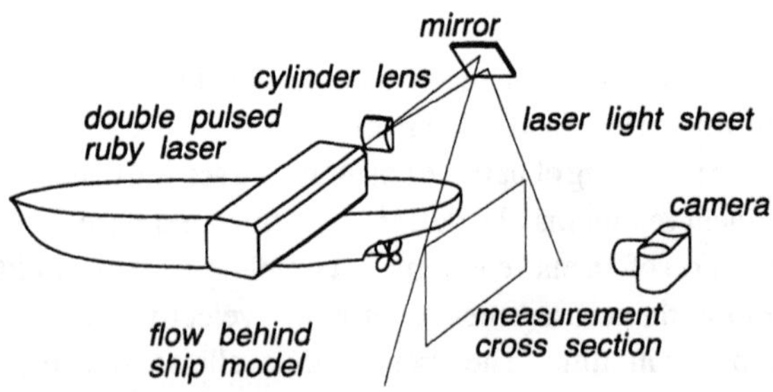

Figure 2. Typical PIV-measurement. The transverse velocity field behind a ship model is measured in a cross section of the flow behind the model.

Figure 2 illustrates a typical PIV measurement using the most common method and equipment. The figure shows a measurement of the area behind the stern of a ship model, e.g. in a tow tank. The laser defines a thin light sheet normal to the model axis. The recording optics define the extent of the light sheet, which will be recorded. The recording itself may be a photographic film using a 35 mm camera or the 6×6 or 6×9 cm format. The result of the PIV measurement is a 2-D array of velocity vector projections onto the measurement plane, measured at the instant when the laser was fired. Usually some time will pass before the laser and the recording equipment are ready for the next shot, so conventional time records of the time history of the flow cannot be made with present day equipment. Some systems based on video recording, however, have a limited capacity for time history recording (at the video frame rate).

THE PIV METHOD

The velocity of a flowing medium is measured by recording the displacement of small particles carried with the flow and subsequently analyzing the particle displacements recorded.

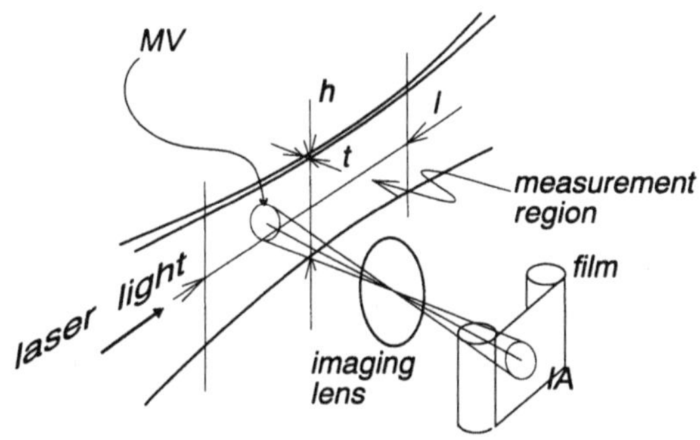

Figure 3. Definition of measurement region and measurement volume.

Particles embedded in a region of the flow are illuminated by two or more short laser pulses fired with a known time separation. The particle positions are recorded by means of a photosensitive medium (photographic film or matrix photodetector). The intersection volume between the region illuminated by the laser and the field of view of the recording camera or detector determines the region of the flow, which is measured, which we shall denote the measurement region, see Figure 3. The smallest resolvable area of the film, the interrogation area (IA), projected onto the measurement region, determines the measurement volume (MV), i.e. the volume of the flow, which contributes information to a single velocity measurement.

The particle displacements in the object plane Δx, Δy are found from the displacements in the image plane ΔX, ΔY on the film:

$$\Delta x = \frac{1}{M}\Delta X \text{ and } \Delta y = \frac{1}{M}\Delta Y. \tag{1}$$

Knowing the magnification M of the imaging and the time separation Δt between the laser pulses, the velocity projections on the measuring plane u_x, u_y may be found:

$$u_x = \frac{\Delta x}{\Delta t} \text{ and } u_y = \frac{\Delta y}{\Delta t}. \tag{2}$$

The results (the primary velocity data) are usually presented in a vector diagram, as shown in Figure 4.

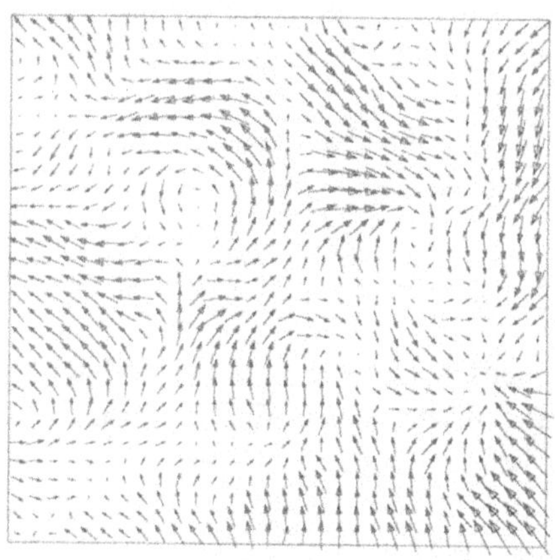

Figure 4. Presentation of the primary velocity data.

PIV CHARACTERISTICS

The PIV-instrument may be described in terms of an optical transducer with the following transducer properties:

- spatial/temporal resolution
- velocity range
- characteristic transfer function
- disturbances from other flow quantities
- errors: systematic errors (bias) and fluctuation errors (noise)

Spatial Resolution

The spatial resolution is very important, since it determines the ability of the system to measure small spatial scales, which are important for the creation or dissipation of the

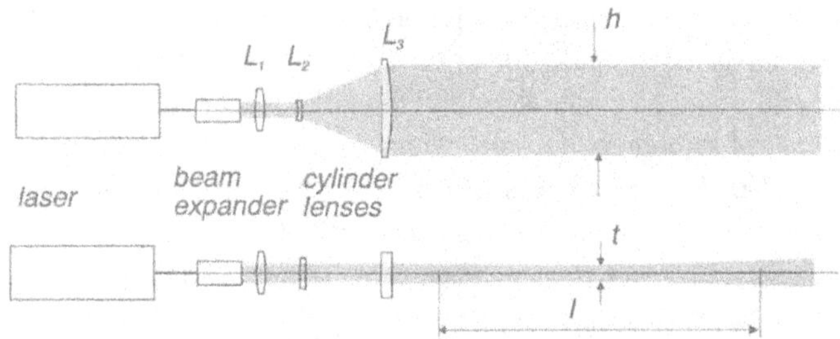

Figure 5. Focusing optics for a laser sheet.

turbulence, and because the resolution determines the ability of the system to measure spatial gradients, which in turn are important for the measurement of e.g. vorticity. The spatial resolution is determined by the thickness of the laser sheet and the quality of the imaging system.

The laser sheet is formed by focusing a TEM_{00} mode laser beam by means of a lens L_1, see Figure 5. The focal length f_1 of lens L_1 determines the thickness of the laser beam in the measurement region. Cylindrical lenses L_2, L_3 with focal lengths f_2 and f_3 expand the laser beam to a sheet (actually a cross section in the shape of an elongated ellipsoid). It is important for a uniform illumination of the particles that the thickness of the laser sheet is approximately constant throughout the measurement region. Before the laser beam enters the cylindrical expansion, its original diameter d_l may have been expanded by a standard beam expander by a factor E_1. The laser wavelength is λ.

The thickness of the sheet t, defined as the diameter at which the intensity is one half of its value at the center, is then given by:

$$t \cong \sqrt{\frac{\ln 2}{2}} \frac{4}{\pi} \frac{f_1 \lambda}{d_l E_1} = \sqrt{\frac{\ln 2}{2}} d_f, \tag{3}$$

where d_f is the usual $1/e^2$-intensity diameter of a focused laser beam.

The length of the focused sheet l, again defined by the points at which the intensity is one half its value at the center, is:

$$l \cong \frac{8}{\pi} \frac{f_1^2 \lambda}{d_l^2 E_1^2} = \frac{\pi}{2\lambda} d_f^2, \tag{4}$$

and the height of the sheet h, using the same definition, is:

$$h \cong tE_2 = t\frac{f_3}{f_2}. \tag{5}$$

Thus, a thin laser sheet (small t) may be obtained by using a lens L_1 of a short focal length f_1, since t is proportional to f_1, or by using a large expansion E_1 of the original laser beam, but this results is obtained at the cost of a significantly reduced length l, since l is proportional to f_1^2 and inversely proportional to E_1^2. One must choose a reasonable compromise between good spatial resolution, i.e. small t, and constant thickness of the laser sheet, i.e. large l.

Example:

Choosing a long focal length lens f_1=1000 mm, a laser diameter of d_l=1 mm and an expansion of E_1=1, we find from the above formulas:

t=0.37 mm, l=1273 mm.

These values seem reasonable for a measuring distance of 1 m. Thus the creation of a usable laser sheet presents no problems in this case!

Temporal Resolution

Limited by the speed at which you can rewind the film or transfer the image electronically!

Velocity Range

The velocity range is determined by the imaging conditions and Δt. Let us try to see how far we can get with a common 35 mm camera. Assume for the moment, that the film can resolve the image - we shall later test this assumption and see if there is enough light in the image plane to expose the film.

The size of the particle image is determined by two factors: The geometric magnification and the diffraction in the optical aperture of the camera. Figure 6 illustrates the imaging process. We find:

Diameter of geometrical image, d_g:

$$d_g = M\, d_p, \tag{6}$$

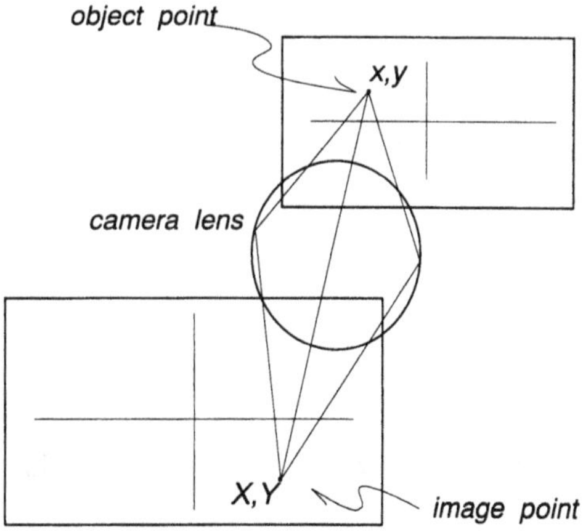

Figure 6. Imaging optics.

where d_p is the diameter of the particle. Geometrical optics is of cause an approximation, which does not take into account diffraction in the receiver aperture. Diffraction of a point image creates a blur spot of diameter d_s:

$$d_s = 2.44(M+1)\,F\,\lambda, \tag{7}$$

where $F=f/D_A$, M is the magnification, f is the focal length of the camera lens and D_A is the diameter of the camera aperture.

One can define an effective diameter by:

$$d_e = \sqrt{d_g^{\,2} + d_s^{\,2}}. \tag{8}$$

This is a useful measure of the image size.

Example:

We consider a standard camera objective with an f-number of 8 and the use of a ruby laser:

F=8, λ=0.6943 μm, M=0.25, d_p=10 μm.

From the formulas above we then find:

d_g=0.25 ×10 (μm) = 2.5 μm, d_s=2.44 (0.25+1)×8×0.6943 (μm) = 16 μm,
d_e=$(2.5^2 + 16^2)^{1/2}$ (μm)= 17 μm.

We see that the image size is largely determined by diffraction (in this case).

Displacement

Using the symbol D for the length of the displacement in the image plane, we may define the minimum measurable displacement D_{min} as the displacement equal to the effective image diameter: $D_{min} \equiv d_e$. We then find for the displacement in the object plane:

$$d_{min} = 1/M\;D_{min} \tag{9}$$

If the velocity range is specified as $R \equiv u_{max}\,/\,u_{min}$, the ratio between the maximum and minimum velocity in the flow, the maximum displacement, is given by:

$$D_{max} = R\,D_{min}. \tag{10}$$

The pulse separation must be chosen accordingly:

$$\Delta t = D_{min}\,/\,u_{min} = D_{max}\,/\,u_{max}. \tag{11}$$

Example:

Using imaging parameters as defined above and a minimum velocity u_{min} of 10 ms^{-1} and a velocity range R=10, we find for minimum displacement, maximum displacement and time separation between pulses:

D_{min} = 17 (mm) /M = 68 mm, D_{max} = 10×68 (mm) = 680 mm, Δt = 6.8 ms.

Film Exposure

To register the image, the energy density on the film must equal the specification from the manufacturer. We estimate the energy density on the film beginning with the energy density incident on the particle in the measurement region:

$$\eta_0 = \frac{E_l}{A} = \frac{E_l}{t\,h} \tag{12}$$

where E_l is the energy of the laser pulse.

The energy scattered from particle is:

$$E_s = \eta_0 \int_{\Omega} \sigma \, d\Omega \, , \tag{13}$$

where σ is the scattering cross section and Ω is the solid angle extended by the camera aperture. E_s is thus the energy scattered into the camera aperture

The energy density in film plane is then:

$$\eta_b = t \frac{E_s}{\frac{\pi}{4} d_e^{\,2}} . \tag{14}$$

The condition for sufficient exposure is $\eta_b > c_1 \eta_f$ or:

$$\frac{E_1}{t\,h} \int_{\Omega} \sigma \, d\Omega \frac{1}{\frac{\pi}{4} d_e^{\,2}} > c_1 \eta_f \, , \tag{15}$$

where η_f is the exposure energy density given by the manufacturer, and c_1 is an illumination factor of order one. To proceed, we must know the amount of light scattered from the particle and the light collection efficiency of the camera lens. One can compute the so-called Lorenz-Mie scattering from a spherical particle, or one can use available tables and curves, e.g. the curves shown in Figure 7 to estimate the light scattering[6].

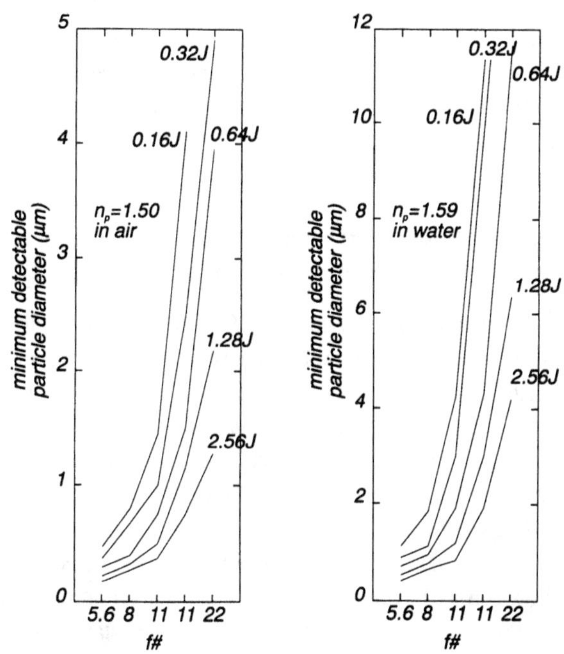

Figure 7. Curves for determining min. detectable particle diameter (from Adrian[6]).These curves are valid for a laser sheet of 100 mm^2, Kodak Technical Pan 2415 and a laser wave length of 0.6943 1m (ruby).

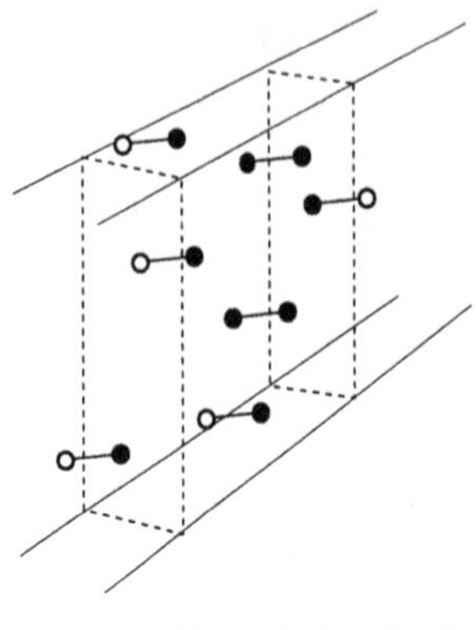

Figure 8. "Out-of-plane bias".

Other Transducer Properties

Characteristic: The connection between particle velocity and the displacement in the film plane is linear using a low aberration optical system. The displacement is the two-dimensional projection onto the measuring plane of the three-dimensional particle motion.

Influence of other parameters in the fluid: As in the case of laser velocimetry (LV) various types of bias exist in PIV. Bias occurs when the number of sampled data points in some way or other is correlated with the velocity.

One sampling error is the so-called out-of-plane bias. It occurs when valid particle image pairs are sampled selectively because one particle image in some of the pairs is missing, e.g. due to the fact that fast particles have larger displacements, and consequently there is a greater probability that one of the images of a pair falls outside the IA. This will bias the measured velocity towards a value, which is too low. Figure 8 illustrates out-of-plane bias; the filled circles indicate particles within the measurement volume. The open circles indicate particles, whose images are not registered inside the interrogation area. Another type of error occurs, if the system is not able to distinguish a displacement. This may happen at very low velocities, because the displacement is of a magnitude comparable to the particle image diameter. In this case low velocity measurements will be selectively rejected, and the measured mean value will be too high. The magnitude of these effects has been computed for known velocity distributions by means of computer simulations of particle image recordings[7].

Image shifting: By moving the image in the film plane with a known velocity (image shifting), e.g. by means of a rotating or scanning mirror[8] (see Figure 9) or by an electro-optic displacement[9], the registered displacements will be added to the common image shift. If the image shift is greater than any of the particle image displacements, the measurable velocity range is shifted to include velocity zero or negative velocities, and the sign of the velocity can thus be determined unambiguously. In a highly turbulent flow the particle image displacements after image shifting will appear to have less spread, which makes it easier to determine the peak location. However, this improved detectability is obtained at the cost of resolution. In a low turbulence intensity flow it may be advantageous to subtract a common image shift from all particle displacements, thereby making the actual differences stand out more clearly.

THE PIV RECORDING

We my consider the chemical changes in a photographic film a signal and treat this signal in the same way we treat an electronic signal. Just like other detectors, the photographic process generates noise. How do we input the signal from the film to the data processor (computer), and what is the effect of the noise?

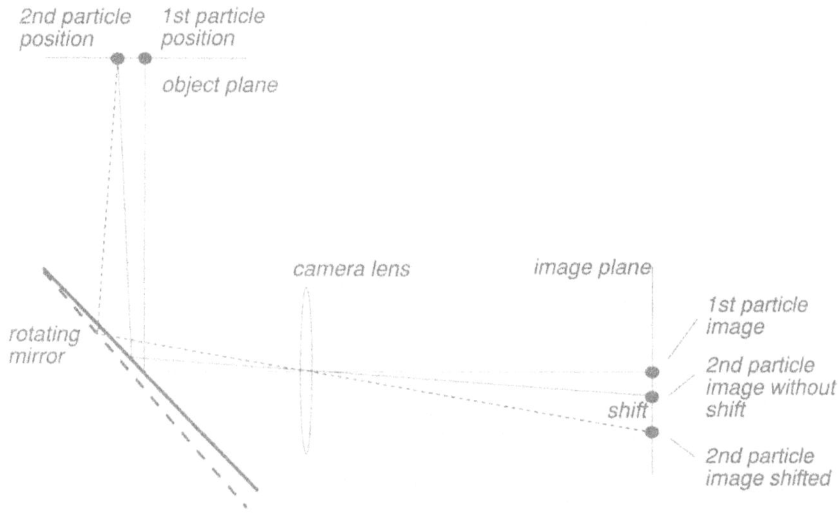

Figure 9. Image shifting.

The Photographic Film as a Detector

The photographic film is an optical quantum detector with a large internal gain (just like the PM tube). The signal can be described by an optical transfer function, and the noise can be associated with different noise mechanisms.

When light strikes a film, the silver halide grains in the film gel are sensitized, see Figure 10, in principle by absorption of a single photon. In reality the quantum efficiency is more like η=0.25.

During development of the film, the silver halide grains are transformed (reduced) to pure silver. This process amplifies the effect of the illumination. The gain factor depends on the grain size; the larger the grains, the larger the amplification. The grains cause noise in the picture (film graininess). The connection between exposure and the film density is often displayed as the so-called Hurtler-Driffield curve or D-log$E$ curve (D=density, E=exposure), see Figure 11.

The entire process can be described by a simplified model, e.g. Kelly's model[10], see Figure 12. The emulsion can be considered a linear optical element with an OTF (optical transfer function), which describes the transfer of the optical image located just in front of the film to the sensitized silver grains. The development process is nonlinear, but it is possible to work in a region, where the connection between the density of the developed image and the incoming optical exposure (energy) is linear. It is important in many applications such as optical transforms and holography to be working in this range.

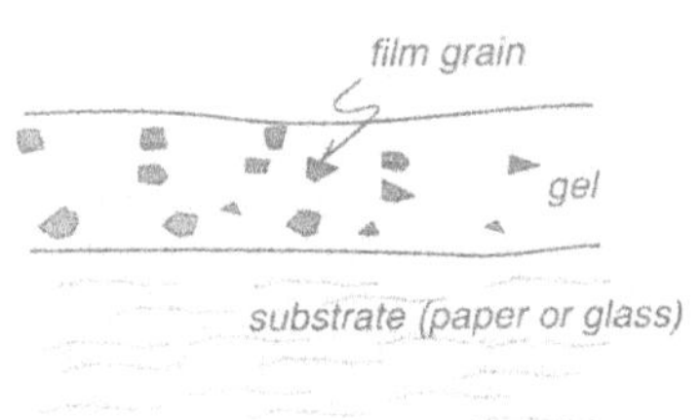

Figure 10. Exposure of photographic film.

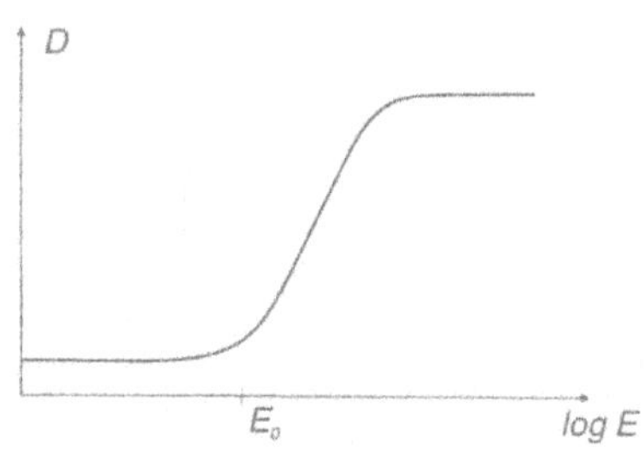

Figure 11. Hurtler-Driffield (H-D) curve.

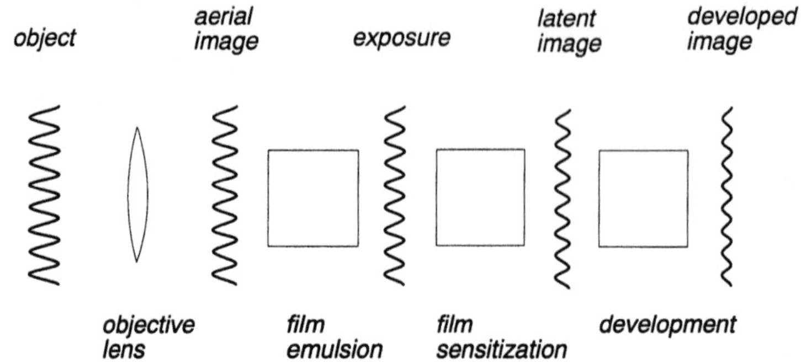

Figure 12. Kelly's model for the photographic process.

The Image Transfer

A photo is of course normally "read" by a person visually, but for optical data processing the information in the film must be translated to an electronic signal. In principle this can take place by scanning the image with a microphotodensitometer. The output of the microphotodensitometer is an analog signal, which after A/D conversion is converted to a digital signal, containing values for the optical density of the individual microscopic areas of the film (the pixels). Due to the finite number of grains seen by the detector aperture of the photodensitometer and the random dispersion of the grains in the emulsion there will be some noise in this signal, see Figure 13. Even when the photodensitometer is replaced by some other type of detector, e.g. scanning with a 2-D photodiode array or by a CCD matrix, the noise characteristics will be of the same kind, and the noise will be analyzed in the same way. The image elements from

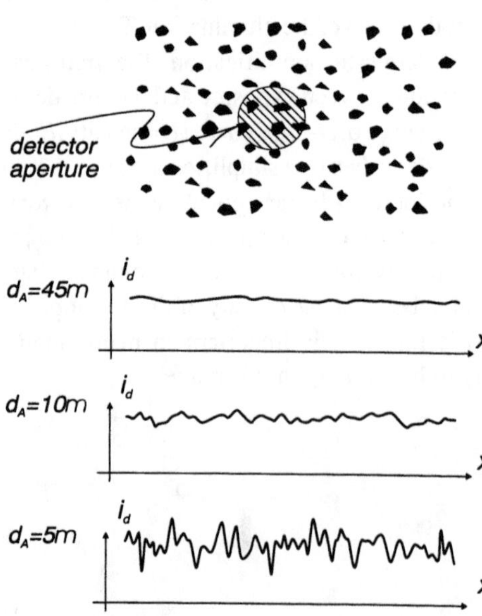

Figure 13. Scanning an image.

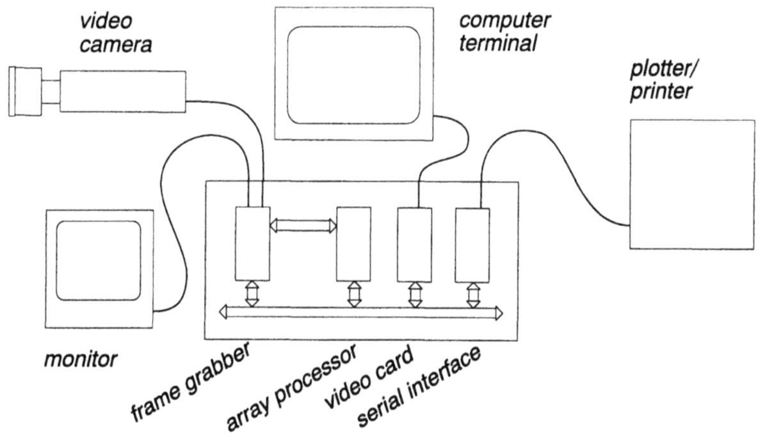

Figure 14. System for digitizing and processing PIV images.

a scan (e.g. 512x512 elements with gray scale values in 256 levels) are stored in the computer for later processing.

A relatively low cost system for recording and processing of images is shown in Figure 14. The video signal from the video camera is connected to a plug-in card in a PC, a so-called frame grabber. This card contains an A/D-converter circuit, synchronizing circuits and image memory. The system also includes an array processor card, which allows relatively fast floating point calculations, e.g. FFT computations.

The PIV "Signal"

The "signal" consists of the silver grains in the film emulsion. This signal is converted to electronic form by detecting the light intensity, which is reflected from or transmitted through a small area of the film. This is the smallest resolvable area of the film and it is called a pixel. The whole image is read out by scanning neighboring pixels sequentially.

The film may be considered a quantum detector with large internal gain:

η = quantum efficiency (approximately 0.25)

G = gain (order of 10^3 - 10^6)

σ = rms noise (depends on the gain, grain size and pixel size)

The signal is digitized and stored in the computer memory. The digitized area of the film can consist of 512x512 8-bit words (pixels), see Figure 15. Some systems use 256x256 or 128x128 pixels. This area corresponds in the PIV signal processing to an interrogation area (IA) of the PIV-image (a single velocity data point).

A typical PIV image can consist of 100x100 such IAs. The information content in one PIV image is thus of the order of: $512 \times 512 \times 8 \times 100 \times 100 \approx 2 \times 10^{10}$ bits.

In practice we can expect that several PIV images of the same measuring plane will be recorded, partly in order to make sure that a good quality image is obtained, partly to make sure that the measurement is statistically significant. Ideally 10 or 100 images from the same plane might be needed.

Often it will be necessary to go on and investigate neighboring planes in the flow and/or repeat the measurements under other conditions. Evidently, there is a need for an efficient method of processing the data!

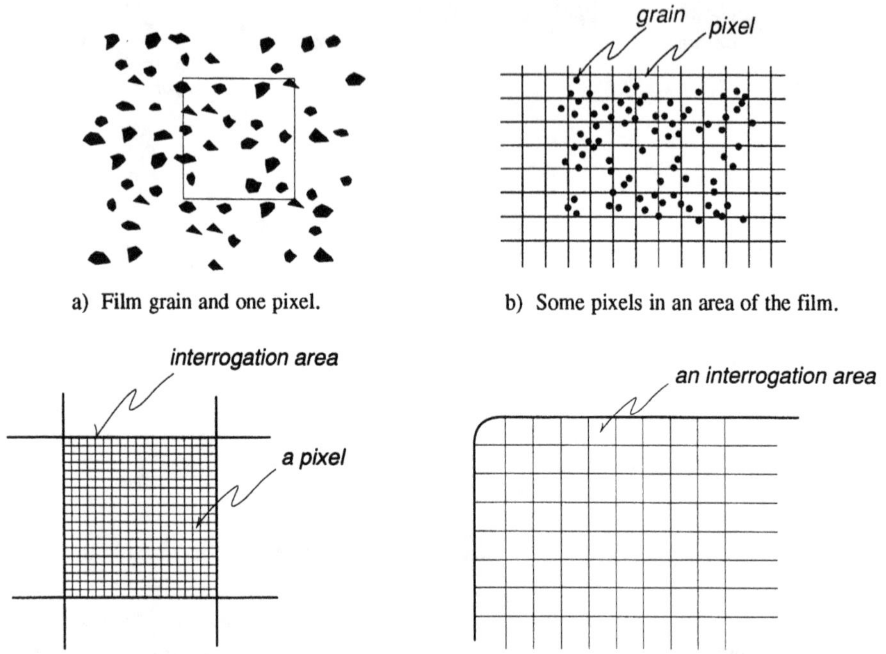

a) Film grain and one pixel.

b) Some pixels in an area of the film.

c) One interrogation area (e. g. 256×256 pixels)

d) PIV photo (e. g. 100×100 Interrogation areas).

Figure 15. The subdivision of the film into interrogation areas and pixels.

Analysis of the Noise

The noise in the digitized image is analyzed in the same way electronic noise is analyzed, only the signal in this case is two dimensional. The signal is written:

$$i(\boldsymbol{x}) = cI\tau(\boldsymbol{x}), \qquad \tau'(\boldsymbol{x}) = \tau(\boldsymbol{x}) - T, \qquad T = \lim_{A\to\infty} \frac{1}{A}\int_A \tau(\boldsymbol{x})d\boldsymbol{x}, \tag{16}$$

where $i(\boldsymbol{x})$ is the photocurrent, I is the intensity of the illumination, c is a constant of proportionality, $\tau(\boldsymbol{x})$ is the transmission through the interrogated part of the film, T is the mean transmission and $\tau'(\boldsymbol{x})$ is the fluctuating part of the transmission. $\boldsymbol{x}$ and A indicate the position in and the size of the reading aperture. I and T are assumed to be constant inside the part of the film, which is interrogated, but this may not be true in general.

We can now compute the correlation function and spectrum:

The transmission through the aperture A, described by the aperture function $w(\boldsymbol{x})=1/A$, can be written:

$$\tau'_A(\boldsymbol{x}) = \int_A \tau'(\boldsymbol{x}')w(\boldsymbol{x}-\boldsymbol{x}')d\boldsymbol{x}' \tag{17}$$

The autocovariance of the detected signal is:

$$R_A(\Delta\boldsymbol{x}) = \langle \tau'_A(\boldsymbol{x})\tau'_A(\boldsymbol{x}+\Delta\boldsymbol{x})\rangle \tag{18}$$

where <> indicates ensemble mean.

The spectrum follows from the Wiener-Khintchine theorem:

$$S_A(v) = \frac{1}{\pi}\int_A R_A(\Delta x)e^{-i2\pi v\cdot\Delta x}d\Delta x = S(v)W(v), \tag{19}$$

where $W(v)$ is the Fourier transform of the aperture function. The spectrum of the detected signal is thus the noise spectrum multiplied by the Fourier transform of the aperture function. If the spectrum of the noise is known and knowing the aperture function, we can find the variance by the inverse Fourier transform:

$$R_A(0) = \int S_A(v)dv = \int S(v)W(v)dv \le \int S(v)dv = \sigma^2. \tag{20}$$

Assuming a constant noise spectrum (white noise), $S(v) = C$, which is a reasonable assumption for the many small grains, we get for the variance:

$$\sigma^2 = C\int W(v)dv = \frac{C}{A}. \tag{21}$$

As expected we get lower noise with a larger aperture, but at the cost of a lower resolution.

PIV SIGNAL PROCESSING

We want to find the particle displacement, $\Delta \boldsymbol{x}_i = (\Delta x_i, \Delta y_i)$, for the ith particle located at $\boldsymbol{x}_i$ in the object plane, so that we can compute $u_{x,i} = \Delta x_i / \Delta t$ and $u_{y,i} = \Delta y_i / \Delta t$. We must find $\Delta \boldsymbol{X}_i$ in the corresponding point $\boldsymbol{X}_i$ in the image plane, see Figure 16.

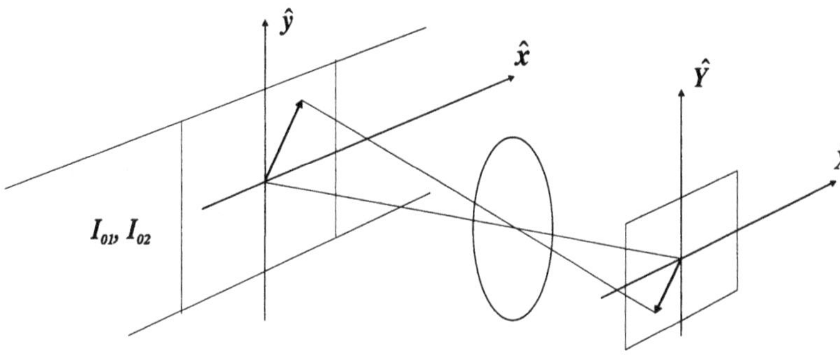

Figure 16. Coordinate systems used by imaging the particles onto the film and the subsequent reading of the film.

The following functions are involved in the imaging:

$I_{01}(\boldsymbol{x}), I_{02}(\boldsymbol{x})$ - illumination of the measurement region by first and second exposure.

M - magnification from object plane to image plane.

$\tau_0(\boldsymbol{X} - \boldsymbol{X}_i)$ - normalized transmission of the ith particle image (ass. all particle images identical).

$\tau(\boldsymbol{X})$ - transmission of the film in the point $\boldsymbol{X}$.

$g(\boldsymbol{x},t) = \sum \delta(\boldsymbol{x} - \boldsymbol{x}_i(t))$, sampling function for particles.

$I_I(\boldsymbol{X} - \boldsymbol{X}_I)$ - illumination in the interrogation area; $\boldsymbol{X}_I$ - center of interrogation area

$I(\boldsymbol{X})$ - light field after passage of (or reflection from) the film

Now we can write:

$$\tau(X) = \int \{I_{01}(x)g(x,t) + I_{02}(x)g(x,t+\Delta t)\}\tau_0(X - X_I)dx \tag{22}$$

This describes the transmission as a convolution of the illuminated particle sampling function and the particle transmission function. We get the intensity just behind the film as the product of the illuminating intensity and the transmission function:

$$I(X) = I_I(X - X_I)\tau(X) \tag{23}$$

Finally, we can form the autocovariance of the intensity field just behind the image:

$$R(S) = \int_A I(X)I(X+S)dX \tag{24}$$

$R(S)$ is the autocovariance function of the transmitted (reflected) light in the interrogation area, A. By introducing a statistical description of the particle motion through the moments of $g(x,t)$ and by making assumptions about the imaging (through $\tau_0(X - X_i)$) and illumination (through $I_{01}(x)$, $I_{02}(x)$ and $I_1(X\text{-}X_1)$), we can investigate $R(S)$. (For further details see Adrian[7] and Keane and Adrian[8]).

We can distinguish three cases:

1. Weak seeding: The particles are assumed so well separated that the chance that two particles are present in the interrogation area at the same time is negligible. The signal processing consists in a determination of the separation between the centroids of the particle images. Special software must find the centroid coordinates and compute the projections on the coordinate axes, see Figure 17. It is also possible to project the image onto the x- and y-axes (summing pixels in vertical and horizontal direction) first and then find the distances between the two peaks of the projections, which represent the projection of the particle positions, as illustrated in Figure 18. From this we find directly Δx and Δy. This latter method is probably the fastest, since it involves two 1-D peak finders instead of one 2-D peak finder.

Problem: Due to the random particle positions we can not be sure to have only one pair of images in any one IA and at the same time be sure to have images in all neighboring IAs. Therefore it is necessary to seed more heavily in order to obtain a continuous velocity information. However, this means that on the average more than one particle pair is present in each IA. It is no longer possible to determine uniquely a single particle image pair, and it becomes necessary to use more sophisticated forms of signal processing.

2. Medium seeding: The seeding concentration is so high that there are particle image

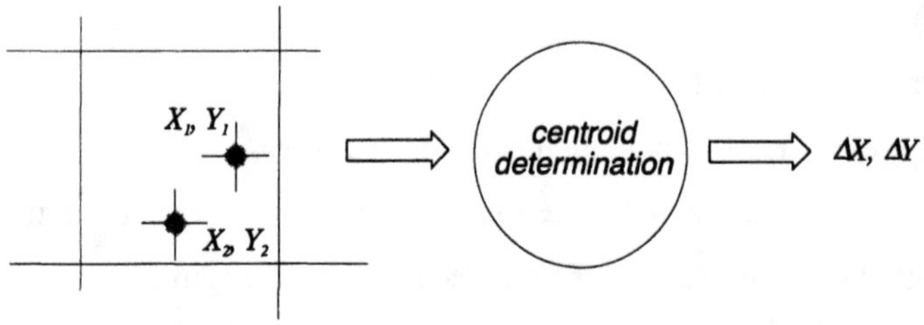

Figure 17. By weak seeding the most direct signal processing is simply to find the distance between the two images of the particle. This can be done by determining the distance between the two maxima or between the two centroids.

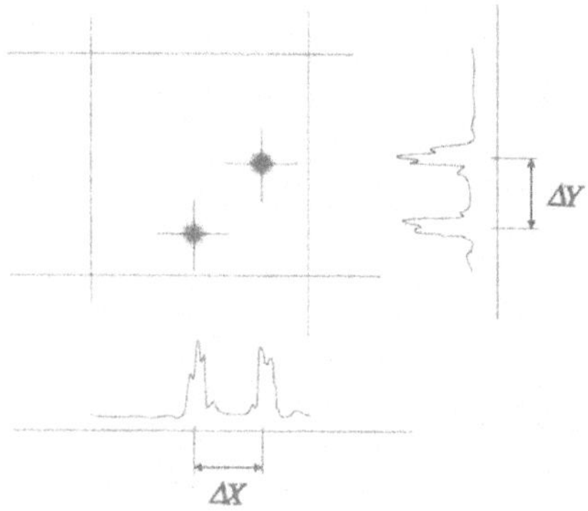

Figure 18. By projection of the image onto the two coordinate axes the signal processing is reduced to finding the distance between two peak values of two 1-D signals. The compression can be done optically or electronically.

pairs in all IAs with great probability. This assures continuous velocity information. But now there are many image pairs in each IA. How do we treat this signal?

a) 2-D FT:.

The FT of a double image with a fixed displacement is a complex function. The modulus or the square of this function is a fringe pattern, as illustrated in Figure 19. When discussing PIV this is known as the Young's fringe pattern. The most efficient way of finding the period in a fringe system is by Fourier analysis. The second FT computes the period in the fringe pattern. As there is one fixed period we get a PS consisting of two side peaks plus a center peak. We are thus led to perform two FTs and a squaring followed by a peak finder.

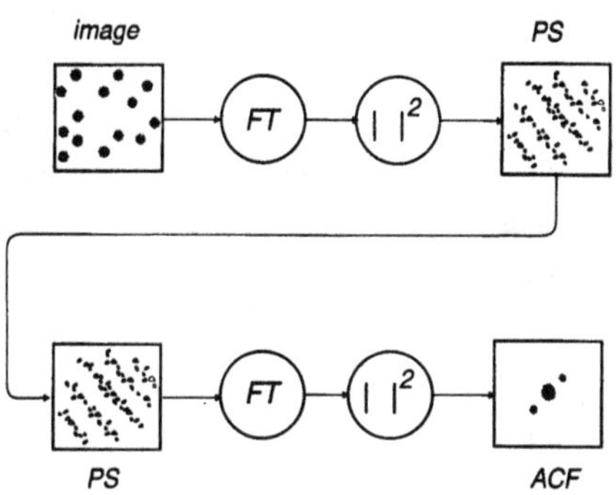

Figure 19. Signal processing by two consecutive power spectra.

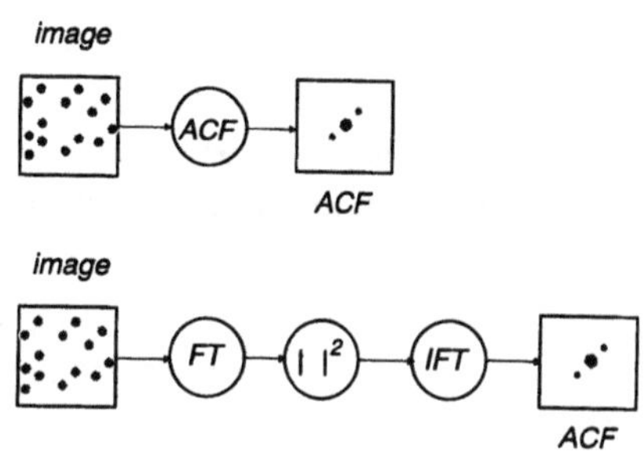

Figure 20. Processing by direct ACF or by use of the Wiener-Khintchine theorem.

b) 2-D ACF:.

We may also try to compute the ACF directly from the image from the definition of the ACF, Figure 20. The autocovariance function of the displaced set of points consists of two peaks on either side of a center peak. The positions of the side peaks indicate the displacement of the object. The most efficient way of computing the ACF is through the W-K theorem, where

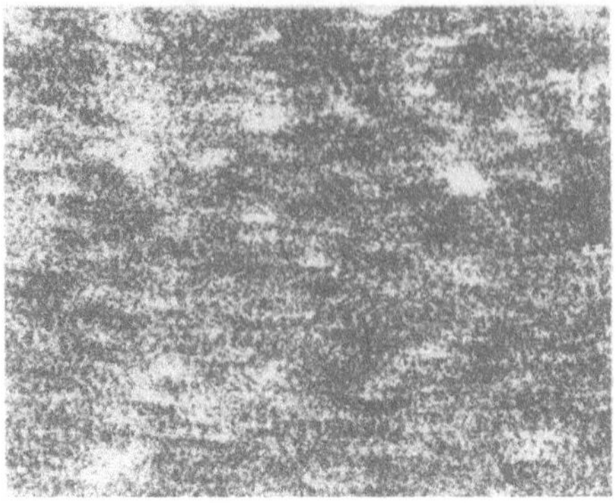

Figure 21. Speckle velocimetry. Section of double exposed speckle pattern.

one first finds the power spectrum by an FT followed by a squaring, then through an inverse FT finds the ACF. Apart from minor details we are led to the same computations as above.

3) Heavy seeding

The seeding concentration may be made so high that the optics are no longer able to resolve the individual particles. Instead a speckle pattern is recorded on the film. The spatial distribution of the speckles is determined by the distribution of particles. A displacement of the particle swarm will result in a corresponding displacement of the speckle pattern in the image plane. The optimum signal processing is again a 2-D ACF of the image in the IA. Figure 21 shows a double exposed speckle pattern.

Data Processing

The purpose of the measurement may be to find the spatial structures of the flow or perhaps more traditional statistical values for the flow. In both cases many images are required in order to assure the statistical significance of the measurement, as illustrated in Figure 22. Let us compare the processing time of PIV to the traditional laser velocimeter, LV.

After processing the signal, fluid mechanical computations are performed on the resulting velocity data, e.g. computation of velocity gradients, vorticity etc. These computations are not especially time demanding, so the time consuming part of the process is the signal processing, i.e. the conversion of displacement information into velocity information.

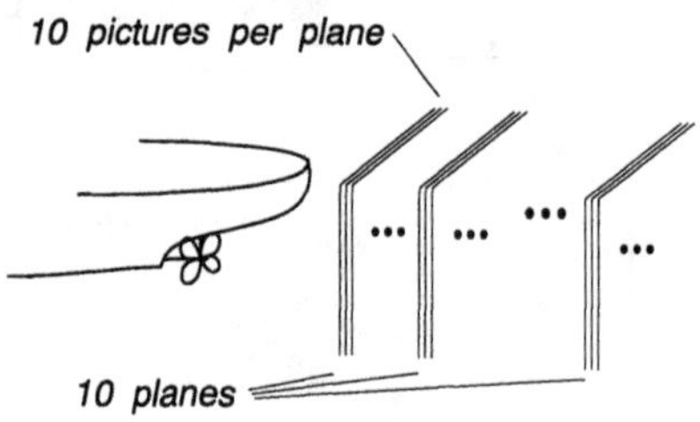

Figure 22. Problem: Measure 100×100 points in the flow field behind a model ship propellor. Make 10 measurements in each of 10 planes.

LV samples a time record in each point (up to three components in each point). Each record must contain a minimum number of samples, $N_{samples}$, (e.g. 10 for accuracy) with a time separation of at least $2T_i$ (for statistical significance), where T_i is the integral time scale (here estimated to 0.01 s). The total number of data in the set is:

a) For static moments (as e.g. mean, rms): 1 per. $2T_i$ total 10^7 data

b) Dynamic moments (spectrum and ACF): 100-1000 per $2T_i$ 10^9-10^{10} data

The signal processing takes place in real time. The statistical computations on each record take about 1s. The total processing time for static moments including time to traverse from point to point is, assuming a traversing time T_{trav}=0.5s:

$$T_{proc}=N_{planes}N_{points}(N_{samples}\times 2\times T_i+T_{trav})=10\times 10^4(10\times 2\times 10^{-2}+0.5)=19 \text{ h}. \tag{25}$$

Here N_{planes} is the number of measurement planes behind the propeller and N_{points} is the number of measuring points in each plane.

PIV samples 10 double exposed photos with at least $2T_i$ time separation for each measurement plane. The recording time is:

$$T_{rec}=N_{planes}(N_{samples}(2T_i+T_{ekspo}+T_{film})+T_{trav})=10(10(0.02+10^{-5}+1)+10)=200\text{s} \tag{26}$$

where T_{ekspo} is the exposure time for the film and T_{film} is the forward winding time of the camera.

The processing time with an image processor, which can make a 2-D FT in T_a=0.5s (e.g. VAX 780 and Numerix Array processor), and a scanning system with T_{scan}=0.04 s is:

$$T_{proc}=N_{planes}N_{samples}(N_{points}(T_a+T_{scan})+T_{setup})$$
$$=10\times 10(10^4(.5+.04)+100)=1.01\times 10^6\text{s}=152 \text{ h} \tag{27}$$

where T_{setup} is the setup time for the processor.

With our system (at DTH) the corresponding processing time would be 10 times longer. The need for a reduction in processing time is apparent! Two methods are being pursued: Use of parallel computing (transputers) and optical processing.

With an optical processor, which can carry out 25 ACFs per s, the corresponding processing time could be, assuming a processing time of 0 and a scan time of 0.04s:

$$T_{proc}=N_{planer}N_{samples}(N_{points}(T_a+T_{scan})+T_{setup})$$
$$=10\times 10(10^4(0+0.04)+100)=5\times 10^4\text{s}=13.9 \text{ h}. \tag{28}$$

OPTICAL PIV SIGNAL PROCESSING

As mentioned it is desirable to try to increase the speed of processing of PIV signals. Both PS and ACF may be carried out optically. The basis for these processes is the optical Fourier transform (OFT). These methods are still under development and are not available commercially. However, it may be of interest to briefly describe some of the methods.

The Optical Fourier Transform

An OFT may be carried out by means of a single lens. We shall briefly refresh the background for the OFT:

The starting point is Huyghens' principle according to which the field $\Psi(\boldsymbol{x})$ in the plane S with coordinates $\boldsymbol{x}=(x,y)$ may be derived from the field $\Psi'(\boldsymbol{x})$ in the plane S' with coordinates $\boldsymbol{\eta}=(\xi,\eta)$, see Figure 23. The assumption is that each point in the aperture A' emits a spherical wave. These waves are added in phase at the point P (x,y,z) (the Huyghens-Fresnel integral):

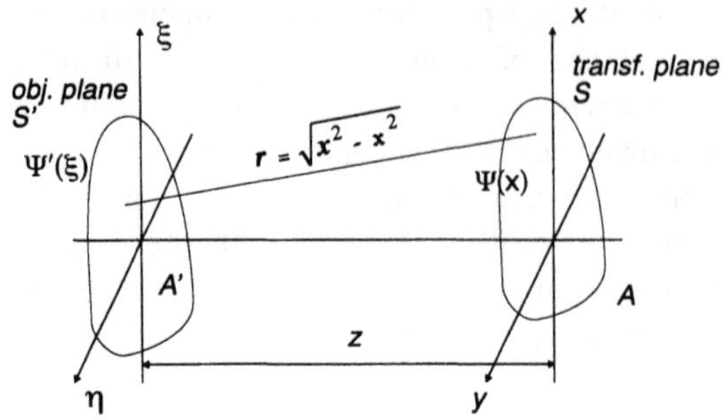

Figure 23. Geometry for optical Fourier transform.

$$\Psi(x) = \int_{A'} \Psi'(\xi)\Lambda(\xi,x)\frac{e^{-ikr}}{r}d\xi \tag{29}$$

$\Lambda(\xi,x)$ is a direction factor for the strength of the spherical wave travelling from ξ to x. Assuming both the aperture A' and the observation plane A to be small compared to the distance z between them, we may replace Λ by a constant $K = 1/ikz$ and $1/r$ may be replaced by $1/z$:

$$\Psi(x) = \frac{1}{i\lambda z}\int_{A'} \Psi'(\xi)e^{-ikr}d\xi \tag{30}$$

Expressing $r(\xi,x)$ by its coordinates, we may write:

$$r = \sqrt{(x-\xi)^2 + (y-\eta)^2 + z^2} \cong R + \frac{\xi^2+\eta^2}{2R} + \frac{\xi x + \eta y}{R} \qquad \left(R = \sqrt{x^2+y^2+z^2}\right) \tag{31}$$

Then we may write for the field:

$$\Psi(x) = \frac{1}{i\lambda z}e^{-ikz}\int_{A'} \Psi'(\xi)e^{-i2\pi(\xi^2+\eta^2)/\lambda z}e^{-i2\pi(\xi x+\eta y)/\lambda z}d\xi \tag{32}$$

If the aperture is a circular lens, and if the observation plane is in the focal plane of the lens, $z=f$, we may introduce the phase delay in the incident wave caused by the lens, as illustrated in Figure 24:

$$\Delta\Phi = -k(\xi^2+\eta^2)/2f \tag{33}$$

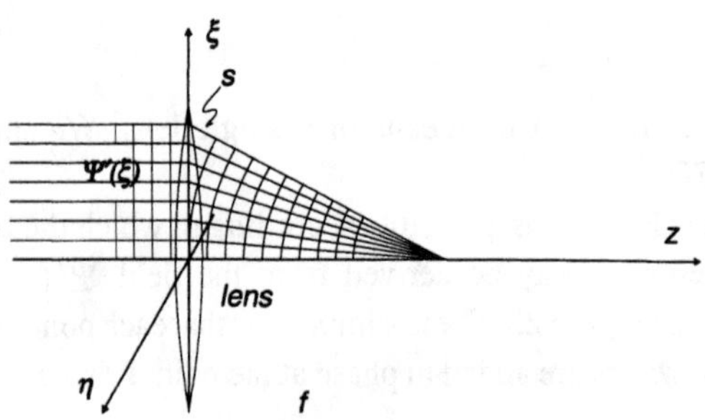

Figure 24. According to its definition a lens introduces a phase delay, which causes all rays to intersect in the focal point.

or a phase factor on the incident wave:

$$e^{-k(\xi^2+\eta^2)/2f}. \tag{34}$$

The wave in the observation plane then becomes:

$$\Psi(x) = \frac{1}{i\lambda f} e^{ikf} \int_{A'} \psi'(\xi) e^{-i2\pi(\xi \bullet x)/\lambda f} d\xi \tag{35}$$

which is exactly the mathematical expression for the 2-dimensional Fourier transform of $\psi'(\xi)$ (apart from the scaling and phase factors in front).

An observer watching the screen S' or a photodetector at S' will detect the intensity at S':

$$I(x) = |\Psi(x)|^2 \qquad \text{(the 2-dimensional "optical power spectrum").} \tag{36}$$

Optical image processing of PIV images may be carried out in several ways.

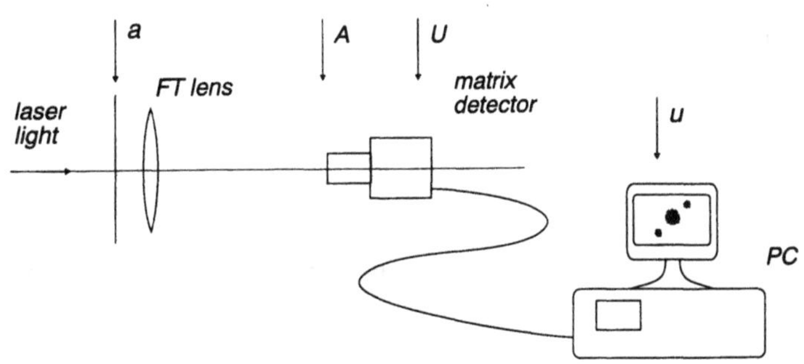

Figure 25. Hybrid optical/numerical processing, the so-called Young's fringe method.

Optical PIV Processors

The main reason for using optical processing on PIV recordings is that the accuracy available in a PIV transparency and the resolution required in a PIV measurement match very well the performance of available optical processing schemes. It is then possible to use the high speed of an optical transform to bring down the total processing time for a PIV measurement. The disadvantage, however, of optical processing is the slow and cumbersome process of reading in and out to the active optical element.

Young's fringe method: The most widely used optical processing scheme is a hybrid method, also called Young's fringe method, in which the interrogation area is Fourier transformed optically, squared to form the power spectrum, which is then detected and stored in a computer. After digitization the power spectrum is Fourier transformed numerically to yield the autocorrelation function. The location of the peaks is detected and converted to velocity information. The process may be described symbolically with reference to Figure 25:

a	Object (the interrogation area of the image)
A	Optical FT of a
$U = c_1 \lvert A \rvert^2$	Squaring performed by the optical detector

$u = \mathrm{FT}\{U\} = c_1 a \otimes a$ Autocorrelation of a ($\otimes$ denotes correlation)

Programmable optical light modulator (spatial light modulator, SLM): An all-optical processing method may be implemented by means of an optical component called a spatial light modulator (SLM). An SLM is a programmable component (e.g. liquid crystal display), whose optical transmission in a number of cells arranged in a 2-D matrix (e.g. 512×512 elements) may be controlled electronically. These components may work by a pure phase modulation or by amplitude modulation.

An optical autocovariance by means of an SLM is shown in Figure 26. The first transform is recorded by video camera no. 1 and the electronic image is transferred to an electrically programmable SLM. The following transform results in the ACF, which is picked up by video camera no. 2. As mentioned earlier we may project this image onto the two coordinate axes and process the two resulting 1-D signals to obtain the two velocity projections.

A simplified analysis of the processor is given by the following:

a	Object (PIV interrogation area)
$A = \mathrm{FT}\{a\}$	Optical FT of a
$I = c_1 \lvert A \rvert^2$	Intensity
$\tau = c_2 I$	SLM transmissitivity is proportional to I.
$U = R\tau = c_2 RI = c_1 c_2 R \lvert A \rvert^2$	Modulated plane wave, R. The modulation τ is proportional to the square of A, as provided by the detector.

$$
\begin{aligned}
u &= \mathrm{FT}\{U\} \\
&= c_1 \mathrm{FT}\{RAA^*\} \\
&= c_1 r * \mathrm{FT}\{AA^*\} \\
&= c_2 r * a \otimes a
\end{aligned}
$$

where $\otimes$ denotes covariance and * means convolution. r is a delta function in the direction of

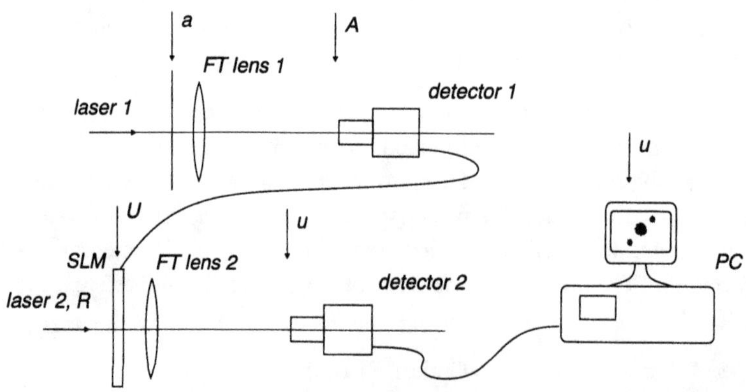

Figure 26. All-optical processor based on a spatial light modulator.

R). Thus the FT of U gives the autocovariance of a, convolved with the FT of the plane wave, i.e. located at the focal point for the plane wave.

Photorefractive crystal: A photorefractive crystal may be used as the active element in an all-optical OFT. A photorefractive crystal is a nonlinear optical crystal in which an optical field can be stored as a volume hologram. The recording in the crystal takes place as a modulation of the refractive index of the crystal. The index in turn is modulated by the local light intensity in the crystal. Thus, if two light beams (an object beam and a reference beam) are made to intersect inside the crystal, an optical interference pattern is created inside the crystal. The intensity of this optical field modulates the local index of refraction and leaves an index modulation, also after the two incident fields are turned off. Another light beam, which is passed through the crystal, will receive a phase modulation from the index pattern. This modulation may be used for optical filtering or further processing of the field.

Physically, the crystal is both photoconductive and electro-optical. The incident light field creates photoelectrons in proportion to the local intensity. These photoelectrons diffuse through the crystal until they are trapped by acceptor sites (impurities or lattice defects). The positive donor ions form a local electric field, which through the electro-optic effect modulates the local index of refraction. The result is a modulation of the refractive index, which in turn will modulate an optical field travelling through the crystal.

Photorefractive materials have certain advantages as light modulators. The effect depends on the total optical energy in the crystal, not on the instantaneous power, since the number of

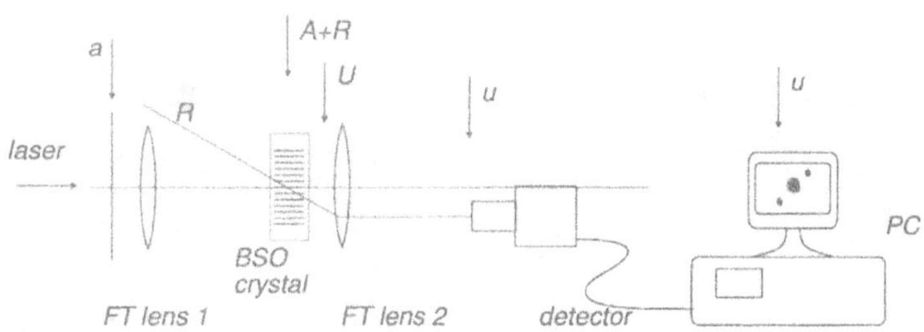

Figure 27. An optical autocorrelator based on a BSO crystal. The crystal stores a hologram of the FT of the object field (the PIV transparency) by means of the reference beam. When the hologram has formed, the reference beam is turned off, after which the FT of the object field (which is still present) is multiplied by itself through the phase modulation in the crystal. The result is an ACF of the PIV image.

charge carriers is proportional to the amount of light. Thus, the photorefractive effect thus can be substantial even for low power lasers. The writing speed depends on how fast the energy is delivered to the crystal. Writing time constants using He-Ne lasers are of the order of seconds, while gratings may be written in the crystal in µs or less with pulsed lasers. Once the hologram is formed, it is stable in darkness, but can be erased by uniform illumination.

Analysis, see Figure 27:

a	Object field (PIV interrogation area)
A+R	FT of a plus plane wave, R

$$
\begin{aligned}
I &= c_1|A+R|^2 = c_1\left(|A|^2+|R|^2+A^*R+AR^*\right)\\
\tau &= c_2\exp[i\beta I] \cong c_2(1+i\beta I)\\
U &= c_3AT \cong c_2c_3A(1+i\beta I)\\
&= c_2c_3\left(A+ic_1\beta\left(A|A|^2+A|R|^2+AA^*R+AAR^*\right)\right)
\end{aligned}
$$

The intensity I is proportional to $A+R$ squared. The transmission of the SLM τ is proportional to I. The hologram in the crystal is read out with the FT of the object field A, while the reference beam is switched off. Thus the field after the crystal U is proportional to the product of A and T.

$$
\begin{aligned}
u &= \mathrm{FT}\{U\}\\
&= \mathrm{FT}\{A\}+i\beta\mathrm{FT}\left\{A|A|^2\right\}+i\beta\mathrm{FT}\left\{A|R|^2\right\}\\
&\quad +i\beta\mathrm{FT}\{AA^*R\}+i\beta\mathrm{FT}\{AAR^*\}\\
&= u_1+u_2+u_3+u_4+u_5 \qquad \text{where}\\
u_4 &= i\beta a\otimes a^*r
\end{aligned}
$$

The output of the processor is the FT of U. It consists of several terms, but one of them, u_4, is proportional to the convolution of the ACF of a and the FT of the reference beam r, i.e. located at the focal point of R.

Other Optical Processing Methods

Filtering in the Fourier plane: By spatial filtering in the Fourier plane, we can extract those areas, whose FT has a given magnitude and direction, i.e. light coming from regions of the object plane having a certain displacement (velocity). By this method we obtain iso-velocity plots. The method has been used with LSV and LSP, but the resolution is not very good due to speckle effects.

CONCLUSION

Particle image velocimetry is still not a well established measurement method. One reason is the technological development, which changes the fundamental assumptions as to what can be done with a given technology. The photographic recording is still much superior in its ablilty to store high resolution images, but CCD chips are catchin up, and it is probably safe to say that no one will miss the wet chemical development process. Also laser technology is under development, and the emergence of small, more robust and cheaper sources of pulsed laser light will help make the PIV method mere practical.

As for processing of PIV images, the most widely used method today is the combined optical-digital procedure of the Young's fringe method. As computers become cheaper and more powerful the completely digital computation of the ACF will probably take over. The all-optical processor is still a possibility, but is now limited by the bottleneck caused by the limited input and output speed and questions about durability and life of the nonlinear optical components. Perhaps both of these options will be replaced by a dedicated electronic processor developed for this specific purpose, and therefore surpassing the other methods in speed and user friendliness.

REFERENCES

1. See the chapter on "Correlation methods of PIV analysis" by R. D. Keane.
2. See the chapter on "Three-dimensional particle velocimetry" by K. D. Hinsch.
3. R. J. Adrian, "Particle-imaging techniques for experimental fluid mechanics", *Annu.. Rev.. Fluid Mech.*, 23:261-304 (1991).
4. P. Buchhave, "Particle image velocimetry - status and trends", Int. J. Heat and Mass Transfer,.
5. K. D. Hinsch, "Particle Image Velocimetry", in: "Speckle Metrology", R. S. Sirohi, ed., Marcel Dekker, Inc., New York, Basel, Hong Kong (1993).
6. R. J. Adrian, and C. S. Jao, "Pulsed laser techniques: Application to liquid and gaseous flows and the scattering power of seed materials", *Applied Optics*, 24:44-52 (1985).
7. R. D. Keane and R. J. Adrian, "Optimization of particle image velocimeters. Part I. Double Pulsed Systems," *Meas. Sci. Technol.*, 1:1202-1215 (1990).
8. R. J. Adrian, "Image shifting technique to resolve directional ambiguity in double-pulsed velocimetry," *Appl. Optics*, 25:3855-3858 (1986).
9. C. C. Landreth and R. J. Adrian, "Electro-optical image shifting for particle image velocimetry", *Appl. Optics*, 27:4216-4219 (1988).
10. G. O. Reynolds, J. B. DeVelis, J. B. Parrent Jr. and B. Thompson. "Physical Optics Notebook: Tutorials in Fourier Optics", SPIE Optical Engineering Press (1989)
11. Adrian, R. J., "Statistical properties of particle image velocimetry measurements in turbulent flow", Laser Anemometry in Fluid Mechanics, Vol. III, Ladoan, Instituto Superior Technico, Lisbon (1988).

CORRELATION METHODS OF PIV ANALYSIS

Richard D. Keane

Department of Theoretical and Applied Mechanics
University of Illinois at Urbana-Champaign
216 Talbot Laboratory
104 S. Wright St.
Urbana, IL 61801

INTRODUCTION

As described in the previous chapter, Particle Image Velocimetry (PIV) uses images of seeded marker particles in a fluid flow to measure two-dimensional or three-dimensional instantaneous velocity fields in experimental fluid mechanics. The seeded fluid flow is illuminated by short duration pulsed sheets of light at precise time intervals to obtain particle images that are recorded on a digital video camera array, on photographic film or on a hologram. The velocity vector field is obtained by analyzing these recorded images to measure the particle displacements and thus determine the velocity by knowing the time interval. Although many methods of PIV interrogation have been explored experimentally, as discussed by Adrian[1], Buchhave[2], Gray[3] and Hinsch[4], the broad range of methods can be reduced to a smaller group of alternative methods based upon the seeding density of particles within the flow field, denoted by C and defined to be the number of seeded particles per unit volume of the fluid. For a sufficiently high seeding density, the local velocity of the fluid in a small volume of the flow can be determined by measuring the particle image displacements within a corresponding digitized region of the image field, known as an interrogation spot. The image density, N_I, is defined from Adrian and Yao[5] to be the mean number of particle images within an interrogation spot of diameter, d_I and can be written as

$$N_I = C\, d_I^2 \Delta z_0 / M^2, \tag{1}$$

for square interrogation spots where the image magnification is M and the light sheet thickness is Δz_0.

As the image density increases, alternative methods of PIV interrogation can be applied. In the low image density case where the mean number of particle images per interrogation spot is low, particle tracking methods can be utilized to measure image displacements or multiple pulses of light sheet illumination can be used to increase the number of images sufficiently to use correlation methods.

In the high image density case, it is natural to measure the average displacement of local groups of particles in an interrogation spot as opposed to tracking individual particle images. The most common forms of analysis of high image density recordings are based on correlation techniques that are performed either numerically using Fast Fourier Transforms on digitized image fields, numerically on the digitized Young's fringe patterns that result from optically Fourier transforming the image field, or in a purely optical correlation processor. For high image density recordings of PIV image fields, the following alternative methods of interrogation are possible:

(a) Digital auto-correlation or cross-correlation analysis of single exposures of double or multiple pulse recordings. Two digital Fast Fourier Transforms are performed on each digitized interrogation spot for auto-correlation analysis while three FFTs are required for cross-correlation analysis.

(b) Young's fringe method. One optical transform and one digital FFT are performed.

(c) Optical auto-correlation. Two optical transforms are performed.

It can be shown that each of these three methods is similar so that the numerical correlation of the digitized image field will be discussed below as an example. Some comparisons between these three methods have been made recently by Keane and Adrian[6] and by Westergaard and Buchhave[7], while Buchhave and Jacobsen[8] have analyzed the optical correlator method of PIV image processing using a spatial light modulator. Furthermore, digital methods of correlation are common due to advances in array processors and digital methods of data acquisition such as CCD cameras.

PIV PERFORMANCE USING CORRELATION METHODS

Originally, in PIV analysis, numerical correlation was performed on digitized images obtained by a frame grabber from a single photographic recording upon which a double exposure of the flow field was made. Despite developing procedures being required to obtain the negative and the positive photographic recording, the advantage of high resolution made photographs an attractive medium. In addition, cinematographic cameras have been used by Lin and Rockwell[9] to examine vortical structures in the near wake of a cylinder by recording image fields sequentially to obtain a time history of the flow, even though the data cannot be interrogated in real time. Moreover, correlation methods can be generalized to digitized images from multiple exposures on a single photographic frame and to direct digital recording of either single, double or multiple exposures on video cameras or CCD arrays. Although large format photographic film such as 100mm x 125mm still offers higher resolution than digitally recorded data, the relative ease of CCD image acquisition and the availability of 2K x 2K CCD cameras has offered an attractive alternative to 35mm recording and numerical correlation of direct digital recordings are now commonplace.

In all correlation analysis, the performance of the PIV method of fluid velocity measurement is determined by

(1) the spatial resolution of the measurements,

(2) the valid data detection rate, and

(3) the accuracy and reliability of the velocity measurements.

In order to optimize the PIV experimental and interrogation procedure, it is important to understand the effect that the experimental parameters and the choice of interrogation method have on these three criteria.

The spatial resolution for auto-correlation analysis is bounded by the size of the measurement volume which, ideally, should be small relative to the length scales of the flow field. For example, in the study of turbulent flow fields, spatial resolution on the order of the Kolmogorov length scale, η, would enable vorticity measurements and measurements of the viscous dissipation of the turbulent energy to be made very accurately.

The valid detection rate is defined as the number of interrogation spots per unit area of the image plane that produce velocity measurements which satisfy certain interrogation criteria. The accuracy of all detections that satisfy the interrogation criteria is used as a measure of the ability of those interrogation criteria to produce valid measurements which agree with a known velocity field. Thus, the valid detection rate is a measure of the technique's ability to measure flow velocity correctly and can only be determined precisely by simulation or by comparison with alternative accurate methods of fluid velocity measurement.

The performance of a PIV system determined using the three criteria above, is dependent upon both the experimental parameters and the interrogation procedure with its accompanying criteria. Hence, to optimize the auto-correlation analysis in PIV, it is necessary to choose the experimental parameters and the interrogation parameters to achieve the highest spatial resolution possible and a satisfactory valid detection rate, for the flow field under consideration.

It can be seen in Figure 1 that the experimental light sheet and the image recording system are designed for a velocity field, $\boldsymbol{u}(\boldsymbol{x})$, with the following parameters chosen to optimize the PIV technique:

(a) the mean volumetric concentration of seeding particles, C,

(b) the thickness of the illuminating light sheet, Δz_0,

(c) the size, d_I, and shape of the interrogation spot,

(d) the mean particle image diameter, d_τ, and

(e) the time interval, Δt, between successive exposures.

The image recording system is designed with a lens magnification, M, which can be chosen to improve spatial resolution at the cost of reducing the field of view. Furthermore, the mean particle image diameter, d_τ, in the image plane is dependent upon M and upon the wavelength of the illuminating light, λ, and the f-number of the camera lens, $f^\#$, for known mean particle diameter, d_p, due to diffraction limitations.

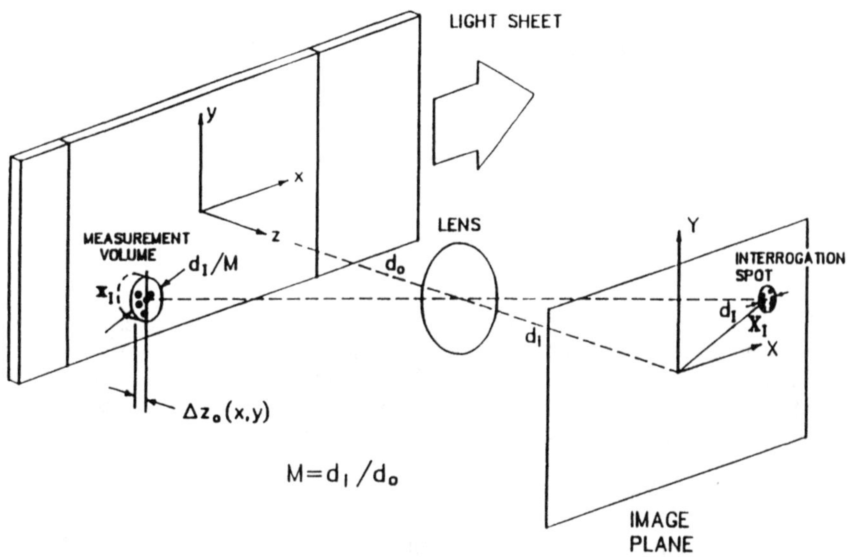

Figure 1. Experimental light sheet and image recording system for planar pulsed laser velocimetry. A measurement volume within the illuminated light sheet corresponds to an interrogation spot within the image plane.

To select the best method of PIV analysis for a given fluid flow and to compare the alternative methods possible using PIV, it is desirable to reduce the experimental and interrogation parameters to a smaller set of dimensionless parameters, from which the most important ones can be selected. For auto-correlation analysis, the most important of these dimensionless parameters can be listed as:

1. Data validation (or detectability) criterion, D_0
2. Particle image density, N_I
3. Relative in-plane image displacement, $|\Delta X|/d_I$
4. Relative out-of-plane particle displacement, $|w_I|\Delta t/\Delta z_0$
5. Relative image displacement variation (gradient effects), $M|\Delta u|\Delta t/d_I$, $M|\Delta u|\Delta t/d_\tau$
6. Relative particle image diameter, d_τ/d_I.

For a particular fluid flow measurement, it is possible to choose the experimental and interrogation parameters by determining optimal ranges for these dimensionless parameters, based on some knowledge of the velocity field, $\boldsymbol{u}(\boldsymbol{x})$. The relationship between these dimensionless parameters and the correlation function will be illustrated below to indicate how optimal choices of such parameters are made.

Spatial auto-correlation analysis of a double-pulse single frame system.

For interrogation of a photograph, an interrogation spot is illuminated by a beam of intensity, $I_I(\boldsymbol{X}\text{-}\boldsymbol{X}_I)$, where $\boldsymbol{X}_I$ is the centre of the interrogation spot. The transmitted light intensity through the interrogation spot is given by

$$I(\boldsymbol{X}) = I_I(\boldsymbol{X}-\boldsymbol{X}_I)\ \tau(\boldsymbol{X}), \tag{2}$$

where $\tau(\boldsymbol{X})$ is the intensity transmissivity of the photographic image. This photographic transmissivity can be written, using the notation of Adrian[10], in terms of the particle image transmissivity, $\tau_0(\boldsymbol{X})$ and the laser light sheet intensities, $I_{01}(\boldsymbol{X})$ and $I_{02}(\boldsymbol{X})$, namely

$$\tau(\boldsymbol{X}) = \sum_i \{I_{01}(\boldsymbol{x}_i)\ \tau_0(\boldsymbol{X}-M\boldsymbol{x}_i(t)) + I_{02}(\boldsymbol{x}_i)\ \tau_0(\boldsymbol{X}-M\boldsymbol{x}_i(t+\Delta t))\}\quad , \tag{3}$$

where Δt is the pulse separation for double-pulsed images. The spatial auto-correlation function of the transmitted light intensity for an interrogation spot is given by

$$R(\boldsymbol{s}) = \int I(\boldsymbol{X})\ I(\boldsymbol{X}+\boldsymbol{s})\ \mathrm{d}\boldsymbol{X}\ , \tag{4}$$

where $\boldsymbol{s}$ is a two-dimensional displacement vector in the correlation plane. It can be seen from the above equations that the correlation function depends upon those particle properties, which determine the intensity of light recorded, such as its diameter and reflectivity. The auto-correlation function for double-exposure images consists of five components (Adrian[10])which are illustrated in Figure 2 and are

$$R(\boldsymbol{s}) = R_C(\boldsymbol{s}) + R_P(\boldsymbol{s}) + R_{D^+}(\boldsymbol{s}) + R_{D^-}(\boldsymbol{s}) + R_F(\boldsymbol{s})\ , \tag{5}$$

where $R_C(\boldsymbol{s})$ is the convolution of the mean intensities in equation (4) while $R_F(\boldsymbol{s})$ is the fluctuating noise component of the correlation estimator. $R_{D^+}(\boldsymbol{s})$, the correlation of all first images shifted by $\boldsymbol{s}$ with subsequent unshifted second images and $R_{D^-}(\boldsymbol{s})$, the correlation of all second images shifted by $\boldsymbol{s}$ with prior unshifted images, are identical in form and shape, with $R_{D^+}(\boldsymbol{s})$ being centred at the positive mean displacement of the images and $R_{D^-}(\boldsymbol{s})$ being centred at the negative mean displacement. The tallest component, $R_P(\boldsymbol{s})$, the correlation of particle images with themselves, has a maximum value when $\boldsymbol{s} = \boldsymbol{0}$ and has a width comparable with d_τ, the particle image diameter. The amplitudes of both $R_{D^+}(\boldsymbol{s})$ and $R_{D^-}(\boldsymbol{s})$, are less than one half that of $R_P(\boldsymbol{s})$ because the correlation between pairs of images in an interrogation spot is at most one half of the correlation of images with themselves. It can be seen that the symmetry of $R(\boldsymbol{s})$ leads to ambiguity in locating the true displacement, unless the direction of motion is known a priori or unless image shifting described below is implemented. The correlation between pairs of images is maximized when all particle images appear as pairs in the interrogation spot, which imposes a limit on the image displacement to minimize loss of image pairs within the interrogation spot. When particles move into or out of the interrogation volume between illuminating pilses, they do not contribute to the correlation between pairs of images, producing only noise peaks in R_C+R_F by correlating with other random particle images. The signal to noise ratio is the ratio of the peak amplitude of $R_{D^+}(\boldsymbol{s})$ to the peak amplitude of the instantaneous noise, R_C+R_F.

The mean image displacement across a given interrogation spot is determined by locating the centroid of the displacement component, $R_{D^+}(\boldsymbol{s})$, namely

$$\mu_{D^+} = \int \boldsymbol{s} R_{D^+}(\boldsymbol{s})\mathrm{d}\boldsymbol{s} \Big/ \int R_{D^+}(\boldsymbol{s})\mathrm{d}\boldsymbol{s}\quad , \tag{6}$$

from which the mean velocity is estimated as $\mu_{D^+}/M\Delta t$ where M is the magnification and Δt is the pulse separation.

In practice, care must be taken to select an area in the correlation plane over which the centroid is determined. It must be small enough to minimize the effects of random noise peaks from R_C+R_F near the signal peak and large enough to span the peak of $R_{D^+}(\boldsymbol{s})$ when a velocity gradient spreads it. Furthermore the effects of background electronic or photographic noise which add a background plateau to the signal must be taken into

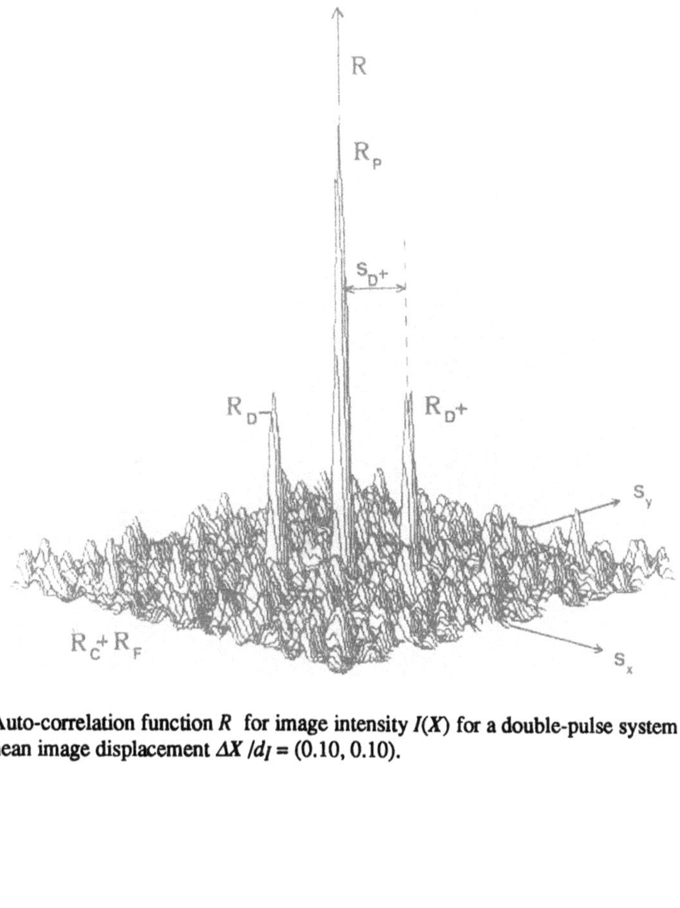

Figure 2. Auto-correlation function R for image intensity $I(\boldsymbol{X})$ for a double-pulse system where N_I = 15 and the mean image displacement $\Delta \boldsymbol{X} /d_I$ = (0.10, 0.10).

account. Alternatively, the mean image displacement, μ_{D^+}, can be approximated to subpixel resolution by fitting a curve or surface to the displacement component, $R_{D^+}(\boldsymbol{s})$ around the peak pixel value and determining the peak location of the smooth curve or surface. These curves or surfaces could be parabolic, Gaussian or could be based on the particle image shape and diameter deduced from the self-correlation peak, $R_P(\boldsymbol{s})$. It has been recommended by Westerweel[11], that the Gaussian estimator performs best in low pixel resolution PIV interrogations and Prasad et al.[12] have examined these alternatives for a range of pixel resolutions.

Signal Peak Search Algorithm

It can be seen that from Figure 2 that the auto-correlation function R is symmetric with respect to rotations about the origin of 180^0, leading to possible directional ambiguity in locating the mean image displacement from $R_{D^+}(s)$ or $R_{D^-}(s)$, unless some knowledge of the flowfield results in a peak search over one half-plane of R. To overcome the problem, image shifting using either electronic, optical or mechanical methods is used to provide all second images with a known induced image shift based upon prior knowledge of the flow field. With an imposed image shift on all second images, the recorded image displacements are modified to lie in a pre-selected half-plane of the correlation function. As shown in Figure 3, the image shift, X_S, has been chosen to ensure that $\Delta X+X_s > 0$ for all particles in the interrogation volume. In this case, the image shift must be larger than the absolute value of the most negative X-component of image displacement causing the recorded images to form image pairs whose correlation peak lies in the right half-plane of R, thus removing directional ambiguity. In order to separate the self-correlation peak, $R_P(s)$, from the signal peak after image shifting, the image shift must be chosen to be larger than the bound given above by some particle image diameters, namely

$$\Delta X + X_s > 3d_\tau \quad . \tag{7}$$

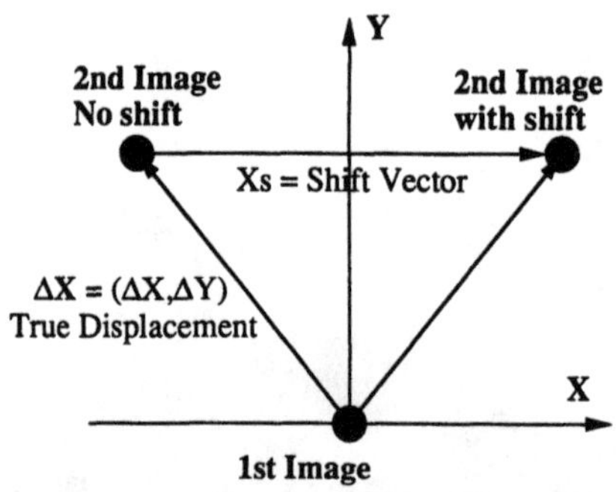

Figure 3. The image shift, Xs, ensures that $\Delta X + X_s > 0$ for all image pairs in the interrogation spot. In this case, the recorded displacement of the particle image will lie in the right half-plane, removing directional ambiguity.

This image shifting can be accomplished by means of a rotating mirror, or by electro-optic methods such as a calcite crystal of specified thickness, producing an image shift from two cross-polarized illumination beams forming the light sheets. Electronic image shifting by a fixed number of pixels within a CCD camera produces a highly accurate image shift.

When directional ambiguity has been resolved by image shifting, the tallest peak in the correlation half-plane corresponding to the image shift is located. The influence of the self-correlation peak is removed by windowing a sufficiently large area around the origin, without affecting the signal peak. The windowed area can be determined from the mean particle image diameter or interactively before interrogation proceeds. Alternatively, a restricted area in the correlation plane can be searched, based on data from neighbouring interrogation spots. While such a restricted search improves the computational efficiency of the algorithm, it becomes less reliable when the velocity field changes rapidly from one interrogation spot to the next.

A detectability threshold, D_0, can be enforced to reduce the probability of invalid measurements. One can require that the ratio, $D(X_I)$, of the tallest peak in R, $R(s_1)$, to the next tallest peak, $R(s_2)$, (excluding R_P) exceeds the prescribed threshold, D_0. The choice is made to minimize the false detection probability while still ensuring an acceptable valid detection rate. If D_0 is too high, then there may be measurements rejected although the signal peak is located because the amplitude of the noise peaks are high and, if D_0 is too low, then there may be noise peaks accepted as valid measurements when the signal and noise peaks are similar. The detectability or data validation parameter, $D(X_I)$, serves as an approximate estimate of the signal to noise ratio for the measurement volume at X_I,

$$D(X_I) = R(s_1) \,/\, R(s_2)\,. \tag{8}$$

Alternative peak locations for the second, third, and even fourth tallest peaks in a half-plane of $R(s)$ can be located to obtain alternative image displacements. Post-interrogation processing of the velocity vector field can be performed to determine vectors which are consistent with the vectors of the neighbouring field in cases where there is an invalid measurement using the detectability criterion.

NOISE AND BIAS SOURCES IN THE AUTO-CORRELATION FUNCTION, R(s)

There are several independent sources of noise in $R(s)$ resulting in loss of accuracy and reliability in obtaining velocity measurements, even in the theoretically ideal analysis of identical particle images. They can be listed as follows:

(a) Discretization in spatial resolution due to finite pixel size,
(b) Discretization of image intensity, usually recorded as 8-bit words,
(c) Electronic CCD noise and photographic grain noise and fog level noise,
(d) Spurious correlations of random image pairs.

To overcome the discretization in spatial resolution, which occurs when the image field is recorded either photographically or using a digital CCD camera, the flow field can be recorded using the highest resolution formats available for either film or CCD formats. Large format film such as 100mm x 125mm or 70mm gives better resolution than 35mm film for the same field of view because the magnification needed is lower and the film grains image a smaller volume of the flow field. High resolution film such as Kodak Technical Pan or TMax film, with resolution of approximately 300 and 180 lines per mm respectively, minimizes film grain noise although it is generally less sensitive than other regular film. In digital recording, large array CCD cameras of array size 1K x1K, 2K x2K and even 5K x5K are now available and allow higher pixel resolution for interrogation spots than earlier 512 x 512 arrays for a specific spatial resolution. The most severe limitation of large array CCD cameras is the time required to transfer a recorded image field before a subsequent image can be recorded.

To ensure adequate pixel resolution in an interrogation spot, it has been shown by Prasad et al[12] that arrays of size 128 x 128 or 64 x 64 are preferable to 32 x 32 arrays for performing digital FFT's in determining sub-pixel image displacements from peaks in the auto-correlation function. The increased number of pixels decreases the bias towards discrete pixel displacements that occurs when the pixel resolution is too coarse.

The selection of seeding particles and their imaging characteristics is important in minimizing the particle image size which is determined by the particle diameter and the diffraction limited spot size of the recording optics. It is important to choose the particle diameter as small as possible to ensure that the particles follow the fluid flow, yet the particles must be sufficiently large to ensure adequate light reflection for image recording. High accuracy and high spatial resolution require very small markers because the uncertainty in image displacement is proportional to particle image size,

$$\frac{\sigma_u}{u_{max}} = \frac{\sigma_{\Delta x}}{\Delta x_{max}} = \frac{C d_\tau}{\Delta x_{max}}, \tag{9}$$

where $\sigma_{\Delta x} = C d_\tau$.

The particle image size and the pixel resolution of the digitized image field are inter-dependent as the spatial resolution improves with small images and there is a pixel bias due to discretization of the image field. The accuracy of the displacement measurement, however, becomes independent of the pixel resolution if there are more than 2-3 pixels per image diameter (Prasad and Adrian[12]). Hence pixel resolution must be sufficiently high to enable small seeding particles to span a number of pixels to overcome the pixel bias. Westerweel[11] has provided an optimal model for PIV analysis when such high pixel resolution is not available.

A second source of bias in the auto-correlation function occurs when the image density is so low that spurious correlations of random image pairs become larger in amplitude than the true displacement peak height. For a uniform flow with velocity, **u**, the mean value of the displacement correlation component, $R_{D^+}(\mathbf{s})$, evaluated at its peak value can be written as

$$<R_{D^+}(\mathbf{s}_{D^+})|\mathbf{u}> = \frac{1}{2}<R_P(\mathbf{0})>F_O(w_I\Delta t)\exp[-4|M\mathbf{u}_I\Delta t|^2/(d_I^2+d_\tau^2)], \tag{10}$$

where $F_O(w_I\Delta t)$ is the factor representing loss of signal due to out-of-plane motion causing loss of image pairs and $\exp[-4|M\mathbf{u}_I\Delta t|^2/(d_I^2+d_\tau^2)]$ represents the loss of signal due to loss of in-plane image pairs from the interrogation spot for the model chosen for particle image transmissivity and interrogation spot intensity (Keane and Adrian[13]). It can be seen that the displacement peak becomes smaller and less likely to be detected as the velocity increases due to increased loss of image pairs. It becomes smaller as the image density decreases as the self-correlation peak $<R_P(\mathbf{0})>$ is dependent upon the number of images in the interrogation spot. Hence, in regions of low seeding, there will be a bias in the measured data, due to spurious noise peaks being detected. Furthermore it can be seen that there is a small bias effect due to in-plane loss-of-pairs above, particularly if the particle image size, d_τ, is significant in comparison to d_I, giving further reason to choose small seeding particles.

A third source of bias in the auto-correlation function occurs when the velocity field has variations across the interrogation spot caused by a spatial velocity gradient. Using gaussian or top hat models for particle image intensities, Keane and Adrian[14] have shown that the peak and centroid of the displacement correlation function, $R_{D^+}(\mathbf{s})$, are located at

$$s_{D^+} = \left(Mu_I\Delta t - Mu_I\Delta t\left(\frac{M\Delta u_x\Delta t}{d_I}\right) - Mv_I\Delta t\left(\frac{M\Delta u_y\Delta t}{d_I}\right) - Mw_I\Delta t\left(\frac{M\Delta u_z\Delta t}{\Delta z_0}\right), \right.$$
$$\left. Mv_I\Delta t - Mu_I\Delta t\left(\frac{M\Delta v_x\Delta t}{d_I}\right) - Mv_I\Delta t\left(\frac{M\Delta v_y\Delta t}{d_I}\right) - Mw_I\Delta t\left(\frac{M\Delta v_z\Delta t}{\Delta z_0}\right)\right) \tag{11}$$

where

$$(\Delta u_x, \Delta u_y, \Delta u_z) = \left(\left|\frac{\partial u}{\partial x}\right| \frac{d_I}{2M}, \left|\frac{\partial u}{\partial y}\right| \frac{d_I}{2M}, \left|\frac{\partial u}{\partial z}\right| \frac{\Delta z_0}{2M} \right)_{x_I} \tag{12}$$

and

$$(\Delta v_x, \Delta v_y, \Delta v_z) = \left(\left|\frac{\partial v}{\partial x}\right| \frac{d_I}{2M}, \left|\frac{\partial v}{\partial y}\right| \frac{d_I}{2M}, \left|\frac{\partial v}{\partial z}\right| \frac{\Delta z_0}{2M} \right)_{x_I} \tag{13}$$

represent the variation in the u and v components of the velocity vector caused by a three-dimensional velocity gradient. This gradient bias is determined here for velocity fields which have constant gradient (first order terms only) and is independent of the interrogation spot size or the light sheet thickness. The bias can be reduced by decreasing the pulse separation, Δt, at the cost of reducing the particle image displacement, $M\mathbf{u}\Delta t$ and its accuracy as well.

OPTIMAL AUTO-CORRELATION ANALYSIS

It has also been shown by Keane and Adrian[14], that, in auto-correlation analysis, the valid detection probability depends upon the mean number of true pairs of particle images that are present in an interrogation spot. The mean number of pairs of images left in an interrogation spot after losses due to in-plane and out-of-plane motion are calculated, can be written as $N_P F_I F_O$, where N_P is the mean density of particle image pairs in the image plane for multiple pulse analysis and F_I and F_O are the reduction factors to account for in-plane loss-of-pairs and out-of-plane loss-of-pairs respectively. For double-pulse analysis, $N_P = N_I$, whereas for multiple pulse experiments with n pulses, $N_P = (n-1)N_I$.

The dependence of the valid detection probability on the density of the effective particle image pairs is shown in Figure 4 for a range of in-plane and out-of-plane velocities and for a range of double, triple and quadruple pulse image recordings. The success rate is independent of the image displacement and the number of pulses when considered as a function of $N_P F_I F_O$ and a detection probability of at least 95% is possible when there are at least 8 pairs of images per interrogation spot when $D_O = 1.2$. To achieve this many pairs, the seeding density must be chosen to take account of the loss factors, F_I and F_O, when determining N_P.

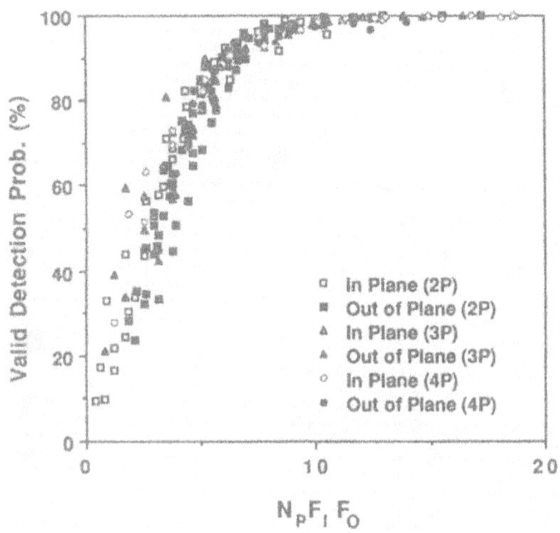

Figure 4. For general velocity fields, the valid detection probability for auto-correlation of double- or multiple-pulse systems for a range of image displacements, $|\Delta X|/d_I$, depends only upon the effective mean number of image pairs, $N_P F_I F_O$, for variable number of pulses, n, and $D_O = 1.2$.

While the velocity fields considered in Figure 4 were constant, velocity variation within the interrogation volume broadens and distorts the mean Young's fringes and diminishes the mean correlation peak in amplitude while broadening it. For individual realizations, where the particle images are randomly positioned, velocity variations lead to splintering and broadening of the signal peak from individual displacements based on image locations in the velocity field. Thus in the presence of a significant velocity gradient, the image density must be increased to ensure adequate signal to noise ratio and to overcome the effects of splintering and fragmentation of the signal peak in determining the centroid of the signal peak.

To optimize PIV performance using auto-correlation analysis, the following procedure and criteria are recommended to ensure high success rates above 95% for interrogations.

(a) Select the interrogation spot size, d_I, and the light sheet thickness, Δz_0, to give adequate spatial resolution of the flow field. The image magnification, M, is chosen to give adequate pixel resolution for the interrogation spot size.

(b) Select the particle image diameter, d_τ, to be sufficiently small to provide spatial accuracy in measuring displacement and select the image density, N_I, (or N_P) > 15, independent of the number of pulses.

(c) Choose Δt to be sufficiently large to obtain adequate separation of the particle image pairs, with image shifting, if necessary, to remove directional ambiguity. Choose Δt to be sufficiently small to minimize bias due to in-plane velocity gradients and to minimize loss of image pairs due to in-plane and out-of-plane motion, namely:

(1) $M|\mathbf{u}|\Delta t/d_I \leq 0.25$, (In-plane loss of pairs)

(2) $|w|\Delta t/\Delta z_0 \leq 0.25$, (Out-of-plane loss of pairs)

(3) $M|\Delta \mathbf{u}|\Delta t/d_I < 0.05$, (Velocity gradient spreading)

(4) $M|\Delta \mathbf{u}|\Delta t/d_\tau < 1.0$, (Velocity gradient splintering)

(5) $1.0 \leq D_0 \leq 1.2$. (Detectability criterion)

In flow fields which have low turbulence and low velocity gradients, the "one-quarter" rules for in-plane and out-of-plane displacements above as (1) and (2), are generally sufficient to determine Δt once the spatial resolution and image density have been determined. However, in flow fields with large velocity gradients, further changes in spatial resolution and seeding density may be necessary to satisfy restrictions caused by velocity gradient effects. Finally, the velocity gradients may be so large that PIV fails to provide adequate spatial resolution of the small scale structures of the turbulence, in which case supplementary analysis using particle tracking techniques based on coarse scale PIV vector fields may be successful as suggested by Keane and Adrian[15].

POST-INTERROGATION ANALYSIS

Post-interrogation analysis of PIV vector fields is employed to remove bad vectors, if none of the measured vectors for a given interrogation spot is judged to be acceptable according to criteria based on neighbouring vectors, and to fill in spaces where no vectors were acceptable from the detectability criterion. When several alternative peaks are recorded in the correlation plane for a given interrogation spot, alternative vectors can be considered to replace spurious vectors. In his recent thesis, Westerweel[16] has undertaken a comparison of alternative criteria to determine what criteria for data validation are robust and successful. The most successful criterion suggested is based on the median value of the eight vectors from surrounding interrogation spots with an acceptable variation about this median value being chosen. Meinhart et al.[17] have used this criterion to choose from three alternate peaks at each location or choose no vector at all.

In the case of no measured vector being acceptable, there are interpolation schemes available to fill regions where there are no valid vectors. However, care must be taken in interpolation as vorticity calculations and other higher order measurements are strongly dependent upon the interpolation scheme used.

Similarly, smoothing of the data field can be performed using a smoothing kernel, which is chosen to filter the vector field and remove small scale structures.

Successful auto-correlation analysis has been undertaken to examine a large range of turbulent fluid flows and to determine the coherent structures in them. Liu et al.[18] have studied high Reynolds turbulent channel flow with high spatial resolution of the scales of motion. The vortical structures and stagnant regions in the unsteady aerodynamics of high speed flow over airfoils have been investigated by Raffel and Kompenhans[19] using photographic film recording of sub-micron particles to follow the airflow and good comparisons have been made with numerical codes. Similar high speed flows over an airfoil have been measured by Lourenco and Krothapalli[20], using high resolution digital data acquisition methods. Other applications include turbulent thermal convection, two phase or multiphase flows, free surface flows and breaking waves and measurements inside rotating machinery, such as turbines and internal combustion engines.

SPATIAL CROSS-CORRELATION ANALYSIS

Auto-correlation analysis is based upon the calculation of the most likely local displacement of particle images by correlating the image intensity field of an interrogation spot with itself. This calculation, while based upon the chosen spatial resolution in determining the size of the interrogation volume, does not depend upon, and does not utilize any prior knowledge of the flow field under consideration to improve the signal strength of the correlation function. Increasing the number of image pairs measured for a given interrogation volume can be achieved by correlating its interrogation spot with a region in the image plane in which the images of particles in the initial interrogation volume are located for the second exposure. This alternative to auto-correlation analysis is known as cross-correlation analysis where separate first and second interrogation spots are chosen to optimize the signal strength by increasing the number of image pairs contributing to the signal. The first and second interrogation spots are not necessarily taken from the same image field as is the case in video camera imaging, where successive images can be recorded on successive frames . Willert and Gharib[21] used such an approach to cross-correlate successive digitized recordings of flow in a vortex ring. Cross-correlation of successive photographic images of a turbulent open channel water flow have been used by Utami et al.[22] to study the turbulent structure of open channel flow.

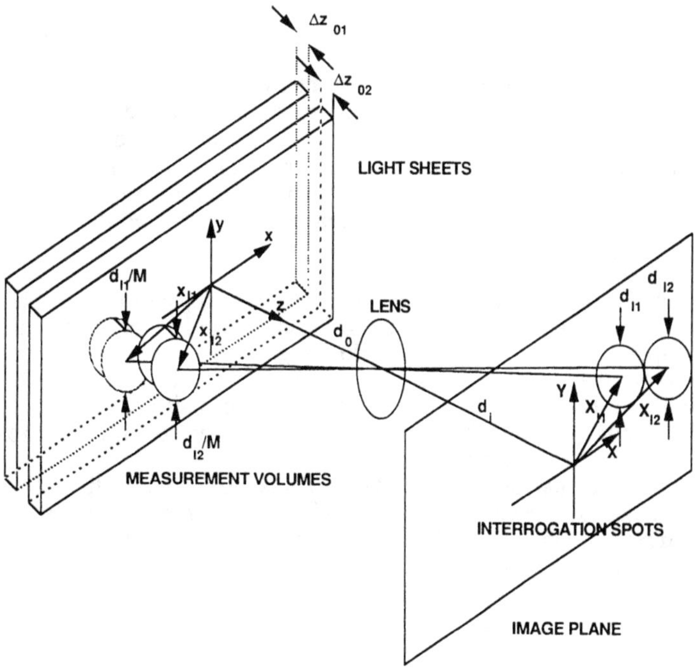

Figure 5. Experimental light sheets and image recording system for pulsed laser velocimetry to ensure maximal image pairing using cross-correlation analysis. Two successive measurement volumes within the illuminated light sheets correspond to the first and second interrogation spots within the image plane.

The most general experimental light sheet and image recording system for cross-correlation of a double-pulse single exposure recording is shown in Figure 5. The relative separation of the light sheets is chosen to maintain illumination of pairs of images in the presence of out-of-plane motion and the second light sheet can be thicker than the first to allow for a variation in the out-of-plane velocity. The cross-correlation function of the two intensities from the interrogation spots is defined to be

$$C(s) = \int I_1(X) I_2(X+s)\, dX \,. \tag{14}$$

Cross-correlation methods of interrogation are superior to auto-correlation methods for double-pulse PIV systems applied to a range of velocity fields for the following reasons:

(a) Size and location of interrogation spots.

The versatility in selecting the size and relative separation of interrogation spots permits higher spatial resolution by more efficient pairing of particle images. Furthermore, high spatial resolution can be obtained in a preferred direction, such as in near-wall shear flows, while maintaning sufficient image pairs for an adequate signal-to-noise ratio The second interrogation spot is located at the local mean image displacement from the first spot and can chosen to be slightly larger to enhance image pairing when there is some uncertainty about the separation and the velocity variation. The spatial resolution can be improved, if necessary, while maintaining the same effective image pair density as auto-correlation, by reducing the size of the first interrogation spot.

(b) Seeding density.

The seeding density for successful cross-correlation analysis can be lower than for auto-correlation analysis while maintaining an adequate detection probability or detection rate, because there is an increased efficiency in image pairing with fewer or no losses of image pairs. Because more particle image pairs can be recorded for a given seeding density in cross-correlation, either the seeding density can be lower or the interrogation spot size can be reduced.

(c)Increased dynamic range of the velocity measurements.

The dynamic range of the measurements can be improved over auto-correlation methods, because the relative window offset can be monitored and varied continuously to the approximate local mean image displacement. This flexibility allows the image displacement to have a dynamic range larger than the interrogation spot size,whereas in auto-correlation, the maximum image displacement was restricted to being approximately one quarter of the spot size.

(d)Velocity gradient bias.

Optimal choice of light sheet thickness and relative separation and optimal choice of interrogation spot sizes and relative separation result in no loss of image pairs due to out-of-plane or in-plane motion. Thus, the velocity gradient bias in velocity measurements present in auto-correlation analysis, can be eliminated. However, if a single light sheet is used, there will still be a bias due to out-of-plane loss of pair effects but no velocity bias from in-plane motion, provided that the second interrogation spot is chosen to eliminate in-plane image pair loss, i.e. by setting $F_I = 1.0$.

Cross-correlation analysis using digital FFT's is more computationally intensive than auto-correlation analysis because it requires digitized data from two interrogation spots for calculation and require three digital FFT's comparedto only two FFT's on one interrogation spot in auto-correlation analysis, namely

$$C(s) = |F^{-1}\{F\{I_1(X)\}^* F\{I_2(X+s)\}\}| \quad . \tag{15}$$

It can be seen from Keane and Adrian[13] that the cross-correlation function can be decomposed in a similar manner to the auto-correlation function but the displacement component, $C_{D^+}(s)$, now has a increased amplitude while that of $C_{D^-}(s)$ has been diminished, when the interrogation spots are chosen to minimize loss of in-plane image pairs,

$$C(s) = C_C(s) + C_P(s) + C_{D^+}(s) + C_{D^-}(s) + C_F(s)\,. \tag{16}$$

The optimization of PIV using cross-correlation is based on selecting experimental and interrogation parameters to maximize the valid detection probability using the same three criteria as those above.

OPTIMAL CROSS-CORRELATION ANALYSIS

The valid detection probability for a range of relative interrogation spot sizes and separations is illustrated in Figure 6 in terms of the effective particle image density, $N_I F_I F_O$, for both (a) single-exposure double-frame (as in CCD or video camera digital images) and (b) double-exposure single-frame cross-correlation analysis (for photographic imaging, for example) where $D_O = 1.2$.

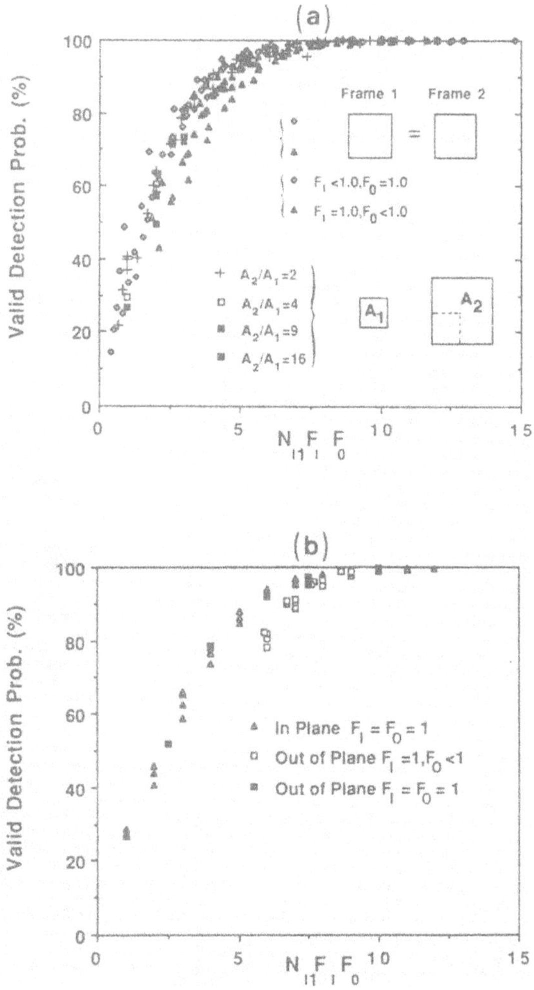

Figure 6. Valid detection probability as a function of mean effective image density, $N_{I1}F_I F_O$ for cross-correlation analysis of (a) single-exposure, double-frame systems and (b) double-exposure, single-frame systems for variable relative interrogation spot sizes A_2/A_1 within the image plane.

Although the results do not appear to be any different from auto-correlation analysis results in Figure 4, the chief advantage of cross-correlation lies in being able to choose F_I = 1.0 and possibly F_O = 1.0 and making optimal use of seeded particle images. In order to achieve a valid detection rate of 95%, the effective particle image density need be no larger than 6 for single exposures (such as CCD or video recording) and need be no larger than 8 for double exposures. The valid detection rate is insensitive to the size of the second window provided that $F_I = 1.0$, but an unnecessarily large second window reduces available pixel resolution in the first window and reduces computational efficiency.

As was described for auto-correlation analysis of a variable velocity field, the velocity variation within an interrogation volume broadens and distorts the mean Young's fringes and diminishes and broadens the mean correlation peak of the cross-correlation function in a similar fashion. Hence, an increase in image density is required to maintain signal strength when splintering and broadening of an individual realization occurs. However, the statistical bias in the centroid of $\langle C_D^+(\mathbf{s})|\mathbf{u}\rangle$ against high velocities does not occur when there are no image-pair losses. A simple shear velocity gradient of u versus y of increasing magnitude is illustrated in Figure 7 to show how the bias diminishes as F_I approaches unity. The bias in the measured mean velocity velocity for an in-plane shearing velocity, decreases as F_I approaches unity, independently of the mean image displacement and image density. However, it should be remembered that recording a single light sheet of the fluid flow will produce a bias due to any out-of-plane motion, which cannot be removed by cross-correlation analysis.

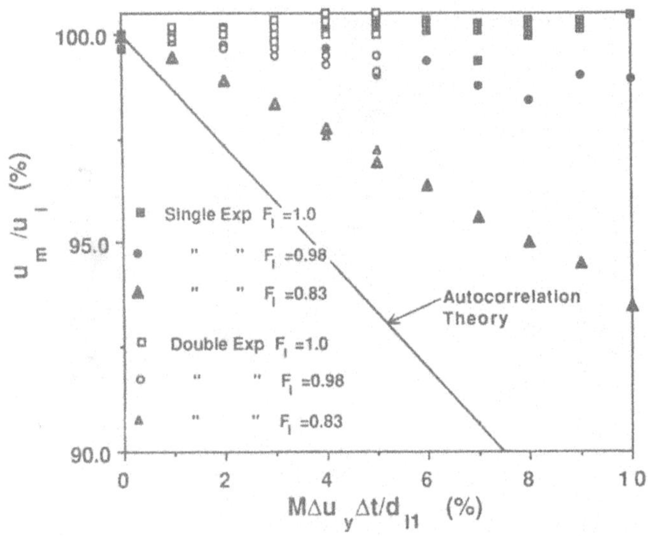

Figure 7. Measured velocity component, u_m, relative to the true velocity component, u_I, for single-exposure and double-exposure cross-correlation systems for variable velocity fields for which $\partial u/\partial y \neq 0$, as a function of the relative variation in the image displacement, $M\Delta u\Delta t/d_I$. The analytical result from equation (11) is included.The bias is reduced as the loss of image pairs due to in-plane motion is reduced in each system and F_I increases.

In summary, to optimize PIV performance using cross-correlation analysis, the following procedure, similar to that for auto-correlation, is recommended. However, in the case of cross-correlation, the spatial resolution is determined by particle displacement as well as by the dimensions of the first interrogation spot. In practice, as the particle image displacement is usually smaller than the interrogation spot size, the spatial resolution is determined by the spot size.

(a) Select the first interrogation spot size, d_{I1} and light sheet thickness, Δz_0 to give the desired spatial resolution.

(b) Select $N_{I1} > 10$, assuming there will be some out-of-plane image losses and some velocity gradient.

(c) Choose Δt to minimize pair losses due to out-of-plane motion and to give adequate image displacements with image shifting, if necessary.

(1) $|w|\Delta t/\Delta z_0 \leq 0.25$, (Out-of-plane loss of pairs)

(2) $M|\Delta u|\Delta t/d_{I1} < 0.05$, (To maintain SNR)

(3) $M|\Delta u|\Delta t/d_{\tau} < 1.0$, (Velocity gradient splintering)

(d) Choose the location, X_{I2} and size, d_{I2} of the second interrogation spot to ensure no loss-of-pairs due to in-plane motion.

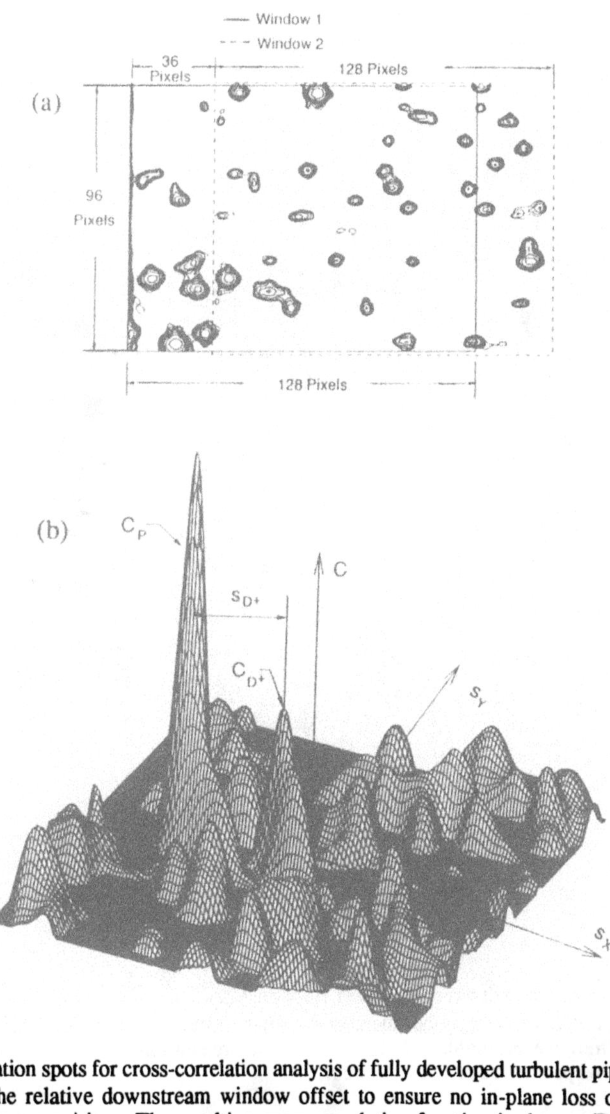

Figure 8. Interrogation spots for cross-correlation analysis of fully developed turbulent pipe flow with Re_D = 50,000, showing the relative downstream window offset to ensure no in-plane loss of image pairs, for marginal particle image pairing. The resulting cross-correlation function is shown. (Reproduced by kind permission of C. D. Meinhart).

Cross-correlation analysis has been used successfully by Meinhart et al.[17] in the study of fully developed turbulent pipe flow where Re_D = 50,000 to obtain high resolution velocity vector fields. The interrogation spots for a sample interrogation and the resulting correlation function are shown in Figure 8. The spatial resolution in the wall-normal direction is enhanced by using only 96 pixels in the radial direction and the relative separation of the interrogation spots is monitored continuously and adjusted to be approximately equal to the local mean image displacement. The instantaneous total velocity field and the fluctuating velocity field, obtained by removing an ensemble averaged mean velocity are illustrated in Figure 9 where the streamwise spatial resolution is 214μm and the wall-normal (radial) spatial resolution is 160μm with 25% overlap of interrogation spots in each direction.

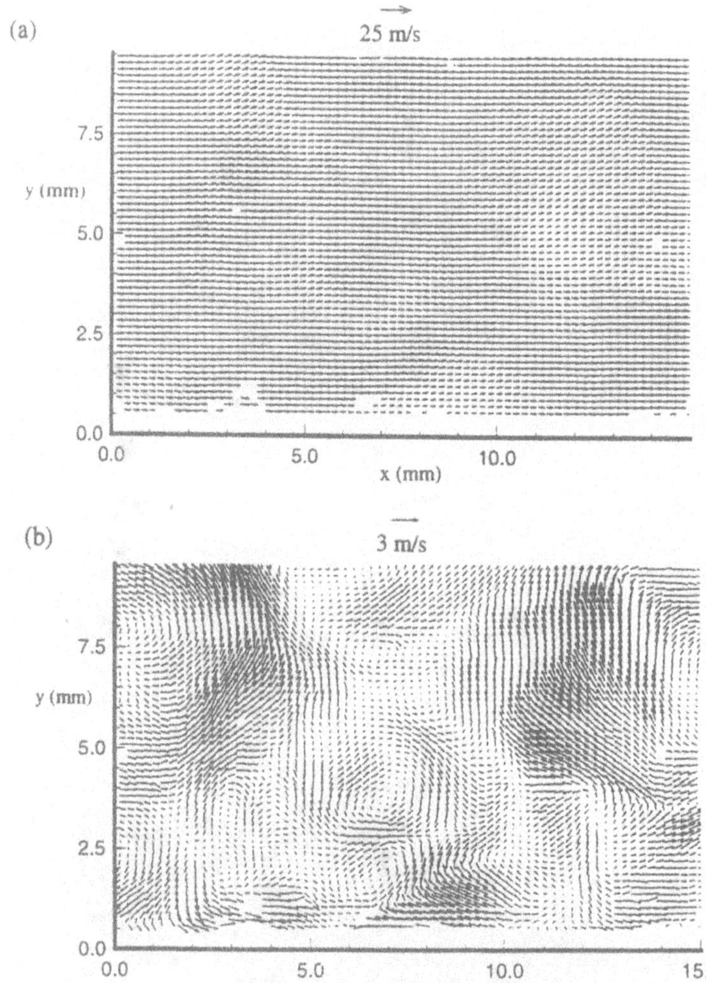

Figure 9. Instantaneous velocity vector field of fully developed turbulent pipe flow with Re_D = 50,000, showing (a) the total velocity and (b) the fluctuating velocity vectors determined by removing line averaged mean velocities from the ensemble. Enhanced spatial resolution occurs in wall-normal direction. (Reproduced per kind permission of C.D. Meinhart).

ALTERNATIVE METHODS FOR LOW IMAGE DENSITY ANALYSIS

(a) Particle Tracking Velocimetry. Particle tracking velocimetry (PTV) is the PIV technique that can be employed when the particle concentration becomes too low for correlation methods to be used successfully. In the wall boundary layers of some flows, particle seeding becomes low, enabling PTV techniques to succeed when correlation methods do not have a high detection rate. Particle tracking is usually performed using multiple illuminations of the flow field to produce a string of particle images on either a single exposure or on successive exposures. Particle tracking has been employed to measure both two-dimensional and three-dimensional flow fields using multiple recording cameras. Recent three-dimensional PTV experiments have been performed by Kent et al.[23] in studying flow in an internal combustion engine.

In general, the separation between particles in the fluid is sufficiently large to permit the tracking of each particle image individually, because the spacing between successive images of a given particle is less than the mean separation of particle images in the recorded image plane.

(b) Multiple pulse auto-correlation analysis. If the low particle seeding density can be overcome by multiple exposure images to produce a sufficiently high effective image pair density, $N_P F_I F_O$, then correlation analysis can be employed, as an alternative to particle tracking methods. Successful multiple-pulse PIV analysis has been undertaken by Cenedese and Paglialunga[24] in a study of vortical motion with ten successive exposures recorded and by Arroyo et al.[25] in the study of Rayleigh-Bènard convection in which as many as 40 exposures were imaged.

Although particle tracking algorithms have been applied to low seeding density experiments, it is possible to apply them to high image density flows, when a knowledge of the local mean flow, based on PIV correlation methods provides a measurement for a given interrogation spot. The local mean image displacement can then be used as an initial approximation to determine individual particle displacements from particle images in the interrogation spot. The spatial resolution of the resulting PTV data can improve the PIV spatial resolution by a factor of 3 or 4.

CONCLUSIONS

Particle Image Velocimetry (PIV) is a practical, robust and accurate experimental method for measuring two- or three-dimensional velocity vector fields in two or three dimensions of a wide range of fluid flows. The spatial resolution of PIV that has been achieved using correlation methods is as low as 160μm, which is comparable to that achieved by LDA or hotwire anemometry and the accuracy of PIV is comparable to that of LDA in measuring turbulent flow fields.. However, with such high spatial resolution, even with the largest format imaging available being 100mm x 125mm film, there is a reduced field of view for such high magnification recording. Improved spatial resolution is expected when particle tracking techniques are implemented to enhance correlation methods.

New developments in PIV are underway in improving digital (CCD) methods of data acquisition to obtain larger fields of view and greater pixel resolution imaging with 2K x 2K cameras being currently available. Three dimensional velocity fields in two and three dimensions have been measured using stereographic imaging and holographic techniques. These techniques will be discussedin greater detail in the following chapter.

The development and availability of high speed parallel array processors has enabled correlation methods to be used to interrogate ensembles of recorded image fields to yield at least 100 vectors per second and produce statistical data based on many realizations interrogated within a reasonable time. Such high speed processing will become more crucial as high resolution three-dimensional spatial measurements of three-dimensional velocity vectors will require such efficiency to enable statistical analysis of ensembles of recordings to be undertaken.

Acknowledgement

The support of the National Science Foundation through Grant ATM 89-20605, TSI Inc., Argonne National Laboratory and the organizing committee of the Summer School on Optical Diagnostics for Flow Processes is gratefully acknowledged. The advice and direction of Professor R. J. Adrian in preparing the material for the Summer School is greatly acknowledged also.

REFERENCES

1. R.J. Adrian, Particle-imaging techniques for experimental fluid mechanics. *Ann. Rev. of Fluid Mech.,* 23: 261 (1991).
2. P. Buchhave, Particle image velocimetry - status and trends, *in*: "Experimental Heat Transfer, Fluid Mechanics, and Thermodynamics," J.F. Keffer, R.K. Shah, and E.N. Ganic, eds., Elsevier Science Publishing Co., (1991).
3. C. Gray, The evolution of particle image velocimetry, *in*: "Proceedings of IMechE symposium on Optical Methods and Data Processing in Heat and Fluid Flow," London (1992).
4. K.D. Hinsch, Particle image velocimetry, *in*: "Speckle Metrology," R.S. Sirohi, ed., Marcel Dekker, New York (1993).
5. R.J. Adrian and C.S. Yao, Development of pulsed laser velocimetry (PLV) for measurement of turbulent flow, *in*: "Proceedings of the 8th Biennial Symposium on Turbulence," X. Reed, G. Patterson, J. Zakin, eds., University of Missouri Press, Rolla, 170, (1984).
6. R.D. Keane and R.J. Adrian, Optimization of particle image velocimeters, Part I: Double-pulsed systems, *Meas.Sc. and Tech*., 1:1202 (1990).
7. C.H. Westergaard and P. Buchhave, PIV: Comparison of 3 auto-correlation techniques *in*: "Laser Anemometry: Advances and Applications - Proceedings of the Fifth International Conference," J.M. Besem, R.Booij, H.W.H.E. Godefroy, P.J. de Groot, K. Krishna Prasad, F.F.M. de Mul, E.J. Nijhof, eds., Proc. SPIE 2052, 535 (1993).
8. P. Buchhave and M.L. Jacobsen, Optical correlator for PIV image processing, *in*: "Proceedings of the 3rd International Conference on Laser Anemometry, Advances and Applications", BHRA/Springer Verlag, 609 (1990).
9. J.-C. Lin and D. Rockwell, Cinematographic system for high-image-density particle image velocimetry, submitted to *Exp. in Fluids* (1993).
10. R.J. Adrian, Statistical properties of particle image velocimetry measurements in turbulent flow, *in*: "Laser Anemometry in Fluid Mechanics - III," R. Adrian, T. Asanuma, D. Durao, F. Durst, J. Whitelaw, eds., Instituto Superior Tecnico, Lisbon 115 (1988).
11. J. Westerweel, Analysis of PIV interrogation with low pixel resolution, *in*: "Optical Diagnostics in Fluid and Thermal Flow," Soyoung S. Cha, James D. Trolinger, eds., Proc. SPIE 2005, 624 (1993).
12. A.K. Prasad, R.J. Adrian, C.C. Landreth and P.W Offutt, Effect of resolution on the speed and accuracy of particle image velocimetry interrogations, *Expt. in Fluids* 13:105 (1992).
13. R.D. Keane and R.J. Adrian, Theory of cross-correlation analysis of PIV images, *J. of App. Sc. Res.* 49:191 (1992).
14. R.D. Keane and R.J. Adrian, Optimization of particle image velocimeters, Part II: Multiple-pulsed systems, *Meas. Sc. Techn.* 2: 963 (1991).
15. R.D. Keane and R.J. Adrian, Prospects for super-resolution with particle image velocimetry, *in*: "Optical Diagnostics in Fluid and Thermal Flow," Soyoung S. Cha, James D. Trolinger, eds., Proc. SPIE 2005, 283 (1993).
16. J. Westerweel, "Digital particle image velocimetry; theory and application," Ph.D. thesis, Delft University Press, Delft (1993).
17. C.D. Meinhart, A.K. Prasad and R.J. Adrian, Parallel digital processor system for particle image celocimetry, *Meas. Sc. Tech.* 4:619 (1993).
18. Z.-C. Liu, C.C. Landreth, R.J. Adrian and T.J. Hanratty, High resolution measurement of turbulent structure in a channel with particle image velocimetry, *Exp. in Fluids*, 10:301 (1991).

19. M. Raffel and J. Kompenhans, PIV measurements of unsteady transonic flow fields above a NACA 0012 airfoil, *in*: "Laser Anemometry: Advances and Applications - Proceedings of the Fifth International Conference," J.M. Besem, R.Booij, H.W.H.E. Godefroy, P.J. de Groot, K. Krishna Prasad, F.F.M. de Mul, E.J. Nijhof, eds., Proc. SPIE 2052, 527 (1993).
20. L.M. Lourenco and A. Krothapalli, Application of on-line particle image velocimetry to high speed flows, *in*: "Laser Anemometry: Advances and Applications - Proceedings of the Fifth International Conference," J.M. Besem, R.Booij, H.W.H.E. Godefroy, P.J. de Groot, K. Krishna Prasad, F.F.M. de Mul, E.J. Nijhof, eds., Proc. SPIE 2052, 683 (1993).
21. C.E. Willert and M. Gharib, Digital particle image velocimetry, *Exp. in Fluids*, 10:181 (1991).
22. T. Utami, R.F. Blackwelder and T. Ueno, A cross-correlation technique for velocity field extraction from particulate visualization, *Exp. in Fluids*, 10: 213 (1991).
23. J.C. Kent, N. Trigui, W.-C. Choi, Y.G. Guezennec, R.S. Brodkey, Photogrammetric calibration for improved three-dimensional particle-tracking velocimetry (3D PTV), *in*: "Optical Diagnostics in Fluid and Thermal Flow," Soyoung S. Cha, James D. Trolinger, eds., Proc. SPIE 2005, 400 (1993).
24. A. Cenedese and A. Paglialunga, Digital direct analysis of a multi-exposed photograph in PIV, *Exp. in Fluids*, 8:273 (1990).
25. M.P. Arroyo, T. Yonte, M. Quintanilla and J.M. Saviron, Particle image velocimetry in Rayleigh-Bénard convection: photographs with a high number of exposures, *Optics and Lasers in Engineering*, 9:295 (1988).

VISUALIZATION OF COHERENT STRUCTURES

Jens Juul Rasmussen and Bjarne Stenum

Optics and Fluid Dynamics Department
Risø National Laboratory
DK-4000 Roskilde, Denmark

ABSTRACT

The visualization of coherent structures is illustrated in two-dimensional flows as rotating and stratified fluids. The visualization is based on the development of dye concentration for measurements of the gross properties of the structures, and on a particle tracking method, which provides detailed knowledge of the flow field.

INTRODUCTION

The existence of coherent vortical structures is a characteristic feature of two-dimensional or quasi-two-dimensional flows. Such structures have long life times, much longer than their own internal rotation period. They are dominating the large scale dynamics and are naturally of great importance in connection with the transport of mass, heat and momentum. They are likely to form after external forcing and seem to be the universal outcome of any regular as well as irregular forcing in two–dimensional systems. They also appear spontaneously in two–dimensional turbulence due to self–organization processes characterizing the two–dimensional flows.

There are several exampels where the two–dimensional approximation is relevant; in this contribution rotating[1,2] and stratified fluids[3,4,5] (e.g. fluids with a finite density gradient in the vertical direction) will be considered. In geophysical flows in atmospheres and oceans the two–dimensionality is brougth about by the combined effects of the planetary rotation and the density stratification. In these systems vortical structures are abundant and play a significant role for the large–scale dynamics. The most famous example of a naturally occuring vortex is probably the Great Red Spot of Jupiter, which appears as a giant vortex with an extension of approximately 25.000 km (i.e. more than twice the diameter of the earth). This vortex have been observed for more than 300 years. Other cases where the dynamics can be considered two–dimensional comprise: magnetized plasmas, where vortical structures formed in low–frequency turbulent fluctuations play an important role in connection with the plasma

Optical Diagnostics for Flow Processes
Edited by L. Lading *et al.*, Plenum Press, New York, 1994

confinement, and flows in an essential two–dimensional domain such as soap films and other types of films or membranes. The understanding of the dynamics of coherent vortical structures in two–dimensional flows is thus not only of fundamental interest; but it has several pratical implications as well.

Although considerable progress in the understanding of the dynamics of two–dimensional vortical structures has been made during the last couple of decades, there is still a lot of unsolved fundamental problems. These include, for instance, the understanding of the formation process of vortices from arbitrary initial forcing, the mutual interaction of vortices, their interaction with boundaries and the large–scale transport properties of vortices. In order to obtain a detailed understanding of these problems, investigations combining both analytical, numerical studies and laboratory experiments are essential. To benefit from such combined investigations it is necessary to be able to visualize the vortical structures in a quantitative way, which ultimately demands detailed information about the total flow field, i.e., essentially the velocity field from which the vorticity and the streamfunction can be derived. This is of cause possible for the numerical results and here only the question about the presentation arises. In that connection we emphasize that the amimation by means of video techniques is of great help in understanding the detailed dynamical evolution.

In this contribution we discuss relative simple means of visualizing and quantifying two–dimensional vortices in laboratory flows. First we describe a simple dye concentration development, which gives the gross properties of the structures: paths and bulk velocity. Then we describe a particle tracking facility, which makes possible the quantitative measurements of whole field quantities. This tracking facility is developed at the University of Cambridge[6] and is purcased under the trade name DigImage. The experimental investigations of the vortical structures are mainly performed in rotating fluids or in stratified fluids. Aside from being of interest for the fundamental dynamics of vortices these investigations are also as indicated above of direct geophysical interest and in addition rotating fluids have several industrial applications[2] .

COHERENT VORTICAL STRUCTURES IN TWO–DIMENSIONAL FLOWS

Before we discuss the details of diagnosing vortical structures we shall review some features of two–dimensional flows and discuss the different types of vortical structuteres. The motion of a two–dimentional, incompressible fluid is governed by the Navier–Stokes equation

$$\frac{\partial \omega}{\partial t} + \vec{u} \cdot \nabla \omega = \nu \nabla^2 \omega \, , \tag{1}$$

where

$$\nabla \times \vec{u} = \vec{\omega} = \omega \hat{z} \, , \qquad \vec{u} = \nabla \psi \times \hat{z} \, , \qquad \nabla^2 \psi = -\omega \, . \tag{2}$$

Here, $\vec{u}$ is the velocity, ω the scalar vorticity, ψ the streamfunction and ν the kinematic viscosity. Equation 1 expresses, in the inviscid case, $\nu = 0$, the conservation of the vorticity on each fluid element. This feature makes the two–dimensional dynamics significantly different from the usual three–dimensional dynamics. In three dimensions one has an extra degree of freedom for the development of a vortex, it may be stretched, whereby the vorticity is changed. This implies that for two–dimensional systems (inviscid) in addition to the conservation of the total energy (the energy density being proportional to $\frac{1}{2}v^2$) also the so–called total enstrophy (the enstrophy density being proportional to $\frac{1}{2}\omega^2$) is conserved. The simultaneous conservation of energy and enstrophy introduces restrictions on the evolution of the turbulent energy spectrum

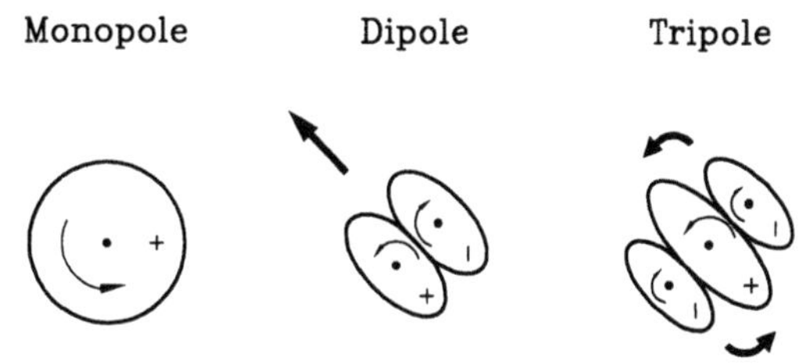

Figure 1. Classification of vortical structures. The "thin" arrows indicates the direction of the swirling velocity inside the structures, while the "thick" arrows indicate the bulk motion of the structures.

and leads to the so-called inverse energy cascade, by which the energy cascades to smaller and smaller wavenumbers[7,8] , while the enstrophy is cascading to larger and larger wavenumbers. As a result the energy initially distributed over both large and small scales eventually evolves into large scale energy containing vortical structures, while the enstrophy gets concentrated in thin shear layers on the vortex boundaries. In this way a self-organization of the flow[9] takes place. These features have been clearly demonstrated in numerical simulations of two-dimensional flows[10] and also in experiments, as we shall see below.

The vortical structures can be represented as solutions of the Navier-Stokes equation, Eq. 1, in the inviscid case. All such solutions is found from $\vec{u} \cdot \nabla \omega = 0$, which by the introduction of the streamfunction reduces to

$$\omega = f(\psi), \tag{3}$$

where f is any integrable function. Note that the solutions found from Eq. 3 are the solutions in the frame of reference moving with the structure. These solutions can be classified in terms of the spatial moments of the vorticity distribution about their centers. The simplest structures are illustrated in Fig. 1. The *monopolar vortex* is the simplest and most common structure. It is a localized, mostly circular symmetric, structure with the swirling motion of just one orientation, i.e., it may be characterized by a single set of closed, concentric streamlines. For the present homogeneous model the center of the vortex is fixed at the same position. The vorticity distribution is also symmetric, but its sign may change with radius. Simple monopoles (with just one sign of the vorticity) are stable, while the stability in general depends on the vorticity distribution. Also monopoles with ellitical shape can be found, they will rotate around their "center of mass". The *dipolar vortex* consists essentially of two closely packed monopolar vortices of opposite vorticity. The dipole contains a net linear momentum which makes it translate, in contrast to the monopole, which only contain angular momentum. If the dipole consists of two identical vortices (with opposite sign of the vorticity) it will move on a straight path along the direction of the (anti-)symmetry line between the two vortices, and we refer to it as a balanced dipole. If one of the vortices is stronger than the other, the dipole will have a finite angular momentum

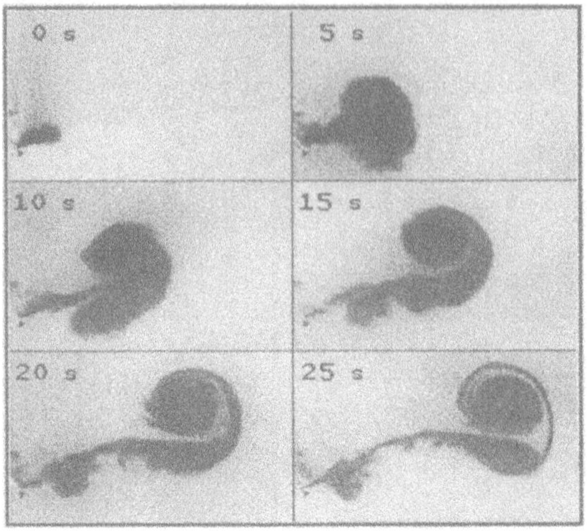

Figure 2. Generation of symmetric dipolar vortices by a deflected jet of dyed water. Shown is the development of the dye concentration. There is 5 s (one half rotation period) between the frames.

and will move on a circular trajectory with the radius determined by the ratio between the angular and dipolar momentum. Such a dipole will be refered to as an unbalanced dipole. The *tripolar vortex* consists of a core vortex of elliptical shape and two satellite vortices of opposite vorticity. The tripole contains angular momentum, but no linear momentum and it rotates steadily around the center of the core vortex. All these structures have been observed in natural flows and are easily generated in laboratory experiments in rotating as well as stratified flows. A small viscosity does not change the dynamics of these free vortical structures substantially.

For the simple linear relationship between vorticity and streamfunction, i.e. $f(\psi) = \lambda^2\psi$ analytical expressions for structures with continuous vorticity have been found. However, there is no natural law saying that f is bound to be linear and in the investigations of vortical structures it is essential to be able to deduce the function f. This may be done from so-called scatterplots, where the vorticity is plotted versus the corresponding streamfunction (measured in the frame of reference moving with the vortical structure) obtained in the same points of space.

EXPERIMENTAL ARRANGEMENTS

Rotating Fluid

The setup for the experiments in rotating fluids consist of a square tank with a flat bottom. It has a side length of 1 m and is filled with water with a depth of 15 cm. The tank is placed on a rotating table. The rotation rate was relatively low and the depth variation of the free surface had negligible influence on the vortex propagation. In the experiments reported her the rotation rate was $\Omega_0 = 0.63$ s^{-1}. The dipole vortices are created by injecting a jet of water into the mid-plane of the water from a nozzle with an inner diameter of 1 mm for a short period of typically 5 s. After a short transient period of a couple of rotation periods following the jet injection, the solid body rotation causes the flow to organize into nearly two-dimensional motion and a dipolar vortex is formed as shown in Fig. 2, where the vortex has been visualized by dye, i.e. the injected jet is dyed. Our setup is somewhat similar to that described by Flierl et al.[11] who investigated the dynamics of free unbalanced dipoles moving on curved paths. We have been able to excite balanced dipoles moving on straight paths by deflecting the jet on a vertical plate placed about 5 cm from the nozzle. By changing the direction of the nozzle-plate system we can control the propagation direction of the dipole vortex rather acurately. The size and velocity of the dipole can be controlled by regulating the jet inlet time and the pressure of the jet. The tank is surveyed by a video camera mounted in the rotating frame approximately 2 m above the surface of the water. The video signal is transmitted to a video tape recorder (VTR) via a set of slip rings and recorded on video tape for further analysis.

Stratified fluid

Experiments in a stratified fluid were performed in a rectangular PVC tank of dimensions 0.6 m x 0.8 m x 0.5 m. A linear salt stratification is produced by the two–tank method[12] , typically with a buoyancy frequency between 1 and 2 Hz in the fluid of depth about 20 cm. Dipole vortices are produced by horizonthal injection of a turbulent jet. The density of the injected fluid matches that of the fluid at the level of the injection nozzle. The vertical motion in the jet is suppressed due to the stratification, forcing the flow to be two–dimensional. Subsequently the flow gets organized in well-defined, "flat" dipole vortex structure. Once it is formed it moves slowly (slower than the dipole formed in a rotating fluid) with roughly preserved shape. Also here can the size and velocity to some extend be controlled by the injection duration and the pressure of the injected jet. A thin (1 cm) horisonthal light sheet for illumination of tracer particles in a layer is produced by means of a conventional 250 W quartz slide projector, and evolution is recorded on the VTR via a video camera surveying the horisonthal plane via a mirror.

RESULTS WITH DYED FLUID

The simplest way of visualizing is to follow the evolution of a dye concentration. The dye may be introduced into the vortex when it is well–formed or it may be introduced into the fluid that forms the vortex as for example in Fig. 2, where the injected jet is dyed. Coherent structures are characterized by the ability of trapping the fluid and the exchange of fluid with the surroundings is limited. Thus, the development of the dye concentration gives a good measure of the gross properties such as stability, propagation path, size and velocity of the vortical structures. The dye acts as a passive tracer and also the exchange with the surrounding fluid may be visualized, e.g. entrainment and detrainment of fluid. In our investigations we have mainly used conventional food colours for dying the flow and we find that red and green gives the best contrasts. One advances by using food colour is that it slowly bleaches if a small amount of clorine is added to the water (approximately 1/2 l clorine pr. 100 l water makes the colour bleach in approximately 10 minutes). This makes it possible to perform many experiments

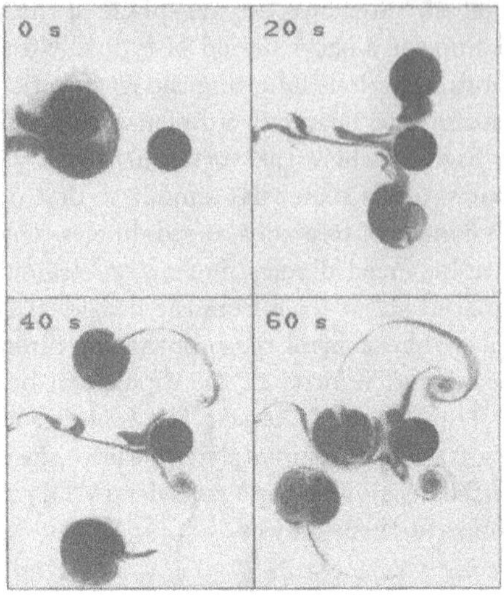

Figure 3. Interaction of a dipolar vortex with a cylinder of radius 5 cm, the dipole radius is 10 cm. The dipole is visualized by the development of the dye concentration. The time T = 0 corresponds to the time where the dipole is fully developed. To emphasize the structure the background has been removed by the DigImage system.

without having to change the water and clean the tank. Others colours that give a larger contrast are for instance fluorescine and terasil blue, however, these cannot be bleached and only a couple of experiments can be performed before the water has to be changed.

The frames shown in Fig. 2 are the directly grapped from the video recordings by means of the DigImage system witout any image processing. The system has the capability of image enhancement as, for instance, background removal etc. Also intensity plots in false coulors can be made. In Fig. 3 we show the interaction of a dipole with a cylindrical obstacle, which is visualized by the dye concentration inside the dipole, note also the dye trailing the dipole which indicates the detrained fluid due to the slow damping of the dipole. The results are in detailed agreement with results from a numerical solution of the Navier-Stokes equations (Eq. 1) in annular geometry[13,14] using an initial dipole with parameters determined from the experimentally measured dipole in Fig. 3. The numerical solution of Eq. 1 is based on a fully dealiased, spectral scheme based on Fourier-Chebyshev expansions, which allow very high resolution of boundary layers.[15]

Qualitatively, the interaction of dipoles with curved surfaces is not too different from that seen in collisions with flat walls, as it has previously been reported by several authors[4,5,15,16] . We summarize here the main aspects of this interaction: As the dipole approaches the cylinder it induces opposite-signed wall vorticity layers. These thicken due to viscosity. At minimum approach, these wall layers roll tightly into secondary vortices and couple with opposing primary lobes. The wall vorticity production continues, feeding the secondary vortices through vortex sheets, as the former are being advected by the primaries. At a critical instant, detachment of the combined structures occurs. These structures moves on curved paths and collide again in front of the cylinder.

Although we have only given examples of the dye evolution in rotating fluids also in the stratified fluid vortices may be visualized by adding dye.

PARTICLE TRACKING

Detailed information of the internal dynamics in the coherent vortical structures as well as their mutual interactions and interactions with solid walls cannot be obtained by means of studies with dye. Therefore, it is important to be able to measure total velocity fields instantaneously. By such measurements detailed comparisons with numerical studies are also made possible as well.

Total two–dimensional velocity fields in fluid experiments can be obtained by tracing a large number of small particles by means of the so–called particle imaging techniques.[17] In this method small particles are added to the fluid, illuminated by a light sheet and subsequently the velocities fields are determined by analysis of images of the particles in the fluid. A frequently used method for the image processing, when the particle density is high, is to use a correlation technique to determine the the displacements of groups of particles on photographic images obtained by pulsed illumination. This method is usually refered to as the "particle image velocimetry" (PIV) method. However, at it present state this method is time consuming and not convenient for studies of the time evolution of a fluid systems, although it provides a very high spatial resolution. A thorough review of PIV techniques has been given by Adrian[17] and elsewhere in these proceedings.

In order to measure two–dimensional velocity fields in our laboratory experiments we have installed a setup based on particle tracking velocimetry (PTV) at particle densities sufficiently small to allow location and tracking of the individual particles. This system, DigImage[6] , allows tracking of a large number (about 4000) of individual particles and is a simple but efficient tool for measurements of velocity fields, as compared to PIV at high particle densities. Futhermore the spatial resolution is high enough for studies of large scale coherent structures. Small particles are added to the fluid illuminated by a light sheet and the experiment is imaged by means of a CCD Super VHS camera. Initially a commercial Sony CCD–650E camera was utilized. Later this camera has been replaced by a black and white EEV P47580 Super photon CCD camera. The video signal is passed to a video tape recorder (VTR) (Panasonic, model AG–7350). In the subsequent image processing and analysis the video frames are digitized one by one by means of a frame grabber (Data Translation 2862) with a resolution of 512 x 512 pixels at 256 grey levels. For the colour camera only the luminance component (Y) was supplied to the frame grabber.

The VTR can be controlled by the DigImage software. It is impossible to acquire high quality images from a VTR in the paused mode. Therefore, audio markers are added to the video tape in order to control positioning of the tape. The images are

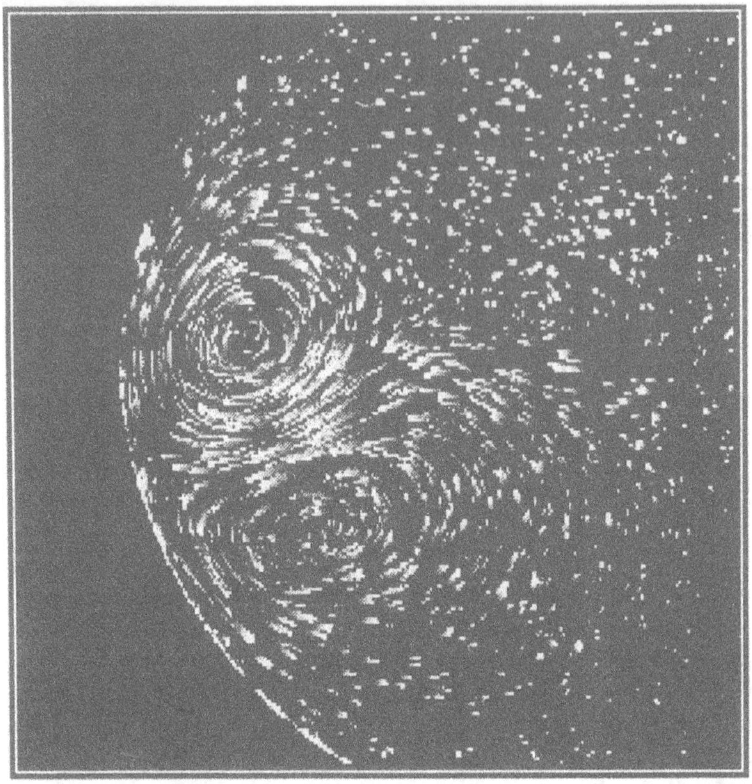

Figure 4. Particle streaks for a dipole excited by a jet in a stratified fluid. The dipole is moving towards a concave wall indicated by the semi-circle.

acquired when the tape is rolling forward in the play mode according to the preset time intervals. Four frames are digitized in one circle. Subsequently, the tape is repositioned backwards for the next acquiring circle. The images are mapped to a reference pixel coordinate system via a number of fixed fluorescent reference points, and a pixel to world coordinates transformation is performed in accordance with an initial calibration. In the digitized images the particles are localized and matched to one another from one frame to the next. In this way the particle tracks and hence, for example, the velocity–field and its evolution can be determined. The particle tracking for about 500 frames can be performed automatically in few hours on a usual PC (486DX, 50 MHz).

The size of the added particles depends on the area of interest in the experiments. The particle location has the best accuracy for particles covering slightly more than one pixel in the images. In such images the particles can be located with subpixel accuracy. The particles are located as the mass centroid of regions fulfilling a preset threshold. Before the particle location takes place some image enhancement as subtraction of background illumination or filtering may be performed. After the particle location the particle pairing, based on minimization of an association function,[6] is performed and the path data are stored in a file. Based on this data file the velocity fields and their derivatives can be determined since the time intervals are known as well. Practically the velocities are determined by fitting the slope to a given path and finally the velocities are associated to regular grid points. This method based on the conventional video technique can only be applied for moderate flow velocities since the interlaced video frames are superpositions of two fields obtained with a time difference of 0.02 s. For

Velocity map at t =400.00s

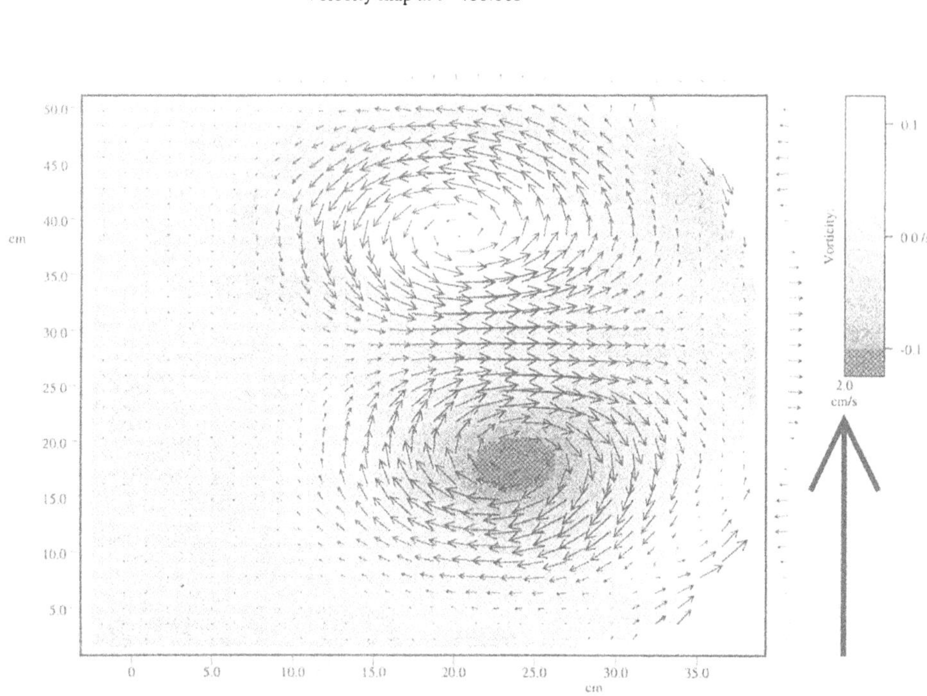

Velocity map at t =800.00s

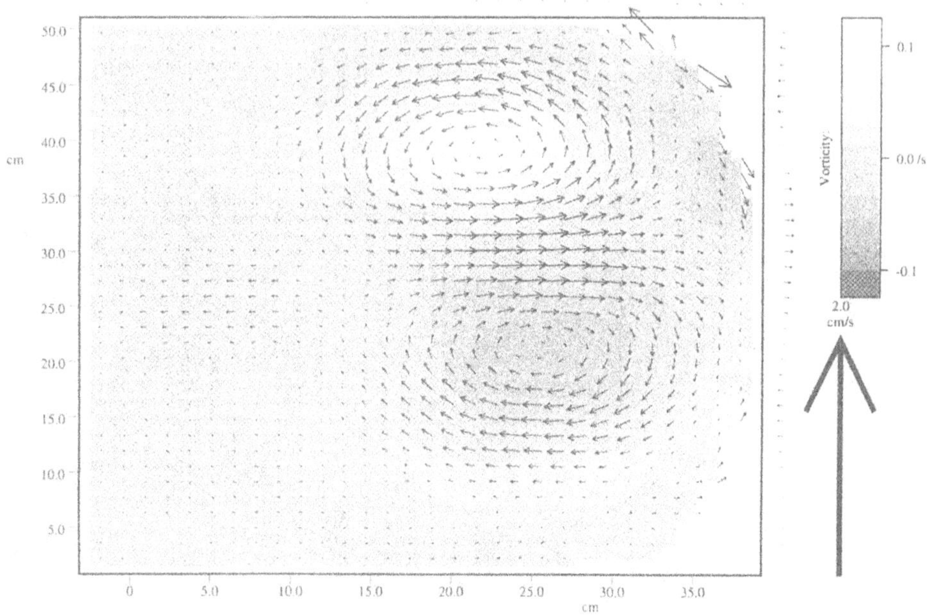

Figure 5. The evolution of the velocity field and the vorticity of the dipole in Fig. 4. At the right–hand–side of the figures are indicated the scales, grey scale for the vorticity and an arrow for the velocity. The tiny arrows on the top and the right–hand–side of the frame indicate the values of the averaged velocity. The time is in seconds after the injection of the jet.

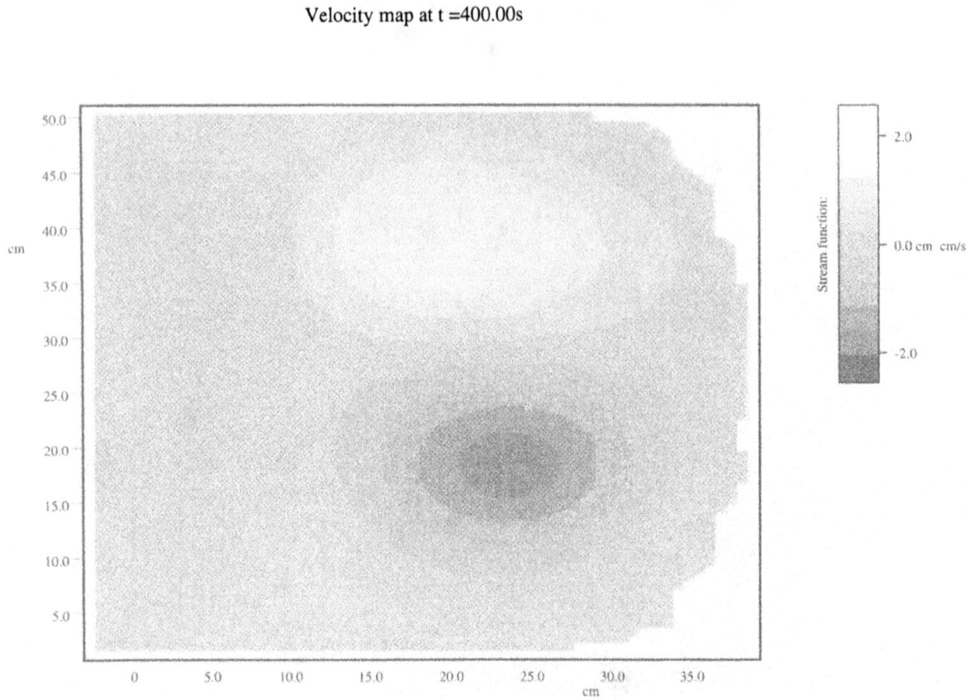

Figure 6. The streamfunction for the dipole in Figs. 4 and 5 at T= 400 s. The grey scale for the value of the streamfunction is indicated at the right–hand–side of the figure.

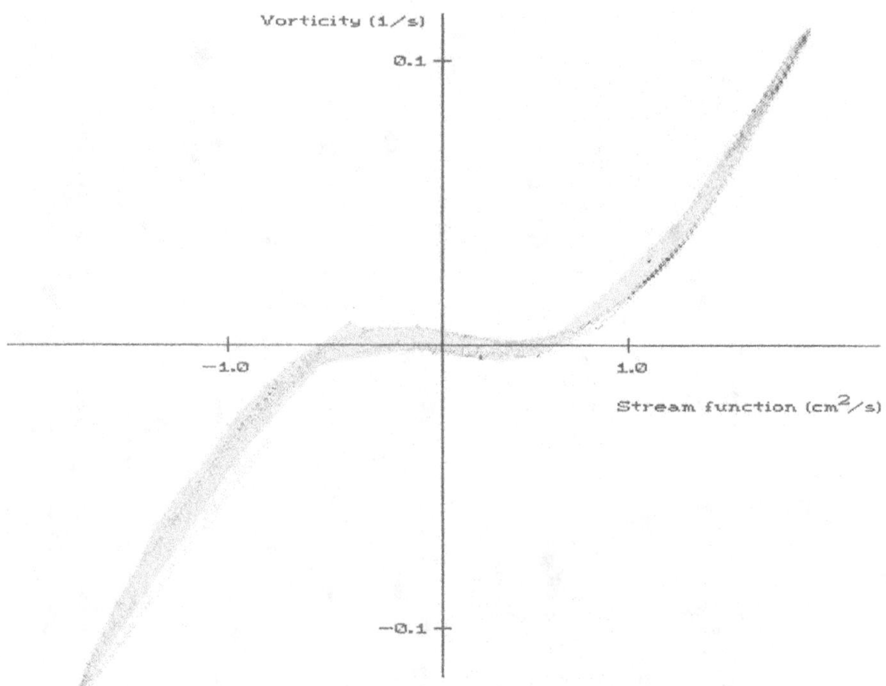

Figure 7. The scatter–plot of the relation between streamfunction and vorticity for the dipole in Figs. 4, 5 and 6 at T= 400 s.

large particle velocities only single fields can be used for the particle tracking. However, this implies a reduction in the spatial resolution.

In order to perform detailed studies of the dipole wall collisions in the rotating tank setup we have applied the DigImage particles tracking velocimetry facility. Small opaque particles with densities smaller than that of the fluid have been added to float on the surface and illuminated from the top. However, it turned out that particles floating on the surface have a tendency to make clusters in time scales of the same order as the experimental time scale, due to the surface tension. Although, a large number of different particles and reduction of the surface tension with chemicals have been applied the clustering was found to influence the particle tracking significantly.

In stratified fluids, on the other hand, the particles are floating in one layer in a depth corresponding to the density of the particles. Therefore, the particles are not influenced by the surface tension, and we avoid the clustering.

In Fig. 4 we show an example of the streak lines of a dipole in the stratified fluid obtained by means of DigImage, and in Fig. 5 we see the development of the velocity field and vorticity at two different stages of a dipole collision with a concave wall. In Fig. 6 we show the corresponding streamfunction at T = 400 s and finally in Fig. 7 we have shown the scatter-plot illustrating the functional relationship between the vorticity and streamfunction in the central part of the dipole. This scatter plot is not corrected for the translation velocity of the dipole, this would also be rather small since the dipole is almost at rest due to the collision with the wall. The scatterplot may be interpretated as the dipole is startening to split into two half due to the interaction with the wall similar to what is observed in Fig. 3.

ACKNOWLEDGMENTS

The DigImage system has been developed at Department of Applied Mathematics and Theoretical Physics, University of Cambridge, England by Stuart Dalziel and is commercialized through Cambridge Environmental Research Consultants Ltd. The software supports a personal computer, a commercial Super VHS video tape recorder (Panasonic, model AG–7350) and frame grabber (Data Translation 2862). The system was financially supported by a grant from the Danish Natural Science Research Council.

The helpful assistance of Mogens Nielsen in the construction of the experimental devices and of Lars Bækmark in running our computer systems is gratefully acknowledged.

REFERENCES

1. E.J. Hopfinger and G.F.J. van Heijst, "Vortices in rotating fluids", Ann. Rev. Fluid Mech. **25**, 241 (1993).

2. E.J. Hopfinger, editor, "Rotating Fluids in Geophysical and Industrial Applications", CISM Courses and Lectures - No. 329, Springer–Verlag, Wien - New York (1992).

3. G.J.F. van Heijst and J.B. Flor, "Dipole formations and collisions in a stratified flow", Nature **340**, 212 (1989).

4. G.J.F. van Heijst and J.B. Flor, "Laboratory experiments on dipole structures in a stratified fluid", in: "Mesoscale/Synoptic Coherent Structures in Geophysical Turbulence", J.C.J. Nihoul and B.M. Jamart, eds. Elsevier, Amsterdam, pp. 591 - 608 (1989).

5. S.I. Voropayev, Ya.D. Afanasyev and I.A. Filippov, "Horizontal jets and vortex dipoles in a stratified fluid", J. Fluid Mech. **227**, 543 (1991). S.I. Voropayev and Ya.D. Afanasyev, "Two-dimensional vortex-dipole interactions in a stratified fluid", J. Fluid Mech. **236**, 665 (1992).

6. S.B. Daziel, "Decay of rotating turbulence: some particle tracking experiments", Applied Scientific Research **49**, 217 (1992)

7. G.K. Batchelor, "The Theory of Homogeneous Turbulence", Cambridge University Press, Cambridge (1953); G.K. Batchelor, "Computation of the energy spectrum in homogeneous two-dimensional turbulence", Phys. Fluids **12** (Suppl. II), 233 (1969).

8. R.H. Kraichnan and D. Montgomery, "Two-dimensional turbulence", Rep. Prog. Phys. **43**, 547 (1980).

9. A. Hasegawa, "Self-organization processes in continous media", Advances Phys. **34**, 1 (1985).

10. J.C. McWilliams, "The emergence of isolated coherent vortices in turbulent flows", J. Fluid Mech. **146**, 21 (1984); B. Legras, P. Santangelo and R. Benzi, "High- resolution numerical experiments for forced two-dimensional turbulence", Europhys. Lett. **3**, 811 (1988); D. Montgomery, X. Shan and W.H. Matthaeus",Navier-Stokes relaxation to sinh-Poisson states at finete Reynolds numbers", Phys. Fluids A **5**, 2207 (1993).

11. G.R. Flierl, M.E. Stern and J.A. Whitehead, "The physical significance of modons: laboratory experiments and general integral constraints", Dyn. Atmos. Oceans, **7**, 233 (1983).

12. J.M.H. Fortuin, "Theory and application of two supplementary methods of constructing density gradient columns", J. Polymer Sci. **44**, 505 (1960).

13. E.A. Coutsias, J.P. Lynov, A.H. Nielsen, M. Nielsen, J. Juul Rasmussen and B. Stenum, "Vortex dipoles colliding with curved walls" in: "Future Directions of Nonlinear Dynamics in Physical and Biological Systems" P.L. Christiansen et al., eds. Plenum Press, New York, pp. 51-54 (1993).

14. J.P. Lynov, E.A. Coutsias and A.H. Nielsen, "A spectral algorithm in the vorticity-streamfunction formulation for two-dimsional flows with no-slip walls", in: "Computational Fluid Dynamics '92" Vol. 1, Ch. Hirsch et al., eds, Elsevier Science Publishers B.V., Amsterdam, pp.413-420 (1992).

15. E.A. Coutsias and J.P. Lynov, "Fundamental interactions of vortical structures with boundary layers in two-dimensional flows" Physica D., **51** 482 (1991).

16. P. Orlandi, "Vortex dipole rebound from a wall", Phys. Fluids, A **2**, 1429 (1990); P. Orlandi, "Vortex dipoles impinging on cicular cylinders", Phys. Fluids A **5**, 2196 (1993).

17. R.J. Adrian, "Particle-imaging techniques for experimental fluid mechanics", Ann. Rev. Fluid Mech. **23**, 261 (1991).

THREE-DIMENSIONAL PARTICLE VELOCIMETRY

Klaus D. Hinsch

FB 8 - Physik
Carl von Ossietzky Universität Oldenburg
Postfach 2503
D-26111 Oldenburg, Germany

INTRODUCTION

In recent years, the problems tackled by experimentalists and theoreticians in fluid dynamics have grown considerably. Sophisticated flow facilities like large size or high speed wind tunnels are the basis for ambitious experiments, and powerful computers allow modeling of very complex flow fields. Interest has turned to nonstationary flows, their spatial structures and their development in time. An ultimate aim is to measure and understand the evolution of turbulence.

The metrological tools for the experimental study of flow fields have also seen an impressive improvement. Much early information about flows has been accumulated by mostly qualitative visualization techniques[1]. Smoke or dye line patterns have produced impressive milestones in the history of flow analysis[2]. Attention then turned to the development of anemometers that provide quantitative velocity records of high accuracy at a selected location in space. Thermal probes like hot wire anemometers (HWA), on the one hand, have become a routine tool in velocimetry. Their spatial resolution, however, is limited by the finite dimensions of the probe. Furthermore, in delicate flows, the anemometer may introduce disturbances. Optical probes, on the other hand, allow noncontact remote measurements within a very small sample volume. Thus, the favorite choice in single point measurements is the laser Doppler anemometer (LDA). It is noninterfering, but it relies on the presence of small particles in the flow. The technique assumes, of course, that the particle motion faithfully resembles the fluid motion. In gaseous flows this requires very small particles and sets limits to the largest velocity gradients acceptable.

Fluid velocity, the characteristic parameter of the flow, is a three-dimensional quantity. A complete specification calls for its three components u, v, and w in the x-, y-, or z-direction, respectively. Single point anemometers like HWA or LDA have been modified for 3-D operation. Yet, such a measurement provides only data from a single location in space. Many flow configurations of interest, however, are of three-dimensional nature. Especially the structural character of a flow, a topic of current interest, is only revealed by its spatial features. When the flows are very voluminous or measuring time is expensive, the effort or the cost to scan the flow with a single point probe are enormous and may be even prohibitive. Even more, when the flows are nonstationary there is no time to move a single point probe through the flow field.

In view of these problems, much recent effort has been dedicated to the development of flow velocimetry in several dimensions. We may classify techniques for flow velocimetry in terms of the dimensional size of the data that are covered (Fig. 1). A technique can be labeled a (k,l,m)-method where k=1,2,3 indicates the number of velocity components measured, l=0,1,2,3 gives the number of dimensions in space, and m=0,1 indicates whether a measurement yields velocity at only a single instant in time or a complete time record. A few examples: Conventional one-dimensional LDA is a (1,0,1)-method, particle image velocimetry PIV in its usual form a (2,2,0)-method. The utmost in flow velocimetry is a (3,3,1)-method where the velocity vector is measured throughout a complete volume as a function of time. Holographic cinematography can yield such data.

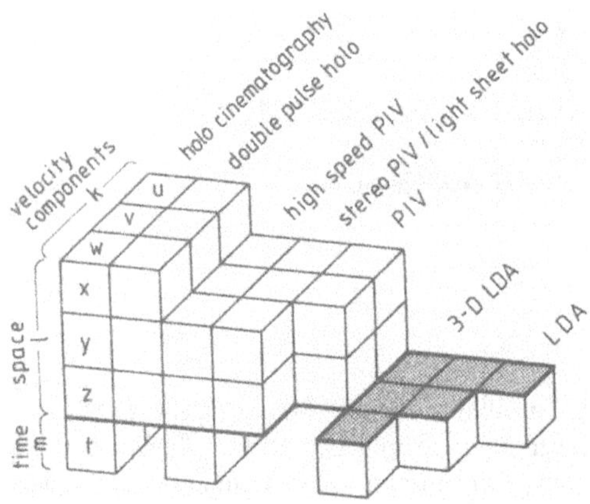

Figure 1. Techniques for optical flow velocimetry and the dimensions of data space.

In this chapter we describe efforts to explore the dimensions of fluid dynamics beyond particle image velocimetry (PIV) in its common 2-D configuration. We present only methods that rely on the detection of small tracer particles in the flow. There are other interesting and promising more-dimensional approaches in fluid analysis like tomography, visualization of fluorescent tracer molecules or global Doppler that are treated elsewhere.

LIGHT SHEET TECHNIQUES

Classical PIV: a 2-D slice from the flow

Let us recall briefly the great step forward in whole-field techniques that has become known as PIV[3]. It can be a good starting point for extensions into more dimensions, because profitable use can be made of existing powerful PIV interrogation schemes. Furthermore, the thorough understanding of the imaging of tiny particles is prerequisite for all methods to follow.

A sheet from a powerful light source, usually a laser, is directed along a plane in the fluid (Fig. 2). It illuminates tiny tracer particles of suitable properties within the flow. The particles are photographed at right angles to the sheet. When the light source is pulsed repeatedly, a series of particle images are recorded. Since the repetition period of the light pulses is known, magnitude and direction of the displacement of the particle images yield the velocity.

PIV realizations can be classified by the number of pulses and frames that are employed[4]. In the most simple and commonly used version two pulsed images are superimposed on a single photographic frame. The evaluation thus calls for the determination of the separation of the pairs of particles. Usually this quantity is averaged over finite interrogation regions that must be much smaller than the size of the relevant flow structures in the transversal x- and y-dimensions.

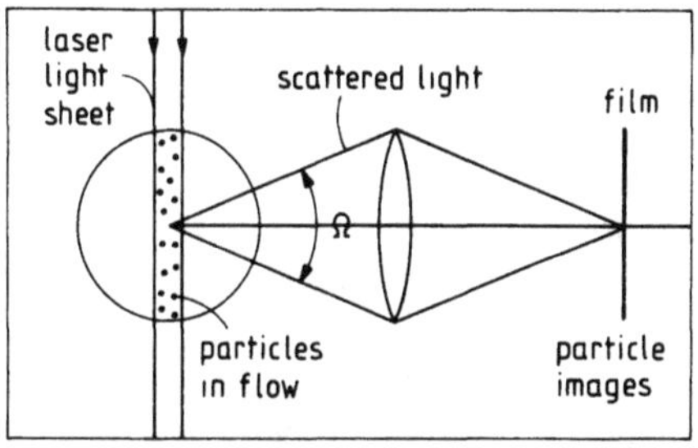

Figure 2. Optical setup for PIV (Particle Image Velocimetry).

PIV has found successful applications in flows where the light sheet can be oriented parallel to a prevailing flow direction. This guarantees a sufficiently small out-of-plane velocity component and ensures that most particles stay in the light sheet between exposures. Unpaired particles contribute to noise and reduce the efficiency of the method. Usually a light sheet depth of 1 to 2 mm is appropriate which is made to fit the focal depth of the recording camera system. Thus a PIV study usually ends up with values of u and v at the interrogation locations (x,y,z_0) where z_0 is given with the accuracy of the light sheet depth.

It should be mentioned that PIV makes poor use of the light scattered by the particles. As is demonstrated by the example in Fig. 3, particles suited for flow tracer applications scatter light mostly into the forward direction. At 90° observation the light intensity is down to about 0.1 percent. Thus, powerful laser illumination (ruby, Nd:YAG or copper vapor for pulsed applications, Ar-Ion for scanned beam light sheets) is required. Contrary to the forward direction, however, where the light is concentrated into a narrow lobe, the mean scattered light does not change very much within a large angular region around the direction of observation. This is a feature important for particle imaging.

Particle velocimetry calls for small and high quality particle images. Basically, the resolution in optical imaging is limited by diffraction and is determined by the angular spread

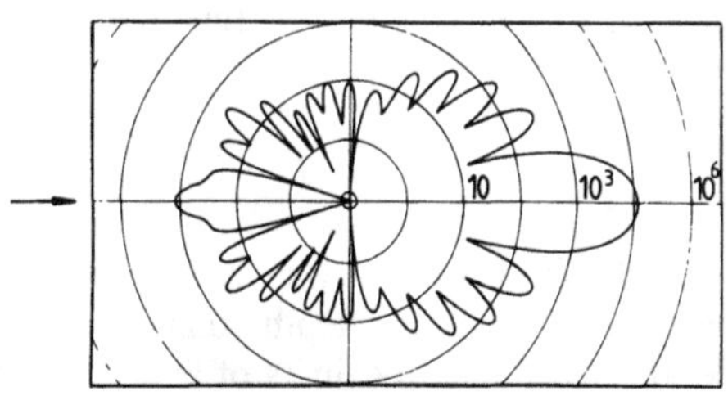

Figure 3. Scattering of 0.633 µm wavelength laser light by a 3-µm diameter water droplet in air (with permission after [28]).

of the light that contributes to image formation (angle Ω in Fig. 2). Omitting nonessential numerical factors, we obtain for the transversal resolution length (λ being the wavelength of light)

$$d_t \approx \lambda/\Omega, \tag{1}$$

and for the longitudinal resolution length

$$d_l \approx \lambda/\Omega^2. \tag{2}$$

When a particle is smaller in size than the resolution length it will not be resolved in the image, when it is larger its geometrical image has to be convolved with a point spread function whose width is given by the resolution length.

Due to the wide scattering around 90°, the angle Ω in PIV is determined by the aperture of the imaging system. Large aperture optics is therefore required for small diffraction limited particle images. At a magnification M=1 an f/2.8 lens yields some 6λ transversal and 30λ longitudinal resolution. Due to depth of focus considerations (particles throughout the complete light sheet depth must be recorded properly) the lens is usually stopped down to a compromise value. This is to make sure that particle images are almost identical irrespective of depth position.

More than a decade of intensive PIV development has produced many different processing routines. These range from digital particle tracking to optical or digital correlation analysis. In the latter, the cumbersome identification and pairing of particles is avoided by processing an ensemble of particles within the interrogation area. Details depend on the particle density, single or multiple frame techniques and preferences of the researcher concerning computerized or optical processing. In any case, there exists powerful hard- and software that is also available when 3-D problems can be decomposed into 2-D subtasks.

For later use in stereo extensions of PIV let us recall the typical steps in the partly optical correlation processing of a PIV double exposure record. The photographic recording (or a negative contact copy thereof) is interrogated with a narrow focused laser beam to produce a diffraction pattern known as Young's fringes system. The fringe pattern is usually read into a computer that performs a two-dimensional Fourier transformation of the pattern, ending up with the autocorrelation function of the input negative. This consists of a central peak accompanied by two side peaks at a distance equal to the wanted particle separation. Identification of the correlation peaks and determination of their position thus concludes the velocity measurement. The final result of a flow analysis with this method usually is a field of small velocity vectors plotted at the interrogation locations.

The limits of particle image velocimetry in the form presented are obvious: a two-dimensional cut from the flow volume is extracted, only the in-plane components of the velocity are determined, and this at a single instant in time that is set by the pulsing of the laser. It is, of course, assumed that the pulse separation is much smaller than the time scales in the flow. Current progress in the generation of powerful laser pulses with repetition rates tailored to specific needs allows PIV to conquer the time domain. Thus, the design of high-speed PIV setups is possible.

Stereoscopic PIV

We mentioned that the light sheet is directed in such a way that the main flow direction falls within the plane selected. Otherwise, the depth component of the flow would produce too many unpaired particle images. Within the limits of light sheet depth and depth of focus, however, a displacement in z-direction could be tolerated. As a matter of fact, its measurement opens the next dimension in the conquest of space.

There have been various proposals to measure the displacement in depth by modifications of the approved PIV setup. Examples are color coding by probing the flow with adjacent light sheets of different color[5] or by recording the particles with a coded aperture[6]. Here, we will introduce stereo-photography, a promising concept that receives considerable attention[7].

In stereo photography an object is viewed from two directions to extract its three-dimensionality. The underlying principle is easily summarized with the aid of Fig. 4. Let us consider three particles, numbered 1 to 3, distributed in space. We are registering the relative separations between particles as seen from two different directions. These separations are identical for all particles that lie within planes normal to the bisector of the viewing directions. One such plane is the reference plane in the illustration. The separations differ,

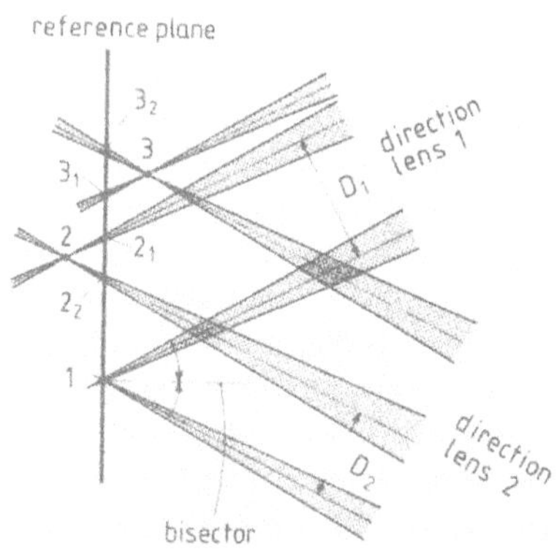

Figure. 4. Schematic of stereo photography of particles and the determination of the depth coordinate by back-projection.

however, when the particle combination includes a partner from a position behind (particle 2) or in front (particle 3) of the reference plane. For the pair of particles 1 and 2, for example, D_2 is smaller than D_1, because the second particle is behind the reference plane. It would be vice versa had we combined particle 1 and particle 3. From a combination of position information in two stereo photographs it is therefore possible to extract the depth information of the particle field. Obviously, the sensitivity of the method increases with the angle between the viewing directions. It should be mentioned that the stereo effect, i.e., that a camera registers only the projection in the viewing direction introduces an error in ordinary PIV when viewing extends over a large angle[8].

A stereo version of particle photography in PIV faces basic problems. There are two configurations to view the light sheet at an oblique angle and yet maintain focused particle images:

Angular displacement method. The imaging systems point at the particle field such that their optical axes form the required stereo angles with the light sheet (Fig. 5a). Camera lens and image plane must then be tilted to intersect in a common line (Scheimpflug condition). This results in a geometrical distortion of the image with space variant magnification.

Translation method. The imaging systems are aligned at right angles to the light sheet and view the interesting section of the particle field far off-axis (Fig. 5b). In this case the images are displaced off the usual imaging region and the backplanes of the cameras must be translated. While this translation can be avoided with an appropriate arrangement of mirrors[9] (Fig. 5c), the image is usually subject to large aberrations due to the off-axis conditions. This is critical when high quality particle images are required.

Several studies of three-dimensional flows using the translation method have been reported[7]. We have explored the capability of the angular displacement method and employed the Scheimpflug arrangement. There exist proper camera systems that allow the separate alignment of lens and image plane. The resulting two double exposure recordings may be processed by any of the techniques familiar from PIV. In the interrogation of the stereo images, however, the geometric distortions must be accounted for. The magnification of corresponding regions in the object field differs from left to right image. To be independent of system parameters, a calibration with a precision grid in object space is advisable. The interrogation can then be carried out on points of the distorted non-Cartesian grid in the image spaces to render regular sampling in object space.

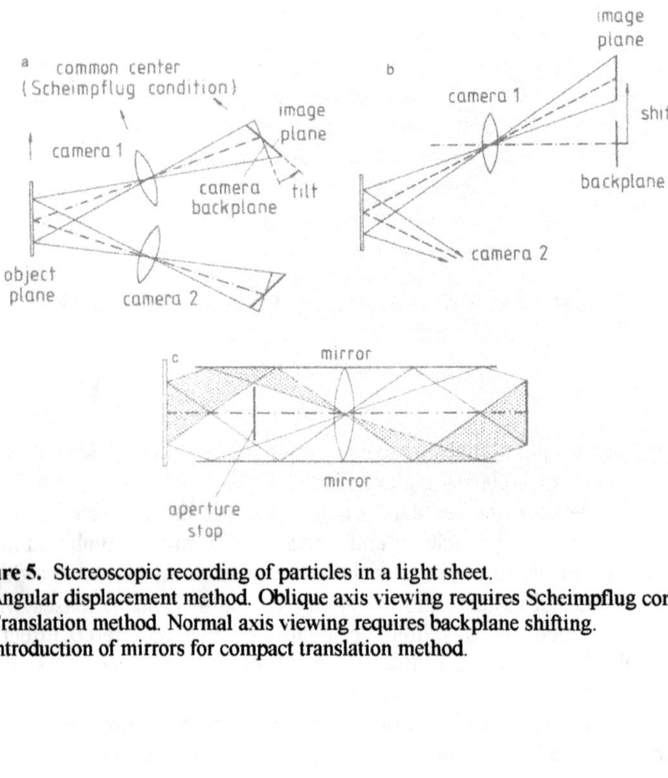

Figure 5. Stereoscopic recording of particles in a light sheet.
a. Angular displacement method. Oblique axis viewing requires Scheimpflug condition.
b. Translation method. Normal axis viewing requires backplane shifting.
c. Introduction of mirrors for compact translation method.

Distortions can be avoided when a telecentric imaging system is used in the camera. This may consist of two lenses of focal length f_1 and f_2, respectively, separated by a distance $s=f_1+f_2$. When the object is placed, say, in front of lens No. 1 there results a magnification $M=f_2/f_1$ that is independent of object position.

We have seen that the performance in determining the depth coordinate of a particle is greatly improved by a stereo arrangement. Basically, this can be attributed to the utilization of scattered light from widely separated angles. In a single lens system the aperture sets a practical upper limit to the angle Ω that determines depth resolution. It can only be increased further by employing two seperate lenses. In the usual stereo configurations, however, this advantage has to be paid for by two separate images that have to be matched for evaluation. We have tried to overcome this by a technique that we have called back-projection stereo. Its principle is explained by returning to Fig. 4. Assume that we have a means to back-project the recorded images into object space. The back-projected light is registered in a reference plane within the light sheet. Thus, for particle 1 (in the reference plane) both the back-projected images overlap in focus at the former particle position in the reference plane. However, for particle 2 or 3 (behind or in front of the plane) there results a pair of images each (2_1,2_2 and 3_1,3_2, respectively). The intrapair separation is a measure of the depth position, the left-to-right sequence of the images yields the distinction between the space behind or in front of the reference plane.

Let us apply this reasoning to double pulse particle image velocimetry. Let particle 2 and 3 represent the first and second exposure of a particle traveling with the flow. Instead of the ordinary two images we now get four back-projected images for a particle pair, and from the separations between 2_1 and 3_1 and 2_2 and 3_2 both the in-plane as well as the out-of-plane displacement can be derived. When, for example, the particle moves parallel to the reference plane these separations are the same. When one of the particles is caught in the reference plane, there are only three images; when both are in the reference plane we find the usual two images. The additional particle images produce corresponding peaks in the autocorrelation function from which the wanted information can be obtained.

There are several possibilities for the necessary back-projection. We have experimented with photographic recording of the particles in the image plane and back-imaging after production of a contact copy. Fig. 6 demonstrates results obtained in a model environment of 20-μm particles embedded in plastic. The cameras were viewing a light sheet of 2 mm depth symmetrically at 20° each to the normal. The object was displaced by 1 mm in-plane and 0.5 mm out-of-plane. Due to the two predominant particle separations, the fringe system displays a beat modulation and the resulting autocorrelation output shows two side peaks. Unfortunately, the performance of this method is rather poor in real flows, because of photographic thresholding and repeated introduction of aberrations.

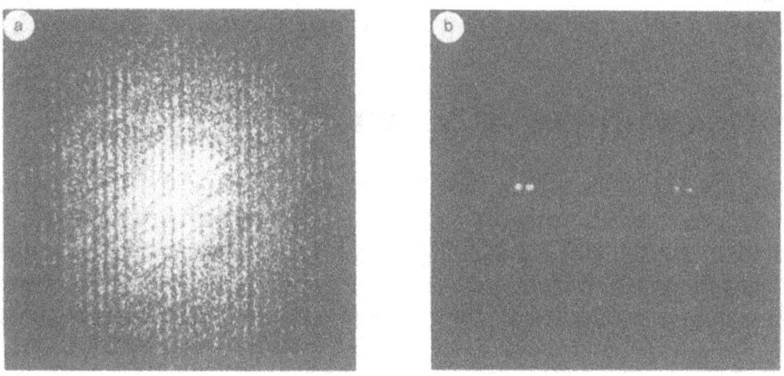

Figure 6. Interrogation of a transparency of a back-projected stereo pair of 20-μm particles in plastic. The out-of-plane displacement shows as a beat in the Young's fringes system (a) and twin-peak side orders in the autocorrelation (b). The central order was suppressed.

Our present solution is a direct stereo-photographic back-projection system. The optical setup for such an operation makes use of two unit magnification telecentric imaging optics as introduced above. Such a system is completely symmetric as it consists of two identical lenses separated by the double focal length. Therefore, half of the system can be omitted and a mirror can be introduced in the central plane of symmetry to reverse the imaging (Fig. 7). Let the reference plane in the flow intersect the optical axis at a distance equal to the focal length. Then, any object point P in the reference plane is imaged back onto this plane (point I). Any point in front of (or behind) the reference plane is imaged symmetrically behind (or in front of) the plane. The back-projection configuration of Fig. 4 is therefore realized with an inverted image of object space. Identical conditions hold for both the imaging directions. The introduction of a beam splitter parallel to the reference plane produces the image I' that can be collected on a photographic plate for further processing. For discrimination between the left and right images, the introduction of two distinctive apertures is considered to label the images.

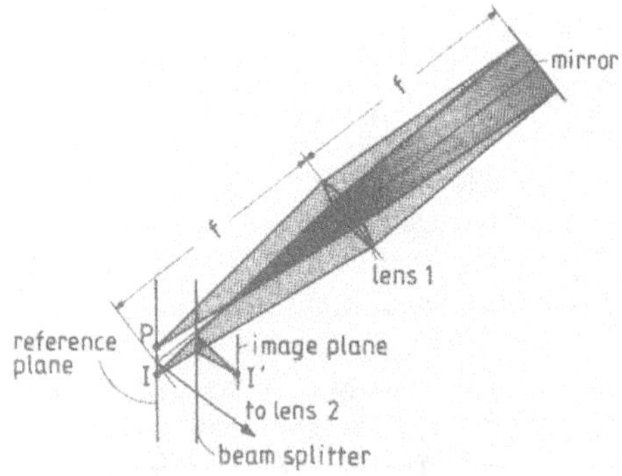

Figure 7. Back-projection of particle images by introduction of a mirror in the symmetry plane of a unit magnification telecentric optical system of 4f length.

Proper pulsing of the light source and cinematographic recording of the particle images could extend the stereoscopic techniques into the time domain arriving at a (3,2,1)-technique.

HOLOGRAPHIC PARTICLE VELOCIMETRY

For true imaging of three-dimensional space, holography is the technique of choice. Let us explain the principle in a setup for holographic recording of a particle field (Fig. 8a). All particles of interest in the flow are illuminated with a properly expanded laser light wave. The complex light field scattered by the particles is stored by superposing a reference wave from the same laser source and recording the interference pattern on a photographic plate. Due to

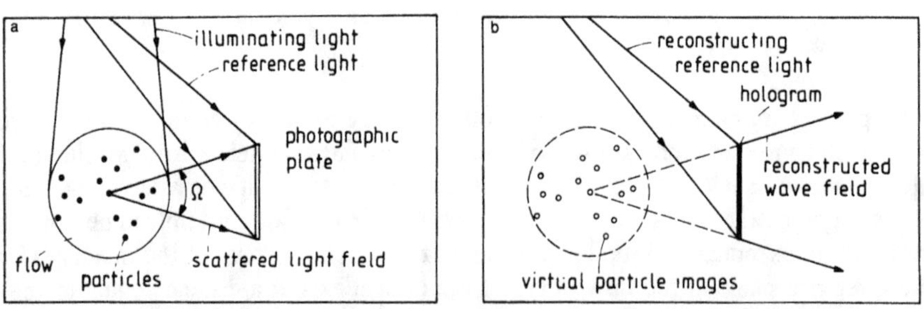

Figure 8. Off-axis holography of a field of tracer particles in a flow.
a. Recording.
b. Reconstruction of the virtual image.

the oblique angle between reference wave and object light this is called off-axis holography. The coherence length of the laser must be adequate to guarantee interference over all path length differences encountered. The developed pattern (the hologram) is used to reconstruct an image of the object. For this purpose the hologram is illuminated with the original reference wave that reconstructs a virtual image of the particle field (Fig. 8b). When looking or photographing through the hologram the particle configuration can be seen as it was during the recording of the hologram.

A very useful feature of holography is that also a real image of an object can be obtained without the need for a lens system. For this purpose, the hologram must be illuminated with the complex conjugate of the reference wave. For a divergent spherical reference wave this is a convergent wave from the opposite direction (shown in Fig. 9a). In this case the hologram produces backward traveling object waves that form a real image of the light sheet. With a CCD-array, for example, a plane anywhere through the field could be examined for particles. If aberrating media like windows or low quality lenses have distorted the object wave on its way to the holographic recording plate, these distortions can be compensated during the backward path of the conjugate wave provided the distorting media are still in the old position (Fig. 9b). Then the particle images are formed by the original ideal wavefront.

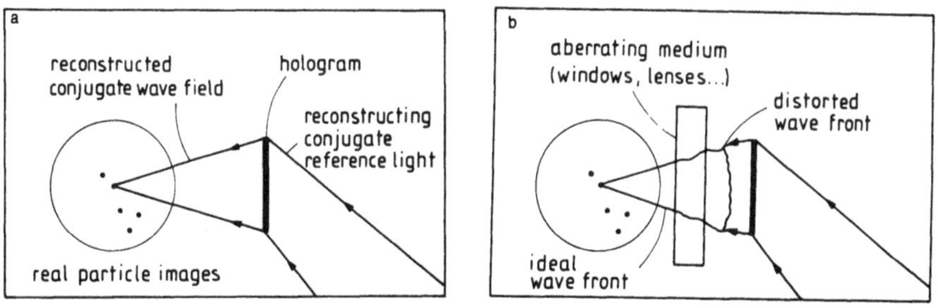

Figure 9. Generation of a real image of a tracer particle field by reconstruction with a phase conjugate wave.
a. Ordinary reconstruction.
b. Compensation of aberrations.

For good particle image quality the reconstructing wave has to meet special requirements. The geometry of the recording wave has to be reproduced and the wavelength must be the same. In some lasers, this poses problems when a pulsed source is used for the recording and continuous illumination is wanted for viewing. The 10% wavelength difference between ruby and He-Ne laser, for example, is detrimental to a faithful reconstruction. For high fidelity in our ruby laser experiments, we have reconstructed the images with repeated pulses from the same laser. Nd:YAG laser systems are better, because there exist pulsed and CW lasers at the same wavelength. Another requirement is that distortions in the photographic material due to shrinkage, for example, must be avoided. As to resolution, usually the angle Ω is now given by the size of the hologram.

The real power of holography lies in the almost unrestricted recording depth. But already holography of a light sheet can be utilized for various tasks because the depth position of the particles is preserved. When the stereoscopic techniques, for example, are applied to a holographic virtual image, a single camera system would suffice. The stereo images could be taken of the frozen-in particle distribution, one after the other. In another application, the generation of real images could substitute the lens-and-mirror arrangement in back-projection stereoscopy. A hologram recorded in either arm and illuminated with the conjugate reference wave would reproduce the desired back-propagating waves. Furthermore, it has been proposed to use a holographic real image of the particles in a light sheet as input for a 3-D autocorrelation evaluation analogous to 2-D Young's fringes processing[10]. The resulting autocorrelation peaks are arranged in 3-D space and their separation provides the 3-D displacement.

It should be mentioned that holographic particle imaging in the off-axis configuration has been used under still a different aspect. When operated with a separate reference beam for each of the pulses the first and second image can be distinguished thus removing the directional ambiguity that is inherent in simple double pulse recordings[11]. A dual pulsed Nd:YAG laser system, for example, supplies suitable pulses that are aligned to travel identical object illumination paths but different reference paths[12,13]. Upon reconstruction, first and second exposure can be viewed separately.

In-line particle holography

Off-axis holography is easily disturbed by vibrational noise, requires good coherence, and relies on an accurate reproduction of the reference wave for reconstruction. For small objects like the flow tracing particles, the so-called in-line arrangement is much simpler, provides better stability and claims better signal-to-noise ratio[14].

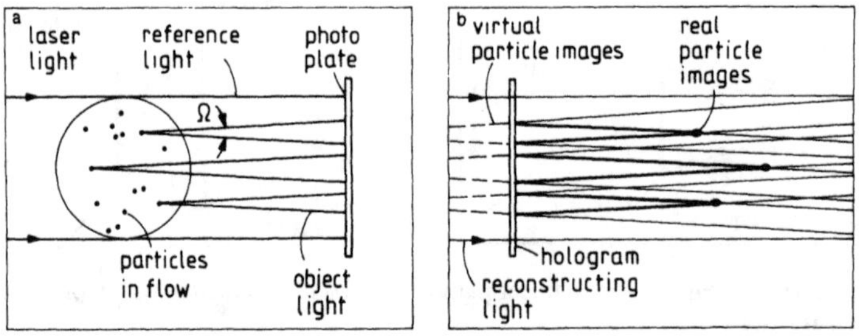

Figure 10. In-line holography of particle fields.
a. Recording.
b. Reconstruction.

Fig. 10 illustrates a schematic of in-line particle holography. There is no extra reference wave, because the tiny particles leave enough of the illuminating plane wave light undisturbed so that it can serve this purpose. In-line holography relaxes many of the restrictions mentioned for off-axis holography. Its application in particle studies has a long history[15,16]. Since forward scattering is employed, the light is utilized much more efficiently, but the angular range Ω of the recorded light is usually smaller than in sidewise imaging. Because Ω determines the resolution, this can become a serious disadvantage. For reconstruction, the hologram is repositioned in the illuminating wave. Due to the on-axis geometry, both the virtual and real image are reconstructed simultaneously. Usually a microscopic analysis of the real particle images is carried out for evaluation. These images, however, are disturbed by the unused portion of the reconstructing wave and by noise from the virtual images[14].

An in-line hologram obtained from 20-µm particles embedded in a plastic block of 10 mm depth is shown in Fig. 11. The diffraction patterns of single particles can be distinguished. The high particle density that is also typical of flow conditions, however, results partly in overlapping light fields tending to produce laser speckle.

To understand the resolution limits in an in-line arrangement and provide for improvements, let us apply equations (1) and (2). Most registered light falls within the central lobe of the diffraction pattern of the particle which is given approximately by $\Omega \approx \lambda/d_p$. Thus

$$d_t \approx d_p \tag{3}$$

$$d_l \approx d_p^2/\lambda\,. \tag{4}$$

Clearly, in many practical cases where d_p is considerably larger than λ axial focusing according to Eq. (4) will be much lower than transverse focusing as in Eq. (3). It has been shown that the depth of field that can be covered with an in-line hologram is some 100 times the longitudinal focus length[17]. Thus, about 100 different planes can be distinguished. There

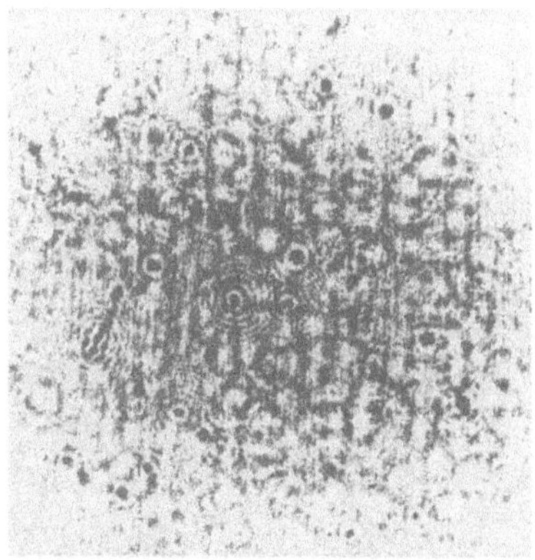

Figure 11. In-line hologram of 20-μm particles.

are many applications of in-line particle holography that take the real images as they are and improve on resolution by sophisticated processing of the images[18,19,20]. But the in-line setup can also be modified directly for improvements. Early solutions introduced an auxiliary illumination beam to enhance the scattered object light[21]. Fig. 12 shows a solution[12,17], where two traditional in-line setups are combined at right angles so that the inaccurate depth position from, say, hologram 1 can be improved by an additional evaluation from hologram 2. In other approaches multibeam illumination was introduced to increase the aperture by recording several scattered fields, each of extent Ω, on the same hologram[12]. Thus highly effective forward scattering is combined with good axial focusing by a large aperture.

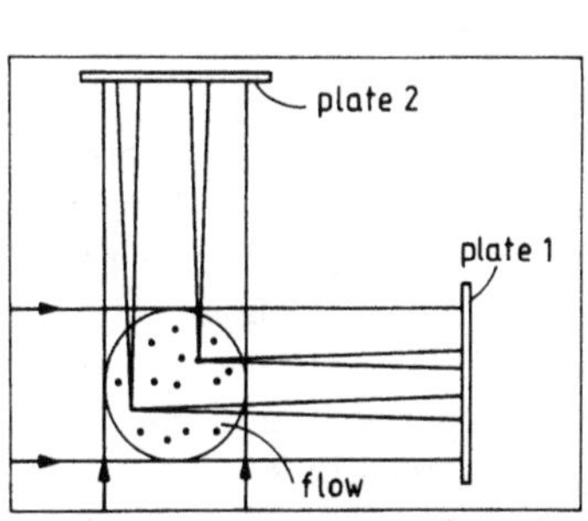

Figure 12. Improvement of axial resolution for in-line holography of particles by combination of two holograms recorded at right angles.

Off-axis particle holography

The true power of holography for 3-D particle imaging in flow studies is illustrated by a recently presented off-axis arrangement that comprises many of the unique features holography offers[13]. The recording setup is shown in Fig. 13a. The illuminating light is produced by a Q-switched Nd:YAG laser that is pulsed twice at instants t_1 and t_2. Let us go through the optical arrangement to explore its various components. The pulses are spatially separated so that the first pulse forms, say, reference wave R_1 and the second pulse R_2. For the illuminating light both pulses are superposed. The light scattered by the particles is collected under two near-forward directions with relay optics that form particle images close to the holographic plate. A beam-bending prism is introduced in each imaging branch so that both object waves can be recorded on the same holographic plate. Thus, two holograms of the particle fields are recorded with different reference waves.

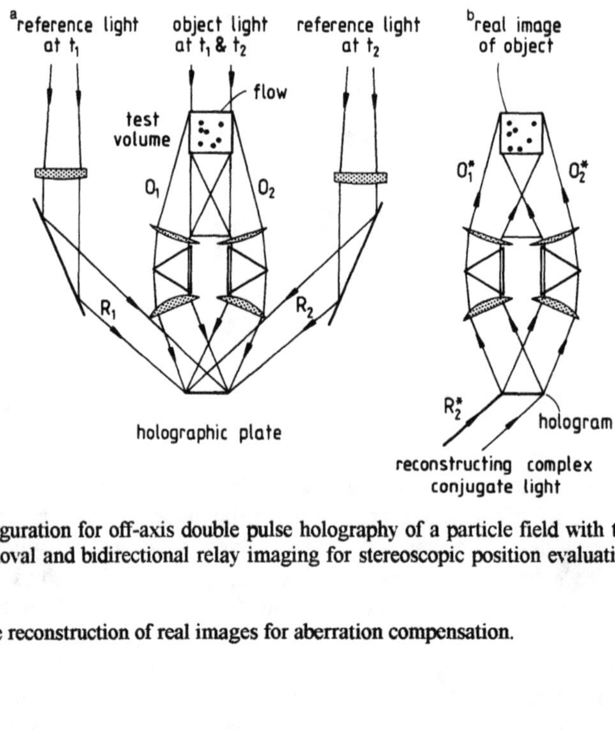

Figure 13. Configuration for off-axis double pulse holography of a particle field with two reference beams for ambiguity removal and bidirectional relay imaging for stereoscopic position evaluation (with permission after [13]).
a. Recording.
b. Phase conjugate reconstruction of real images for aberration compensation.

In the reconstruction process (Fig. 13b) one complex conjugate reference wave is applied after the other to produce the backward traveling object wave at the times of first and second exposure. These are directed backwards through precisely the old imaging optics thus canceling all prior aberrations during the recording. Both the temporal states of the particle field can thus be reconstructed consecutively by applying either reference wave. A light sensitive detector like a CCD target, for example, is placed somewhere in the real image. Now the situation is analogous to the back-projection stereo configuration that we have described earlier. Depending on the depth postion of the particle with respect to the plane of the CCD we register identical images from both directions $O_1{}^*$ and $O_2{}^*$ (when the particle lies within this plane) or separated images that indicate the depth coordinate by the amount of separation and the left-right sequence. By moving the CCD the whole volume can be scanned.

The work shown gives a good example of how holography is suited to study arbitrarily complicated 3-D flows. In a series of planes in depth the 3-D velocity vector has been evaluated to yield very detailed information about the flow. Quite some effort, however, must be accepted for this technique and it might be useful in some types of flows to look for a less sophisticated solution.

Multiple light sheet holography

We have introduced an alternative approach in the holographic exploration of flows by recording a set of light sheets that sample the flow field in depth[22]. If the flow displays a predominant direction that determines the orientation of the light sheet, the velocity often changes only gradually in depth. If the spacing of the sheets is adjusted to the depth scale of the flow structures, much valuable information about the three-dimensionality of the flow can be obtained with a system of just a few light sheets. At the same time, favorable properties of ordinary PIV are preserved. First of all, optimum use is made of the available light by redirecting the sheet several times through the flow (Fig. 14). Because the particles scatter only little, the quality of the light sheet stays sufficient for several passes. If needed, suitable beam shaping optics can improve the sheet dimensions after multiple passes through the measuring field. Second, each light sheet is available for any of the well established or developing evaluation techniques including three-dimensional aspects. Finally, a sophisticated balance of the limited coherence of the laser and the coherence requirements for holography ensures that each sheet can be reconstructed separately eliminating even most of the crosstalk from defocused particle images in other sheets.

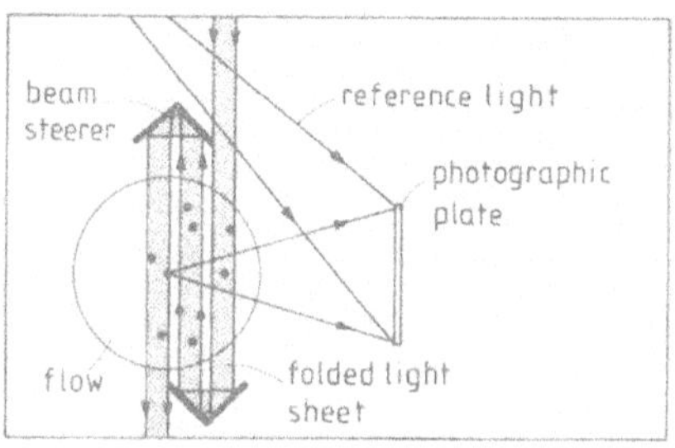

Figure 14. Holographic particle velocimetry with a folded light sheet.

Fig. 15 gives a schematic of a setup serving this purpose in a wind tunnel study. We are looking at the 0.25 m x 0.25 m cross section of the tunnel in the measuring region. Two prism beam steering devices produce four light sheets at 40 mm spacing. The scattered light is incident on a holographic plate together with four reference beams from different directions in space. All of these waves are present at the same time, but provisions have been made that light from each sheet interferes only with one reference beam each. For this purpose the limited coherence length of the laser light is used. Wavelength selective devices in the laser produce a coherence length that is just sufficient for the holographic recording of a single light sheet together with a reference beam that matches in optical path length. Take the examples shown in Fig. 16: The solid curve gives the coherence function, i.e., the quality of a recorded hologram, for the superposition of object light and reference light No. 1. Sheet No. 1 is adjusted such that it falls within the coherence length. Sheet and reference beam No. 3 (broken curve) are adjusted so that they interfere with each other but not with No. 1, and this concept is repeated for the other sheets and reference beams. By reconstructing with different reference beams, particles in one sheet can be evaluated without disturbing light from any other plane in depth.

As an example we present the flow field in two planes through the von Kármán vortex street behind a circular cylinder at Re = 190. Ideally, this is a two-dimensional structure. Often, however, it is observed that the vortices are not shed parallel to the cylinder axis

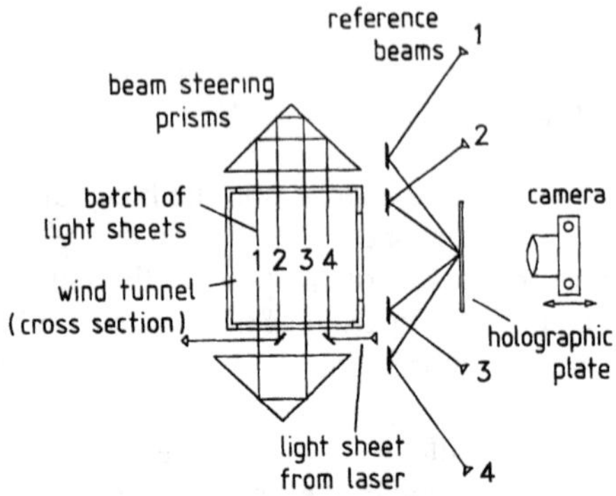

Figure 15. Multiple light sheet holography with controlled coherence in a windtunnel. Cross-sectional view of tunnel and optical beam configuration.

which has been attributed to effects from the termination of the vortex axes at the ends. The slight three-dimensionality of this phenomenon is ideally suited to demonstrate the capability of our PIV method. Already two planes reveal important features of the flow structure. In

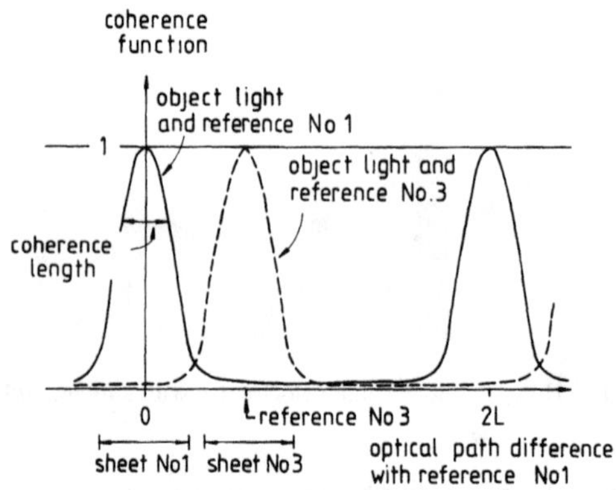

Figure 16. Coherence function for multiple light sheet holography.

our experiment, light originated from a 10-J Q-switched double pulse ruby laser (optical resonator length 0.7 m) that was modified for about 0.2-m coherence length. For this purpose one etalon in conjunction with the output mirror was removed. The original second etalon was exchanged for a model of specifically designed Q-factor to yield the desired coherence. Velocity vector plots of both light sheets that are spaced 40 mm in depth are shown in Fig. 17. The cylinder is positioned 55 mm upstream to the right and halfway up the frame. The vortex street is evident in both the planes. A clearly visible spatial shift in the periodic structure is evidence of the oblique vortex shedding. It amounts to about 9 mm which yields an angle of about 13°.

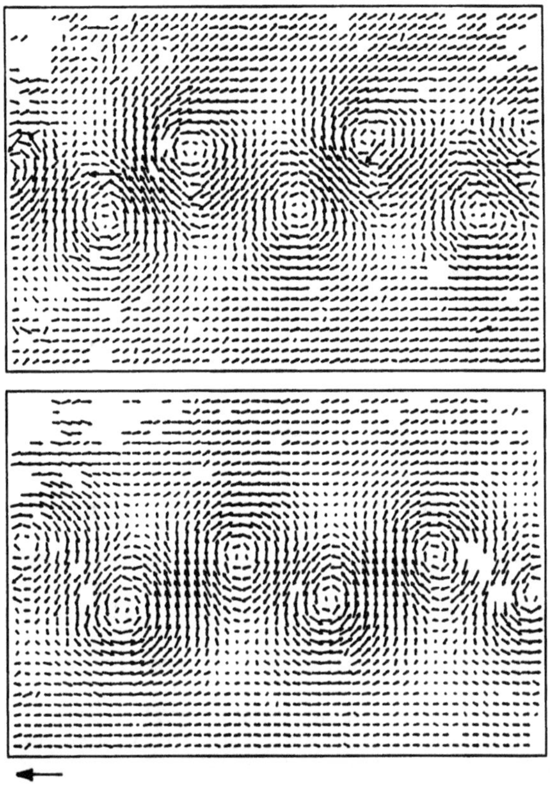

Figure 17. Double light sheet holography of oblique vortex shedding behind a cylinder. Velocity maps from two planes normal to the cylinder axis and spaced 40 mm. Mean velocity of 0.47 m/s has been subtracted. Arrow is 0.2 m/s.

CONCLUSIONS

Optical flow velocimetry is gradually conquering all dimensions. We have seen that a hologram can store most of the spatial data about the particles that track a flow field. Even more, the holographic techniques can be turned into cinematographic tools when the laser light source is operated at the appropriate repetition rate. In recent studies on cavitation bubble dynamics, for example, small holograms were recorded on rotating holographic plate material at 20 kHz rate with a dye laser pumped by a copper vapor laser[23]. In other investigations, rotating drum cameras served a similar purpose[24,25]. It should be mentioned that all optical recording and evaluation techniques described in this paper are finding powerful competitors in video equipment and digital image processing[26,27]. These techniques are limited in depth and resolution, but take advantage of advancing video and computer technology. The details of such techniques, however, are beyond the scope of this paper.

Nowadays, powerful equipment reveals the spatial and temporal structures of complex flows in experimental investigations. The amount of experimental data, of course, grows tremendously, and the evaluation and visualization of multi-dimensional records presents new challenges. Thus, each problem deserves special consideration to select the appropriate level in the dimensional hierarchy shown in Fig. 1.

ACKNOWLEDGEMENT

Cordial thanks are due to H. Hinrichs for many useful suggestion and assistance in compiling the experimental material presented in this paper.

REFERENCES

1. W. Merzkirch. "Flow Visualization" (2nd edition), Academic Press, Orlando 1987.
2. M. Van Dyke. "An Album of Fluid Motion," Parabolic, Stanford 1982.
3. K.D. Hinsch, Particle image velocimetry (PIV), in: "Speckle Metrology," R. S. Sirohi, ed., Marcel Dekker, New York (1993).
4. R.J. Adrian, Particle-imaging techniques for experimental fluid mechanics, Ann. Rev. Fluid Mech. 23:261 (1991).
5. A. Cenedese and A. Paglialunga, A new technique for the determination of the third velocity component with PIV, Proc. 4th Int. Symp. Appl. Laser Anemometry to Fluid Mechanics, Lisbon, Paper 3.14, (1988).
6. C.E. Willert and M. Gharib, Three-dimensional particle imaging with a single camera, Exp. Fluids 12:353 (1992).
7. A.K. Prasad and R.J. Adrian, Stereoscopic particle image velocimetry applied to liquid flows, Exp. Fluids 15:49 (1993).
8. S.K. Sinha, Improving the accuracy and resolution of particle image for laser speckle velocimetry, Exp. Fluids 6:67 (1988).
9. M.P. Arroyo and C.A. Greated, Stereoscopic particle image velocimetry, Meas. Sci. Technol. 2:1181 (1991).
10. J.M. Coupland and N.A. Halliwell, Particle image velocimetry: Three-dimensional fluid velocity measurements using holographic recording and optical correlation, Appl. Opt. 31:1004 (1992).
11. J.M. Coupland et al., Particle image velocimetry: Theory of directional ambiguity removal using holographic image separation, Appl. Opt. 26:1576 (1987).
12. F. Hussain et al., Holographic particle velocimetry: prospects and limitations, in: Holographic Particle Image Velocimetry, E.P. Rood, ed., ASME FED 148:1 (1993).
13. R.J. Adrian et al., An HPIV system for turbulence research, in: Holographic Particle Image Velocimetry, E.P. Rood, ed., ASME FED 148:21 (1993).
14. H. Meng et al., Intrinsic speckle noise in in-line particle holography, J. Opt. Soc. Am. A 10:2046 (1993).
15. B.J. Thompson, Holographic methods for particle size and velocity measurement - recent advances, in: Holographic Optics II: Principles and Applications, G.M. Morris, ed., Proc. SPIE 1136:308 (1989).
16. C.S. Vikram. "Holographic Particle Diagnostics," SPIE Milestone Series, Bellingham (1990).
17. L.P. Bernal and J. Scherer, HPIV measurements in vortical flows, in: Holographic Particle Image Velocimetry, E.P. Rood, ed., ASME FED 148:43 (1993).
18. P.R. Hobson, Precision coordinate measurements using holographic recording, J. Physics E 21:139 (1988).
19. M. Dadi et al., A study by holographic velocimetry of the behavior of free small particles in a flow, Exp. Fluids 10:285 (1991).
20. J.A. Liburdy, Holocinematographic velocimetry: resolution limitations for flow measurement, Appl. Opt. 26: 4250 (1987).
21. H. Royer, Holographic velocimetry of submicron particles, Opt. Commun. 20:73 (1977).
22. K.D. Hinsch et al., Holographic and stereoscopic advances in 3-D PIV, in: Holographic Particle Image Velocimetry, E.P. Rood, ed., ASME FED 148:33 (1993).
23. W. Lauterborn, High-speed off-axis holographic cinematography with a copper-vapor-pumped dye laser, Opt.Lett. 18:4 (1993).
24. F. Eisfeld, High-speed photography, high-speed cinematography and high-speed holography as tools to investigate fast flows, in: "Flow Visualization VI," Y. Tanida and H. Miyashiro, eds., Springer-Verlag, 419, Berlin (1992).
25. G.A. Ruff, High speed in-line holocinematography for dispersed-phase dynamics, Appl. Opt. 29:4544 (1990).
26. T. Dracos and N.A. Malik, 3D particle tracking velocimetry - its possibilities and limitations, in: "Flow Visualization VI," Y. Tanida and H. Miyashiro, eds., Springer-Verlag, 785, Berlin (1992).
27. N. Kasagi and Y. Sata, Recent developments in three-dimensional particle tracking velocimetry, in: "Flow Visualization VI," Y. Tanida and H. Miyashiro, eds., Springer-Verlag, 832, Berlin (1992).
28. Invent GmbH, Datasheet on program STREU, Erlangen 1993.

FLOW PATTERN IDENTIFICATION AND NEURAL NETWORK PROCESSING

F. Carosone, A. Cenedese

Department of Mechanics and Aeronautics
University of Rome "La Sapienza"
via Eudossiana, 18
00184 Rome Italy

INTRODUCTION

A heavy computational burden is needed to determine a velocity field from images produced by different flow visualization techniques, whereas human beings perform a fast scene analysis and pattern identification. This consideration has lead to develop computational architectures that mimic the one found in the human brain, in the hope of achieving human-like performances in real-time image analysis. These architectures are called artificial neural networks. They share with humans the ability of recognizing patterns through noise and distortion, automatically, as a result of their structure and not by using human intelligence embedded in the form of ad hoc computer programs. This paper provides an introduction to the field of artificial neural networks and shows an application to the analysis of the images obtained with PIV (Particle Image Velocimetry) technique. It will be seen that artificial neural networks can perform both low-level (e.g. filtering, edge detection) and high-level (e.g. pattern identification) analysis of flow fields.

PATTERN IDENTIFICATION AND NEURAL NETS

A pattern is an ideal model of real objects (or of a group of objects linked by their mutual relations). Humans and animals can easily abstract a pattern from many imperfect samples: all samples are grouped together by a recognition of similarity. Similar samples are considered distorted versions of one pattern, and they constitute one class. From this point of view, pattern identification may be regarded as a classification: when a new sample is considered similar to those belonging to a class, it is assigned to that class and it is said to be identified[1].

The task can be formulated in the following way. A pattern is represented by a number of features (numerical, symbolic) and a data structure (vector, tree, graph), that indicates the morphology of the mutual relations among these features. In the simplest case, a pattern is a vector whose N components are continuous valued real numbers, that represent the measurements of features useful to characterize that pattern. The final goal is to divide the N-dimensional vector space into decision regions, where each region codes for a class[2].

Neural-net classifiers are adaptive and non-parametric (i.e. make no assumption on the shape of the probability density function of the features). They produce better results than Gaussian classifiers in many cases[1]. The latter ones process all input data to estimate the probability density function parameters. Bayes' decision theory[3] can then predict the most likely decisional boundaries. On the contrary, adaptive non-parametric classifiers use discriminant functions to define the same borders. Neural-net classifiers can 'learn' the discriminant functions directly from examples.

Their behavior can be described in terms of input-output matches[4]. An input vector is presented to the net. The net first produces an output, different from the desired response. Then, according to this difference, the net changes its internal structure so that it accomplishes the right match. A 'learning algorithm' specifies how to do the job. When a new input vector is shown, and the above procedure is repeated, the network learns the new input-output match without 'forgetting' the first example. Therefore the net learns how to reproduce a series of input-output matches with no explicit knowledge of the underlying rule (or set of rules) that produces those matches. This feature may be useful when the underlying rules are long to code by an algorithmic approach (best-fitting curves, non-linear field). The job is performed properly, provided that examples are chosen with care.

This kind of learning (called "supervised") implies that a teacher tells the net the right answer for each input vector. In fact, a label is associated to the input vector to specify the desired output. In other cases ("unsupervised learning"), the net is provided with an auto-organization mechanism that changes its structure so that it groups similar inputs and enhances the difference between different inputs. In fact, the net maps all inputs to a vector space where similarity is computed as distance: similar patterns are placed at neighboring locations.

The learning procedure teaches the net how to perform a specific rule of classification (in terms of a series of input-output matches to be accomplished). After this phase, the net structure is 'frozen', so that it is not allowed to change anymore. The net is then checked on new data ('test phase'): a new set of examples (i.e. input vectors), different from the previous but still generated by the same underlying rule, is shown to the net. If the net produces the right classification for each of them, it is settled fit to work on the real-world problem. However, the same net (i.e. the same architecture plus the same learning algorithm) could be trained to be adapted to a completely different task, just with a different set of examples used during the learning procedure[5]. This flexibility has been a reason of success for neural networks over the past few years. Their application has shown to be useful for decision theory problems, when the decision 'rules' are complex, difficult to formulate explicitly (e.g. non-linear), or time-varying.

Finally, all neural networks possess an inherent parallel architecture. This means that they can be naturally layered down on parallel hardware, and this makes them suitable for visual pattern recognition specifically. As a matter of fact, it is a belief that massive parallelism is essential for high performance image analysis. This is the basis upon which neural networks are successfully applied to the field of image recognition.

ARTIFICIAL NEURAL STRUCTURES

The architecture of artificial neural networks (ANN) is motivated by the computational features found in biological nervous systems: a large number of relatively simple non-linear processing units, and a high degree of connectivity among them. Processing units are called 'neurons', reminiscent of biological neural nets. The connection between two neurons is called a 'synapse'. The synaptic strength may be varied. Typically, it gets higher when the link is often used, it gets lower when this happens seldom. Thus, neural nets adapt themselves to a specific task by changing their structure through the synaptic strengths. ANN mimic this biological feature by associating a real value *W* to each connection, which indicates the link 'weight'. Signal amplification (or attenuation) through a link is achieved by a simple product of the signal value by the link weight. The learning algorithm is a set of rules to modify the net weights, as closely as possible to the biological model[5]. It is remarked that the learning algorithm does not depend on the specific application: it is a net property[6]. Every net is defined by its neurons, its architecture, and its learning algorithm[4].

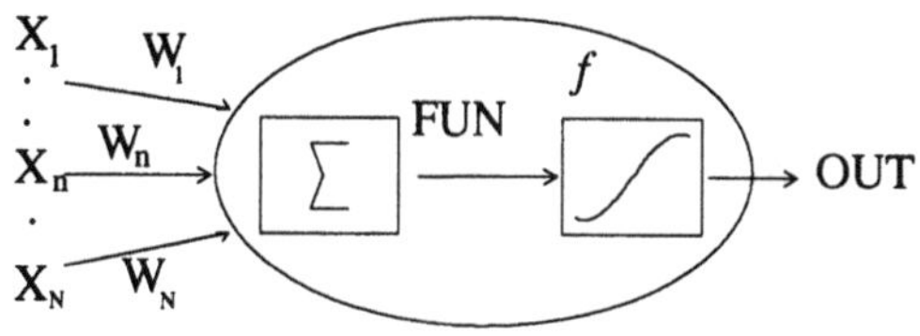

$$FUN = \sum_{n=1}^{N} W_n \cdot x_n \qquad OUT = f(FUN)$$

$$(FUN = FUN(W_n, x_n))$$

Figure 1. Neuron structure.

Each neuron performs a summation on its weighted inputs and then passes it through a non-linear function called 'activation function' (fig.1). In some cases, artificial neurons compute more complex functions of weight and input values, before the activation function. This is the same process as in the human brain neurons, though artificial activation functions are just rough approximations of the real ones. In the simplest case, a threshold (fig.2) is used:

$$\begin{array}{ll} OUT = 1 & \text{if fun > threshold value T} \\ OUT = 0 & \text{otherwise} \end{array} \qquad (1)$$

The activation function is also often a sigmoid (fig.2), more biologically plausible:

$$OUT = \frac{1}{1 + e^{-fun}} \qquad (2)$$

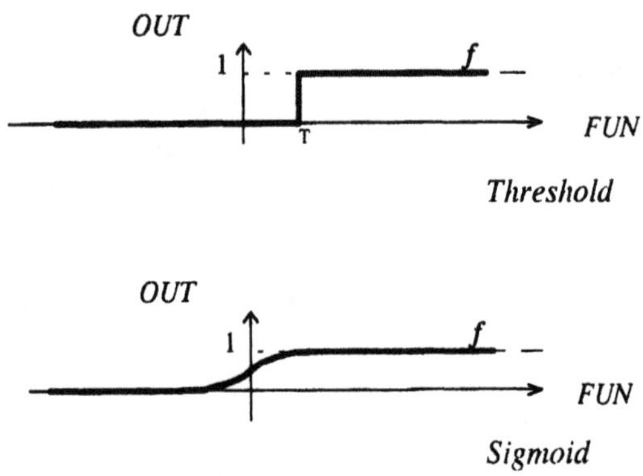

Figure 2. Activation functions.

The network structure typically consists of some layers of neurons (for instance the Perceptron architecture in fig.3). An input layer consists of nodes that perform no computation and that serve only to distribute the inputs in parallel to the following hidden layer. Input values may be continuous or binary. Information is then propagated through the network via every link. A hierarchical structure is often chosen: more layers (typically, but not always, three) are used. Connections are from the input layer to the first 'hidden' layer, and from each layer to the following one, but not vice-versa. Such nets are named *feed-forward nets*. On the other hand, *recurrent nets* have connections through weights extending from the outputs of a layer to the inputs of the previous or the same layer.

The learning algorithm indicates how to change the link weights W according to the OUT values of all neurons. It depends on the particular architecture, and details will be covered in the next paragraphs for each network.

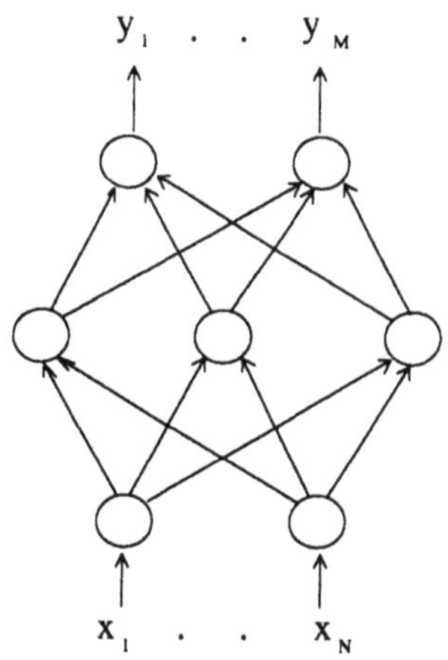

Figure 3. Perceptron architecture.

When neural nets are used as classifiers[1], the output layer consists of as many neurons as the expected classes. Each neuron codes for a class. The neuron whose output value *OUT* is maximum, specifies that the input pattern belongs to the class associated with that neuron. For single pattern recognition, the output layer is reduced to a single neuron with a threshold activation function: its output is then a binary value that tells whether a pattern has been recognized or not.

THREE LAYER PERCEPTRON

Net Architecture

The Perceptron net is likely the most widely used artificial neural network[6]. Its architecture is shown in fig.3. It is composed by three layers. The first one is an input layer, whose N nodes just distribute information through the net: they are called nodes (and not neurons) since they perform no computation. They accept N continuous values. Connections are feed-forward. The second layer is the 'hidden' layer. Each neuron in this layer is linked to all input nodes. Its non-linear transformation is a sigmoidal function. This choice accomplishes the requirements of the Kolmogorov theorem[7], that in its easiest form states that a three-layer network with a non-linear continuous non-decreasing activation function can compute any continuous function of N variables[8]. The state of activation OUT_h of each hidden neuron (fig.1) is calculated by:

$$OUT_h = f(\Sigma_n W_{nh} \cdot x_n - \theta_h) \tag{3}$$

where x_n are the input values, W_{nh} denotes the weight associated to the link from node n to neuron h, θ_h is a neuron threshold value, and f is the sigmoidal function (fig.2).

The third layer is the output layer. Each neuron in this layer is structurally identical to those in the hidden layer: the activation function is still a sigmoid. Each output neuron codes for a class, when the net is used for classification. The neuron in the output layer with the maximum state of activation is selected. The network is initialized with all weights set to random values in the range [0,1].

The Back-Propagation Algorithm

The Back-Propagation is the standard learning algorithm for the Perceptron net. Although many variants have been investigated so far, the base version is proposed here[6]. Each example consists of a *training pair*: a continuous valued input vector $\boldsymbol{x}$ (whose N components are as many as the input nodes) and a desired output vector $\boldsymbol{d}$ (whose M components are as many as the output neurons). For classification, typically, each output neuron represents a class, so a desired classification c of a pattern is specified by:

$m = 1,....M$ number of classes

$$d_m = 1 \quad \text{if } m = c, \text{ for selected class} \qquad (4)$$
$$d_m = 0 \quad \text{if } m \neq c, \text{ for all others}$$

When $\boldsymbol{x}$ is presented to the net, this provides an output vector $\boldsymbol{y}$ according to the relationship [3], applied to hidden and output neurons. The net output $\mathbf{y}$ is then compared with the desired output $\boldsymbol{d}$ and a measure of difference is calculated at each output neuron:

$$\delta_m = y_m \cdot (1 - y_m) \cdot (d_m - y_m) \qquad m = 1,.....M \qquad (5)$$

Weights towards output neurons are updated by:

$$W_{hm\,new} = W_{hm\,old} + \eta \cdot \delta_m \cdot OUT_h \qquad (6)$$

where η is a gain term, OUT_h is the output of neuron h in the hidden layer and W_{hm} is the link weight from h to m. The Back-Propagation algorithm starts at the output neurons and then works back to the first layer. The measure of difference for neurons in the hidden layer is:

$$\delta_h = OUT_h \cdot (1 - OUT_h) \cdot \Sigma_k \delta_k \cdot W_{hk} \qquad h = 1,......H \qquad (7)$$

where δ_k and W_{hk} refer to all neurons 'above' neuron h. Finally, weights from input nodes to hidden neurons are modified, using the same relationship [6] as before.

After all weights have been updated, a new training pair ($\boldsymbol{x}$, $\boldsymbol{d}$) is shown to the net. The weight change from the new training pair affects the previous values of weights, so that the whole set of training pairs must be shown several times to the net, until weights don't change significantly for all training data. This condition is called convergence. For instance, the number of different training pairs might be of the order of one hundred in an application, but the whole training set might need some ten thousands examples until convergence. It must be noted that the Kolmogorov theorem states that a solution exists, but assures nothing about the convergence time; moreover, the Back-Propagation algorithm has neither a physiological nor a mathematical foundation. On the other hand, the Perceptron with this learning algorithm has worked well in many practical applications.

Weights are not modified anymore during the test phase. For each input vector $\boldsymbol{x}$ of unknown classification, the network selects the most appropriate output neuron by assigning the maximum state of activation to it. The neuron indicates the class to which the input pattern belongs. Since the maximum valued neuron is selected, the network compensates for minor corruptions of the input patterns.

Application to Trajectory Recognition

A Perceptron network with Back-Propagation has been applied to trajectory recognition from PIV (Particle Image Velocimetry) images[9]. A typical PIV image is shown in fig.4. The flow past a bluff body in a water channel is seeded with scattering particles and then illuminated[10].

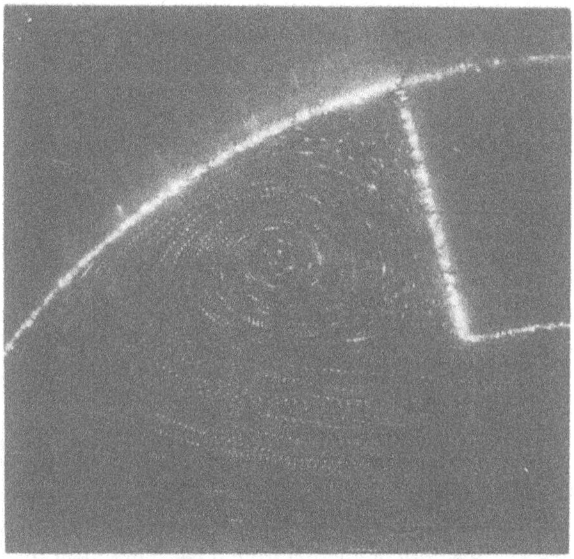

Figure 4. PIV multiexposed image.

A video camera takes a movie, frame by frame, of the flow evolution. Each frame is analyzed by a suitable software: particle images are recognized and identified by the co-ordinates of the particle barycenters, stored in an array with a time label. Subsequent frames were superimposed in fig.4. PIV technique yields the velocity field from the distances between successive locations at which a particle is found (in subsequent frames), since the time interval is known. Particles belonging to the same area in three subsequent frames were considered. If the time interval between two images is sufficiently small, trajectories may be postulated nearly linear. Moreover, acceleration may be expected to be not greater than a certain amount. These considerations were translated in the following procedure to match triplets of particle images according to likely trajectories.

Let A, B, C (A', B', C' and so on) denote three subsequent images of the same particle (fig.5). Then the lengths and the angular orientations of both vectors **AB** and **BC** can be calculated. For the optimum trajectory (refer to fig.5),

$$|AB| \approx |BC| \qquad \text{(no acceleration on trajectory)}$$
$$\alpha_{AB} \approx \alpha_{BC} \qquad \text{(linear trajectory)} \qquad (8)$$

Therefore the ratio between corresponding quantities should be 1 in the optimal case, and close to this value for those triplets that are to be identified as trajectories (how far from 1, it depends on the specific flow conditions and must be taken into account through an adequate training of the network).

The ratio between the lengths ($|BC|/|AB|$) and the ratio between the angular orientations (α_{BC}/α_{AB}) were the actual inputs to the Perceptron net. Two input nodes received these ratios and fed the hidden layer in parallel. The output layer consisted of one neuron, trained to produce an output value equal to 1 when a triplet indicated a plausible trajectory.

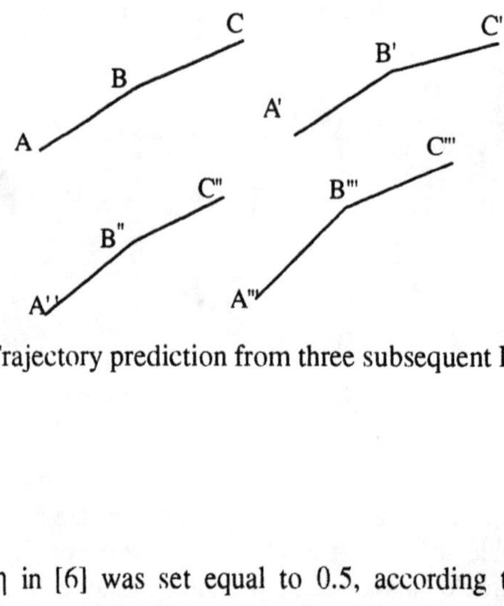

Figure 5. Trajectory prediction from three subsequent PIV frames.

The gain term η in [6] was set equal to 0.5, according to literature[2,5,6]. (The higher is η, the faster is the training, but values exceeding 0.5 often produce instability). The training set consisted of about one hundred right triplets (i.e. satisfying the constraints in [8]) and one hundred wrong triplets: a casual seek through the whole training set accessed one of these examples at a time. About 30000 examples were needed to make the net learn to discriminate trajectories from spurious triplets.

Doubts remain about a totally supervised learning: a different training set has to be generated for each specific flow condition. Yet, it seems not practical for the operator to provide each triplet in the training set with a label indicating whether that pattern has to be accepted or not. Therefore, a more complex network, with a combined supervised/unsupervised learning, is needed to reduce the amount of supervised training data. A non-supervised network will be discussed in the next paragraph.

THE SELF-ORGANIZING MAP

Net Architecture

The Self-Organizing Map[11] is perhaps the one, among neural networks, with the strongest mathematical background. The net is characterized by a big accuracy and short training times[1]. It is often used in conjunction with other nets. It consists of one layer of neurons fed by a linear array of N input nodes. These accept N continuous values $x_1 ... x_n ... x_N$ and send them in parallel to every neuron in the layer. Connections are feed-forward (fig.6). The weights associated to these links will be named W_{nm} from input node n to neuron m (m=1,...M).

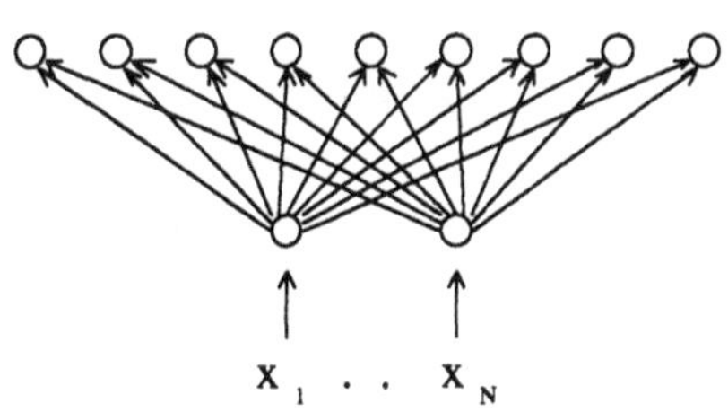

Figure 6. Self-Organizing Map architecture.

The network job may be divided in two phases. In the first phase, each neuron m produces the following output FUN_m:

$$FUN_m = \sqrt{\sum_{n=1}^{N} (W_{nm} - X_n)^2} \tag{9}$$

where $\boldsymbol{x}$ is the input vector and $\boldsymbol{W}_m$ is the weight vector associated to neuron m, that is the vector whose elements are the connective weights from all input nodes towards that neuron. It is to be noted that each neuron does not perform a simple sum of its inputs, but a more complex function FUN_m, that represents the Euclidean distance between the vectors $\boldsymbol{W}_m$ and $\boldsymbol{x}$ in $\Re^N$.

In the second phase, the minimum output value is selected:

$$FUN_{winner} = \min_m \{FUN_m\} \tag{10}$$

The activation function is a threshold, so that in the end:

$$\begin{aligned} OUT_m &= 1 \qquad \text{if h = winner} \\ OUT_m &= 0 \qquad \text{if h} \neq \text{winner} \end{aligned} \tag{11}$$

Hence, only one neuron 'fires': it is called the winner. The firing of a neuron indicates that the corresponding vector $\boldsymbol{W}_m$ has the minimum distance from the input vector $\boldsymbol{x}$ in $\Re^N$. Thus, the network job has a geometric interpretation: the net possesses M internal vectors ($\boldsymbol{W}_m$, as many as the neurons in the layer) and compares them with a co-dimensional input vector $\boldsymbol{x}$ to select which of them is nearest neighbor to $\boldsymbol{x}$.

Application to pattern classification is straightforward. When $\boldsymbol{x}$ is a feature vector (that is, a vector containing the feature values of a sampled pattern), each $\boldsymbol{W}_m$ represents an internal reference pattern, stored in the network, that is the average exemplar of a class. Similarity between two patterns is computed as distance in $\Re^N$. The net has M internal reference patterns and selects which is most similar to the input $\boldsymbol{x}$. Weight vectors in the net are actually random in the beginning: it's the learning algorithm that teaches the net to store the average exemplars of each class in the weight values.

The Vector Quantizer (VQ) Learning Algorithm

The Self-Organizing Map can be trained both in a supervised and in a non-supervised way. Although the supervised learning produces more accurate results, the non-supervised version is presented here for two main reasons. First, a supervised algorithm would need a label (indicating the right classification) for each input vector, which might be not practical sometimes. On the contrary, a non-supervised algorithm is completely automatic. One more difference lies in the expectations on the two learning algorithms. A neural network trained by a supervised algorithm (like the Perceptron net in the previous paragraph) is asked to take decisions (i.e. a teacher tells the net what's wrong and what's right on a number of examples, then the net is asked to tell right from wrong on another set of examples). A non-supervised network is not expected to take decisions, but just to change its internal structure according to the arriving inputs. A job like this might be: keep the eyes wide open in faint light, then almost close them when a sun ray suddenly strikes the eye. That is: adaptive control.

Many biological examples exist about the process of auto-organization of groups of neurons in higher animals to become selective detectors of specific features: for instance, thirty years ago Hubel and Wiesel[12] found that neurons in the region V1 for vision in the human brain, are sensitive to line orientation. Region V4 is specialized in color perception, while region V5 responds to motion[13]. In all cases, neurons are arranged in ordered structures where close neurons respond to similar modalities of sensory signals[14]. Moreover, many of these maps are not completely formed at birth, but emerge in the first months of a child's life by experience. The non-supervised VQ learning makes the artificial network behave like these human cortical maps, in that it changes its structure by experience to let neighboring neurons become sensitive to specific features. The whole algorithm consists of one learning equation:

$$W_{m\,new} = W_{m\,old} + \eta \cdot (x - W_{m\,old}) \tag{12}$$

where W_m is the weight vector associated to neuron m, x is the input vector, and η is the learning parameter, that is scalar valued and decreases monotonically with time. Typically, it starts at about 1 and then vanishes according, for instance, to a linear schedule.

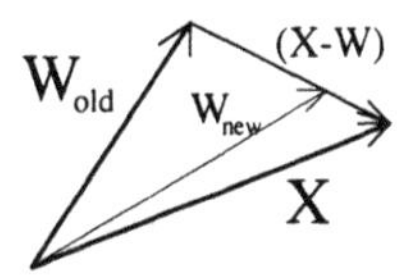

Figure 7. Geometrical representation of VQ learning rule.

The update equation has a geometrical interpretation. W_m and x are co-dimensional vectors in $\Re^N$. Fig.7 shows the two dimensional case. The learning equation pushes the weight vector towards x along the vector that represents the difference between them. Thus η represents an 'adaptation gain' (as similar parameters used in stochastic approximations) that decides how close W_m gets to x.

The updating equation is not applied to all weight vectors. First a winner neuron is selected in the layer, according to the procedure described in the previous section. Then only weight vectors towards neurons in a neighborhood of the winner are updated. The neighborhood is initially as wide as half the layer and then shrinks linearly with time (i.e. as the number of examples presented grows). After about one thousand examples only nearest neighbors are updated. In fact, a wide neighborhood corresponds to initial coarse resolution to bring the net weights to a global topological order. Narrower neighborhoods then produce an improved spatial resolution of the map, for more accurate results.

Normalization of both input and weight vectors is not necessary in principle, but recommended in practice to reduce all vectors to the same range and improve accuracy.

Kohonen and LaVigna[15] demonstrated the mathematical properties of the learning algorithm, that can be related to statistical analysis. Briefly, assume that the training set consists of a large number of vectors x. The Self-Organizing Map can extract any underlying joint-probability density function of the training set. The learning algorithm moves the weight vectors in the net to locations in $\Re^N$ that divide it in M regions of different size, in such a way that the probabilities of the input vectors x are equal in every region. The equal-probability method of dividing the input domain is one of the ways to code a probability density function[16].

Say P distinct patterns have to be recognized (P<M), and a big number of imperfect, distorted, noise-corrupted samples of these patterns is available. Each sample is a vector, whose components represent N measurements of the selected features of classification. It may be assumed that the joint-probability density function of the whole

training set is a linear combination of P gaussian-like functions of N variables. After the training phase, the weight vector distribution is more dense about the peaks of the joint-probability density function and one can easily evaluate them. These peaks are assumed to be the more plausible values of the measured features for each class, that is the exemplary patterns that were mentioned in the last section. It is to be noted that, in this case, a group of neurons is associated to one class.

Application to Particle Recognition

A PIV image contains particle images but also 'blobs' produced by the partial overlapping of two particles, especially if high seeding densities are used in the working fluid. The aim of a particle recognition system is to detect particles and output their centroid co-ordinates[10]. In a previous release, the recognition software discarded the blobs according to a maximum-area inequality, and preserved particles only, for the centroid calculations. A Self-Organizing Map improves the performance of the recognition system by accepting blobs as 'double particles' so that an algorithm may be activated to search for two centroid locations in the blob.

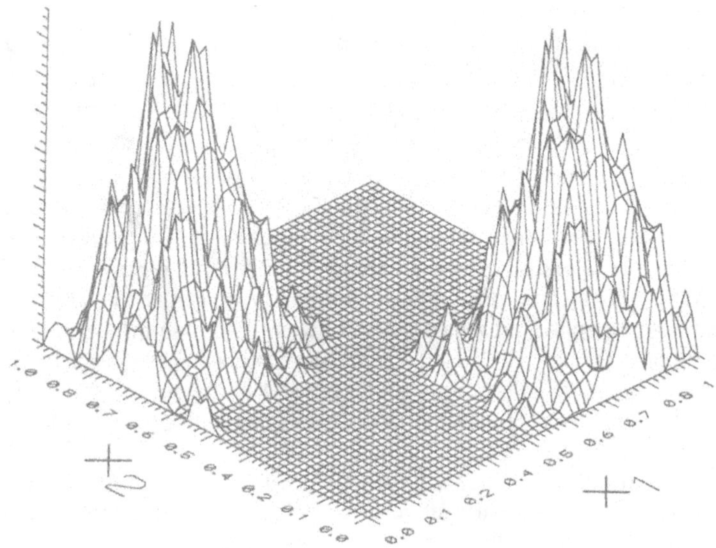

Figure 8. Joint Probability density function of training data for particle recognition.

The recognition system consists of three units. The first unit pre-processes the raw sensor images, performs a thresholding procedure and produces a boolean image. The second unit analyzes the boolean image: it collects connected sets of lighted pixels, treats them as one object and calculates some geometrical features (i.e. pixel area and a roundness parameter). Its output is objects that are to be considered as candidate particles. The third unit is a Self-Organizing Map that examines the candidate particles on the basis of their shape and size: in the learning phase, it extracts a reference particle and a reference double-particle. In the test phase, it classifies all candidate particles according to the reference pattern they are more similar to.

The actual input of the net was a two-dimensional vector whose components were the measured values of area and roundness for each candidate particle. The training set may also consist of digitally simulated images, containing half particles and half blobs. The joint-probability density function of the training set is shown in fig.8. Two peaks are clearly seen, corresponding to the most frequent pairs of the measured features. These values represent the reference particle and the reference double-particle.

The training set consisted of about one thousand examples. Two input nodes received the measured values of area and roundness of each example. 16 neurons were used in the layer, arranged in a linear array. The learning parameter η was initially set to 0.1. The net behavior is illustrated in figg.9 - 10.

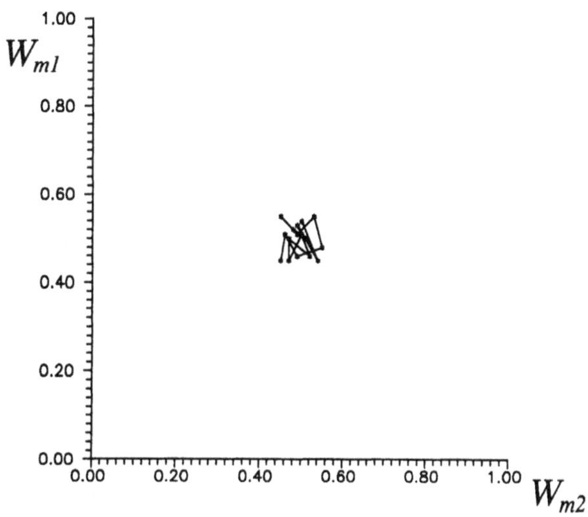

Figure 9. Network weights after random initialization.

Weight vectors are plotted as points in the square $[0,1] \in \Re^2$. Weight vectors of neighboring neurons are connected by lines. Fig.9 shows the initialization of weight vectors: they assume random values with uniform probability density within the range:

$$0.45 < W_{m1} < 0.55 \quad , \qquad 0.45 < W_{m2} < 0.55 \qquad (13)$$

As training progressed, the line connecting the weight vectors moved around in the square, to give a one-dimensional approximation of the two-dimensional probability density function of the input data. The line position after about 1000 examples is illustrated in fig.10: a comparison with fig.8 shows that weight vectors are distributed about the peaks' locations, and that they tend to approximate two circumferences whose radius are the standard deviations about the mean values (calculated for check by statistical analysis, assuming the probability density function to be the sum of two gaussians). Hence, although the reduction of the data structure was considerable (i.e. a function of two variables compressed to a line), essential information about the joint-probability density function was preserved.

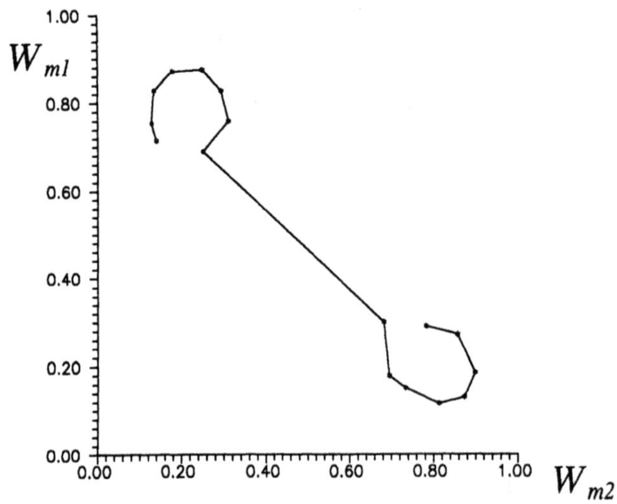

Figure 10. Network weights after 1000 examples.

Finally, a non neural algorithm, by simple evaluation of the distances between weight vectors after training, divided them into two groups: in the test phase, different classification was given to the examples that made different groups 'fire': 'particles' for the neurons of one circumference, 'double-particles' for the neurons of the second circumference. The examples classified as 'particles' were sent to the centroid estimation algorithm, while the examples classified as 'double-particles' were sent to the two-centroids estimation algorithm.

Application to Flow Segmentation

The velocity field obtained by the PIV technique, may look like fig.11. The flow segmentation process consists of decomposing the whole image into P domains of rather uniform flow features. Fig.12 shows the result of the domain decomposition: three regions are enhanced: region A and region B of rather uniform flow, and region C of recirculating flow. Flow segmentation is useful because it yields a criterion to validate trajectories: for instance, in region A, a sampled velocity is discarded if either its magnitude or its orientation is too different from the mean one in that region. The main problem for flow segmentation lies in that the number P of regions is not known a priori.

A Self-Organizing Map can perform the domain decomposition in P unknown regions by assuming, as inputs to the net, the normalized co-ordinates of the sampled velocity co-ordinates along the X and Y axis. The training set is recycled until about 1000 examples are given to the network.

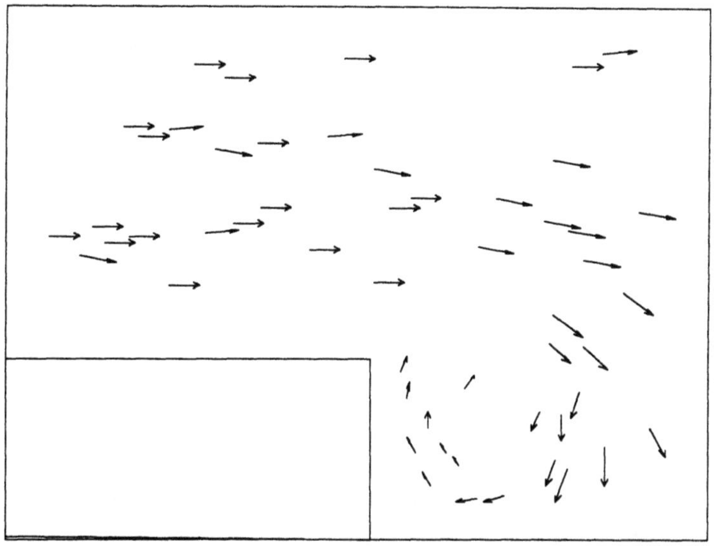

Figure 11. Velocity field close to a bluff body, after PIV elaboration.

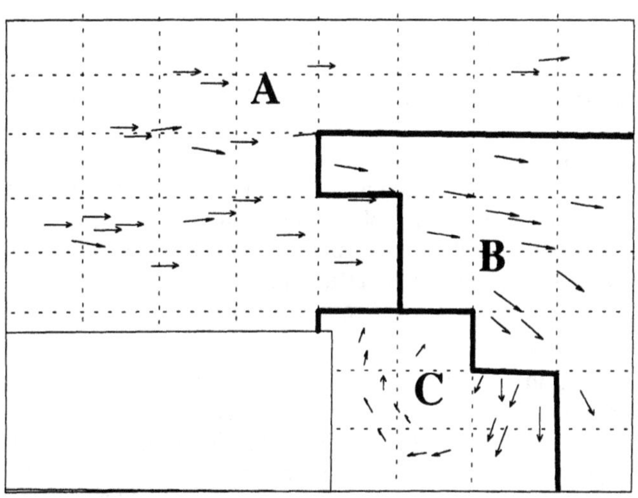

Figure 12. Domain decomposition after flow segmentation.

For the velocity field of fig.11, the joint-probability density function exhibits two peaks, corresponding to the mean values of the velocity co-ordinates in region A and B. The situation is then analogous to the problem of particle recognition.

The Self-Organizing Map finds the two peaks, and a non neural algorithm separates the groups of neurons associated to each peak, by different labels. The number of neurons and the learning parameter is the same as before. In the test phase, weights in the net are fixed, and each velocity sample is processed through the net, makes a neuron fire and give a label to the pair of spatial co-ordinates (x, y) of the location where the velocity was detected. The spatial co-ordinates x and y vary with a coarse resolution, that divides the image into 8×8 squares. Regions A and B are detected as connected domains of squares with the same label. C is detected as a region where neighboring squares have different labels.

REFERENCES FOR LOW-LEVEL ANALYSIS

An image is a continuous, analog two-dimensional signal. Once acquired by a video camera, the digitized. signal (1024×1024 pixels) itself is not practical to work with: it requires extensive storage, and an interpretation of the image in terms of this signal requires a cumbersome procedure. First, the foreground has to be distinguished from the background. Then some objects must be recognized in the image. At a higher level, relations among objects are investigated to lead to a scene description. The last may be regarded as a semantic description using abstract symbols, which is highly compact and efficiently represented on digital computers. Symbols, at every level, are encoded representations of signals. The higher the level, the lower the size of the data structures.

Artificial neural networks tolerate the approximations of a real-world environment. Hence, they seem naturally suited to receive real signals in parallel at their input, and interesting applications are to be found in the field of signal transformation. For such purposes, simulations must be carried out on a suitable parallel hardware. On the other hand, ANN are naturally hostile to serial inference and programming, and their application to symbolic data processing may reveal cumbersome related to an algorithmic approach. These attitudes collocate neural networks at the basement of the hierarchical structure described above for image processing.

Application to low-level analysis is straightforward in principle. Interesting fields might be non-linear filtering techniques (e.g. for edge detection), image coding, foreground/background thresholding, object detection in the foreground. Yet, neural nets should have a huge size to deal with raw signals in real time- or, at least, huge compared to the nets that have been presented in this paper and that, in fact, work at a more symbolic level. These have few tens of neurons, while real-time processing on the image bitmap would require, for instance, 10^6 nodes in the input layer (for a 1024×1024 image) that distribute signal information in parallel towards further layers. Only a massively parallel hardware can make these applications feasible.

Research is in progress where the hardware is available: basically, Transputer-based or SIMD machines. However, concepts are not far from the examples provided in this paper. For example, Ahalt et al.[17] give results of a thresholding procedure using the FSCL (Frequency Sensitive Competitive Learning), derived from the Kohonen algorithm to train the Self-Organizing Map. Accuracy is the same as the optimal vector quantization algorithm, the Linde-Buzo-Gray (LBG) algorithm, and better than the popular Pulse Code Modulation methods.

Another trend is represented by VLSI (Very-Large Scale Integration) circuitry. This is neural based analog circuitry. The ultimate product would be a chip with parallel layout, used as a pre-processor for image analysis. As a reference, Kobayashi et al.[18] proposed a smoothing-contrast-enhancement filter, using a neural network that requires connections for each neuron with every second order nearest neighbor (in addition to the immediate neighbor connections). Such a non-linear filter produces edge images out of a picture, so that objects can be detected by an edge-tracking procedure, that saves computational cost with respect to scanning the whole objects. Wiring complexity is critical to implement vision chips: in fact, only neighbor neurons can actually be connected. This has lead to an interest towards the networks that require connections among just neighboring neurons, i.e. cellular networks, first explored by Roska[19].

Attention must be finally payed to Optical-Hardware Implementation[3]. Although it is still far from commercial use, it will hopefully overcome the limit of VLSI circuitry, and allow the hardware implementation of three-dimensional fully connected neural networks. In view of these future options, research is carried on by means of digital simulations on workstations. These can validly test the actual capability of the neural approach to real world problems.

CONCLUSIONS

Real-time image analysis needs a parallel hardware. Most algorithms that were developed on digital computers consist of a set of sequential rules and, even if they may be parallelized, cannot suit such a hardware as well as artificial neural networks do. Inspired by biological neural networks, they are intrinsically parallel data processors.

Artificial neural networks can be used for pattern identification, using their well-known capabilities of learning from examples and performing classification tasks. In a training phase, they learn a set of patterns from examples (i.e. distorted versions of the patterns). In a test phase, they can match each new example to the most similar pattern.

Two popular artificial neural networks were presented in this paper: the Perceptron net (with the Back-Propagation learning algorithm) and the Self-Organizing Map (performing the Vector Quantization learning). Their features are complementary, in a sense, and many other neural networks are intermediate between these two. The Perceptron uses a supervised learning. Its training time is long (typically, it may requires some ten thousands examples to identify a pattern), but the net is extremely fast in test. The Self-Organizing Map uses a non supervised learning, its training is fast (typically, some thousand examples), but the net is slower than the Perceptron in the test phase.

Applications of both networks to flow image analysis were shown. In particular, the Self-Organizing Map was applied to the recognition of particles from PIV images. The Perceptron net was used to identify the trajectories traced by these particles. Finally, one more Self-Organizing Map was used for the image decomposition in domains of locally homogeneous flow: this process can support an easier procedure of trajectory tracking. The neural networks were tested by means of digital simulations on a workstation: their accuracy resulted to be similar to previous methods. Hence, neural networks have proved to be alternative to more traditional image analysis techniques.

Further research is in progress about the applications of neural networks to low-level image analysis. Signal pre-processing (e.g. filtering, thresholding, edge enhancement) needs large size networks (i.e. composed by thousands of neurons). Thus, attention must be payed to an actual parallel hardware implementation, using parallel computing machines, VLSI vision chips, or Optical hardware.

ACKNOWLEDGMENTS

The EEC sponsored all projects within PATRANS program. We wish to thank C.Del Gracco, A.Paglialunga, M.Terlizzi for the kind information and collaboration about the Perceptron network. We are also grateful to F.De Gregorio and G.Querzoli for the information about the image pre-processing procedures.

REFERENCES

1. R.P. Lippman, "Pattern classification using neural networks", IEEE Communications Magazine, November 1989
2. S.K.Rogers, M.Kabrisky, "An introduction to biological and artificial neural networks for pattern recognition", SPIE Press, 1991
3. B.Kosko, "Neural networks for signal processing", Prentice Hall, 1992
4. R.P.Lippman, "An introduction to computing with neural nets", IEEE ASSP Magazine, April 1987
5. R.Hecht-Nielsen, "Neurocomputing", Addison-Wesley Publishing Company, 1990
6. P.D.Wassermann, "Neural Computing", Van Nostrand-Reinhold, 1989
7. R.Hecht-Nielsen, "Kolmogorov's mapping neural network existence theorem", Proc.Int.Conf. on Neural Networks, 3, pp.11-13, IEEE Press, 1987
8. B.Moore, T.Poggio, "Representation properties of multilayer feedforward networks", presented at the 1988 Annual meeting of the Int.Neur.Network Soc., September 1988
9. C.Del Gracco, A.Paglialunga, M.Terlizzi, Internal Communications.
10. A.Cenedese, G.P.Romano, "PIV: a new technique in flow velocity measurements", Excerpta of the Italian contribution to the field of hydraulic engineering, 5, 1990
11. T.Kohonen, "The Self-Organizing Map", Proc.IEEE, September 1990
12. P.Churhland, T.Seinowski, "The computational brain", MIT Press/Brandford Books, 1992
13. S.Zeki, "A vision of the brain", Blackwell Scientific Publications, (in print).
14. E.I.Knudsen, S. du Lac, S.D.Esterly, "Computational maps in the brain", Ann.Rev.Neurosci.,10, pp.41-65, 1987
15. A.LaVigna, "Nonparametric classification using learning vector quantization", Ph.D.Thesis, University of Maryland, 1989
16. J.Bendat, A.Piersol, "Random data: analysis and measurement procedures", Wiley & Sons, 1971
17. S.Ahalt, P.Chen, A.Krishnamurthy, " Performance analysis of two image vector quantization techniques", Proc.IJNN, July 1989
18. H.Kobayashi, T.Matsumoto, T.Yagi, T.Shimmi, "Image processing regularization filters on layered architecture", Neural Networks, vol.6, n.3, Pergamon Press, 1993
19. T.Roska, "Proceedings of the 1990 IEEE International Workshop on Cellular neural networks and their application", CNNA, 1990

TOMOGRAPHIC METHODS IN FLOW DIAGNOSTICS

Arthur J. Decker

National Aeronautics and Space Administration
Lewis Research Center
Cleveland, Ohio 44135

INTRODUCTION

Flow visualization often displays the integral transforms of flow properties rather than the properties themselves. Interferometry is an example. A double-exposure hologram of two index-of-refraction fields $n_0(x, y, z)$ and $n(x, y, z)$, recorded, for example, from light propagating parallel to the z-axis, yields, in the refractionless limit, interference phase measurements given by

$$\Delta\phi(x,y) = \frac{2\pi}{\lambda}\int (n(x,y,z) - n_0(x,y,z))dz \tag{1}$$

One measurement of the interference phase given by (1), at one point x, y, represents one sample point of the so-called x-ray transform of $n(x, y, z) - n_0(x, y, z)$. The calculation of $n(x, y, z) - n_0(x, y, z)$ from such sample points is an example of computed tomography. High speed detector arrays, powerful small computers, fast parallel processors, and a body of research conducted in the 1970's and particularly in the 1980's make tomography a viable tool to be considered for flow diagnostics.

There are at least three integral transforms of interest to tomographers: the Radon transform, the fan-beam or cone-beam transform, and the x-ray transform. We shall consider only the Radon and x-ray transforms and shall note in particular how the two transforms differ.

The Radon transform, first described by Johann Radon in 1917 (ref. 1), is defined for any number of dimensions n in Euclidean space $\mathbb{R}^n$. The Radon transform R $f(\mathbf{x})$ of a function $f(\mathbf{x})$ where $\mathbf{x} = (x_1, x_2, ..., x_n) \in \mathbb{R}^n$ is the set of all integrals of $f(\mathbf{x})$ over all the hyperplanes of $\mathbb{R}^n$.

The x-ray transform, by contrast, is always a straight-line integral. It is also defined for any number of dimensions n. Consider all planes through the origin. A plane through the origin is defined by the coordinates θ of its normal. Consider a particular plane through the origin and any straight line perpendicular to that plane. The x-ray transform P $f(\mathbf{x})$ of a function f(**x**) is the set of all integrals along all straight lines perpendicular to all planes through the origin. In fact, the definition of the x-ray transform is unaffected by the location of the plane perpendicular to θ. One can imagine a plane in three dimensions being translated along θ and perpendicular to θ until it coincides with an interferogram of a fluid. The assumption is that the entire fluid projects onto the interferogram in this manner. The continuum of interference phase measurements (except a possible reference offset and normalization) is the x-ray transform at θ of the index of refraction of the fluid. Defining the complete x-ray transform requires recording and measuring interferograms for the continuum of directions on the unit hemisphere.

The x-ray and Radon transforms are the same in two dimensions. Medical tomography using x-rays is done typically by stacking two-dimensional slices recorded through a patient. It's common and correct to refer to the linear projections recorded for a slice as samples of the Radon transform. The terminology may mislead flow diagnostics professionals who may need to use three-dimensional tomography at times. Figure 1 summarizes the difference between the Radon and x-ray transforms in $\mathbb{R}^3$. An integral of a function over the plane shown in the figure is a sample point of the Radon transform, and an integral along the line perpendicular to the plane is a sample point of the x-ray transform.

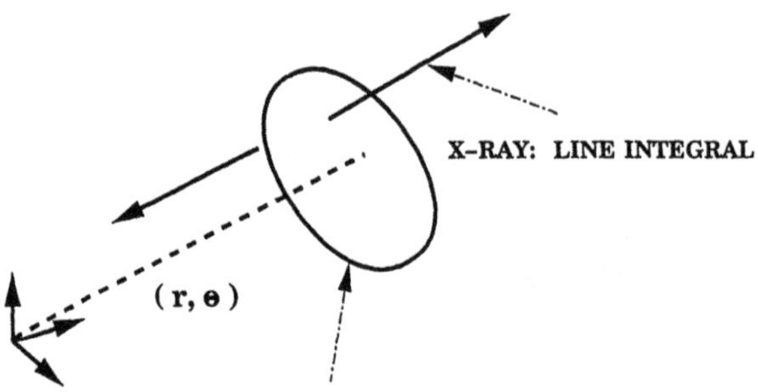

Fig. 1. The three-dimensional Radon and x-ray transforms. Radon transform is integral over plane and x-ray transform is integral along line.

Computed tomography consists of estimating a function or property from measured or theoretical samples of the integral transforms. (Theoretical objects and their transforms are called phantoms.) Figure 2 shows a division of tomography into three major efforts: fast measurements of the transforms; reliable, accurate algorithms or processors for inverting the transforms; and utilization of the computations. In medical tomography, each effort has its own dedicated experts. Tomography is a legitimate branch of applied mathematics and has its own full-time practitioners. The methods employed depend strongly on the measurement technology and the end uses. Tomography is underdeveloped for flow diagnostics. Because this is only a brief overview, it's important to mention some good references.

Perhaps the best general reference is F. Natterer's **The Mathematics of Computerized Tomography** (ref. 2). This book contains excellent discussions and examples of the effects of errors due to sampling, ill-posedness, and missing data. The book also discusses the principal algorithms for inverting the transforms. This reference was written by a mathematician for mathematicians, and the terminology may be difficult for other readers.

S.R. Deans' **The Radon Transform and Some of its Applications** (ref. 3) discusses the principal algorithms in notation that are probably more familiar to engineers. The book has an extensive list of references. Unfortunately, there are no practical discussions of the effects of errors, and such discussions are indispensable for the application of computed

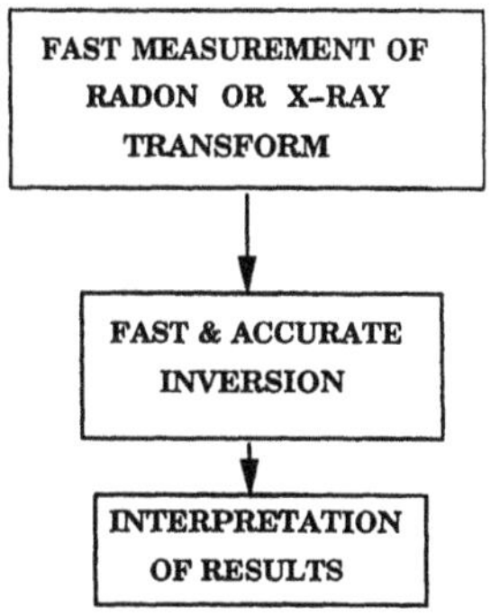

Fig. 2. Organization of tomography into three major subfields.

tomography to flow diagnostics, especially for the incomplete data cases. Dean's and Natterer's books are excellent as a pair.

The book, **Solution of Ill-posed Problems**, by A.N. Tikhonov and V.Y. Arsenin (ref. 4) is not a tomography book, but it does explain the origin and control of ill-posedness (the amplification of errors depending on their spectral content).

The computational methods that accompany the algorithms of tomography are discussed practically and understandably by W.H. Press, et al. in **Numerical Recipes** (ref. 5).

Tomography has its own journals, although they don't normally discuss flow diagnostics. The **Journal of Computer Assisted Tomography** (ref. 6) discusses medical applications; **Inverse Problems** (ref. 7) contains papers by applied mathematicians and physicists.

The remainder of this overview will discuss: measurements of the integral transforms for flow diagnostics; algorithms for inverting the x-ray transform; three-dimensional tomography using incomplete data from interferograms as an example; computers and artificial neural networks for tomography; and some on-going research.

SAMPLING THE INTEGRAL TRANSFORMS FOR FLOW DIAGNOSTICS

Useful tomography requires the parallel, or at least fast, measurement of thousands of channels of data. Some optical technologies such as phase shifting interferometry with CCD cameras meet this requirement. The next section discusses sampling the Radon transformation in three dimensions.

Sampling the Radon Transformation in Three Dimensions

It is mathematically expedient to assume that the fluid property is non-zero only within a sphere. The theorems of tomography assume that a function has this so-called compact support. It's convenient to normalize coordinates so that the fluid is confined to within a spherical region of unit radius called the unit ball Ω^3, and the property itself is often normalized. These stipulations are kept in mind, if not stated explicitly, in the following discussions.

Hanson, et al. (ref. 8) have discussed at least the concept of measuring various fluid properties from laser-induced fluorescence produced by sheet or planar illumination. Consider the case where effects other than density can be ignored. Imagine firing a pulsed laser sheet. The total light emitted in fluorescence is proportional to the number of fluorescing molecules $N(\theta, r)$ in a thin sheet located about the plane defined by radial coordinate r and orientation θ. We could imagine surrounding the region with a spherical calorimeter with a thin slice through which the laser sheet passes unincumbered. The total energy measured is essentially one sample of the three-dimensional Radon transform of the density $\rho(\mathbf{x})$. In a sense, the number of fluorescing molecules is a sample of the Radon transform of the density in the limit where the sheet has zero thickness. In mathematical terms the energy measured is given by

$$E(\theta, r) = \tau N(\theta, r) = \tau t \int_{x \cdot \theta = r} \rho(x) dx \tag{2}$$

where t is the thickness of the sheet and τ is a proportionality factor.

This concept is extended as shown in figure 3. The laser beam is split into multiple sheets. The sheets are separated by large enough distances that times of flight within sheets can be neglected in comparison with the times of flight between sheets. A fast point detector then reads out the samples of the Radon transformation as a string of pulses. Each value of θ must be sampled in the same manner, for example, by routing subsequent laser pulses into different paths, where each path has a point detector.

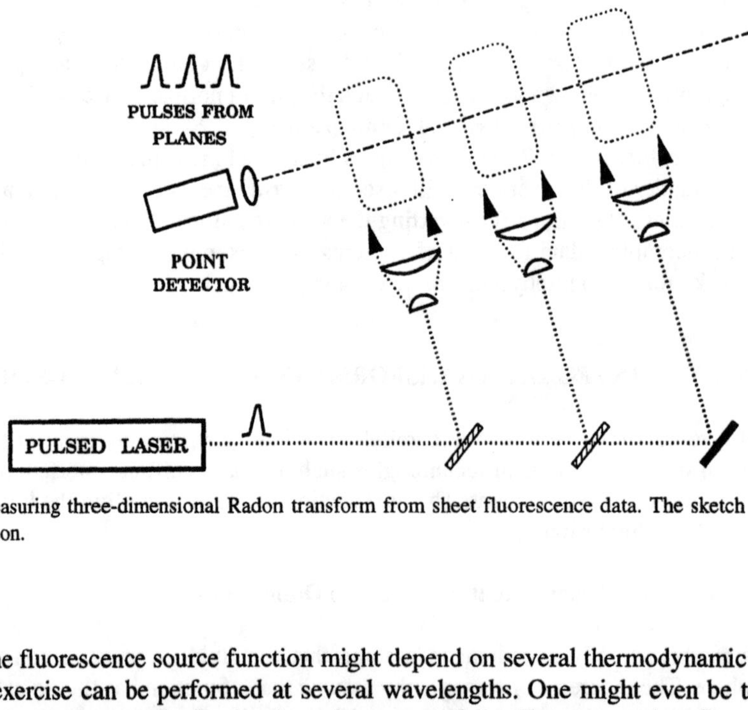

Fig. 3. Measuring three-dimensional Radon transform from sheet fluorescence data. The sketch shows only one direction.

The fluorescence source function might depend on several thermodynamic variables, but the exercise can be performed at several wavelengths. One might even be tempted to extend the exercise to velocity, but here the concept fails. The propagation direction of the beam and component of velocity measured change for each sample direction. The tomography literature generally assumes a scalar field; vector-field tomography is a topic for research.

We will say very little else about three-dimensional Radon transformations. Should you have an appropriate application or practical measurement method, Natterer (ref. 2) provides an adequate discussion. The Radon transform is probably mathematically simpler than the x-ray transform and exhibits a certain locality. This locality is shown by the inverse of (2), which is given by the equation

$$\rho(x) = -\frac{1}{8\pi^2 t}\int_{S^2} \frac{\partial^2}{\partial p^2} N(\theta, p)|_{p = x\cdot\theta = r} d\theta \qquad (3)$$

The density at **x** is recovered effectively from (all) planes in the neighborhood of **x**. The summation over θ, after the differentiation, is an example of back projection.

The Radon transform is not as robust as the x-ray transform for work in three dimensions, particularly if data are limited. The Radon transform involves only three degrees of freedom (two angular and one positional), while the x-ray transform offers four degrees of freedom (two angular and two positional). The Radon transform is more sensitive to data errors (round-off or measurement). For relative error ε, the Radon transform deteriorates as $\varepsilon^{1/3}$, and the x-ray transform deteriorates as $\varepsilon^{1/2}$ (ref. 9).

Sampling the X-ray Transform in Two and Three Dimensions

Visualization data is recorded effectively as an x-ray transform. Figure 4 shows the fluid depicted in figure 3, but with the entire volume illuminated. An afocal imaging system produces approximately a parallel projection of the illuminated volume. An array detector samples the x-ray transform for the direction shown. Array detectors would be required for each direction. Even velocity could be measured in principle for this concept; since the illumination beam has a fixed direction.

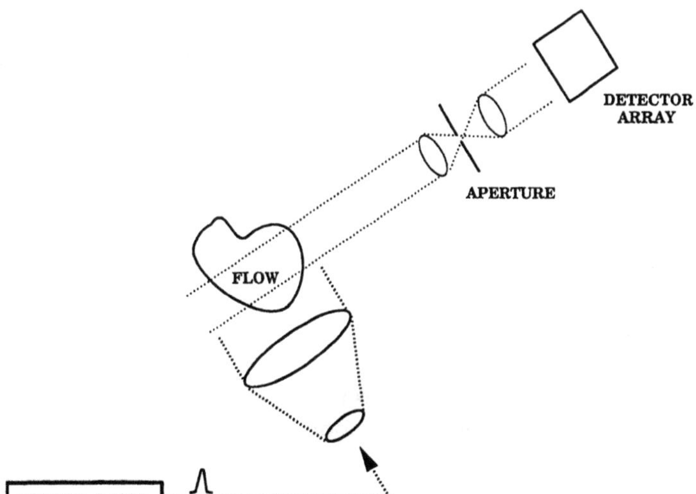

Fig. 4. Measuring three-dimensional x-ray transform from fluorescence data. The sketch shows one direction only.

Diffuse-illumination holographic interferometry is the classical method for generating the x-ray transform in three dimensions. Phase shifting or heterodyne detection can be used for efficient interrogation of the hologram. Later we discuss an example using heterodyne detection. The interference phase, measured in interferometry of fluids in the refractionless limit, is proportional to the x-ray transform of index of refraction as mentioned in the introduction.

Deflectometry (schlieren, moire, Hartmann screens) projects components of the gradient of the index of refraction of a fluid. Deflectometry offers high speed electronic array detection. However, the vector compositions of the components measured vary with direction θ, thereby preventing deflectometry from being used with standard tomographic methods.

The x-ray transform of a fluid property f(**x**) is given, in general, by the equation

$$P_\theta f(x) = \int_{-\infty}^{\infty} f(x + t\theta)dt \tag{4}$$

where θ is the direction in two or three (or even n) dimensions, and **x** here is the coordinate of a point in the projection plane. Naive attempts to invert (4), even from good measurements, may lead to poor results unless correct algorithms are selected and applied correctly.

ALGORITHMS FOR INVERTING THE X-RAY TRANSFORM

S.H. Izen has studied the x-ray transform extensively for n dimensions for both full and missing data (refs. 10 to 13). The work is discussed in a form suitable for applications to flow diagnostics by A.J. Decker and S.H. Izen (ref. 14). Prior to this work, most generalizations of computed tomography to three or more dimensions were applied to the Radon transform. The methods for inverting the x-ray and Radon transforms in two dimensions are, of course, identical.

The two major sources of poor performance are undersampling or discretization and ill-posedness. Systematic errors such as those caused by misregistration of projections also can be serious. The consequences of errors appear as incorrect reconstructions called artifacts (artefacts). Incomplete data due to view limitations or obstructing objects worsens these effects. (By complete data, we mean a uniform sampling of the entire x-ray transform according to the Nyquist criterion.) Natterer (ref. 2) shows prints of the effects of sampling, ill-posedness, and incomplete data. The origins of these effects can be understood in terms of the projection slice theorem.

Projection Slice Theorem

The projection slice theorem states that the n dimensional Fourier transform (or inverse transform) of a property in any plane through the origin of the transform space equals the n − 1 dimensional Fourier transform (or inverse transform) of the x-ray transform in a parallel plane. The inverse Fourier transform of f(**x**) is defined as

$$F_n^{-1} f(x) = (2\pi)^{-n} \int_{R^n} \exp\,(ix \cdot k) f(x)dx \tag{5}$$

Inverse Fourier transforms simplify the notation for the in-depth example to be presented later.

The projection slice theorem in three dimensions is then given by the equation

$$(2\pi) F_3^{-1} f(\eta) = \left[F_2^{-1} P_\theta f\right](\eta) \tag{6}$$

where η represents Fourier transform coordinates on planes perpendicular to θ. The left member of (6) refers to a slice through the three-dimensional object, and the right member refers to the x-ray transform plane.

The direct use of (6) (or its two-dimensional version) in tomography is called a Fourier reconstruction technique. The projection slice theorem produces some immediate insights.

The first insight is that the sampling requirements for the object and the x-ray transform are the same. If **k** for the property distribution is confined essentially to a sphere of radius b , then the distance h between samples of the x-ray transform should satisfy

$$h \leq \frac{\pi}{b} \tag{7}$$

(A function with compact support is only essentially band limited.)

The second insight is that missing data impose no fundamental restrictions on the inversion of the x-ray transform. The Fourier transform of a function with compact support is analytic. The entire transform can be recovered by (error-free) analytic continuation starting from its (error-free) representation in any region. Restrictions are, in fact, imposed by the computational method and by ill-posedness. Analytic continuation is ill-posed as are all procedures for inverting the x-ray transform.

Ill-Posedness and Regularization

Ill-posedness (refs. 4 and 14) can be thought of as the nonuniform response of a computational procedure to the modal content (in any representation) of the errors. Errors appear in measurements, machine precision, or round-off. Ordinary differentiation is ill-posed, for example. If the error contains modes sin(ky) and cos(ky), then the derivative contains corresponding errors that vary as kcos(ky) and –ksin(ky). Anyone who has attempted to differentiate noisy data has seen the effect. The second derivative in (3) for inverting the Radon transform clearly imparts larger effects to errors in the higher frequency components.

The general filtering procedures used to control ill-posedness are called regularization. Regularization is defined in terms of a regularization parameter $\gamma(\varepsilon)$ for error ε, where $\gamma \Rightarrow 0$ as $\varepsilon \Rightarrow 0$. The regularization parameter's form and application depend on the computational procedure for inverting the x-ray transform. Three procedures are discussed briefly: the algebraic reconstruction technique (ART), convolution back projection, and series methods (considered examples of the Fourier reconstruction technique).

Algebraic Reconstruction Technique (ART)

An algebraic reconstruction technique (ART) can be defined to be any iterative technique for inverting the Radon or x-ray transforms (ref. 15). The usual procedure is to discretize the property distribution. For the two-dimensional case, the property value can be assumed to be constant in each finite object element. The object elements are called pixels since the results of the reconstruction are often displayed electronically. Each sample of the x-ray transform then involves only a few pixels (about $M^{1/2}$ where M is the number of pixels). The contribution of pixel m to the j value g_j of the x-ray transform is $a_{jm}f_m$ where the coefficient a_{jm} can be equated to the length of the projection line segment that intersects the pixel (fig. 5). A sample point of the x-ray transform is then given by

$$g_j = \sum_{m=1}^{M} a_{jm} f_m \tag{8}$$

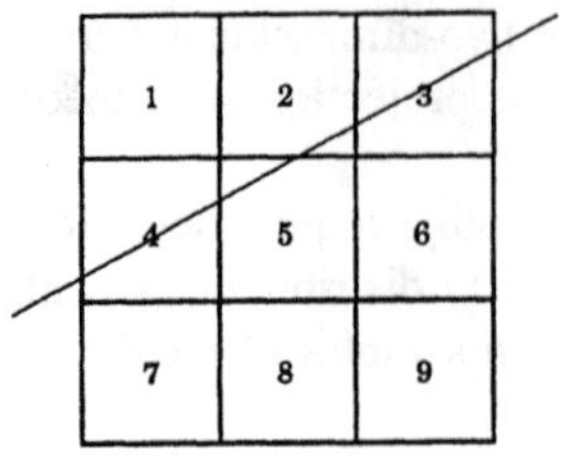

$$a_{j2}f_2 + a_{j3}f_3 + a_{j4}f_4 + a_{j5}f_5 = g_j$$

Fig. 5. Discretization of two-dimensional slice used in one verison of the algebraic reconstruction technique (art). Each pixel (object element) is weighted by relative length of light ray passing through it.

where g_j is measured in the manner described in section 2.2, and the a_{jm} are calculated from geometry. The various sample points of the x-ray transform are then used to build a set of linear equations. Most of the coefficients vanish, thereby yielding a sparse matrix equation that is well suited, in principle, for solution by iterations.

One procedure (ref. 16) follows. Start with a guess F_0 of the M pixel averaged values of the property f. Let a_j be the vector of coefficients for sample point j of the x-ray transform. Execute in order the following N computations for the N measured sample points of the x-ray transform:

$$F_j = F_{j-1} + \frac{\omega}{|a_j|^2}\left(g_j - a_j^T F_{j-1}\right) a_j \tag{9}$$

The parameter ω is called a relaxation parameter and is selected in the interval (0,2). Each pass through the N computations is called one iteration.

Generally, there is an optimum number of iterations. The reconstruction appears to improve until that number is reached, and then artifacts increase with more iterations and destroy the reconstruction. This phenomenon is the manifestation of ill-posedness in ART. The reciprocal of the optimum number of iterations can be thought of as the regularization parameter ($\gamma(\varepsilon) \sim 1/N_{OPTIMUM}(\varepsilon)$).

Although ART does not seem to be regarded as a good technique, that judgement is often made while comparing ART with back projection. C.M. Vest discusses the use of the above approach to ART with holographic interferometry (ref. 17).

Convolution Backprojection

Convolution backprojection is the signature method for performing medical x-ray computed tomography. It is very difficult to meet the requirements of the method in flow diagnostics. Backprojection, by definition, requires projections from all viewing directions. The technique is applied to two-dimensional slices, and the slices are stacked for complete information. Each slice may require a large number of sample points (50,000 to 200,000). Generally, there should be more views than sample points per view. R.M. Lewitt presents a good overview of the technique for the x-ray and fan beam transforms (ref. 18).

A regularized integral form of the algorithm is summarized by the following equations. The coordinates within a slice are denoted by (x,y). The first equation is the backprojection operation itself.

$$f(x,y) = \int_0^{\pi} p^{c}(x \cos \theta + y \sin \theta, \theta)\, d\theta \tag{10}$$

Here, p^c is the projection after convolution given by the equation

$$p^{c}(\hat{s}, \theta) = \int_{-1}^{1} p(s, \theta)\, q(\hat{s} - s)\, ds \tag{11}$$

where $p(s, \theta)$ is a sample point of the x-ray transform at distance s from the origin on a line at angle θ with respect to the x-axis. The function f(x,y) is assumed to have compact support on Ω^2.

The convolving function, including a band limiting window, is given by the equation

$$q(s) = (2\pi)^{-2} \int_{-\pi/\Delta s}^{\pi/\Delta s} |k|\, W(k) \exp(iks)\, dk \tag{12}$$

Equations (10) to (12) must be discretized and an appropriate window W(**k**) supplied for band limiting and regularization (ref. 18).

Two-dimensional backprojection techniques are preferred, if there are no limits on views. One application to flow diagnostics is for external rotating flows as might be associated with a rotating helicopter blade. Multiple holograms can be recorded, for example, at a fixed station as the flow rotates by.

Series Method

A series method is simply an application of the Fourier reconstruction technique to a series representation of a property. Series methods have very useful properties for flow diagnostics. In effect, projection data, model data, and other measurements of a property can be combined easily with a series method.

The simplest approach is to represent the property distribution in terms of orthonormal basis functions $V_{\mathbf{n}}(\mathbf{x})$ as in

$$f(\mathbf{x}) = \sum_{\mathbf{n}} A_{\mathbf{n}} V_{\mathbf{n}}(\mathbf{x}) \tag{13}$$

where **n** represents the indices of the basis functions. The number of indices equals the number of dimensions.

Equation (13) is transformed a term at a time to yield

$$F^{-1} f(\eta) = \sum_n A_n F^{-1} V_n(\eta) \qquad (14)$$

where η represents the coordinates in inverse Fourier space. Each $F^{-1} V_n(\eta)$ can be evaluated analytically. But, by the projection slice theorem (6), each Fourier transformed x-ray sample point has its own Fourier coordinates η_j, and can be equated, except factors of 2π to (14) at the corresponding coordinate.

The computational procedure is to terminate (14) and write an equation in the remaining coefficients A_n for each transformed sample point. The orthonormal functions, in general, will oscillate, and the series is terminated for functions that oscillate too rapidly for the sampling rate (speaking approximately). There are usually many more sample points than coefficients so that a large, overdetermined set of equations is obtained. This set is expressed by the matrix equation

$$Q\,\boldsymbol{a} = \boldsymbol{b} \qquad (15)$$

where **a** is the vector of unknown coefficients, **b** is the generally much larger vector of Fourier transformed sample points of the x-ray transform, and Q is a matrix computed from the transformed orthonormal functions.

A second regularization step (in addition to termination of the series) is required in this case. Computing the generalized inverse of Q is an ill-posed problem that may be severely ill-posed for missing data. Singular value decomposition (SVD) is used to accomplish both inversion and regularization. For SVD, Q is written as

$$Q = UWV^T \qquad (16)$$

where U and V are column orthogonal matrices, and W is diagonal, containing the so-called singular values. In effect, computing the SVD solves the problem, since **a** can be estimated from $V\, W^{-1}\, U^T\, \mathbf{b}$. In fact, doing so may lead to significant errors, if the data contains measurement errors. The so-called condition of Q is measured by the ratio of the largest singular value to the smallest singular value. A large ratio means that Q is ill-conditioned for computing a generalized inverse. Regularization consists of zeroing the singular values which differ by a large factor from the maximum singular value. The effect of zeroing a singular value is to remove the corresponding column vectors in U and V from the calculation. The penalty is the loss of information associated with those components. The benefit is the removal of an error amplification factor proportional to the reciprocal of the small singular value. The reciprocal of the retained singular value ratio should be approximately equal to the error $\boldsymbol{\varepsilon}$.

The series method is used the same way for complete and incomplete data. Equation (13) can also be used to incorporate presumed known values of a property. These known values are called constraints. Constraints might be known from other measurements or they might be inserted from a model. Each known value of f(**x**) produces from (13) one additional equation in the coefficients. These equations are simply appended to the set used to form (15).

The series method does have some significant disadvantages. The orthonormal functions are nonlocal, and errors in computing a coefficient appear throughout the volume. There is a so-called Gibbs phenomenon, which makes it hard to recover edges such as those occurring at shock waves. The use of local functions called wavelets are being investigated to alleviate these problems. A specific example of the use of the series method is presented next.

The series method was used to evaluate the efficacy of computed tomography for the limited angle problem. The limited angle problem, as shown in figure 6, refers to a cone of viewing directions with the cone angle less than 90 degrees. The objective of this project was to evaluate computed tomography for very small cone angles of about 10 degrees. This angle limitation is representative of internal flow diagnostics (the flow diagnostics within the components of a jet engine). A single diffuse-illumination hologram will record projections within a cone angle of about 10 degrees (20 degrees between extreme views).

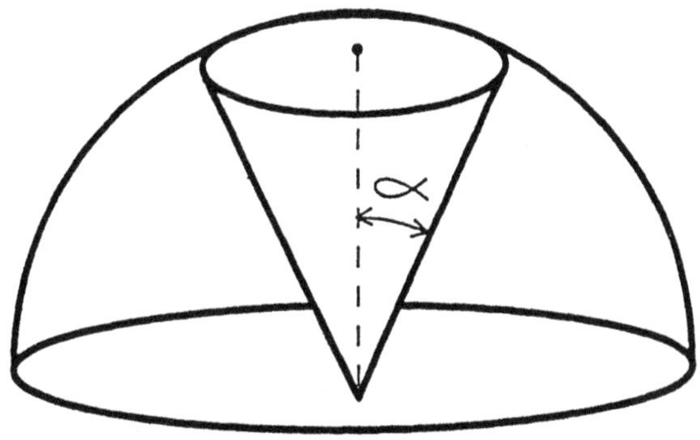

Fig. 6. Limited angle problem in three-dimensional computed tomography ($\alpha < 90$ degrees). Complete data requires sampling directions on entire unit hemisphere ($\alpha = 90$ degrees).

The evolution of this project is discussed in several references (refs. 10 to 14). The detail in those references is beyond the scope of this overview. The objective here is to outline how tomography can be analyzed for a potential application to flow diagnostics.

One decision was to use three-dimensional computed tomography to combat the extreme ill-posedness. Two-dimensional tomography ignores slice-to-slice projections and thereby discards data that can be measured naturally in flow diagnostics. Nevertheless, two-dimensional tomography is computationally much easier. Convolution backprojection produces the best results for full data.

Another decision was to use a standard test object or phantom. The phantom was a ball having a constant index of refraction and unit diameter. The ball was contained in the unit-radius space. The region outside the ball, of course, had a different index of refraction. This phantom represents a spherical shock wave. It is important to note that there were no a priori assumptions about the symmetry of the phantom. Convex polyhedra were also used as phantoms.

A final decision was to assume the use of infinite fringe interferograms measured to an accuracy of 1/50 fringe. This conservative choice probably means that singular values that differ from the maximum by more than a factor of 50 should be rejected.

Interferograms were computed for the phantom to serve as inputs to the tomography routine. The original study used 29 interferograms, where 28 interferograms were arranged in nearest neighbor fashion about a central interferogram. Each interferogram contained 32×32 samples of the computed interference phase. Hence, there were 29,696 computed sample points.

The orthonormal functions $V_n(\mathbf{x})$ of (13) were products of spherical harmonics and Jacobi polynomials, and the inverse Fourier transformed functions $F^{-1} V_n(\eta)$ of (14) were products of Bessel functions of the first kind and spherical harmonics. There is a software package, created during this project, that performs the singular value decomposition for any cone angle, calculates phantoms with and without noise, handles data from real interferograms, and can add constraints (ref. 19).

The termination of the series in (13) depends on the overall maximum degree of the polynomials to be retained. The polynomials are composed of Jacobi and associated Legendre contributions. The number of coefficients is given by

$$N(S) = \frac{(S+1)(S+2)(S+3)}{6} \tag{17}$$

where S is the maximum degree of the polynomials retained. The referenced study (ref. 14) was performed for S = 12 and N = 455. Here, 29,696 equations are solved for 455 coefficients. Required computer resources will be discussed briefly in the next section.

The figures 7 to 9 show the results for a lesser calculation, which was performed for tutorial purposes. The calculation was performed for 29 16×16 samples of the x-ray transform for a total of 7424 samples. The series was terminated at degree S = 8 for N = 165 coefficients.

Figure 7 shows the relative singular-value spectra (diagonal elements of W) for three geometries. Remember that calculating the regularized generalized inverse of Q in (15) depends on the cone angle and accuracy, but not on the phantom. The geometries consist of a 10-degree viewing cone, a pair of 10-degree viewing cones at right angles, and a 90-degree viewing cone. Ill-posedness is seen to range from mild for the 90-degree (full view) case to extreme for the 10-degree case.

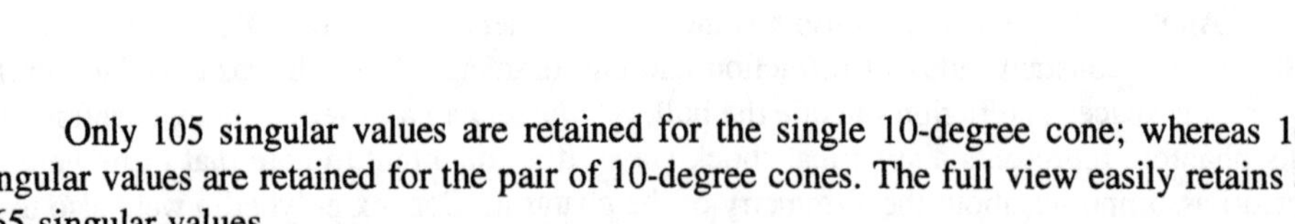

Fig. 7. Relative singular values for the series method and three viewing situations: one 10-degree cone; two 10-degree cones at right angles; and a full-view or 90-degree cone.

Only 105 singular values are retained for the single 10-degree cone; whereas 163 singular values are retained for the pair of 10-degree cones. The full view easily retains all 165 singular values.

Figure 8 shows the density values computed from the phantom interferograms. The density values were computed on a 16×16×16 grid from the coefficients and orthonormal functions. The single viewing cones are defined to be along the z-axis, and only 16 values along a line parallel to that axis are plotted. In other words, the z-axis is perpendicular to the viewing window, and figure 8 shows the performance of tomography along this hard-to-recover direction.

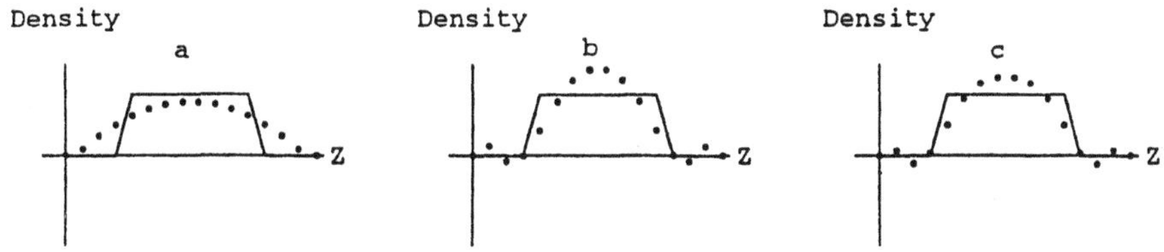

Fig. 8. Relative densities determined for constant index-of-refraction ball for three viewing situations of fig. 7. Solid lines represent phantoms and dots represent their reconstructions by computed tomography. Densities are plotted parallel to viewing cone axis (hard direction to recover in tomography). a) represents one 10-degree viewing cone; b) represents two viewing cones at right angles; and c) represents a full or 90-degree viewing cone. Series were terminated at polynomial degree S = 8.

We conclude that tomography with a 10-degree view limitation is not able to measure enough singular vectors even at S = 8 and cannot recover the edge of the sphere adequately. The 90-degree or full-view case is able to make full use of S = 8 and would easily benefit from higher resolution interferograms and higher order polynomials. The pair of cones makes full use of S = 8 and probably would benefit from more resolution. A major problem with the 90-degree case is the ringing or Gibbs phenomenon at the edges.

The original study included tomography performed on real double-exposure holograms as well (ref. 14). A spherical flask immersed in index matching fluid was used to emulate the ball. The pressure within the flask was changed slightly between exposures to create (with difficulty) an approximately uniform change in the index-of-refraction. Heterodyne interferometry (rather than phase shifting interferometry) was used to measure interference phase. In fact, 49 views were each sampled 32x32 times for a total of 50,176 measurements. The process took about 8 hours. Only 29 views were retained. For this tutorial demonstration, every other measurement was retained and processed by the 10-degree system. Figure 9 compares the results in the central plane for the phantom and the measured data. (The sign depends on which state is chosen as the first exposure.)

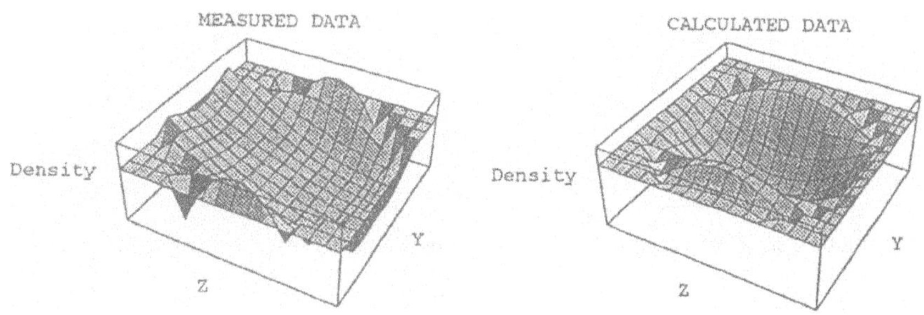

Fig. 9. Comparison of tomography from measured and calculated data for constant index-of-refraction ball. Fringe data are from a double-exposure hologram; change in density between exposures is negative. The tomography was executed for a limited cone angle of 10 degrees; for 29 16x16 interference phase measurements or phantoms; and for a polynomial degree of 8. Density is shown near central plane of ball.

The measured result does not have a dip in the center. Surprisingly, the dip appears for the S = 12 reconstruction, implying that poor performance of the singular value spectrum is not sufficient to reject higher resolutions. Nevertheless, 10-degree viewing cones are probably inadequate for pure tomographic reconstructions.

COMPUTERS AND ARTIFICIAL NEURAL NETWORKS FOR TOMOGRAPHY

The availability of workstations with large memories makes tomography viable for general flow diagnostics. The software used for the original study mentioned in the previous section was developed on an SGI 4D/25 workstation with 16 megabytes of RAM. However, the S = 12 case, in the original study, was performed with a supercomputer and about 200 megabytes of RAM. The tutorial demonstration at S = 8 and with 16×16 interferograms was performed on an SGI 4D/35 with 128 megabytes of RAM. The results of a single SVD for the tutorial demonstration occupy about 20 megabytes of RAM.

The requirements of the application will determine the computational method and the computer resources needed. The series method probably requires the largest memory. A technique such as ART will require less memory. However, a graphics workstation with at least 96 megabytes of RAM would be a reasonable tool for evaluating computed tomography for an application to flow diagnostics.

Another approach to tomography is to use artificial neural networks to calibrate the procedure (ref. 20). An artificial neural network (fig. 10) is simply a collection of interconnected nonlinear processing elements or nodes. The connections for one node of a feed forward net are shown in figure 10. The classical feed forward net is arranged in layers of nodes. Each layer receives inputs from the previous layer only and sends outputs to the next layer only. The input connections to a node are weighted; the weighted inputs are summed; the sum is passed through a nonlinear function; and the output of the nonlinear function is fanned out to the next layer. A network with at least an input layer, an output layer, and one layer in between (a hidden layer) can be calibrated or trained to perform arbitrary mappings between the input and output layers. The training or calibration procedure consists of using an algorithm to adjust the weights in response to the exemplars in a training set of input-output pairs.

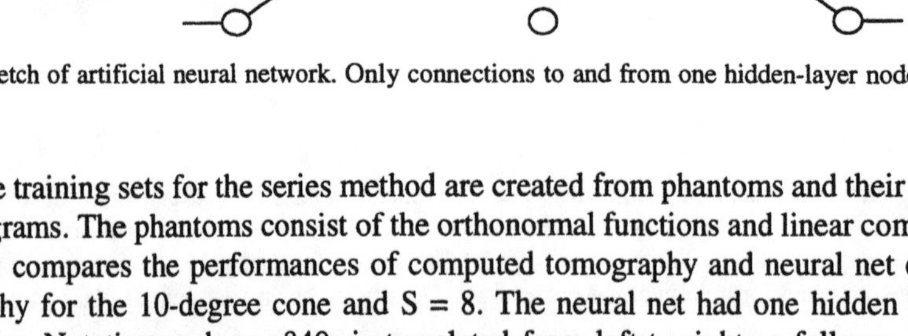

Fig. 10. Sketch of artificial neural network. Only connections to and from one hidden-layer node are shown.

The training sets for the series method are created from phantoms and their calculated interferograms. The phantoms consist of the orthonormal functions and linear combinations. Figure 11 compares the performances of computed tomography and neural net calibration tomography for the 10-degree cone and S = 8. The neural net had one hidden layer with three nodes. Notation such as e840r is translated from left to right as follows: "e" stands for eigenmode; S = 8 for overall degree of polynomial; $m_0 = 4$ for polar index; $m_1 = 0$ for azimuthal index; and "r" for real part. The phantom 11 is a linear combination of the first ten azimuthally symmetrical orthonormal functions. The phantom "sphere" is the ball of figures 7 to 9. Neural net calibration tomography and computed tomography perform similarly.

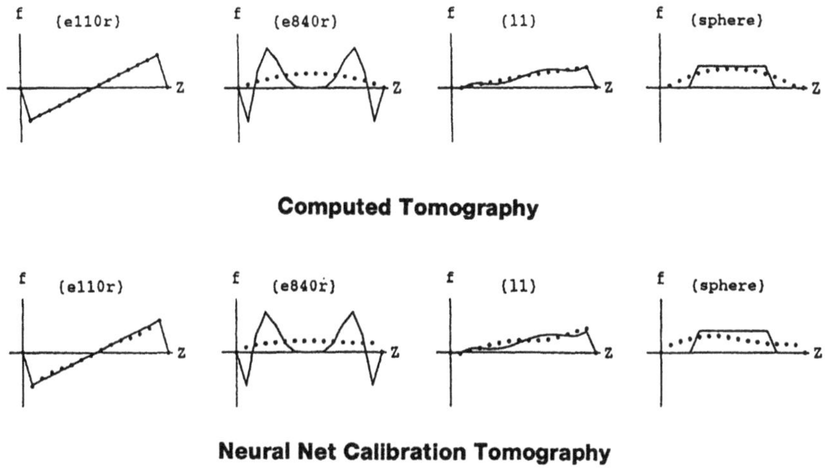

Fig. 11. Comparison of performances of computed and neural-net calibration tomography. Phantoms (solid lines) and their reconstructions (dots) are plotted as in fig. 8.
"e" refers to an orthonormal function; 11 is a linear combination of 10 functions; and "sphere" refers to a constant index-of-refraction ball.

The times required for SVD and neural-net training or calibration are about the same in software. The potential advantage of neural networks is using dedicated parallel processors that are now becoming commercially available. The training and recall speeds of neural net calibration tomography can be many orders of magnitude faster when implemented with the parallel processors.

SOME RESEARCH

We mentioned that tomography of vector fields is not adequately developed in a mathematical sense. There are high speed measurement methods available in deflectometry that possibly suggest performing such studies.

The exterior problem is solvable with the same general approaches already discussed. Measuring a property distribution exterior to an airfoil is an example of an exterior problem. Evidently, the exterior problem is likely to be more ill-posed than the limited angle problem we have discussed. Some demonstrations would be useful.

The series methods are convenient for combining transforms, model data, and other measurements and for handling incomplete data problems. These methods have demonstrated several defects: they use nonlocal orthonormal functions and require substantial computer resources. The nonlocal problem is being addressed by investigating the use of wavelets for tomography (ref. 21). Wavelets are local functions that can also be used to detect shockwave edges via a method called multiresolution analysis.

Clearly, demonstration applications of different methods of tomography for a variety of flow geometries and properties would be useful.

CONCLUDING REMARKS

The necessary analytical, detector, and computer technologies are available for tomography as a tool for flow diagnostics. These available technologies do not necessarily mean that tomography is suitable for all applications. It is probably desirable to perform computer experiments using phantom data before committing resources to hardware. Fortunately, small computers are now well equipped for this kind of analysis.

REFERENCES

1. J. Radon, "Über die Bestimmung von Funktionen durch ihre Integralwerte längs gewisser Mannigfaltigkeiten," *Berichte Sächsische Akademie der Wissenschaften, Leipzig, Math.-Phys. Kl.* **69**, 262–267 (1917).

2. F. Natterer, *The Mathematics of Computerized Tomography* (Wiley, New York, 1986).

3. S.R. Deans, *The Radon Transform and Some of its Applications* (Wiley, New York, 1983).

4. A.N. Tikhonov and V.Y. Arsenin, *Solutions of Ill-Posed Problems* (Winston & Sons, Washington, 1977).

5. W.H. Press, B.P. Flannery, S.A. Teukolsky, and W.T. Vetterling, *Numerical Recipes* (Cambridge University Press, New York, 1986).

6. *J. Comput. Assist. Tomogr.* (Raven Press, New York).

7. *Inverse Probl.* (IOP Publishing Ltd, Bristol).

8. R.K. Hanson, J.M. Seitzman, and P.H. Paul, "Planar Laser-Fluorescence Imaging of Combustion Gases," *Appl. Phys. B* **50**, 441–454 (1990).

9. Ref. 2, pp. 92–95.

10. S.H. Izen, "A Series Inversion for the X-ray Transform in n Dimensions," *Inverse Probl.* **4**, 725–748 (1988).

11. S.H. Izen, "Inversion of the *k*-plane Transform by Orthogonal Function Series Expansions," *Inverse Probl.* **5**, 181–202 (1989).

12. S.H. Izen, "Inversion of the X-ray Transform from Data in a Limited Angular Range," in *Signal Processing, Part II: Control Theory and Applications,* F.A. Grünbaum, J.W. Helton, and P. Khargonekar, eds. (Springer Verlag, New York, 1990), pp. 275–284.

13. S.H. Izen, *An Application for a Limited Solid Angle X-ray Transform,* Vol. 113 Contemporary Mathematics (American Mathematical Society, Providence, R.I., 1990), pp. 151–170.

14. A.J. Decker and S.H. Izen, "Three-dimensional Computed Tomography from Interferometric Measurements within a Narrow Cone of Views," *Appl. Opt.* **31**, 7696–7706 (1992).

15. Ref. 2, p. 137.

16. Ref. 2, pp. 137–144.

17. C.M. Vest, *Holographic Interferometry* (Wiley, New York, 1979), pp. 311–329.

18. R.M. Lewitt, "Reconstruction Algorithms: Transform Methods," *IEEE Proc.* **71**, 390–408 (1983).

19. S.H. Izen, *RECON3D Fully Three-Dimensional Tomographic Reconstructions*, Manual Version 0.855 (Case Western Reserve University, Cleveland, Oh., 1991).

20. A.J. Decker, "Neural Networks for Calibration Tomography," to be published in *SPIE Proc. 2005, Optical Diagnostics in Fluid and Thermal Flow* (1993).

21. S.H. Izen, "Frames for the Radon Transform," to be published.

Inelastic Methods

INELASTIC SCATTERING LASER DIAGNOSTICS; CARS, PLANAR LIF AND PLANAR LII

Douglas A. Greenhalgh

Centre for Photonics and Optical Engineering
School of Mechanical Engineering
Cranfield University
Cranfield
Bedford MK43 0AL, UK

INTRODUCTION

Optical diagnostics using lasers are now an established part of flow measurement. Techniques which employ elastic Mie scattering, such as LDV, PDPA and PIV, have been extensively developed over the past twenty years. In parallel, considerable research to develop optical methods for temperature and gas species measurement has taken place[1-7]. Temperature and species measurements are almost all based on inelastic light scattering and utilise either fluorescent emission or Raman scattering. Because these techniques are spectroscopic their development has been pursued by a separate research community to that involved with Mie scattering techniques. Over the past few 15 years many of these laser spectroscopic techniques have developed into practical methods with a broad range of application to fluids, especially reacting flows and practical devices.

The material summarised here was originally presented as three lectures to a Summer School on "Optical Diagnostics for Flow Processes" at RISO, Denmark, in 1993. The material focuses on three techniques, Coherent Anti-Stokes Raman Scattering (CARS)[2,3,5], Planar Laser Induced Fluorescence (Planar LIF of PLIF)[1,4] and Planar Laser Induced Incandescence (PLII)[12,13,22,23]. This is not intended to belittle the importance of many other inelastic techniques[3], especially Raman[6,7], rather it reflects the interests of the author and highlights the two most important and successful of the inelastic techniques. CARS has been widely demonstrated in a variety of practical and often hostile systems including gas turbine combustion, ic engines and furnaces, these applications have been widely reviewed[2,5]. Recent applications also include coal furnaces[8,9] and rockets[10,11]. PLIF offers the considerable advantage of full field data and numerous applications of PLIF to practical systems have been recently reported[4,12-21]. PLII is another inelastic method which has received much recent attention, this technique offers considerable potential for imaging soot in flames[12,13,22,23].

Firstly an introduction to the basic principles of inelastic scattering is provided, this introduction is aimed primarily at engineers who are familiar with laser methods but are perhaps less familiar with the essentials of spectroscopy. Though general, the introduction

Optical Diagnostics for Flow Processes
Edited by L. Lading *et al.*, Plenum Press, New York, 1994

is naturally slanted towards the CARS and LIF techniques. The principles of CARS, LIF and LII are then discussed, references to some of the recent advances in these technologies and the use of these methods is illustrated, where appropriate, with selected examples.

BASIC PRINCIPLES OF INELASTIC SCATTERING

Elastic scattering from molecules is normally termed Rayleigh scattering. For a medium undergoing a net motion with respect to the source/detector it is usual to find the scattered wavelength fractionally shifted from the incident wavelength even though the scattering process is elastic, this shift is due to the Doppler effect. Doppler wavelength shifts are very small and only of order 1 part in 10^7or less. Inelastic scattering is the term applied to all processes in which a photon, after interacting with a molecule, is scattered to a new wavelength. The most common processes are Raman and fluorescence, in these processes the wavelength shifts vary from 1 to 100 nm. During inelastic scattering, the molecule undergoes a change in internal energy, energy is therefore either dumped in or extracted from the scattering molecules. For non-linear techniques there are exceptions to this, in CARS, for instance, there may be two consecutive energy changes during the scattering which cancel the overall change in internal energy of the scattering molecule. Therefore the nature of inelastic scattering is specifically governed by the internal energy system of the scattering molecules.

The internal energy of a molecule is shared between electronic energy, vibrational energy and rotational energy. When the average distribution between these types of internal energy, for an ensemble of molecules, is equal the system is said to be in thermal equilibrium. All these internal energy modes are quantised, that is the molecules can only contain a precise quanta of energy which is determined by the energy levels associated with each energy mode. The internal energy of a single molecule is defined by its occupation of specific electronic, rotational and vibrational levels. An ensemble of molecules will exhibit a distribution of differing energies with a mean energy equivalent to the system temperature. Typically a simple molecule will posses only a few electronic states or levels and for diatomic molecules both the vibrational and rotational energy levels form simple ladders. Polyatomic species have a much more complex internal energy structure, the general principles are, however, similar. The next three sub-sections describe the basics of internal energy, for simplicity only diatomic molecules are considered. The reader who is interested in more complex systems should consult specialist references[24,25].

Rotational Energy

Molecules can rotate only at precise frequencies, successively higher frequencies forming a ladder of energies, see fig. 1. which is defined by:

$$E_r = hc\,B\,\mathrm{J}\,(\mathrm{J}+1) \tag{1}$$

where h is Planck's constant and c is the speed of light. J is the rotational quantum number and takes only integer values; note that as J increases the energy levels increase quadratically. B is the so called rotational constant and is related to the moment of inertia, I, of a given molecule by:

$$B = h\,/\,(8\pi^2\,cI). \tag{2}$$

The moment of inertia is determined by the mass of the atoms which constitute the molecule and the separation distance of the two atoms for a diatomic species. Strictly, molecules are not rigid rotors, therefore as J increases the molecule size increases, due to centrifugal effects, see fig. 1. A more precise form of eqn. 1 includes a constant, D, to correctly model this effect:

$$E_r = hc\,[B\,J(J+1) - D\,J^2\,(J+1)]. \tag{3}$$

The energy of a photon of light of frequency ν (cm^{-1}), or wavelength λ (nm) where $\nu = 1.0\times10^7 / \lambda$, is $h\nu$. Photon energies corresponding to rotational energies are such that the wavelength is in the microwave region.

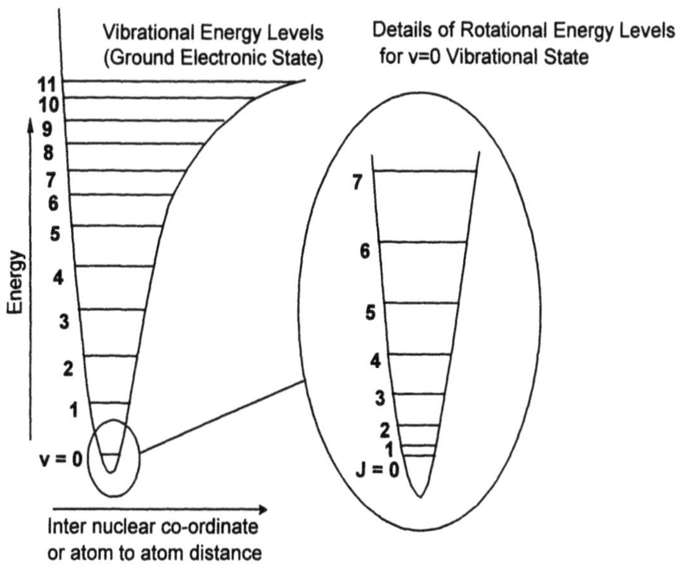

Figure 1. Schematic of vibrational and rotational energy levels for a diatomic molecule.

Vibrational Energy

Motion of an atom relative to another in a molecule occurs through specific vibrational motions, these vibrations are also quantised:

$$E_{\mathrm{v}} = h\omega_{\mathrm{v}}\,(\mathrm{v} + \tfrac{1}{2}) \tag{4}$$

where **v** is the vibrational quantum number and ω_{v} is the vibrational frequency. However as the vibrational frequency increases the molecule is distorted to a larger size, this leads to an anharmonic variation in the ladder which is normally modelled by including higher order terms of $(\mathrm{v} + \frac{1}{2})$ in eqn. 4. Vibrational energies correspond to photons energies in the infra-red from typically 20 to 3 μm.

Rotational -Vibrational Coupling

Molecules vibrate and rotate simultaneously, coupling of these motions leads to changes in the rotational constant, B, as a function of vibrational level, v:

$$B_v = B_e - \alpha_e(v + \tfrac{1}{2}), \quad (5)$$

the subscript "e" refers to the equilibrium position of the two nuclei. α_e is the correction constant to B for vibrational-rotational interaction.

Electronic Energy Levels

The motion of electrons is also quantised, however unlike vibrations and rotations, the energy levels do not form a simple ladder other than in the simplest system which is the hydrogen atom. The low lying electronic states of OH shown in fig. 2 are an example of a typical system .

The symbols for the electronic energy levels indicate the character of the state, for instance spin and symmetry. For a more detailed treaties on the physical implications of these symbols and electronic structure see reference works such as Herzberg[24] and Huber and Herzberg[26].

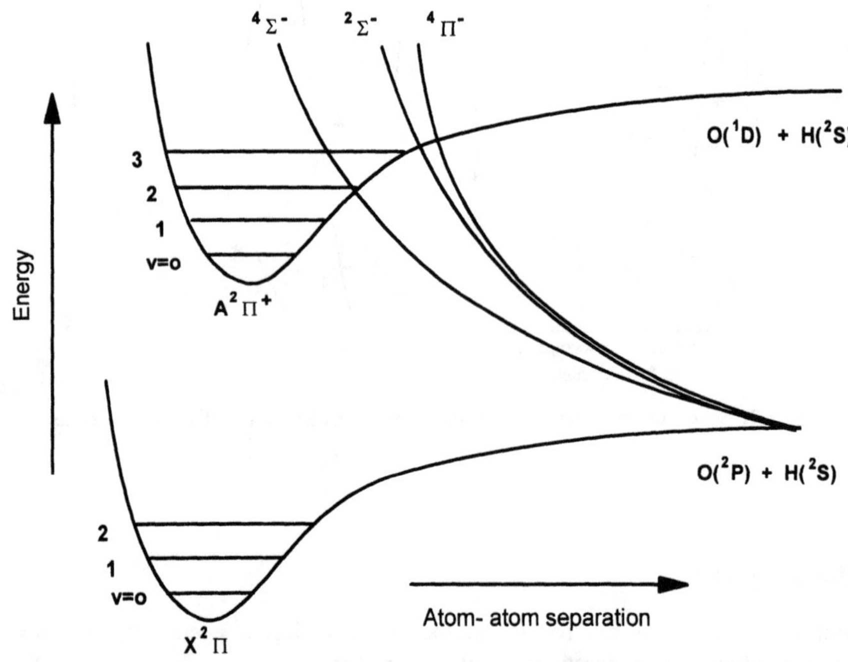

Figure 2. Energy level diagram for the OH radical; electronic and vibrational levels.

Population of Energy Levels

The population of individual energy levels, for an ensemble of molecules, is given by the Boltzmann equation:

$$N_j = [g_j \, d_j \exp(-\omega_j hc/kT)] \, Q^{-1}. \quad (6)$$

where N_j is the number of molecules in the jth state of frequency ω_j or energy $E = \omega_j hc$, g_j and d_j the nuclear spin and molecular degeneracy factors of the jth state, T the temperature and Q the partition function. The rotational degeneracy factor, d_J, is (2J+1). Neglecting centrifugal effects, within a rotational manifold the fractional population of a rotational level is:

$$N_J = N[g_J(2J+1)\exp(-BJ(J+1)hc/kT)](Q_{rot})^{-1}. \quad (7)$$

The nuclear spin degeneracy factor, g_J, is important only for homonuclear diatomics such as N_2, for other molecules such as CO or OH g_J is unity. For N_2, g_J is 6 for even J levels and 3 or odd J levels. The rotational partition function, Q_{rot} ,is:

$$Q_{rot} = \sum_{J=0}^{\infty} (2J+1)\exp(-hcBJ(J+1)/kT). \quad (8)$$

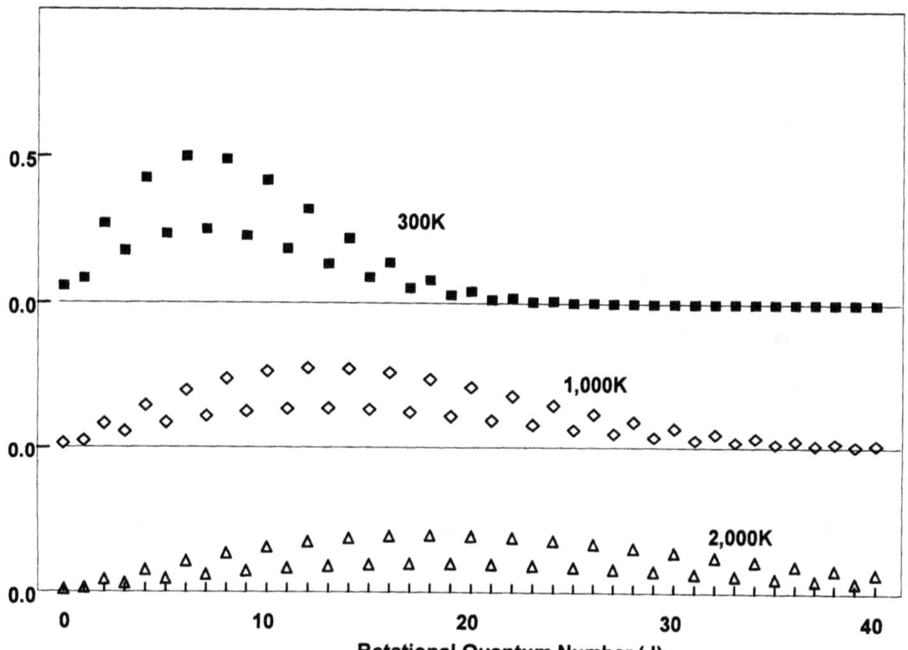

Figure 3. Plot of the fractional rotational population for nitrogen, B≈2.0cm^{-1}.

The vibrational partition function is:

$$Q_{vib} = \sum_{v=0}^{\infty} \exp(-vhc\omega_v/kT). \quad (9)$$

If anharmonicity is neglected, mathematically this approximates to:

$$Q_{vib} = (1\text{-exp}(-hc\omega_v/kT)^{-1}. \tag{10}$$

Thus the fractional population of an individual vibrational level is:

$$N_v = N \exp(-vhc\omega_v/kT)] (Q_{vib})^{-1} \tag{11}$$

and the overall fractional population of the jth ro-vibrational level **v**, J is given by the product of the rotational and vibrational fractional populations, namely eqns. 7 and 11:

$$N_j = N_J \bullet N_v. \tag{12}$$

The key feature of population factors is that they are a function of temperature, indeed it is precisely this feature which enables the measurement of temperature by inelastic scattering. The form of the temperature dependencies for rotations and vibrations are very different, see figs. 3 and 4.

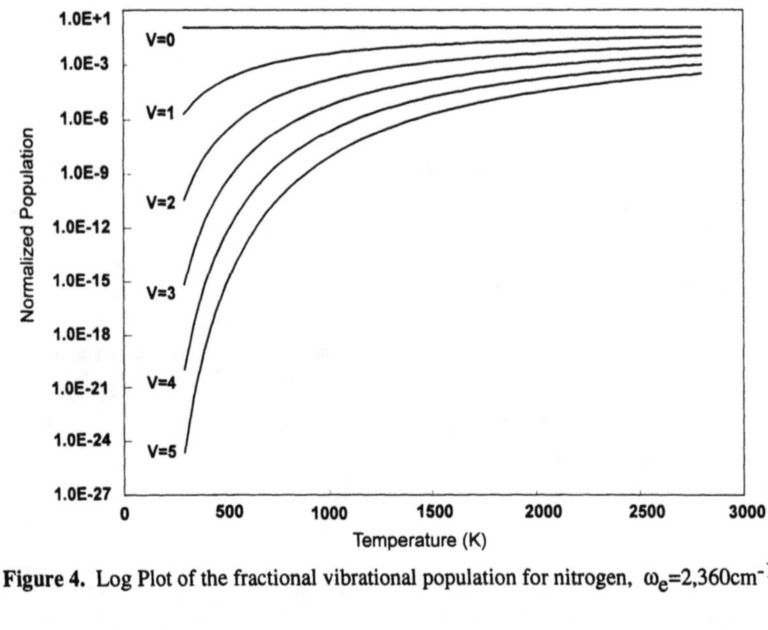

Figure 4. Log Plot of the fractional vibrational population for nitrogen, ω_e=2,360cm^{-1}.

Spectral Structure and Selection Rules

Spectral lines arise from transitions between differing electronic, vibrational and rotational levels. Not all transitions are possible and transitions are only allowed between electronic states for certain electron spins and certain electronic symmetry. Normally only a few electronic levels of a given molecule are important and detailed selection rules may be determined from reference texts[24,25].

The selection rules for vibrational transitions are as follows:

$$\Delta v = 0, \pm 1, \pm 2 \ldots \text{ etc.} \tag{13}$$

where the most important transitions for diagnostics are $\Delta v = +1$. Transitions between rotor states, J, have the following selection rules which are independent of Δv:

$$\Delta J = 0, \ \pm 1, \ \pm 2 \ \text{only}. \tag{14}$$

$\Delta J = 0$ transitions are Q-branch transitions and are only observed if $\Delta v = \pm 1$, the O-, P-, R- and S-branches are respectively ΔJ = -2, -1, +1, and +2. In CARS only O-, Q- and S-branch spectra are allowed, the most important branch is the Q-branch because it is the strongest features. Q-branches are broadened on the high J side because *B* varies with vibrational state, v, and therefore the lines are spaced by $\alpha_e J(J+1)$, see eqn. 5. In LIF the allowed spectral transitions form P-, Q- and R- branches. A schematic example of the structure of a LIF spectrum, in this case the main band of OH is shown in fig. 5. The term (0,0) indicates that the transitions starts and finishes in the v = 0 state of the upper and lower electronic states respectively. Note that only the P_1, Q_1, and R_1 branches are illustrated. In this case the subscript refers to the splitting of each J state due to an unpaired electron spin in the OH radical, an almost identical set of lines P_2, Q_1, and R_2 also exist. As for vibrational spectra, the Q-branch line spacing increases approximately quadratically starting from the band head near to 32,500 cm^{-1}; this characteristic of Q-branches arises due to differences in rotational constants *B* between different electronic levels though in this case for the same vibrational state.

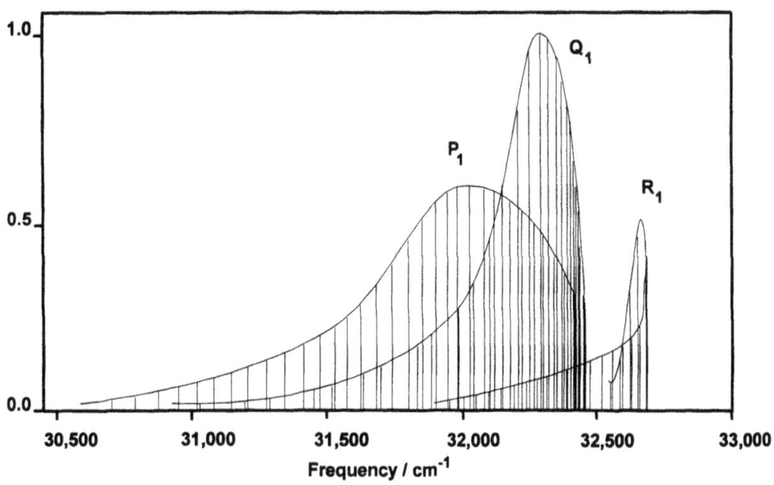

Figure 5. Rotational structure in (0,0) band of the $A^2\Sigma$ to $X^2\Pi$ of OH.

Spectral Line Broadening

In practice the line shape of individual spectral lines is not a delta function as portrayed by fig. 5, broadening of the lines occurs through three basic mechanisms,

Firstly the instrument or laser probe will have a finite resolution or linewidth and this will contribute to the observed spectral width. Secondly, pressure causes linear broadening of the linewidths. Thirdly, molecular motion contribute to broadening through the Doppler effect. Doppler broadening results in a 1/e half width linewidth, $\tilde{\Gamma}_D$, of:

$$\tilde{\Gamma}_D = \frac{\nu_o}{c}\sqrt{\frac{2kT}{m}}. \tag{15}$$

Which in turn gives rise to a Gaussian profile for the line:

$$G_D(\nu) = \frac{\sqrt{\pi}}{\tilde{\Gamma}_D} \exp\left(\frac{-\Delta^2}{\tilde{\Gamma}_D^2}\right). \tag{16}$$

where Δ is the frequency difference from line centre. For CARS spectra of molecules such as nitrogen, with a Raman spectrum around 2,360cm^{-1}, the Doppler width varies from about 0.002cm^{-1} at 300K to around 0.007cm^{-1} at 2,700K. For LIF of the OH radical, which involves optical frequencies around 32,500 cm^{-1}, the Doppler width is much greater and of order 0.084 cm^{-1} at flame temperatures.

Pressure broadening gives rise to a half-width half-maximum linewidth, Γ, and a Lorentzian line shape:

$$G_L(\nu) = \frac{\Gamma}{\pi}\left[\frac{1}{\Delta + \Gamma^2}\right] \equiv \frac{\Gamma}{\pi}\mathrm{Im}\left[\frac{1}{\Delta^2 + i\Gamma^2}\right]. \tag{17}$$

Pressure broadening is of order 0.05 cm^{-1} atms.$^{-1}$, though it can be much lower at high temperatures and significantly greater for highly polar molecules in a polar bath gas.

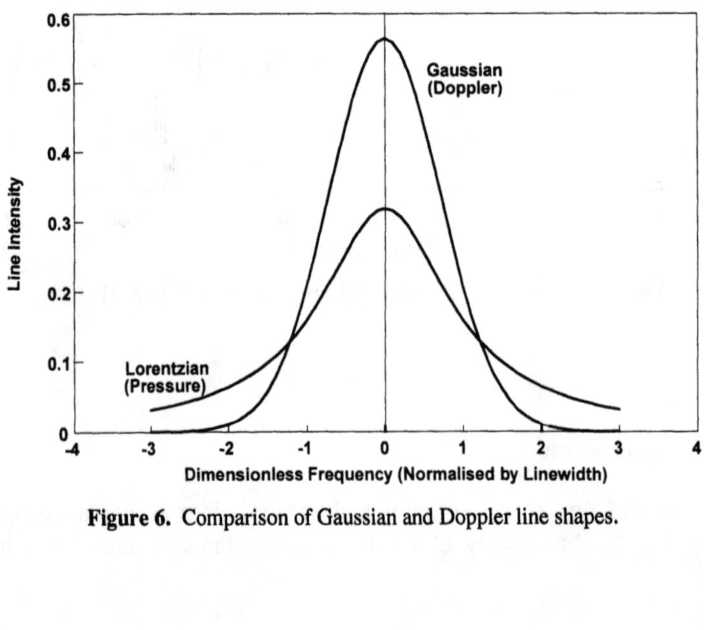

Figure 6. Comparison of Gaussian and Doppler line shapes.

These two line shapes are quite different, this is illustrated in fig. 6. Note that the Lorentzian lines though sharper closer to line centre decay more slowly in the wings.

COHERENT ANTI-STOKES RAMAN SCATTERING (CARS)

When a medium is subject to high laser intensities, it is possible to induce coherent, inelastic scattering from the medium. Coherent elastic scattering is known as Degenerate Four-Wave Mixing, the inelastic process is known as Coherent Anti-Stokes Raman Scattering or CARS.

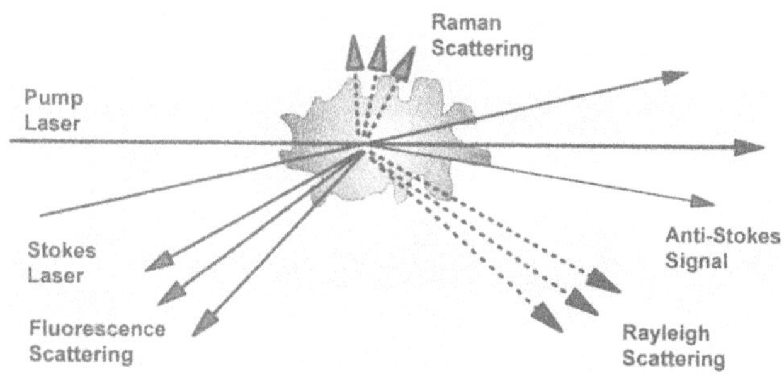

Figure 7. The spectral shapes of the various contributions to $\chi^{(3)}$.

Unlike Raman, Rayleigh or fluorescence, the anti-Stokes signal from CARS is a coherent, laser-like beam. This unique feature of coherent methods enables their application as diagnostic methods in the presence of a highly luminous or fluorescent background.

For diagnostic applications, the most common geometry for CARS is the so called broadband folded BOXCARS configuration. This geometry utilises a broadband Stokes laser with narrow band pump laser to generate a CARS signal from a small intersection region of the laser beams, see fig. 8 below. The main advantage of this geometry is that it minimises the generation volume, thereby maximising spatial resolution. In addition the signal beam is spatially separate from the three input laser beams. The high intensities required for CARS are normally supplied by a frequency doubled, Q-switched Nd:YAG laser of say 100-500 mJ. Part of this output is used to drive a broadband dye laser. The main application of CARS is for temperature measurements, recently some work on the development and application of CARS to species measurement has also occurred[2,5]. In this work, however, only temperature measurements are considered.

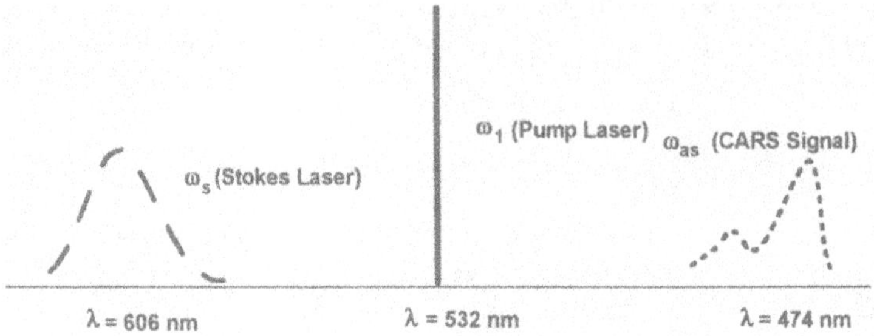

Spectral relationship of the CARS signal to the pump and Stokes lasers

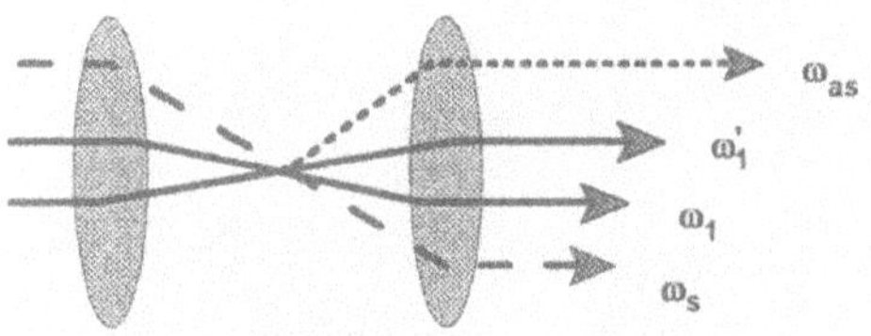

Layout of laser and CARS signal beams in the folded BOXCARS geometry

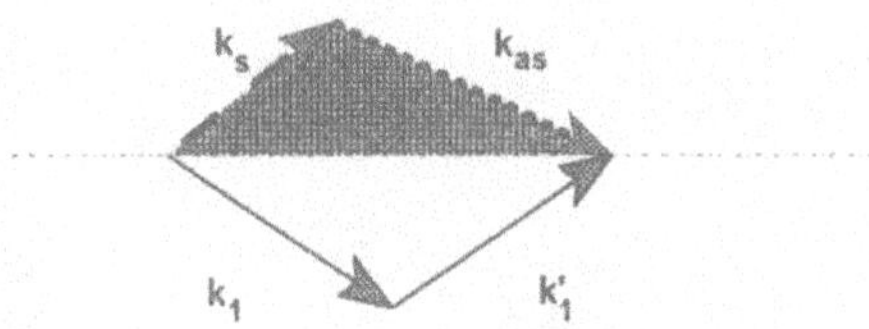

Phase matching diagram for folded BOXCARS

Figure 8. Schematic of the spectral relationships, beam geometry and phase matching in broadband BOXCARS of the nitrogen Q-branch.

Basics of CARS

To understand the basics of CARS we need to consider the coherent processes which arise from a non-linear interaction between intense laser beams in a gas. Intense electric fields induce a polarisation in the medium according to:

$$\mathbf{P} = \chi^{(1)}.E + \chi^{(2)}.E^2 + \chi^{(3)}.E^3.....etc. \tag{18}$$

In CARS we are concerned only with the $\chi^{(3)}$ term where three electric fields mix to yield a fourth. The CARS or anti-Stokes signal is formed by mixing of three laser beams of two different frequencies, ω_1 and $\omega_{s,}$. The frequency relationship to the signal is:

$$\omega_{as} = \omega_1 - \omega_s + \omega'_1. \tag{19}$$

Note that two components of the pump laser at ω_1 are required as shown in fig. 8. Also note that if ω_1 is a narrow band laser then there is essentially a unique relationship between an frequency component of the broad band laser and the CARS spectrum. To a first approximation each frequency component of ω_s, generates an anti-Stokes or CARS spectral component.

The oscillating electric field of a single laser may be described as:

$$\mathbf{E}(\omega_1) = \tfrac{1}{2} A_1^{(o)}\, e^{i(k_1 z - \omega t)} + c.c. \qquad (20)$$

thus the induce third order polarisation for CARS is:

$$\mathbf{P}(\omega_{as}) = \tfrac{3}{8} \chi(-\omega_{as}; \omega_1, \omega_1', \omega_s) A_1^{(o)}\, A_1'^{(o)}\, A_s^{(o)}\, e^{(ik_1 z + k_1' z - k_s z - \omega_1 t - \omega_1' t + \omega_s t)} + c.c. \qquad (21)$$

The use of eqn. 3 in Maxwell's wave equation, assuming a charge-free medium and a slowly varying electric field, after integration gives the main CARS scattering equation:

$$I_{as} = \left(\frac{4\pi\omega_{as}}{c^2 n_{as}}\right) I_1 I_1' I_s |6\chi^{(3)}|^2 l^2 \,\mathrm{sinc}^2(\Delta k l) \qquad (22)$$

where $\mathrm{sinc}(\Delta kl) = \sin(\Delta kl)/(\Delta kl)$, Δk is the phase mismatch between the wave vectors, k, of the laser beams ($\Delta k = k_{as} + k_s - k_1 - k'_1$) and l is the interaction length of the laser beams. The wave vector magnitude is $k_1 = n_1\omega_1/c$, n_1 being the refractive index. To achieve proper phase matching the laser beams must be correctly vectored so that full three dimensional phase matching is achieved, see fig. 8. For this condition the CARS signal is then generated with a unique wave vector and therefore has the properties of a laser-like beam. For gases n is close to unity therefore in broadband CARS it is possible to achieve good phase matching for all the frequencies of the broadband laser. Note from eqn. 22 that the CARS signal scales as the third power of the lasers, or I_1^2 and I_s, and as the square of the interaction length of the laser beams l^2.

The CARS signal also scales as $|\chi^{(3)}|^2$ and this term contains the essential description of the CARS spectrum; for low pressure gasses $\chi^{(3)}$ is given by:

$$\chi^{(3)} = \frac{(\mathbf{v}+1)N}{\hbar} \sum_j \left(\omega_j - \omega_1 + \omega_s - i\Gamma_j\right)^{-1} \Delta\rho_j \alpha_j^2 \qquad (23)$$

where N is the number density of the gas, ω_j the frequency of the jth Raman transition of the probed species, $\Delta\rho_j$ the population difference between the lower and upper energy levels associated with the jth transition and α_j^2 the polarizability matrix element transition which is related to the Raman cross section, $d\sigma/d\Omega$, by:

$$\alpha_j = \sqrt{\frac{d\sigma}{d\Omega}_j}\left(\frac{c}{\omega_s}\right). \qquad (24)$$

From the complex equation, eqn. 23, it is apparent that CARS spectra are formed from both real and imaginary terms. Ro-vibrational Raman type transitions contribute both real and imaginary terms, see fig. 9, however CARS signals are also generated from two-photon electronic transitions (e.g. $\omega_1 + \omega_1$ is of order of $\omega_{electronic}$). The two-photon contributions are typically small because the laser frequencies are non-resonant. Contributions from distant resonances, both electronic and Raman, only contribute an essentially flat Re. $\chi^{(3)}$ part which is termed the non-resonant background, χ_{NR}, see fig. 9.

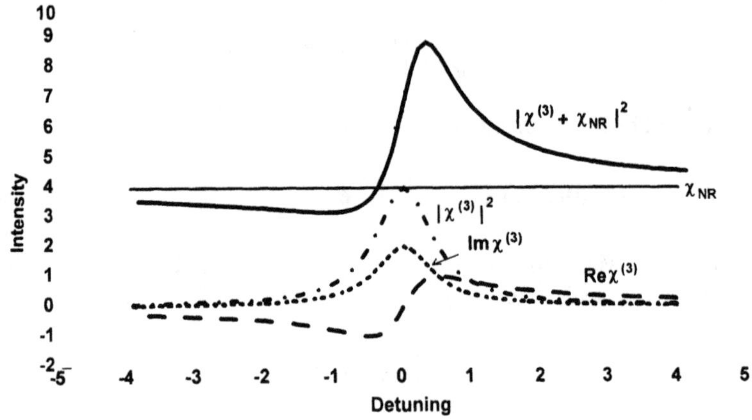

Figure 9. The spectral shapes of the various contributions to $\chi^{(3)}$.

The final shape of a typical spectrum can be calculated from eqns. 22-24, the temperature dependence is principally contained in eqns. 7-12 and the line positions are derived from spectroscopic constants for rotation and vibration. The individual linewidths, which are both J and *T* dependant, can be modelled with simple scaling law models[2] for temperature and pressure effects. For pressure broadened linewidths at low pressure, eqn. 23 explicitly includes a complex Lorentz form. For lower pressures Doppler broadening effects should be included using a Voigt or other appropriate function[2]. At high pressure eqn. 23 is replaced by:

$$\chi^{(3)} = \frac{(\mathbf{v}+1)N}{\hbar}\left[\alpha \cdot \mathbf{G}^{-1} \cdot \Delta\rho \cdot \alpha\right] \tag{25}$$

where α is a vector of trace scattering coefficients, see eqn. 24, $\Delta\rho$ is a diagonal matrix of population difference between the upper and lower energy levels for each transition contributing to the spectrum and, **G** is a matrix whose elements are:

$$G_{jk} = \partial_{jk}(\omega_j - \omega_1 + \omega_s + \Delta_j + i\Gamma_j) + (1 - \delta_{jk}) i\gamma_{jk} \tag{26}$$

Where Δ_j and Γ_j are the pressure shifts and linewidths of each line, δ_{jk} is the Kronecker delta function (i.e. it takes the value 0 unless $j=k$ when it take the value 1) and γ_{jk} are coefficients which describe the rate of collisional exchange of energy between a molecule from one rotor state J to a second state J'. The γ_{jk}'s may be modelled by the same simple scaling laws used to model Γ_j's[2], further details are given elsewhere[2].

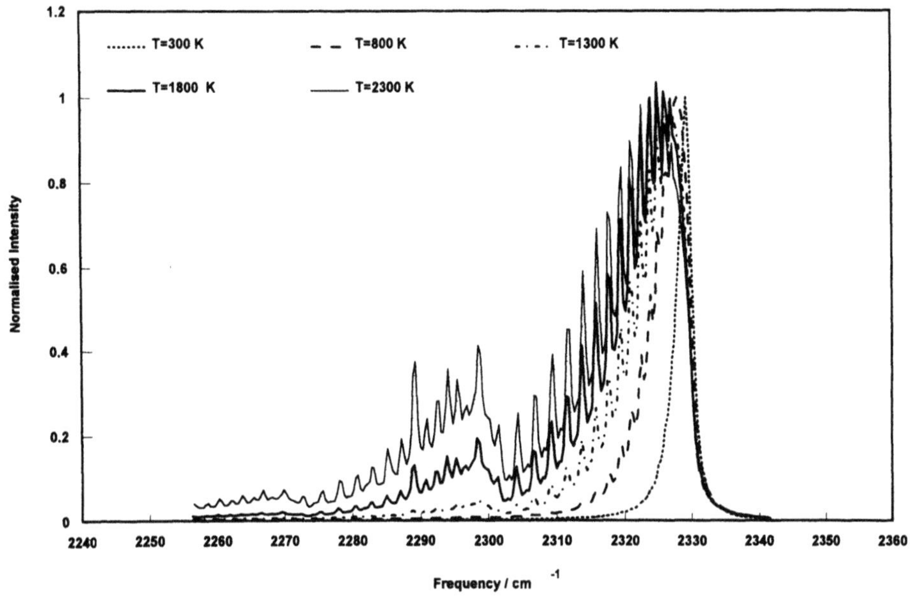

Figure 10. The variation of CARS spectral shape for the Q-branch of nitrogen at 1 atmosphere, 78% concentration and with an effective overall spectrometer/laser resolution of about 1 cm^{-1}.

Temperature Measurement by CARS

Probably the most important application of CARS is for thermometry, for combustion applications temperature is estimated from the shape of the nitrogen Q-branch spectrum, the effect of temperature on the shape is shown in fig. 10. As temperature rises, first we see the broadening of the **v** =0 band to lower frequency and the appearance of structure due to the even J lines. The odd J lines are 4 times weaker due to the square of the nuclear spin degeneracy, see eqn. 7. The J structure is resolved at lower wavelengths because the higher J lines in the spectrum of the Q-branch are shifted to lower frequency and are further apart due to the quadratic spacing of Q-branch lines. At temperatures, above 1100 K, Q-branches of higher **v** levels are seen.

In practice, temperatures are derived by comparing a measured spectrum with a calculated spectrum using a least squares fitting procedure. Since the computation of CARS spectra is complex, sophisticated computer codes are required. For engineering applications it is normal to use a commercial code such as CARP3[27] which is marketed by Epsilon Research Ltd[28].

Raw CARS spectra should first be pre-processed to remove any offset which may arise from the multichannel detector or from scattered light, also the shape of the nitrogen CARS spectrum should be normalised by the CARS spectrum of the broadband dye or Stokes laser. A CARS spectrum of the latter is easily measured by using a non-resonant medium such as carbon dioxide or propane, both species produce strong, spectrally flat, signals in the region of the nitrogen Q-branch spectrum. After recording the non-resonant spectrum this is used to normalise the raw spectrum on a channel by channel basis. The pathway for these various steps of data analysis is illustrated in fig. 11.

Excellent accuracy for ambient pressure CARS thermometry is possible and accuracy's for mean or many pulse spectra of order 1% or better have been reported[2,5]. Specific demonstrations of accuracy at high pressure and temperatures up to 900K and 20 bar[29], at flame temperatures of over 2,000K and 10 bar[29,30] and at near atmospheric pressure to temperatures of 3,500K[31] have been conducted using a variety of validation techniques.

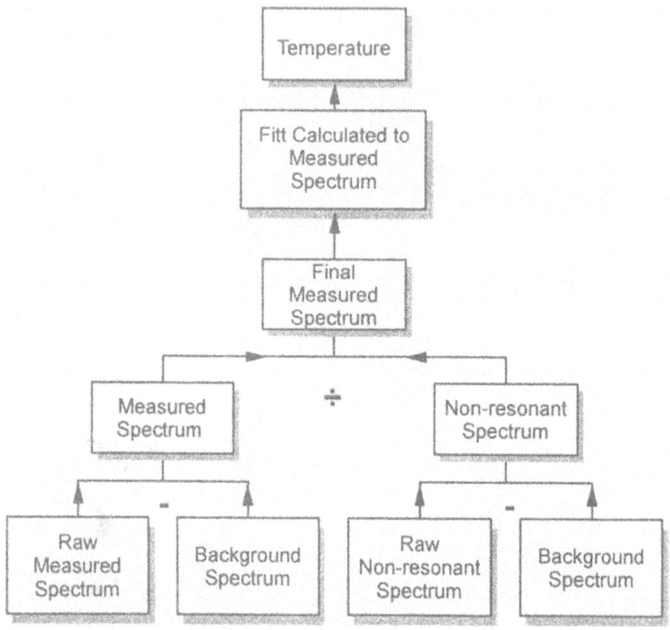

Figure 11. Pre-processing and spectral analysis steps for the determination of temperature from a measured CARS spectrum.

Noise Problems in CARS

The problem of spectral noise in broadband CARS has proven to be an important limitation for accurate single pulse measurements. The noise arises from two main sources, mode noise[2,32-35] and from detector noise[2,36,37]. Mode noise arises because a broadband source must exhibit independent spectral fluctuations each with 100% noise[32]. In the absence of a resonator or cavity, the number of fluctuations or quasi-modes is inversely proportional to the laser pulse length[32]. In a laser the actual number of longitudinal modes is constrained by the total laser bandwidth and the mode spacing. For a typical broadband laser with a cavity length $l_c = 33$ cm this corresponds to a mode spacing of $1/2nl_c$ or .015 cm^{-1}. Therefore over a resolution width of 1 cm^{-1} there will be N=66 modes each with noise of 100%, detecting all this modes together $\sqrt{N}$ averages the noise to 12%. A typical broadband laser is 130 cm^{-1} wide, therefore there are of order 10^4 modes across the width. Mode noise is a primary source of pulse to pulse noise in CARS, also due to the random phases of each mode[33,35] and different mode selection effects[34,35,38] mode noise can only be partially eliminated by normalising the spectrum with a non-resonant reference spectrum on a shot-to-shot basis. In practice mode noise produces uncertainties of order 5-10% on a single pulse temperature measurement.

Most practical CARS systems use a multi-mode Nd:YAG laser combined with a broadband dye laser. This combination gives a superior pulse to pulse repeatability for temperature measurements at atmospheric pressures because the additional laser linewidth (typically 0.3 to 1.0 cm^{-1}) increases the local averaging of the dye laser mode noise[37,38].

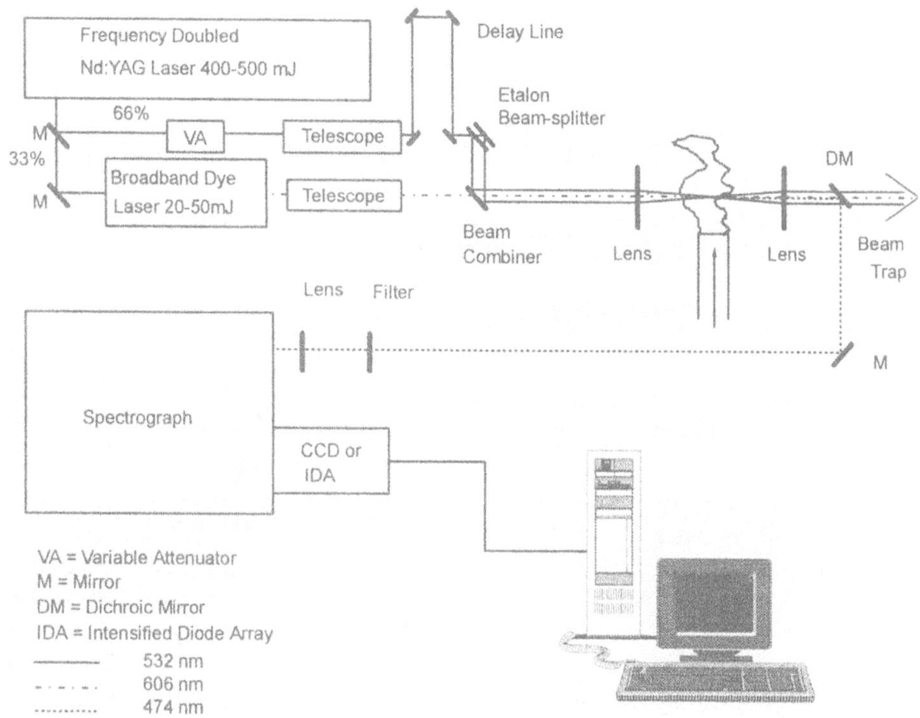

Figure 12. Typical layout for broadband CARS instrument.

Single-mode Nd:YAG lasers, however, produce a much more stable output and will give a superior performance at high pressures where the molecular linewidth is sufficiently large such that good averaging of the dye laser noise occurs through the molecular linewidth. Single mode Nd:YAG lasers, when used with ASE type devices[32,39,40] as substitutes for the dye laser, should give superior performance by reducing the phase noise because the mode structure of ASE devices is essentially random rather than periodic[33]. Future CARS systems are therefore likely to utilise high power single mode Nd:YAG lasers and ASE broadband Stokes sources.

Broadband CARS Equipment

For virtually all practical thermometry, a broadband CARS system is used as discussed above, schematically a typical system is illustrated in fig. 12. For engineering applications a 400 to 500 mJ, frequency doubled Nd:YAG laser is required as the main laser. Approximately 30% of the output of this laser is used to drive a broadband dye laser, see fig. 12. Various optics are used to ensure that the laser beams are all reasonably matched for size and that they are all brought to a tight focus in the sample. For efficient signal generation it is important to produce high laser intensities, however the peak intensity must not produce breakdown, also non-linear pumping of the molecule by stimulated Raman should be avoided. Although long beam interaction lengths produce strong signals, which is achieved by using a relatively small beam crossing angle (say 2 to 5°), it is important to minimise the probe volume size if measurements in high gradient systems are required, see next section.

Experience has shown that a spectrometer with a resolution of 1 cm^{-1} or better is required, a 1 or 1.5 meter spectrometer with a 2,400 l/mm grating is suitable. Most existing CARS instruments use intensified diode array (IDA) multi element detectors, however a modern back-thinned CCD chip offers a usable dynamic range of about 330 compared to the range of an IDA which is 80. These figures for dynamic range are estimated on the basis that Poisson noise will dominate the measured temperature error for CARS signals with a peak intensity of less than 200 measured photo-electrons[36].

Thermometry in Severe Gradients or Flame Fronts

A second important problem for CARS is the measurement of temperatures in sharp temperature gradients as found in flame fronts[42-44]. The source of problem is two fold. Firstly, CARS measurement volumes are formed from the intersection of three laser beams at a shallow angle. The diameter of the beams is typically 100 to 250μ. For an intersection half angle of say 2° and for a beam diameter of 200μ this will give an interaction length of circa 5-6 mm. With care this can be reduced to 100μ × 100μ × 2 mm by using large crossing angles and single mode lasers which have the best beam quality. The limitations in beam diameter are set both by beam quality and the need to avoid very intense fields at the focus which can stress the medium and cause breakdown or optical Stark effects[44] or saturation effects[45,46]. Single mode lasers are preferable because, unlike multi-mode lasers, which have rapidly fluctuating electric fields due to mode beating, they have a smooth temporal shape with lower peak intensities for an equivalent average power. The limitation in volume length is set by the need to generate an adequate signal level which must be sufficient to avoid significant Poisson or detector noise. Because the signal scales as interaction length, l^2, strong signals are only generated when the interaction lengths is in the range 1 to 5 mm or more.

The thickness of a flame is of order 200μm, for a premixed flame this distance will typically divide zones of gas which differ by a temperature of order 2,000K and a density ratio of 6 to 7, see fig. 13. Due to the number density and fractional population dependence of CARS $[N\ \Delta\rho_j]^2$ the signal varies considerably in intensity from the cold unburnt gasses in a flame to the hot burnt gasses. The variations in CARS signal strength and spectral signature are shown in fig. 14, changes in amplitude of order 100 to 500 are possible for the range of temperatures encountered in a typical flame. This effect causes the CARS derived temperature to be too low, indeed for extreme cases this error can reach 1,500 K or greater. Fig. 15 shows a spectrum where the measurement volume spans

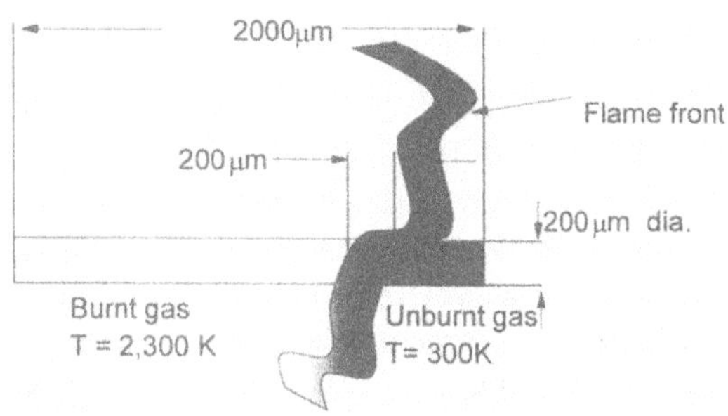

Figure 13. Schematic of a CARS measurement volume in the vicinity of a flame front.

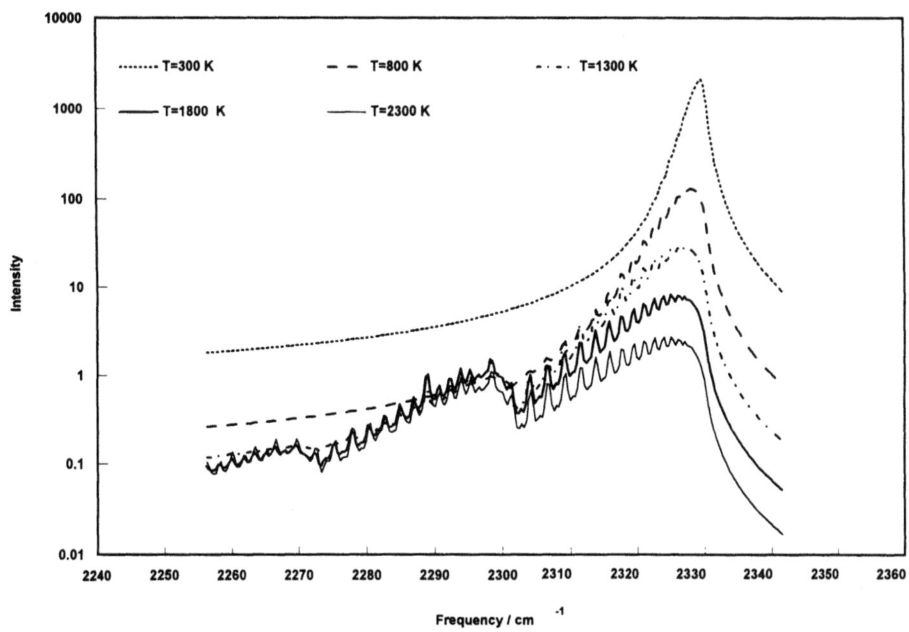

Figure 14. The variation of CARS spectral shape for the Q-branch of nitrogen at 1 atmosphere, 78% concentration and with an effective overall spectrometer/laser resolution of about 1 cm^{-1}.

a premixed flame and contains only 5% cold reactant gas at 300 K, the bulk of the measurement volume is at 2,300 K.

Fitting a single temperature spectrum to a measured spectrum results in a serious error. The magnitude of the error is greatest when the volume of cold gas is small but finite. Recently, a procedure for evaluating two temperature CARS spectra has been published[43] which is based on two facets of the CARS signal. Firstly, when the proportion of cold gas is 80% or greater it is possible to independently estimate the temperatures of the hot and cold zones. As shown by fig. 15, the peak of the CARS nitrogen spectrum around 2,330 cm^{-1} is dominated by the cold contribution whereas the region around 2,300 cm^{-1} is dominated by the hot contribution. Secondly, the CARS signal strength is very sensitive to temperature, see fig. 14, and an estimate of the size of the cold zone can be made on the basis of signal strength. From these two parameters the hot and cold zone temperatures may be derived and there relative sizes estimated, from these three pieces of data either a "volume" weighted or "Favre" (density) weighted temperature may be calculated. For spectra where the hot zone is less than 80% of the measurement volume its temperature cannot be determined separately and it must be assumed equal to the appropriate adiabatic flame temperature or estimated from the reaction profile of a flame derived from a laminar flame calculation [43].

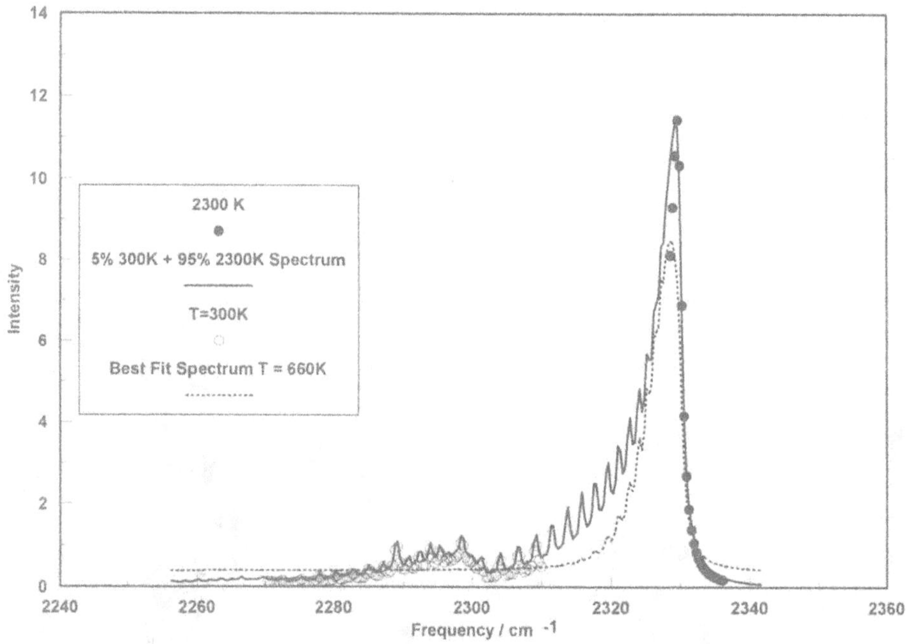

Figure 15. A CARS nitrogen spectrum formed from a zone containing both cold (300 K) and hot (2,300K) gas with a true volume mean temperature of 2,200 K. See text for further details.

PLANAR LASER INDUCED FLUORESCENCE (PLIF)

Planar or 2D fluorescence imaging[1,4,7,21] is a powerful tool for the mapping of species in a turbulent reacting flow for a variety of reasons. Firstly, instantaneous images may be generated with pulsed lasers which generate 10-40 nanosecond pulses, this time scale is fast enough to freeze all fluid motion. Through the analysis of instantaneous images, the structure of flames and mixing contours may be examined in considerable detail. PLIF is also attractive for engineering applications because the final images are easy to interpret and can be readily related to the mechanical geometry of the system. LIF is also a relatively efficient scattering process, therefore high quality images are possible even from trace species such as radicals. Finally, for low and modest laser intensities, LIF signals are linear with both laser power and the number density of scattering species, these features greatly facilitate the processing of images to produce quantitative data.

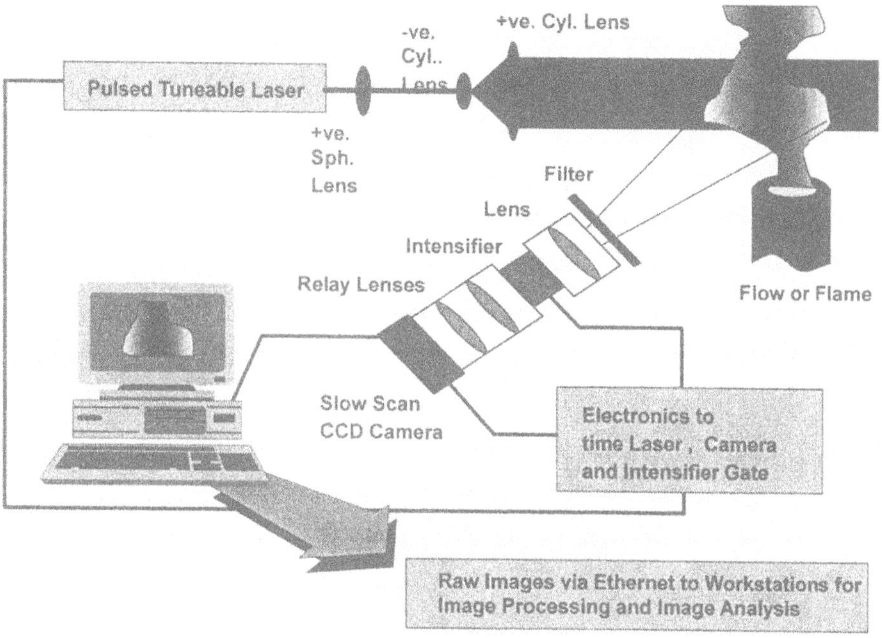

Figure 16. Schematic for a typical PLIF experimental set up.

The basic layout of a PLIF experiment is shown in fig. 16. The essential items are a thin intense laser sheet which is tuned to a molecular absorption and a sensitive camera system. Typically a gated intensifier is used in conjunction with a CCD camera so that faint images can be recorded in the presence of high background light conditions through the use of a fast ≈50-100 nanosecond gate.

Basic Principles of LIF

The term fluorescence is usually applied to the radiation emitted by an atom or molecule when it decays spontaneously from a higher to a lower energy level or state. In LIF the molecule is pumped to a higher level by a photon from a laser tuned to an electronic resonance. This process is illustrated on the far left of fig. 17, and for low intensities it is controlled by the product of laser intensity I_ν, the number density of absorbers in the lower state, N_l, and the Einstein coefficient for the rate of absorption of photons, B_{lu}. Here l, denotes the lower electronic level and the u upper level. Once the molecule is excited there are several competing pathways for de-excitation of the upper level, six of these processes are illustrated in fig. 17.

Fluorescence can arise from spontaneous emission direct from the upper level to which the species is excited by the laser, the rate of photon emission is then the product of the number density of the species in the upper level, N_u, and the Einstein coefficient for spontaneous emission, A_{ul}. More typically, relaxation and energy exchanges processes in the vibrational and rotational levels in the upper state, plus the opportunity to relax to other rotational and vibrational levels in the lower electronic state, cause the fluorescence to be shifted to longer wavelengths as shown by the pathway $A_{u'l'}$ on the right of fig. 17. These relaxation process can be purely internal to the molecule but they are usually driven by molecular collisions with the bath gas. Other important competing processes for N_u include: stimulated emission to the original lower state by the laser at a rate $I_v \cdot N_u \cdot B_{ul}$, further excitation by the laser to higher states or to an ionising level, predissociation (Q_{pre}) of the molecules and radiationless or electronic quenching of the system (Q_{elec}). Electronic quenching, or just quenching as it is normally referred to, also arises from collisions with other bath gas molecules. Quenching cross-section for different species with a target species varies widely, therefore quenching is dependant on the local gas composition. Typically all processes are present in all systems, for practical purposes though, quenching and rotational and vibrational relaxation are the most important.

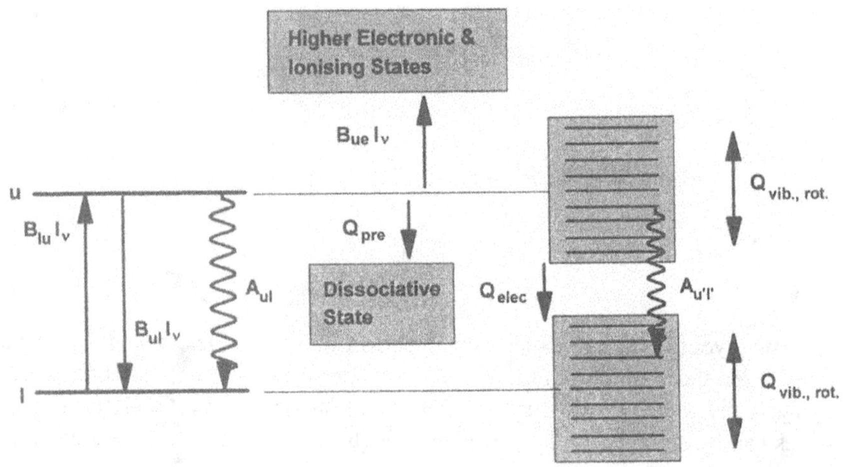

Figure 17. Diagram of the important energy transfer processes in LIF, see text for explanation of the parameters.

Although over simplistic for a practical quantitative treatment, it is instructive to consider a rate analysis for an idealised "two level" system because the principles of saturation and the role of predissociation may be illustrated. In this treatment only the results of a semi-classical analysis[3,4] are used, for a more detailed quantum mechanical analysis the reader should consult Demtröder[47]. The steady-state rate of fluorescence emission, *Rf*, for a simple two level system is:

$$R_f = N_l B_{lu} I_v \frac{A_{ul}}{(A_{ul} + Q_{elec} + Q_{pre})} \frac{1}{\left(1 + I_v / I_v^{sat}\right)} \tag{27}$$

where I_v^{sat} is the saturation intensity parameter for a specific transition. Saturation occurs when further increases in laser intensity result in a detectable depopulation of the upper state due to stimulated emission down to the lower state. In the limit $I_v >> I_v^{sat}$ a constant number density of excited states, N_u, is achieved irrespective of laser intensity.

Neglecting predissociation, in the high intensity limit saturated fluorescence reduces eqn. 26 to:

$$R_f^{sat} = N_l B_{lu} I_\nu \frac{A_{ul}}{(1+const.)}. \tag{28}$$

For low laser intensities, $I_\nu << I_\nu^{sat}$, the saturation term (the final term in eqn. 26) reduces to unity. The attraction of saturation is that the effects of electronic quenching, which is bath gas composition dependent, is eliminated. In practice this is very difficult to achieve because most lasers have an approximately Gaussian spatial intensity distribution and this cause both saturated, partially saturated and non-saturated fluorescence to be simultaneously generated and detected. Since many simple systems, such as OH and CH, are readily saturated it is essential to know the degree of saturation in a given experiment if the measured data are to be quantified. Where possible, for accurate quantitative measurements, it is preferable to work in the low intensity limit and to either have an accurate knowledge of the quenching or, better, to choose a system where quenching is essentially constant for all sensible variations of local gas composition.

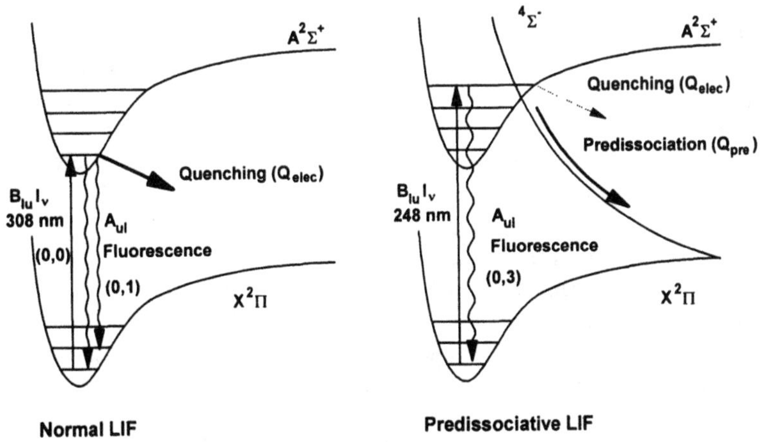

Figure 18. Energy level diagram for OH LIF strategies using tuneable excimer lasers.

An alternative strategy is to choose a energy levels within the molecular system where predissociation is possible such that if $Q_{pre} >> Q_{elec}$ the rate of fluorescence emission becomes:

$$R_f^{pre} = N_l B_{lu} I_\nu \frac{A_{ul}}{(A_{ul}+Q_{pre})}. \tag{29}$$

and therefore changes in quenching due to local gas composition are eliminated. The main advantage of predissociative LIF[17,48,49] over saturated LIF is that no specific laser spatial intensity distribution is required, the main disadvantages is that if $Q_{pre} >> A_{ul}$ the rate of non-fluorescent depopulation of N_u becomes so great that the LIF signal is also considerably reduced. A schematic of predissociative LIF, for the OH radical, using a tuneable KrF excimer lasers is shown in fig. 18. To achieve predissociative LIF the OH radical is excited at a shorter wavelength than usual, 248 nm, such that the v=3 vibrational state of the $A^2\Sigma^+$ is populated. At this vibrational energy the $A^2\Sigma^+$ state crosses over with the dissociative $^4\Sigma^-$ state and most of the excited OH radicals fall apart to H and O atoms before collisions are possible, hence no quenching. The technique is certainly practical in spite of very weak signals, using tuneable KrF excimer lasers it has been possible to obtain OH PLIF images in both spark ignition and diesel engines[48,49]. Also, to achieve predissociative LIF it is important for the detector to collect only fluorescent emissions which originate from the v=3 level[17]; emissions from the v=0, 1 or 2 levels of the $A^2\Sigma^+$ state are subject to quenching and these levels are rapidly populated if the v=3 level is pumped.

The normal approach to OH LIF with a tuneable excimer is to use the 308 nm XeCl line to excite the (0,0) band and to detect either on-resonance (0,0) fluorescence which is strong or the weaker (0,1) fluorescence at 343 nm[50] which has the advantage of being wavelength shifted from the excitation source. All the above schemes are illustrated in fig. 18. Similar approaches are also possible with Nd:YAG pumped dye laser systems, in this case the (1,0) band can be excited at 285 nm and the fluorescence detected from the (0,0) band around 300-320 nm. For practical applications, where windows and other scattering objects are present, an off-resonance detection scheme is preferable.

PLIF of OH Radicals to Estimate Local Mass Burning Rates

An example of a PLIF image of OH taken in a highly turbulent premixed methane air flame[20] is shown in fig. 19. This image was produced using a Nd:YAG pumped dye laser exciting fluorescence on the (0,0) band at 311.8 nm with detection centred around this wavelength. Two jets of flame are produced and the associated flame fronts are denoted by the most intense white regions which indicate the highest OH concentrations.

Images such as fig. 19 have been used to estimate flame curvature effects on mass burning rate[20], this is achieved from the ratio of the super equilibrium OH at a local point in the flame front to the bulk equilibrium OH concentration in the fully burnt gas. In the fully burnt gas an equilibrium level is achieved which is determined by the adiabatic flame temperature (assuming no radiation loss) and the amount of H_2O present which governs the extent of the reversible reaction $H_2O \rightleftarrows H + OH$ and thereby the concentration of equilibrium OH. Because both temperature and the final H_2O concentration are determined by the fuel type and the AFR, the OH concentration can be calculated. In the flame front, where the mass burning is occurring, fast oxidation chemistry causes an additional or super equilibrium level of OH. This enhanced OH concentration is a measure of the rate of local reaction, therefore the effect of flame curvature on local mass burning rate may be estimated[20,51].

PLIF of Fuel in Combustion

Recently, the use of PLIF for the study of fuel-air mixing has received much attention[1,4,12,16,21] and the method has been successfully applied to a gas turbine[18,21], ic engines[16,17,19,51] and a supersonic jet[52]. PLIF of fuel typically uses a UV laser operating somewhere between 300 and 240 nm, the most common sources are XeCl (308 nm), the 4th harmonic of Nd:YAG (266 nm) and KrF (248 nm). The species detected is either a naturally occurring component[51] or a carefully selected seed such as a ketone or aldehyde which is added as a fluorescent seed to the fuel[16-19,21]. Ketones and aldehydes posses excellent properties as fuel markers: they are insensitive to the local composition variations which typify hydrocarbon-air combustion, they produce strong fluorescent emission in the visible around 420 nm and for a number of the lower molecular weight species the quantum yield of fluorescence has either a simple or no temperature dependence[21].

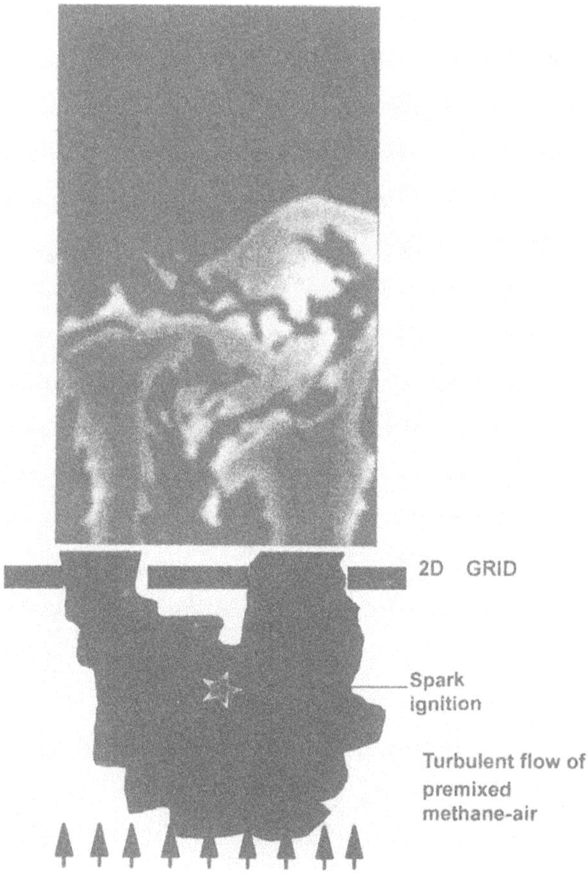

Figure 19. Experimental schematic and PLIF image of OH taken in a highly turbulent atmospheric flame.

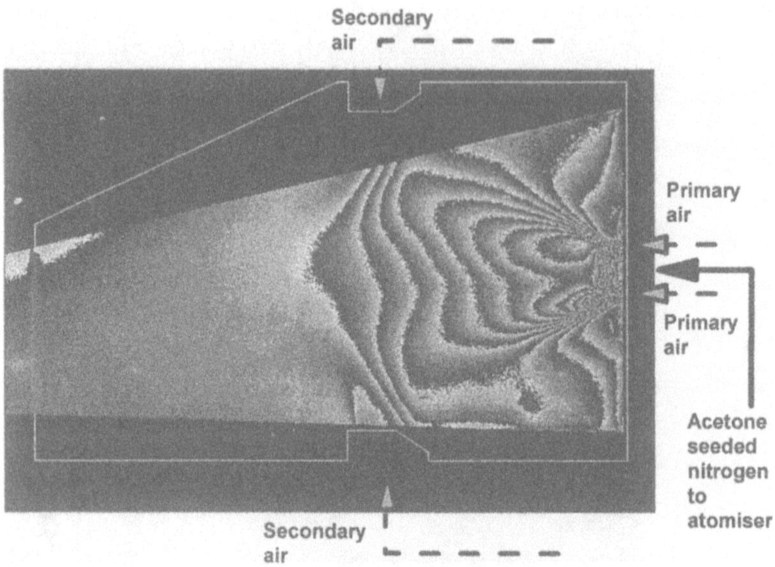

Figure 20. PLIF image of average of fuel fraction in a single sector gas turbine rig. The laser sheet was aligned to bisect the fuel injector. The image is displayed as a repeated linear grey scale.

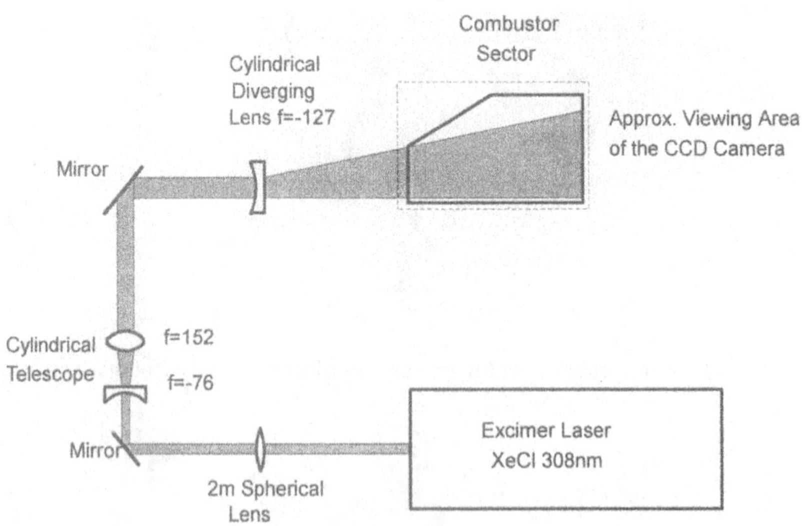

Figure 21. Layout of PLIF experiment for the image in fig. 20.

Two PLIF applications of fuel imaging are illustrated in figs. 20 to 23. Respectively, figs. 20 and 21 show images of fuel fraction taken from a simulated fuel-air mixing experiment in a single sector gas turbine combustor[18,21] and the PLIF geometry for this experiment. The random noise on an instantaneous image is typically of order 5% and the useful dynamic range of order 50. The image in fig. 20 is a mean of 200 laser pulses, therefore the turbulent structure has been averaged out. The noise in fig. 20 is generally

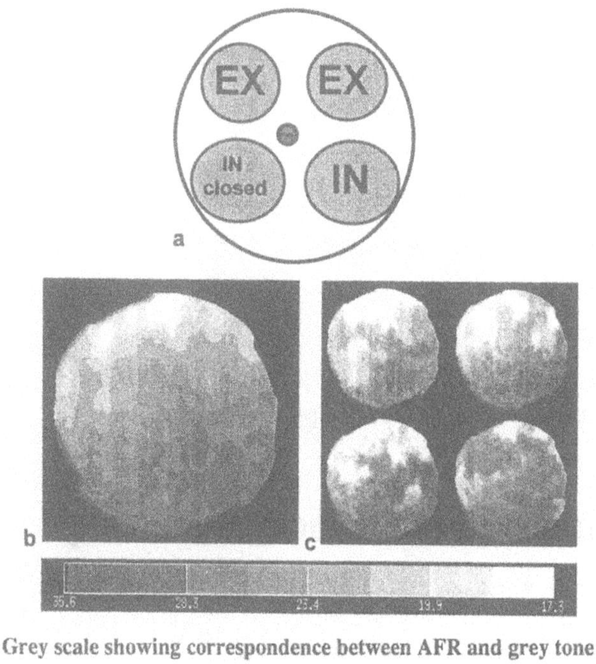

Figure 22. a: Orientation of AFR maps relative to valve geometry, b: Mean AFR map for fuel injection at 60° ATDC on inlent stroke, c: Four single cycle AFR maps as b.

1% or better and the signal levels varied from 500 to 15,000 counts with an almost constant noise level of 10 counts. In fig. 20 the image is displayed as a repeated linear grey scales to better reveal the areas of low fuel fraction. The image clearly highlights a variety of flow features, the most obvious are the hollow nature of the injected fuel cone and the penetration of the secondary mixing jets.

geometry for the engine is shown in fig. 23. In this experiment only one inlet valve operated during the main induction stroke and the fuel was inject whilst that valve was open. This strategy of fuel-air mixing is believed to cause a rich pocket of fuel, in an otherwise lean mixture, to form in the vicinity of the spark plug around the time of ignition The pattern of fuel-air mixing is seen in the PLIF images, fig. 22, and a consistent fuel rich zone on the exhaust valve side of the combustion chamber is evident. The instantaneous PLIF images show distinct changes in the fuel-air mixing from cycle-to-cycle[19].

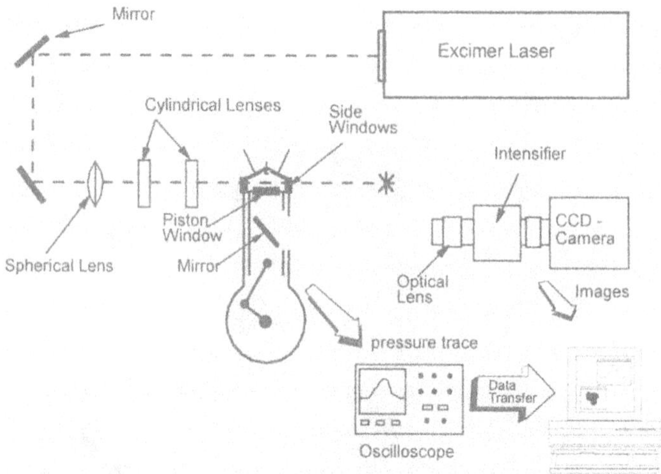

Figure 23. Layout of PLIF experiment for the images in fig. 22.

PLANAR LASER INDUCED INCANDESCENCE (PLII)

Laser induced incandescence (LII) is the process in which soot particles are heated by the incoming laser beam to temperatures well in excess of the bath gas such that they then radiate considerably enhanced incandescence[13,21-23,53,54]. For lasers intensities above a few MW cm^{-2} soot particles rapidly achieve temperatures around 4,000 K. Above a threshold fluence which is of order 10 MW cm^{-2}, the temperature that the particles reach is controlled by vaporisation of C_2 and C_3[55,56]. Increasing the laser fluence will not increase the particle temperature significantly and at very high fluences the particles loss mass rapidly and decrease in size. Large particles achieve slightly greater temperatures because the heating is proportional to volume whereas mass loss is controlled by surface area. The net effect for high laser fluences is a signal which is proportional to soot volume fraction, slightly particle size dependant, yet nearly independent of laser fluence. On this basis a number of soot imaging experiments on soot volume fraction have been pursued[13,21-23]. The original work on LII was by Eckbreth[53] and the first comprehensive model was suggested by Melton[54] which has subsequently been extended by Tait and Greenhalgh[13,21]. Following the latter work, the expected LII signal strength can be calculated from an energy balance for an individual soot particle in a laser beam which gives:

$$K_{abs}\pi a^2 q - 4\pi a^2 (T - T_0)\Lambda - \frac{\Delta H_v}{W_s}\frac{dM}{dt} - q_{rad} - \frac{4}{3}\pi a^3 \rho_s C_s \frac{dT}{dt} = 0 \qquad (30)$$

where the parameters in eqn 29 and the following 8 eqns are:

a	Radius of particle (m)	W_S	Mwt. of solid carbon (kg/mol)
c	Speed of light (m/s)	W_V	Mwt. of vaporised material, taken as 0.036 (kg/mol)
q	Laser fluence (W/m^2)		
t	Time (s)	α	Accommodation coefficient
C_s	Specific heat capacity of carbon ($J\ mol^{-1}\ K^{-1}$)	λ	Wavelength (m)
		ρ_s	Density of solid carbon, 2,260 ($kg\ m^{-3}$)
C_0	Specific heat capacity of flame gases ($J\ mol^{-1}\ K^{-1}$)	ρ_0	Density of flame gases ($kg\ m^3$)
K_{abs}	Absorption coefficient for soot	ρ_v	Carbon vapour density ($kg\ m^3$)
M	Mass of particle (Kg)	λ_{em}	Detected incandescence wavelength (m)
P_0	Pressure of flame gases (Pa)		
$P(\lambda)$	Planck blackbody function	ΔH_v	Heat of vaporisation of carbon, 7.78×10^{-5} ($J\ mol^{-1}$)
R	Gas constant ($J\ K^{-1}\ mol^{-1}$)		
T_0	Temperature of flame gases (K)	Λ	Langmuir heat transfer coefficient
T	Temperature of soot particle (K)	v_v	Carbon vapour velocity ($m\ sec^{-1}$)
T^*	Temperature at which the vapour pressure of carbon is P^* (K)		

The four terms in eqn. 29 are respectively: the rate of heating of the soot particle due to laser absorption, the rate of heat loss by conduction to the bath gas, the rate of heat loss by vaporisation, the rate of heat loss by radiation and the rate of internal energy rise. The term for the heat lose due to conduction needs to account for the fact that the particles are small compared to the mean free path of the gas, thus conduction is controlled by heat transfer across the Langmuir layer, Λ, is given by[57]:

$$\Lambda = \frac{\alpha P_0 (C_0 - \frac{P_0}{2\rho_0 T})}{\sqrt{2\pi P_0 \rho_0}}. \qquad (31)$$

The radiative power of the particle is given by:

$$q_{rad} = \int_0^\infty P(\lambda, T) K_{abs}(a, \lambda) c\pi a^2 d\lambda. \qquad (32)$$

Assuming that most of the light is radiated at wavelengths that are long compared with the absorption length of the soot particle, then $q_{rad} \propto a^3 T^5$, also if the detected wavelengths cover only a few 10's of nm around 600 nm the λ dependence of q_{rad} may be eliminated and the expression further simplified.

Now, if the vapour pressure of carbon is given by the Clapeyron equation then:

$$P(T)=P^{*}\exp\left(\Delta H_{v}\frac{(T-T^{*})}{RTT^{*}}\right) \tag{33}$$

and the rate of mass loss term can be re-evaluated. Given continuity at the particle surface[53] we have:

$$-\rho_{s}\frac{\mathrm{d}a}{\mathrm{d}t}=\rho_{v}v_{v} \tag{34}$$

where v_v is given by the Langmuir evaporation rate:

$$v_{v}=\sqrt{RT/2W_{v}} \tag{35}$$

and from the ideal gas law $P_{v}=\rho_{v}RT/W_{v}$ the rate of mass loss becomes:

$$\frac{\mathrm{d}M}{\mathrm{d}t}=4\pi\rho_{s}a^{3}\frac{\mathrm{d}a}{\mathrm{d}t}=\pi a^{2}(4\rho_{v}v_{v})=\pi a^{2}\left[4\left(RT/2W_{v}\right)^{\frac{1}{2}}P(T)\right]. \tag{36}$$

Hence from eqns 29 to 35 the energy balance can be rewritten as:

$$\begin{gathered}K_{abs}(a)q-4(T-T_{0})\Lambda-4\frac{\Delta H_{v}}{W_{s}}\left(\frac{W_{v}}{2RT}\right)^{\frac{1}{2}}P^{*}\exp\left(\Delta H_{v}\frac{(T-T^{*})}{RTT^{*}}\right)\\-\frac{q_{rad}}{\pi a^{2}}-\frac{4}{3}\rho_{s}C_{s}a\frac{\mathrm{d}T}{\mathrm{d}t}=0.\end{gathered} \tag{37}$$

This equation may then be numerically integrated to find the expected LII signal for the conditions of interest. K_{abs} can be predicted from Mie theory using Lee and Tien's values[58] for the complex refractive index of the soot appropriate to the excitation wavelength used. The signal collected between $\lambda_{max.}$ and λ_{min} is thus:

$$S=\int_{\lambda_{min}}^{\lambda_{max}}\int_{t=0}^{\infty}P(\lambda,T)K_{abs}(a,\lambda)c\pi a^{2}\,\mathrm{d}t\,\mathrm{d}\lambda. \tag{38}$$

Some results from sample calculations from the above model[21] usefully show the principal trends in LII. Typical soot particle are either in or approach the Rayleigh regime, therefore their heating rate is proportional to volume and their energy loss to surface area. Thus the maximum particle temperature increases with particle size and larger LII signals are expected from larger particles. However, mass loss also increases with particle temperature and this reduces the LII signal. As seen in fig. 24 there is a near balance between these effects for 308 nm excitation which yields a signal which is only a slow function of particle size. This should be compared to Mie scattering where the particle size dependence is r^3.

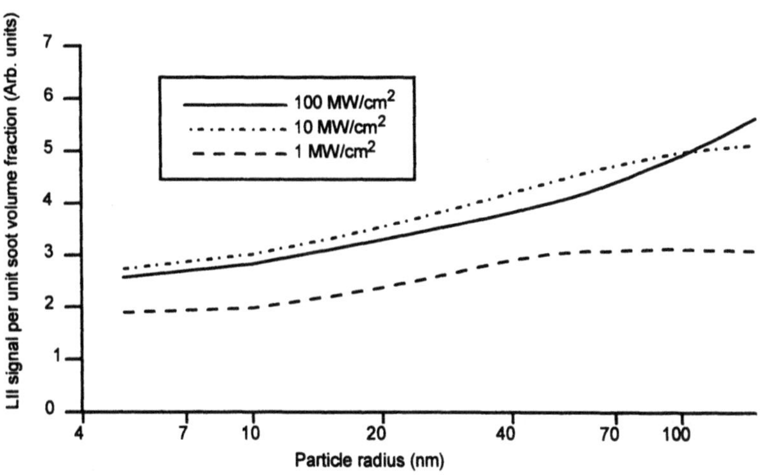

Figure 24. Predicted LII signal versus particle radius for various laser fluences, 308 nm excitation.

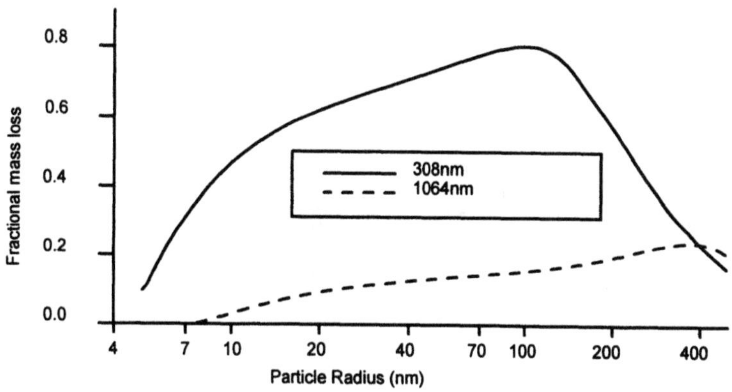

Figure 25. Predicted fractional mass loss at 50 MW cm^{-2} for a 100 nm radius soot particle.

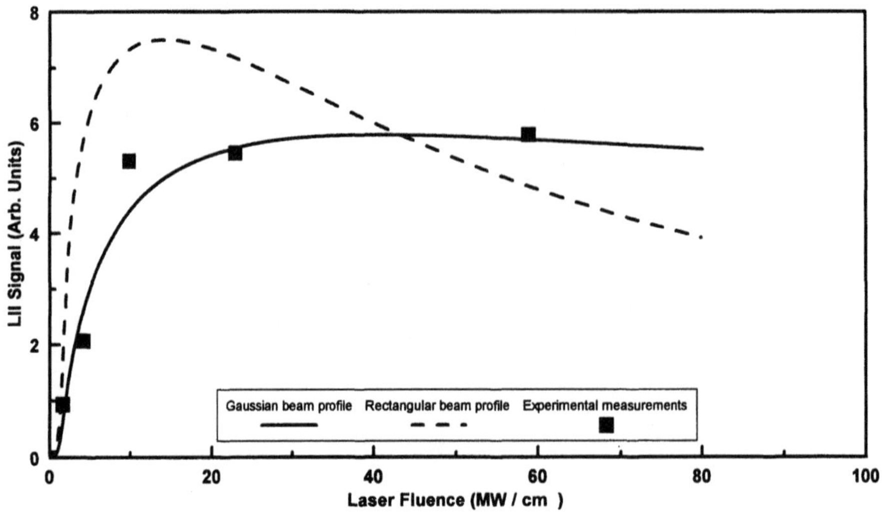

Figure 26. Predicted LII signal versus laser fluence for a 100 nm radius particle.

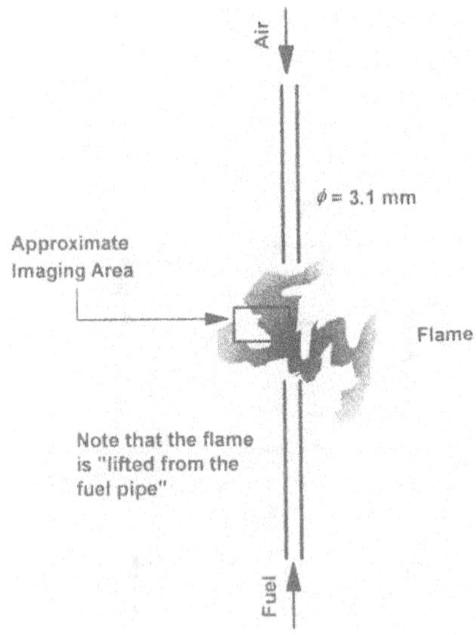

Figure 27. Schematic of a highly turbulent propane-air flame as used to record the image in fig. 25.

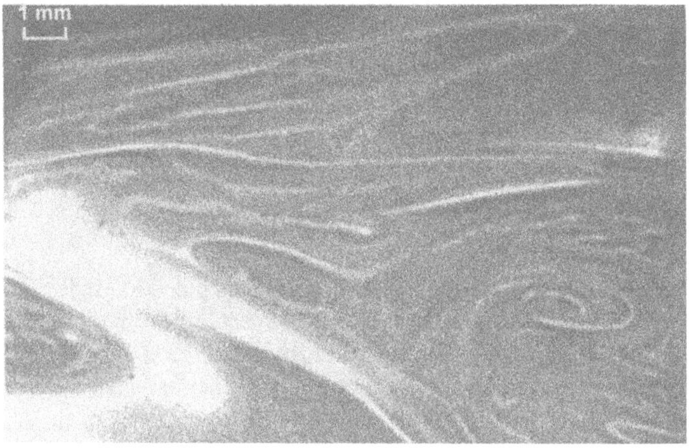

Figure 28. PLII image from a strongly turbulent flame with a spatial resolution of circa 100 μm.

The effect of mass loss is shown in fig. 25, note the very much higher mass loss at the shorter wavelength which is due to the enhanced absorption at this wavelength. Fig. 26 shows the effect of laser fluence on LII signals, for an idealised laser beam of rectangular profile the signal decreases markedly at high laser fluence due to mass loss. For a laser sheet of Gaussian intensity profile across its width the result is a near constant signal from just over 10 to 80 MW cm^{-2} because the enhanced mass loss at higher laser intensity is compensated by sweeping in a larger volume of soot particles.

Figs. 27 and 28 illustrate respectively the geometry of a highly sooting, highly turbulent propane-air flame and a high resolution PLII image of soot from this flame. The image was recorded using a thin laser sheet generated from a XeCl excimer laser, very fine structures of order 100 μm are seen in the soot field which are thought to be formed by rapid vortex mixing[21]. In the image shown in fig. 28 the strongly intermittent nature of soot in a highly turbulent flow is clearly seen.

Although the prime use of LII is as a soot imaging diagnostic it should be noted that the phenomenon gives rise to very strong, spectrally flat, signals and this is a major limitation for the application of PLIF to soot flames[12].

CONCLUDING REMARKS AND FUTURE DIRECTIONS

Inelastic scattering is a growing field in laser diagnostics and techniques such as CARS, and to a lesser extent spontaneous Raman, are now established point methods for temperature measurements. The main advantage of CARS is an ability to measure in hostile flows. The main limitations of CARS thermometry are dye laser noise and spatial resolution. Planar techniques such as PLIF and PLII, though still developing, are rapidly finding wide spread application to practical engineering problems such as fuel-air mixing.

In the future, we can expect further improvements in the planar methods, also with improved and cheaper cameras and lasers the application of PLIF and PLII to 3D visualisation techniques based on multiple time sequenced laser sheets and multiple cameras[4,7] will be possible. The planar methods, PLIF and PLII are undoubtedly set to play a major role in future combustion and fluid flow studies.

ACKNOWLEDGEMENTS

I would like to acknowledge the support of Rolls Royce plc, Honda R&D, Shell TRC, SERC and the CEC who have all supported various aspects of the work presented here. I would also like to acknowledge the assistance of N.P. Tait, N. Farrugia and M. Berckmuller who have contributed various diagrams and other assistance.

REFERENCES

1. D.A. Greenhalgh, Optical Combustion Diagnostics, Fundamental or Practical in "Proc. of The Anglo-German Combust. Symp", British Section of the Combustion Institute, ISBN 0 9520350 0 6, pp. 18-25 (1993).
2. D.A. Greenhalgh, Quantitative CARS Spectroscopy. in "Advances in Non-linear Spectroscopy," R.J.H. Clark and R.E. Hester, J. Wiley & Sons (1988).
3. A.C. Eckbreth, "Laser Diagnostics for Combustion Temperature and Species," Abacus Press (1988).
4. J.M. Seitzman and R.K. Hanson, Planar Fluorescence Imaging in Gases in "Instrumentation for Flows with Combustion," A.M.K.P. Taylor, ed., Academic Press (1993).
5. L.P. Goss, CARS Instrumentation for Combustion Applications in "Instrumentation for Flows with Combustion," A.M.K.P. Taylor, ed., Academic Press (1993).
6. R.W. Dibble, A.R. Masri and R.W. Bilger, *Combst. Flame*, 67:189-206 (1987).
7. M.B. Long, Multidimensional Imaging in Combusting Flows by Lorenz-Mie, Rayleigh and Raman Scattering in "Instrumentation for Flows with Combustion," A.M.K.P. Taylor, ed., Academic Press (1993).
8. D.R. Williams and D McKeown, "CARS Measurement of Temperature in the Coal Fired Tunnel Furnace at RISO National Laboratory, Denmark Winter 1991" AEA Report-EE-0129 (1991).
9. R. Lückerath, M. Woyde, M. Meier and W. Stricker. Temperature Measurements with CARS in a Gas/Coal Dust Fired 350 kW Furnace and in a H_2-Air Ramjet, in "Proc. of The Anglo-German Combust. Symp", British Section of the Combustion Institute, ISBN 0 9520350 0 6, pp. 267-270 (1993).
10. W. Stricker, M. Woyde, R. Lückerath and V. Bergmann, *Ber. Bun.-Ges. Phys. Chem.*, to appear (1993).
11. D.R. Williams, D. McKeown, F.M. Porter, C.A. Baker, A.G. Astill and K. M. Rawley. *Combst. Flame*, 94:77-90 (1993).
12. N.P. Tait and D.A. Greenhalgh. 24th Symp. Int. on Combustion, The Combustion Institute, pp. 1,621-1,628 (1992).
13. N.P. Tait and D.A. Greenhalgh in "Optical Methods and Data Processing in Heat Transfer", IMechE, p.185 (1992).
14. H. Becker, A. Arnold, R. Suntz, P. Monkhouse, J. Wolfrum, R. Maly, W. Pfister. *Appl. Phys.*, B50:473-478 (1990).
15. R. Suntz, H. Becker, P. Monkhouse, J. Wolfrum. *Appl. Phys.*, B47: 287-293 (1988).
16. A. Arnold, H. Becker, R. Suntz, P. Monkhouse, J. Wolfrum. *Opt. Lett.* 15:831-833 (1990).
17. P. Andresen, G. Meijer, H. Schluter, H Voges, A. Koch, W. Hentschel, W. Oppermann, E. Rothe. *Appl. Opt.* 29:2392-2404 (1990).
18. N.P. Tait, D. Bryce and D.A. Greenhalgh. Laser Fluorescence Imaging of Simulated Fuel Vapour Mixing in a Gas Turbine Sector, in "Proc. of The Anglo-German Combust. Symp", British Section of the Combustion Institute, ISBN 0 9520350 0 6, pp. 251-254 (1993).
19. M. Berckmüller, N.P. Tait, D.A. Greenhalgh, K. Ishii, Y. Urata, H. Umiyama, K. Yoshida. In-cylinder Imaging of Fuel Concentration using Planar Laser Induced Fluorescence, IMechE, to be published (1994).

20. N. Farrugia, D.A. Greenhalgh and A. Üngut. PLIF Imaging of OH Radicals to Estimate Mass Burning Rates in Turbulent Premixed Methane-Air Flames. IMechE, to be published (1994).
21. N.P. Tait and D.A. Greenhalgh. *Ber. Bun.-Ges. Phys. Chem.*, to appear (1993).
22. J.E. Dec. Soot Distribution in a D.I. Diesel Engine Using 2-D Imaging of Laser Induced Incandesence, Elastic Scattering and Flame Luminosity. SAE paper 920115 (1992).
23. B. Quay, T.-W. Lee and R.J. Santoro, to be published (1994).
24. G. Herzberg. Molecular Spectra and Molecular Structure, Vol. I. Spectra of Diatomic Molecules. Van Nostrand Reinhold Co. (1950). (also Vols. II and III on Polyatomic Spectra and Infrared and Raman Spectra).
25. B.P. Straughan and S. Walker. Eds. Spectroscopy 3. Halstead, NY. (1976).
26. K.P. Huber and G. Herzberg. Molecular Spectra and Molecular Structure, Vol. IV. Constants of Diatomic Molecules. Van Nostrand, N.Y. (1979).
27. CARP3 is a CARS spectral analysis code developed by AEA Technology, Harwell, Oxon. OX11 0RA, UK.
28. CARP3 is marketed by Epsilon Research Ltd, PO Box 354, High Wycombe, Bucks, HP12 9BZ, UK.
29. F.M. Porter, D.A. Greenhalgh, P.J. Stopford, D.R. Williams and C.A. Baker. *Appl. Phys.* B51:31-38 (1990).
30. M. Woyde and W. Stricker. *Appl. Phys.* B50:519-525 (1990).
31. D.A. Greenhalgh, I.S. Dring, R. Devonshire and F. Boysan. *Chem. Phys. Lett.*, 133:485(1987).
32. D.A. Greenhalgh and S.T. Whittley. *Appl. Opt.*, 24:907-913 (1985).
33. R.J. Hall and D.A. Greenhalgh. *J. Opt. Soc. Am.*, B3:1,637-1,641 (1986).
34. S. Kroll, M. Alden, T. Berglind and R.J. Hall. *Appl. Opt.*, 26:1068 (1987).
35. S. Kroll and D. Sandell. *J. Opt. Soc. Am.*, B5:1,910 (1988).
36. F.M. Porter. PhD Thesis, University of Surrey (1985).
37. D.R. Snelling, G.J. Smallwood, R.A. Sawchuck and T. Parameswaran. *Appl. Opt.*, 26:99 (1987).
38. D.R. Snelling, G.J. Smallwood, and R.E. Mueller. *Appl. Opt.*, 24:2772-2778 (1985).
39. P. Ewart. *Opt. Comm.* 55:124 (1985).
40. P. Snowdon, S.M. Skippon and P. Ewart. *Appl. Opt.* 30:1008 (1991).
41. J.P. Boquillon, M. Pealat, P. Bouchardy, G. Colin, P. Magre and J.P.E. Taran. *Opt. Lett.* 13:722 (1989?).
42. Shepherd, F.M. Porter and D.A. Greenhalgh. *Combst. Flame* 82:106 (1990).
43. D. Bradley, M. Lawes, M.J. Scott, C.G.W. Sheppard, D.A. Greenhalgh and F.M. Porter. 24th Symp. Int. on Combustion, The Combustion Institute, pp. 527-535 (1992)
44. Rahn, R.L. Farrow, M.L. Koszykowski and P.L. Mattern. *Phys. Rev. Lett.* 45:620 (1980).
45. Lucht and R.L. Farrow. *J. Opt. Soc. Am.* 85:1243 (1988).
46. M. Pealat, M. Lefebvre, J.P.E. Taran and P.L. Kelly. *Phys. Rev.* A38:1948-1965 (1988).
47. W. Demtröder. "Laser Spectroscopy". Springer-Verlag, Berlin. (1982).
48. P. Andresen, A. Bath, W. Gröger, H.W. Lülf, G. Meijer and J.J. ter Meulen. *Appl. Opt.* 27:365(1988).
49. A. Arnold, F. Dinkelacker, T. Heitzmann, P. Monkhouse, M. Schäfer, V. Sick and J. Wolfrum. 24th Symp. Int. on Combustion, The Combustion Institute, pp. 1605-1612 (1992).
50. F. Dinkelacker, A. Bushmann, M. Schäfer and J. Wolfrum. Spatially Resolved Joint Measurements of OH and Temperature Fielkds in a Large Premixed Turbulent Flame, in "Proc. of The Anglo-German Combust. Symp", British Section of the Combustion Institute, ISBN 0 9520350 0 6, pp. 295-298 (1993).
51. H. Becker, P.B. Monkhouse, J. Wolfrum, R.S. Cant, K.N.C. Bray, R. Maly, W. Pfister, G. Stahl and J. Warnatz. 23rd Symp. Int. on Combustion, The Combustion Institute, pp. 817-823 (1990).
52. J.C. Hermanson and M. Winter. Imaging of a Transverse, Sonic Jet in Supersonic Flow, AIAA-91-2269 (1991).
53. A.C. Eckbreth. *J. Appl. Phys.* 48:4473 (1977).
54. L.A. Melton. *Appl. Opt.* 23:2201 (1984).
55. E. Rholfing. *J. Chem. Phys.* 89:6103 (1988).
56. P.-E. Bengtsson and M. Alden. *Combust. Flame* 80:322 (1990).
57. E.H. Kennard, "Kinetic Theory of Gases", pp. 312-324, McGraw-Hill, New York (1938).
58. S.E. Lee and C.L. Tien, 18th Symp. Int. on Combustion, The Combustion Institute, p. 1159 (1981).

Index

INDEX

Printed in the USA
CPSIA information can be obtained
at www.ICGtesting.com
LVHW070850181223
766718LV00004B/530

9 781489 912732